PLANETARY REMOTE SENSING AND MAPPING

International Society for Photogrammetry and Remote Sensing (ISPRS) Book Series

ISSN: 1572–3348

Book Series Editor

Zhilin Li
Department of Land Surveying and Geo-Informatics
The Hong Kong Polytechnic University, Hong Kong,
P.R. China

information from imagery

Planetary Remote Sensing and Mapping

Editors

Bo Wu
Department of Land Surveying & Geo-Informatics, The Hong Kong Polytechnic University, Hong Kong, P.R. China

Kaichang Di
Institute of Remote Sensing and Digital Earth, Chinese Academy of Sciences, China

Jürgen Oberst
German Aerospace Center (DLR), Germany

Irina Karachevtseva
Moscow State University of Geodesy and Cartography, Russia

CRC Press
Taylor & Francis Group
Boca Raton London New York Leiden

CRC Press is an imprint of the
Taylor & Francis Group, an **informa** business
A BALKEMA BOOK

CRC Press/Balkema is an imprint of the Taylor & Francis Group, an informa business

First issued in paperback 2021

Typeset by Apex CoVantage, LLC

Library of Congress Cataloging-in-Publication Data
Applied for

Published by: CRC Press/Balkema
P.O. Box 11320, 2301 EH Leiden, The Netherlands
e-mail: Pub.NL@taylorandfrancis.com
www.crcpress.com – www.taylorandfrancis.com

ISBN: 978-1-138-58415-0 (hbk)
ISBN: 978-1-03-209442-7 (pbk)
ISBN: 978-0-429-50599-7 (eBook)

Contents

Planetary Remote Sensing and Mapping – Wu et al.
© 2019 Taylor & Francis Group, London, ISBN 978-1-138-58415-0

Preface

Space exploration expands humankind's understanding of the Earth and the universe, and promotes technology advancement, social progress, and human civilization. Since the early 21st century, a new era in space exploration has begun. The National Aeronautics and Space Administration (NASA) of the United States, The European Space Agency (ESA), as well as space agencies of Japan, China, India, and other countries have sent their probes to the Moon, Mars, and other planets in the solar system. These missions explore the space environment and help us answer long-standing questions on planetary formation and evolution, as well as the origin of life.

Observations from space are required to determine basic geodetic data of planets and satellites, such as size, shape, rotation, and gravity field parameters. Availability of remote sensing data on planetary surface is critical for any planetary exploration mission in terms of safe landing and surface operations. Data from planetary remote sensing and mapping are also important for planetary scientific research such as the mineral abundance, morphological characteristics, and geological processes. The recent planetary exploration missions have acquired and continue to acquire remote sensing data with higher spatial and spectral resolutions, offering new opportunities to explore and improve our understanding of planets and satellites.

This book presents the latest developments and results in planetary remote sensing and mapping from recent planetary exploration missions. The book is composed of six sections and 21 chapters of original research and developments covering fundamental aspects in planetary remote sensing and mapping, including reference systems of planetary bodies, planetary exploration missions and sensors, geometric information and feature information extraction from planetary remote sensing data, planetary remote sensing data fusion, and planetary data management and presentation.

The first section "*Reference Systems of Planetary Bodies*" has three chapters (Chapters 1 to 3), discussing the latest developments regarding the necessity and a technical framework of a new-generation lunar global control network, the basic geodetic and dynamic parameters of Saturn's moon Enceladus, and the physically defined height systems for terrestrial planets and the Moon. The second section "*Planetary Exploration Missions and Sensors*" has four chapters (Chapters 4 to 7). It presents new sensors in planetary missions including the Rosetta/OSIRIS cameras imaging the Comet 67P, the BepiColombo Laser Altimeter for geodetic investigations of Mercury, and new optical sensors for Mars orbital missions. This section also includes a chapter about design challenges of Mars landers and considerations and mission profiles for China's first Mars exploration mission. The third section "*Geometric Information Extraction from Planetary Remote Sensing Data*" includes four chapters (Chapters 8 to 11) and presents the photogrammetric processing techniques for images collected by the representative planetary orbiter cameras, including the High Resolution Imaging Science Experiment (HiRISE) instrument onboard the Mars Reconnaissance Orbiter, the High Resolution Stereo Camera (HRSC) onboard Mars Express, the Narrow-Angle Camera (NAC) onboard the Lunar Reconnaissance Orbiter, and the Mercury Dual Imaging System (MDIS) onboard MESSENGER. The fourth section "*Feature Information Extraction from Planetary Remote Sensing Data*" includes four chapters (Chapters 12 to 15) and emphasizes the studies of geomorphologic features and spectral information for geological and mineral investigations. The fifth section "*Planetary Remote Sensing Data Fusion*" has three chapters (Chapters 16 to 18) and discusses the integration and fusion of multiple-source planetary remote sensing datasets for synergistic use. The last section "*Planetary Data Management and Presentation*" has three chapters (Chapters 19 to 21) and addresses the latest developments in planetary cartography, data management, data visualization, and design of geo-information systems.

With chapters contributed by renowned experts in planetary remote sensing and mapping, the book provides new insights and timely updates on the research and developments with respect to these important aspects of planetary remote sensing and mapping. It is suitable for readers working in the planetary remote sensing and mapping areas as well as for planetary probe designers, engineers, planetary geologists, and geophysicists. It may also present useful reading material for scientists, professionals, educators, and postgraduate students in the fields of remote sensing, photogrammetry, cartography, and geo-information.

This book is supported by the Hong Kong Polytechnic University (Project no. G-YBN8) and the National Natural Science Foundation of China (Project no. 41671426 and Project no. 41471345). This book is also supported by the ISPRS Inter-Commission Working Group III/II: Planetary Remote Sensing and Mapping. We would like very much to thank CRC Press/Balkema, Taylor & Francis Group for their efforts and cordial cooperation in publishing this book.

Bo Wu
Department of Land Surveying & Geo-Informatics
The Hong Kong Polytechnic University
Hong Kong

Kaichang Di
Institute of Remote Sensing and Digital Earth
Chinese Academy of Sciences
China

Jürgen Oberst
German Aerospace Center (DLR)
Germany

Irina Karachevtseva
Moscow State University of Geodesy and Cartography
Russia

Planetary Remote Sensing and Mapping – Wu et al.
© 2019 Taylor & Francis Group, London, ISBN 978-1-138-58415-0

Editors

Bo Wu is an Associate Professor with the Department of Land Surveying & Geo-Informatics of the Hong Kong Polytechnic University, mainly working on Photogrammetry and Planetary Mapping. He worked on NASA-funded projects on Mars and lunar exploration missions when he was a researcher at the Ohio State University in the United States. At the Hong Kong Polytechnic University, he worked on the landing site mapping and selection for China's Chang'E-3 lunar landing mission, and has been leading a team working on the landing site characterization and selection for the Chang'E-4 and Chang'E-5 missions and China's first Mars exploration mission. He served as the Vice President of the Hong Kong Geographic Information System Association, the Co-Chair of the International Society for Photogrammetry and Remote Sensing (ISPRS) Working Group II/6: Geo-Visualization and Virtual Reality, and currently is serving as the Secretary of the ISPRS Inter-Commission Working Group III/IV: Planetary Remote Sensing and Mapping. He received a number of academic awards during his career, including the Talbert Abrams Award and John I. Davidson President's Award from American Society for Photogrammetry and Remote Sensing (ASPRS), the R. Alekseev Award and Gold Medal from the 44th International Exhibition of Inventions of Geneva, and the Duane C. Brown Senior Award from the Ohio State University. He is currently the Associate Editor of the international journal *Photogrammetric Engineering & Remote Sensing* and an Editorial Board member of the *ISPRS Journal of Photogrammetry and Remote Sensing* and *The Photogrammetric Record.*

Kaichang Di is a professor with the Institute of Remote Sensing and Digital Earth (RADI), Chinese Academy of Sciences (CAS). He received his Ph.D. degree in photogrammetry and remote sensing from Wuhan Technical University of Surveying and Mapping (WTUSM) (now Wuhan University) in 1999. He participated in the Mars Exploration Rover mission and contributed to rover localization and mapping for mission operations, when he was a Research Scientist with the Department of Civil and Environmental Engineering and Geodetic Science, Ohio State University (OSU). He is now director of the Planetary Mapping and Remote Sensing Laboratory at RADI. He is involved in tele-operation of the Chang'E-3 rover using visual navigation and environment perception techniques, and has been leading a team to develop new vision-based techniques to support China's future lander and rover missions. He was co-chair of International Society for Photogrammetry and Remote Sensing (ISPRS) Planetary Mapping and Databases Working Group from 2008 to 2016, and he has been chair of ISPRS Planetary Remote Sensing and Mapping Working Group since 2016. He received numerous academic awards during his career, including the first prize of Wang Zhizhuo Innovation Talent Award from WTUSM, the Duane C. Brown Senior Award from OSU, first place of John I. Davidson President's Award for practical papers and first place of ESRI Award for best scientific paper in GIS from American Society for Photogrammetry and Remote Sensing, NASA Group Achievement Award for Mars Exploration Rover mission team, Hundred Talent Program Award from CAS, and first prize of Surveying and Mapping Science and Technology Progress Award from China Society for Surveying, Mapping and Geoinformation.

Jürgen Oberst (Ph.D.: The University of Texas, 1989) is a professor and faculty member of the Technical University Berlin, Institute of Geodesy and Geoinformation Science, and head of the Planetary Geodesy Department of the DLR Institute of Planetary Research. He was/is involved in many past/ongoing planetary science missions specializing in the areas of planetary geodesy and geophysics. He is a Co-Investigator in the team of the HRSC (High Resolution Stereo Camera) on Mars Express and also a member of instrument and science teams of NASA's Lunar Reconnaissance Orbiter and MESSENGER missions. He is also the Science Coordinator for the Laser Altimeter Experiments BELA and GALA on ESA's BepiColombo and the JUICE missions, soon to be launched. Prof. Oberst was winner of a Megagrant awarded by the Ministry of Education and Science of the Russian Federation and consequently Leading Scientist of the Extraterrestrial Laboratory of MIIGAiK (Moscow State University for Geodesy and Cartography) from 2011 to 2015. He was/is PI of planetary science projects supported by the European Union, by ESA, the Helmholtz Association, and the German Science Foundation. Prof. Oberst is member of the IAU/IAG working group on Coordinates and Rotational Elements of the Planets and Satellites as well as Co-chair of the ISPRS Planetary Remote Sensing and Mapping Working Group.

Irina Karachevtseva (Ph.D. in Moscow State University of Geodesy and Cartography (MIIGAiK), 2005) is head of MIIGAiK Extraterrestrial Laboratory (MExLab). She was involved in preparation of Russian mission to Phobos (Phobos-Grunt, 2011) and received the certificate of Helmholtz Association, Germany (2014) for the successful implementation of scientific research within the framework of the Helmholtz-Russia Joint Research Groups (HRJRG-205) focused on Phobos study. Irina Karachevtseva is an initiator and responsible editor of the collective monograph "Atlas of Phobos" (MIIGAiK, 2015), which includes more than 40 maps and the results of studies of this Martian satellite. A set of planetary maps for the Moon, Phobos, and Mercury have been published recently under her leadership. One of her research branches is cartographic support of future Russian landing missions to the Moon: Luna-Glob (Luna-25, 27) and planning of future orbital mission Luna-Resource (Luna-26). Irina Karachevtseva is the co-chair of Planetary Remote Sensing and Mapping Working Group of the International Society for Photogrammetry and Remote Sensing (ISPRS) and co-chair of the Planetary Cartography Commission of the International Cartographic Association (ICA).

Planetary Remote Sensing and Mapping – Wu et al.
© 2019 Taylor & Francis Group, London, ISBN 978-1-138-58415-0

Contributors

J. W. Backer
Astrogeology Science Center
U. S. Geological Survey
Flagstaff, Arizona, USA

M. Bhatt
Physical Research Laboratory
Gujarat, India

A. K. Boyd
Lunar and Planetary Laboratory
University of Arizona
Tucson, Arizona, USA
and
School of Earth and Space Exploration
Arizona State University
Tempe, Arizona, USA

R. Bugiolacchi
Space Science Laboratory
Macau University of Science and Technology
Taipa, Macau

S. Burmeister
Institute of Planetary Research
German Aerospace Center (DLR)
Berlin, Germany

B. Chen
Beijing Institute of Spacecraft System
Engineering
Beijing, China

Y. Chu
Beijing Institute of Spacecraft System
Engineering
Beijing, China

D. Cook
Astrogeology Science Center
U. S. Geological Survey
Flagstaff, Arizona, USA

K. Di
State Key Laboratory of Remote Sensing
Science
Institute of Remote Sensing and Digital Earth
Chinese Academy of Sciences
Beijing, China

J. Dong
Beijing Institute of Spacecraft System
Engineering
Beijing, China

J. Dong
Changchun Institute of Optics, Fine Mechanics
and Physics
Chinese Academy of Sciences
Changchun, China

K. Edmundson
Astrogeology Science Center
U. S. Geological Survey
Flagstaff, Arizona, USA

S. Elgner
Institute of Planetary Research
German Aerospace Center (DLR)
Berlin, Germany

A. Fennema
Lunar and Planetary Laboratory
University of Arizona
Tucson, Arizona, USA

I. Foroughi
Department of Geodesy and Geomatics
University of New Brunswick
Fredericton, Canada

D. Galuszk
Astrogeology Science Center
U. S. Geological Survey
Flagstaff, Arizona, USA

S. van Gasselt
Department of Land Economics
National Chengchi University
Taipei, Taiwan

A. S. Garov
Moscow State University of Geodesy and Cartography (MIIGAiK)
Moscow, Russia

A. Grumpe
Technical University of Dortmund
Dortmund, Germany

D. Guo
Institute of Remote Sensing and Digital Earth
Chinese Academy of Sciences
Beijing, China

K. Gwinner
Institute of Planetary Research
German Aerospace Center (DLR)
Berlin, Germany

B. Giese
Institute of Planetary Research
German Aerospace Center (DLR)
Berlin, Germany

C. Güttler
Max Planck Institute for Solar System Research
Göttingen, Germany

T. Hare
Astrogeology Science Center
U. S. Geological Survey
Flagstaff, Arizona, USA

H. Hargitai
NASA Ames Research Center
Moffett Field, CA, USA

R. Heyd
Lunar and Planetary Laboratory
University of Arizona
Tucson, Arizona, USA

H. Hiesinger
Institut für Planetologie
Westfälische Wilhelms-Universität Münster
Münster, Germany

E. Howington-Kraus
Astrogeology Science Center
U. S. Geological Survey
Flagstaff, Arizona, USA

H. Hu
Department of Land Surveying and Geo-Informatics
The Hong Kong Polytechnic University
Hung Hom, Kowloon, Hong Kong

H. Hussmann
Institute of Planetary Research
German Aerospace Center (DLR)
Berlin, Germany

M. Jia
State Key Laboratory of Remote Sensing Science
Institute of Remote Sensing and Digital Earth
Chinese Academy of Sciences
Beijing, China

Y. Jia
Beijing Institute of Spacecraft System Engineering
Beijing, China

I. Karachevtseva
Moscow State University of Geodesy and Cartography (MIIGAiK)
Moscow, Russia

E. Kersten
Institute of Planetary Research
German Aerospace Center (DLR)
Berlin, Germany

R. L. Kirk
Astrogeology Science Center
U. S. Geological Survey
Flagstaff, Arizona, USA

A. A. Kokhanov
Moscow State University of Geodesy and Cartography (MIIGAiK)
Moscow, Russia

N. A. Kozlova
Moscow State University of Geodesy and Cartography (MIIGAiK)
Moscow, Russia

H. Lin
Institute of Remote Sensing and Digital Earth
Chinese Academy of Sciences
Beijing, China

B. Liu
State Key Laboratory of Remote Sensing Science
Institute of Remote Sensing and Digital Earth
Chinese Academy of Sciences
Beijing, China

Z. Liu
State Key Laboratory of Remote Sensing Science
Institute of Remote Sensing and Digital Earth
Chinese Academy of Sciences
Beijing, China

W. C. Liu
Department of Land Surveying and Geo-Informatics
The Hong Kong Polytechnic University
Hung Hom, Kowloon, Hong Kong

A. Lompart
Technical University of Dortmund
Dortmund, Germany

U. Mall
Max-Planck-Institut für Sonnensystemforschung
Göttingen, Germany

N. Manaud
SpaceFrog Design
Toulouse, France

E. V. Matveev
Moscow State University of Geodesy and Cartography (MIIGAiK)
Moscow, Russia

A. S. McEwen
Lunar and Planetary Laboratory
University of Arizona
Tucson, Arizona, USA

Q. Meng
Changchun Institute of Optics, Fine Mechanics and Physics
Chinese Academy of Sciences
Changchun, China

S. D. Mirchandani
Lunar and Planetary Laboratory
University of Arizona
Tucson, Arizona, USA

M. Mühlbauer
Technical University of Dortmund
Dortmund, Germany

A. Naß
Institute of Planetary Research
German Aerospace Center (DLR)
Berlin, Germany

J. Oberst
Institute of Planetary Research
German Aerospace Center (DLR)
Berlin, Germany
and
Institute for Geodesy and Geoinformation Science
Technical University of Berlin
Berlin, Germany

M. Peng
State Key Laboratory of Remote Sensing Science
Institute of Remote Sensing and Digital Earth
Chinese Academy of Sciences
Beijing, China

J. Ping
National Astronomical Observatories
Chinese Academy of Sciences
Beijing, China

F. Preusker
Institute of Planetary Research
German Aerospace Center (DLR)
Berlin, Germany

W. Rao
Beijing Institute of Spacecraft System Engineering
Beijing, China

B. Redding
Astrogeology Science Center
U. S. Geological Survey
Flagstaff, Arizona, USA

T. Roatsch
Institute of Planetary Research
German Aerospace Center (DLR)
Berlin, Germany

Zh. F. Rodionova
Sternberg Astronomical Institute
Moscow Lomonosov University
Moscow, Russia

D. Rommel
Technical University of Dortmund
Dortmund, Germany

A. P. Rossi
Department of Physics and Earth Sciences
Jacobs-University Bremen
Bremen, Germany

A. L. Salih
Technical University of Dortmund
Dortmund, Germany

P. Schulte
Technical University of Dortmund
Dortmund, Germany

D. Shoji
Earth-Life Science Institute
Tokyo Institute of Technology
Japan

J. Skinner, Jr.
Astrogeology Team
U. S. Geological Survey
Flagstaff, Arizona, USA

F. Sohl
Institute of Planetary Research
German Aerospace Center (DLR)
Berlin, Germany

A. Stark
Institute of Planetary Research
German Aerospace Center (DLR)
Berlin, Germany

G. Steinbrügge
Institute of Planetary Research
German Aerospace Center (DLR)
Berlin, Germany

Z. Sun
Beijing Institute of Spacecraft System
Engineering
Beijing, China

S. S. Sutton
Lunar and Planetary Laboratory
University of Arizona
Tucson, Arizona, USA

R. Tenzer
Department of Land Surveying and
Geo-Informatics
The Hong Kong Polytechnic
University
Hung Hom, Kowloon, Hong Kong

M. Wählisch
Institute of Planetary Research
German Aerospace Center (DLR)
Berlin, Germany

C. Wang
Beijing Institute of Spacecraft System
Engineering
Beijing, China

D. Wang
Changchun Institute of Optics, Fine Mechanics
and Physics
Chinese Academy of Sciences
Changchun, China

Y. Wang
Department of Land Surveying and
Geo-Informatics
The Hong Kong Polytechnic University
Hung Hom, Kowloon, Hong Kong

W. Wan
State Key Laboratory of Remote Sensing
Science
Institute of Remote Sensing and Digital
Earth
Chinese Academy of Sciences
Beijing, China

K. Wickhusen
Institute of Planetary Research
German Aerospace Center (DLR)
Berlin, Germany

C. Wöhler
Technical University of Dortmund
Dortmund, Germany

B. Wu
Department of Land Surveying and Geo-Informatics
The Hong Kong Polytechnic University
Hung Hom, Kowloon, Hong Kong

X. Wu
Institute of Remote Sensing and Digital Earth
Chinese Academy of Sciences
Beijing, China

X. Xin
State Key Laboratory of Remote Sensing Science
Institute of Remote Sensing and Digital Earth
Chinese Academy of Sciences
Beijing, China

Y. Yang
Institute of Remote Sensing and Digital Earth
Chinese Academy of Sciences
Beijing, China

X. Zhang
Institute of Remote Sensing and Digital Earth
Chinese Academy of Sciences
Beijing, China

A.Yu. Zharkova
Moscow State University of Geodesy and Cartography (MIIGAiK)
Moscow, Russia
and
Sternberg Astronomical Institute
Moscow Lomonosov University
Moscow, Russia

A.E. Zubarev
Moscow State University of Geodesy and Cartography (MIIGAiK)
Moscow, Russia

W. Zuo
National Astronomical Observatories
Chinese Academy of Sciences
Beijing, China

Section I

Reference systems of planetary bodies

Planetary Remote Sensing and Mapping – Wu et al.
© 2019 Taylor & Francis Group, London, ISBN 978-1-138-58415-0

Chapter 1

A technical framework for construction of new-generation lunar global control network using multi-mission data

K. Di, B. Liu, M. Peng, X. Xin, M. Jia, W. Zuo, J. Ping, B. Wu and J. Oberst*

ABSTRACT: A lunar global control network provides geodetic datum and control points for mapping of the lunar surface. The widely used Unified Lunar Control Network 2005 (ULCN2005) was built based on a combined photogrammetric solution of Clementine images acquired in 1994 and earlier photographic data. In this research, we propose an initiative and technical framework for construction of a new-generation lunar global control network using multi-mission data newly acquired in the 21st century, which have much better resolution and precision than the old data acquired in the last century. The new control network will be based on a combined photogrammetric solution of an extended global image and laser altimetry network. The five lunar laser ranging retro-reflectors, which can be identified in LROC NAC images and have cm-level 3D position accuracy, will be used as absolute control points in the least-squares photogrammetric adjustment. Recently, a new radio total phase ranging method has been developed and used for high-precision positioning of Chang'E-3 lander; this shall offer a new absolute control point. Systematic methods and key techniques will be developed or enhanced, including rigorous and generic geometric modeling of orbital images, multi-scale feature extraction and matching among heterogeneous multi-mission remote sensing data, optimal selection of images at areas of multiple image coverages, and large-scale adjustment computation, etc. Based on the high-resolution new datasets and developed new techniques, the new generation of global control network is expected to have much higher accuracy and point density than the ULCN2005.

1 INTRODUCTION

As a realization of the lunar reference system, a lunar global control network (LGCN) provides geodetic datum and control points for mapping of the lunar surface, and it is of fundamental importance for both scientific and engineering applications. In history, earth-based telescope observations had been used to establish some local or regional lunar control point networks (i.e., catalogs of landmarks of accurately known coordinates) (Schimerman, 1973). Since the late 1960s, lunar control networks have been established by photogrammetric solutions using images taken by orbiting spacecraft. Well-known global lunar control networks include the Unified Lunar Control Network (Davies *et al.*, 1994), the Clementine Lunar Control Network (Edwards *et al.*, 1996), and the Unified Lunar Control Network 2005 (ULCN, 2005) (Archinal *et al.*, 2006). The widely used ULCN2005 was built based on a combined photogrammetric solution of 43,866 Clementine images acquired in 1994 and earlier photographic data; the resultant 3D positions of 272,931 points have horizontal accuracy of 100 m to a few hundred meters and vertical accuracy of 100 m level (Archinal *et al.*, 2006, 2007).

Since the beginning of the 21st century, many nations/organizations have successfully launched new lunar exploration missions, ushering in a new golden age of lunar exploration. Orbital missions, such as European Space Agency's SMART-1, Japan's SELENE (Kaguya), India's Chandrayaan-1, the United States' Lunar Reconnaissance Orbiter (LRO), and China's Chang'E-1 and Chang'E-2 missions, have acquired large volumes of high-resolution images and high-precision laser altimetry data. In addition, China's Chang'E-3 lander and rover successfully landed and conducted detailed in-situ investigation of the landing site (Liu, Di, Peng *et al.*, 2015).

For comparative and synergistic use of the lunar remote sensing data from multiple missions so as to obtain maximum value for science and exploration, the datasets must be co-registered in a common coordinate reference frame (Kirk *et al.*, 2012). A global lunar control network directly supports such co-registrations. Due to the various uncertainties of the orbits and the imaging sensors, there exist widespread spatial inconsistencies among these new high-resolution data. There also exist considerable differences between the new lunar remote sensing data and ULCN2005. It is highly desirable to construct a new-generation LGCN using the newly acquired multi-mission high-resolution data, to better support lunar scientific research and future lunar exploration missions.

Comparing with the datasets used to construct ULCN2005, the new datasets, if used in combination, are significantly better in terms of image resolution, stereo coverage, laser altimetry precision and point density, and orbit precision. For example, benefiting from lunar gravity field data by the GRAIL mission (Zuber *et al.*, 2013), LRO orbit determination reached an accuracy of ~20 m; the accuracy was further improved to ~14 m after incorporating crossovers of Lunar Reconnaissance Orbiter Laser Altimeter (LOLA) data (Mazarico *et al.*, 2012). Furthermore, the five lunar laser ranging retro-reflectors (LRRRs), which have cm-level 3D position accuracy, can be identified in high-resolution images, such as Lunar Reconnaissance Orbiter Camera (LROC) Narrow Angle Camera (NAC) images (Liu, Di, Wang *et al.*, 2015; Wagner *et al.*, 2012). Therefore, these LRRRs can serve as absolute control points in construction of the new LGCN. All these new capabilities make it feasible to construct a new-generation LGCN with much higher accuracy than that of ULCN2005 and the historical ones.

In this research, we propose an initiative for construction of a new generation LGCN using multi-mission data acquired in the 21st century. The data to be used, technical framework, and key techniques will be elaborated and discussed in the following sections.

2 DATA TO BE USED

Orbital images and laser altimetry data acquired by multiple missions will be used in construction of the new-generation LGCN. Their characteristics and contributions to the network are elucidated below.

2.1 *High-resolution imagery data*

2.1.1 Chang'E-2 stereo images

Launched on 1 October 2010, the Chang'E-2 (CE-2) orbiter carried a high-resolution CCD stereo camera, which acquired images with a spatial resolution of 7 m and 1.5 m respectively at the flight heights of 100 km and 15 km (Zhao *et al.*, 2011). The CE-2 CCD camera consists of two line arrays that are separately fixed on the same focal plane, thus offering forward- and backward-looking stereo images (viewing angle 7.98 and −17.2 degrees respectively) in the same track through push-broom imaging. The two line arrays share the same optical axis with a focal length of 144.4 mm. Each line array has 6144 pixels. By completion of the mission, the CE-2 CCD camera obtained 607 orbits of image data, with 7 m resolution images covering the entire Moon and 1.5 m resolution images covering the preselected landing site of Chang'E-3 (Zuo *et al.*, 2014). Up to now, CE-2 CCD image dataset is the highest-resolution stereo image dataset in the world that covers the entire Moon.

In previous research, Di *et al.* (2014) developed a self-calibration bundle adjustment method that can eliminate the inconsistencies of CE-2 CCD images (back-projection image residuals between images of the same track and neighboring tracks) from more than 20 pixels to subpixel level and also reduce the differences between CE-2 data and LOLA data by 9–10 m. Li *et al.* (2015) have been working on the production of new global topographic mapping products, i.e., digital elevation model (DEM) and digital orthophoto map (DOM), using selected 384 imagery strips; comparing with the LRRRs, the planimetric displacement of the CE-2 products is 21 m – 97 m, the height difference is 2 m – 19 m; comparing with LOLA DEM, the average of height difference is 43 m, and standard deviation is 110 m.

Due to its high resolution and coverage, the CE-2 CCD stereo image dataset will be one of the major data sources in construction of the new generation LGCN. We will select about 400 orbits of CE-2 stereo images out of 607 orbits to cover the entire Moon. The amount of data involved is over 2 TB.

2.1.2 LROC NAC images

NASA's LRO was launched in June 2009 and inserted into a circular (30–50 km), polar orbit. LROC consists of a wide-angle camera (WAC) and two narrow-angle cameras (Robinson *et al.*, 2010). WAC has a spatial resolution of 100 m and swath of 100 km, and NAC acquired images with a spatial resolution of 0.5–2 m and a swath of 5 km. Using data from the wide-angle camera, a near-global terrain model "GLD100" with a resolution of 100 m has been produced (Scholten *et al.*, 2012).

LROC NAC images have been widely used for 3D mapping of the priority sites, e.g., past and future landing sites. So far, LROC NAC images almost cover 99% of the Moon surface, but stereo coverage is only 4% (I. Haase and M. Henrikson, pers. comm., April. 2017). 2D global image mosaics of LRO NAC images have been produced and can be accessed through the Lunaserv map-server, using the "LROC NAC overlay" feature (Estes *et al.*, 2013). Due to its highest resolution among all lunar orbital images, LRO NAC images will be an important data source in construction of the new LGCN. Over 200,000 NAC images will be needed to cover the entire Moon with total amount of close to 50 TB data.

2.1.3 SELENE terrain camera images

SELENE (Kaguya) was launched in September 2007 and carried Terrain Camera (TC) consisting of two line arrays for stereo imaging. It acquired images covering over 99% of the lunar surface with a resolution of 10 m from the nominal altitude of 100 km. Haruyama *et al.* (2012) corrected the models of TC detector distortion and attachment angles and reduced the differences between TC DEMs and the laser altimeter measurements; subsequently, global DEM and DOM were produced and released with a resolution of 1024 pixels per degree (30 m pixel^{-1} at the equator). SELENE TC stereo images have slightly lower resolution that that of CE-2 stereo images, and can be used complementarily in this work.

2.2 *Laser altimetry data*

Launched in June 2009, LOLA acquires five parallel profiles, separated by ~56 m; shots along the track are ~10–12 m apart. The range resolution of LOLA data is 10 cm (Smith, Zuber, Neumann, *et al.*, 2010). LOLA DEMs with different resolution have been produced and released, e.g., a global DEM with a resolution of 256 pixels per degree (118 m pixel^{-1} at the equator) was produced using 6.5 billion LOLA measurements gathered between July 2009 and July 2013 (Smith, Zuber, Jackson, *et al.*, 2010). Currently, the LDEM_1024 is the highest resolution global DEM with a resolution of 1024 pixels per degree (30 m pixel^{-1} at the equator) (Smith, Zuber, Neumann, *et al.*, 2010).

Recently, Barker *et al.* (2016) produced a lunar DEM by co-registration and combining SELENE TC DEM with LRO laser altimetric data. The model, designated as SLDEM2015, covers latitudes within ±60°, at a horizontal resolution of 512 pixels per degree (~60 m at the equator) and a 3 to 4 m root-mean-square (RMS) elevation residuals to LOLA profiles.

As of March 2016, LOLA has obtained ~6.8 billion altimetric measurements and will continue to acquire high-precision altimetric points (Smith *et al.*, 2017). The LOLA data will be one of the major data sources in construction of the new-generation LGCN.

2.3 *Absolute control points*

It is critical and beneficial to have some absolute control points in image block adjustment to ensure high accuracy. The five LRRRs, established in missions of Apollo 11, 14, and 15 and the Lunokhod 1 and 2 missions, reach cm-level accuracy through long-term measurements and can be treated as absolute control points (Wagner *et al.*, 2012). Traditionally, it was very hard to incorporate such absolute control points in lunar mapping and establishment of lunar control network because the image resolutions are not sufficient to identify these LRRRs. With the advent of high-resolution images, particularly the up to 0.5 m resolution LRO NAC images, these LRRRs are clearly discernible (Liu, Di, Wang *et al.*, 2015; Wagner *et al.*, 2012). This makes it feasible and attractive to incorporate these LRRRs as absolute control point in construction of the new LCGN.

Recently, a new radio total phase ranging method has been developed and used for high-precision positioning of Chang'E-3 lander; this shall offer a new absolute control point (Ping, 2016).

3 TECHNICAL FRAMEWORK AND KEY TECHNIQUES

3.1 *Overall framework*

The new control network will be based on a combined photogrammetric solution of an extended global image and laser altimetry network. The five LRRRs and Chang'E-3 lander will be used as absolute control points in the global adjustment. The laser altimetry data will be used as vertical control. Figure 1.1 shows the overall technical framework for construction of the new-generation LGCN. Key techniques are explained and discussed in the following sub-sections.

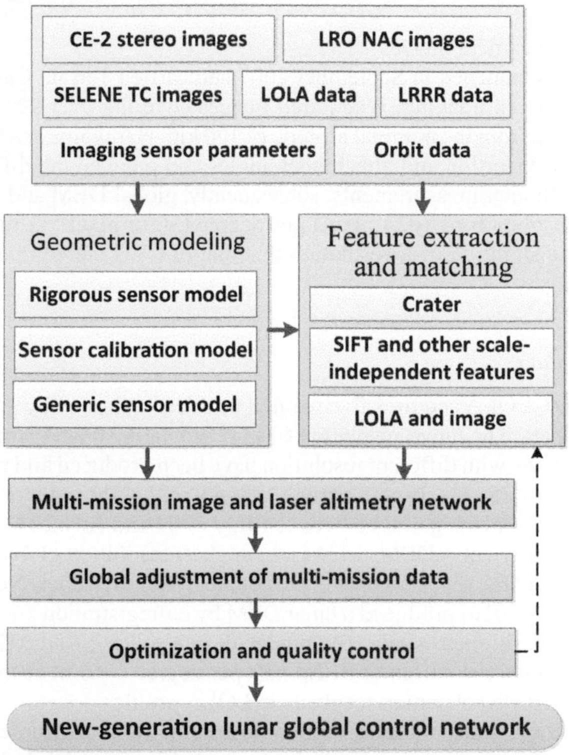

Figure 1.1. Overall technical framework

3.2 *Key techniques*

Systematic methods and key techniques will be developed, enhanced, and integrated, including rigorous and generic geometric modeling of orbital images, multi-scale feature extraction and matching among heterogeneous multi-mission remote sensing data, optimal selection of images at areas of multiple image coverages, and large-scale adjustment computation, etc.

3.2.1 Geometric modeling of orbital images

Geometric modeling of the multi-mission orbital images is the basis for photogrammetric processing of these images. Rigorous sensor models of the orbital images have been established by different groups of researchers based on collinearity equations with exterior orientation (EO) and interior orientation (IO) parameters (Di *et al.*, 2014; Haruyama *et al.*, 2012; Speyerer *et al.*, 2016; Tran *et al.*, 2010; Wu and Liu, 2017; Wu *et al.*, 2014).

IO refers to the transformation from image coordinate system (lines and samples) to the focal plane coordinate system centered at the principal point of the image according to the calibrated IO parameters of the camera. Any optical distortions of the camera lens should be corrected in the IO process. Due to the complexity and difference in design of the imaging sensors, the IO processes for different sensors are often different, e.g., there may be different numbers of IO parameters and the coordinate transformation equations may be different.

EO refers to the coordinate transformation from the focal plane coordinate system to an object space coordinate system, e.g., lunar body-fixed coordinate system. The rigorous sensor model (RSM) can be represented as collinearity equations as either (1) or (2) (Di *et al.*, 2014):

$$
\begin{aligned}
x &= -f\frac{a_1(X-X_s)+b_1(Y-Y_s)+c_1(Z-Z_s)}{a_3(X-X_s)+b_3(Y-Y_s)+c_3(Z-Z_s)} \\
y &= -f\frac{a_2(X-X_s)+b_2(Y-Y_s)+c_2(Z-Z_s)}{a_3(X-X_s)+b_3(Y-Y_s)+c_3(Z-Z_s)}
\end{aligned} \tag{1}
$$

$$
\begin{bmatrix} X-X_s \\ Y-Y_s \\ Z-Z_s \end{bmatrix} = \lambda \mathbf{R}_{\mathbf{ol}}\mathbf{R}_{\mathbf{bo}}\mathbf{R}_{\mathbf{ib}} \begin{bmatrix} x \\ y \\ -f \end{bmatrix} = \lambda \mathbf{R} \begin{bmatrix} x \\ y \\ -f \end{bmatrix} \tag{2}
$$

where (x, y) are the focal plane coordinates of an image point obtained through IO process; f is the focal length; (X, Y, Z) and (X_s, Y_s, Z_s) are the ground point and the perspective center position in lunar body-fixed (LBF) coordinate system, respectively; $\mathbf{R_{ib}}$ is the rotation matrix from the image space coordinate system (ISCS) to the spacecraft body coordinate system (BCS); $\mathbf{R_{bo}}$ is the rotation matrix from the BCS to the orbit coordinate systems (OCS); $\mathbf{R_{ol}}$ is the rotation matrix from the OCS to the LBF; λ is a scale factor; **R** represents the overall rotation matrix from the ISCS to the LBF; and a_i, b_i, and c_i $(i = 1, 2, 3)$ are the elements of the rotation matrix **R**, i.e., the functions of Euler angles $(\varphi, \omega, \kappa)$.

For push-broom imaging sensors, each image line has a different set of EO parameters (Xs, Ys, Zs, φ, ω, κ). The changes of the EO parameters over short trajectories are usually modelled using polynomials, e.g., third-order polynomials. Thus, when a long track of orbital image is processed, EO parameters are first divided into several segments and then fitted separately using third-order polynomials to ensure fitting precision.

The rigorous sensor models are generally complex and different for different sensors. It is preferable to be used in bundle adjustment of images from the same sensor of the same mission. But it would be very complex to use rigorous sensor models in adjustment of images from multi-missions.

Rational function model (RFM) is a commonly used generic geometric model in photogrammetric processing of high-resolution earth observation images. It is a mathematical fitting of rigorous geometric model, and has advantages of platform independence, simple form, and high calculation speed (Liu and Di, 2011; Di *et al.*, 2003). The RFM represents the relationship between image-space coordinates and object-space coordinates using the ratios of polynomials (Di *et al.*, 2003), as shown in Equation (3):

$$r = \frac{P_1(X,Y,Z)}{P_2(X,Y,Z)}$$
$$c = \frac{P_3(X,Y,Z)}{P_4(X,Y,Z)} \tag{3}$$

where (*r*, *c*) are the row and column coordinates of a image point, (*X*, *Y*, *Z*) are the ground coordinates of the point. The three-order polynomial P_i (i=1, 2, 3, and 4) has the following general form:

$$\begin{aligned} P_i(X,Y,Z) = {} & a_1 + a_2X + a_3Y + a_4Z + a_5XY + a_6XZ + a_7YZ + a_8X^2 \\ & + a_9Y^2 + a_{10}Z^2 + a_{11}XYZ + a_{12}X^3 + a_{13}XY^2 + a_{14}XZ^2 \\ & + a_{15}X^2Y + a_{16}Y^3 + a_{17}YZ^2 + a_{18}X^2Z + a_{19}Y^2Z + a_{20}Z^3 \end{aligned} \tag{4}$$

where a_1, a_2 . . . to a_{20} are the coefficients of the polynomial function *Pi*, named as the rational polynomial coefficients (RPCs). The RPCs of the image are derived by least-squares fitting using vast numbers of virtual control points generated by the RSM of the image.

Recently, Liu *et al.* (2014, 2016) studied the feasibility and accuracy of RFM fitting for lunar (LRO NAC, CE-1, CE-2) and Mars (HiRISE, HRSC) orbital images. Experimental results show that the traditional line-based RFM can precisely fit the rigorous sensor models with a RMS residual of 1/100 pixel level for orbiters without exposure time changing; while for orbiters with exposure time changing (e.g., CE-2 and HRSC), the proposed two solutions, time-based RFM or sensor-corrected images with line-base RFM, can also reach such a high-fitting precision (Liu *et al.*, 2016).

3.2.2 Multi-mission image and laser altimetry network

An extended global image and laser altimetry network is essential to a high-precision photogrammetric solution for the LGCN. The network should consist of a sufficient number of evenly distributed tie points (homologous feature points) that link all the images together based on feature extraction and matching among multi-mission remote sensing data. Due to the differences in image resolution (e.g., 0.5 m of LRO NAC vs. 7m of CE-2), pixel aspect ratio, pointing angle, illumination condition, etc., extraction and matching features among multi-mission images are very challenging. We propose feature extraction and matching among heterogeneous multi-mission remote sensing data using crater, scale-invariant feature transform (SIFT), and other scale-invariant features.

Impact craters are the most common surface features on the Moon. Many automated methods have been developed to detect and extract craters from lunar images and/or DEMs (e.g., Kang *et al.*, 2015). Bowl-shaped simple craters can be extracted from images with different resolutions using these methods with some manual checking and editing; the coordinates of the centres and sizes (i.e., radii) of the craters are then obtained through least- squares fitting. Matching of craters between images can be done based on distribution pattern matching with RANSAC-like outlier detection. As a result, images with different resolutions from different missions are matched and the craters' centres are used as tie points to build the image network.

SIFT is a popular algorithm to detect and describe local features, and is invariant to image scale and rotation (Lowe, 2004). Matching of SIFT features (keypoints) is realized using Euclidean distance of their feature vectors. SIFT matching can be applied as a complementary to crater matching in image areas where there are few craters. Other scale-invariant features, such as SURF (Bay *et al.*, 2008) and AKAZE (Alcantarilla *et al.*, 2013) features, will be compared with SIFT and can be used combinedly to extract and match more features.

Matching between image and altimetry data poses another challenge due to their heterogeneous natures. One practical way is to use DEM as a "bridge" between the 3D laser points and the image. The LOLA DEM, or the LOLA+TC merged DEM, can be used to generated simulated images using hill-shading techniques with the sun azimuth and elevation angles same as that of the images to be matched. Then, the simulated images are comparable to the actual images, and the image matching methods described above can be applied. Consequently, the altimetry points are matched to the images.

3.2.3 Optimal selection of images in areas of multiple image coverages

With continual data acquisition by multiple lunar orbiter missions, many areas of the lunar surface have been covered many times by different orbiters or the same orbiter. Optimal selection of images is very important in order to achieve the best geopositioning precision in those areas with multi-image coverages.

In order to automatically identify stereo image pairs for topographic mapping, Becker *et al.* (2015) provide recommended methods and criteria considering image overlap, 3D stereo imaging "strength", as well as similarity in spatial resolution, illumination, and spectral wavelength range.

Recently, we performed an empirical analysis of the geopositioning precision of multi-image triangulation using LROC NAC images at the Apollo-11 and Chang'E-3 landing sites. Experiments with multiple images indicate that utilizing more images produces higher precision than almost all dual-image models; meanwhile, using fewer images can produce better precision than using all available images together (Di *et al.*, 2016; Liu *et al.*, 2017). A progressive selection method has been proposed to find the best image combination for maximum precision. With further validation and improvements, the method can be used in optimal selection of images in construction of the image network.

In some special cases, spacecraft vibrations can cause angular "jitter" and image distortions. Images with jitter effects should be avoided in image selection for control network construction.

3.2.4 Global adjustment of multi-mission data

Global combined adjustment of the multi-mission image and laser altimetry network is the key to the success of the new generation LGCN. Considering the complexities and differences of multiple imaging sensors, we propose to perform global adjustment of multi-mission data based on the generic sensor model, RFM. The RPCs of the RFM are computed by fitting of the rigorous sensor models of the images of different sensors. To ensure the fitted RPCs are sufficiently accurate, we will perform regional bundle adjustment to improve the accuracy of the EO parameters of the participating images.

For regional bundle adjustment, the lunar globe is divided into to regional image blocks with some overlaps between neighboring blocks, and each block consists of multiple strips of images. There are may be tens to hundreds of images involved in a regional adjustment. For each block, images from different missions (i.e., CE-2, SELENE) will be adjusted separately based on rigorous sensor models with LOLA data as vertical constraints. Combined block adjustment of CE-2 stereo images (or SELENE images) and LOLA data can be achieved using the method developed in Wu *et al.* (2014) to improve the EO parameters and can also incorporate self-calibration parameters to improve IO parameters (Di *et al.*, 2014). Since we use LOLA data as a reference in the regional bundle adjustment and since the LRO orbit has been refined, it is not necessary to re-adjust the EO parameters of the LRO NAC images in this stage. But the geometric calibration results (Speyerer *et al.*, 2016; Wu and Liu, 2017) should be considered when generating RPCs for NAC images.

After the regional adjustment, the EO and IO parameters of the participating images are refined such that the inconsistencies between neighboring image strips are eliminated/reduced, and the 3D coordinates of the tie points are more consistent with the LOLA data. Based on the refined EO and IO parameters of the images, RPCs of the images will be obtained through least-squares fitting (Di *et al.*, 2003; Liu *et al.*, 2016).

Inconsistencies among multi-mission images, typically adjusted separately, will also be furthered reduced in the global adjustment. According to previous research and experiments, the regional adjustment should be able to reach a sub-pixel accuracy in image space; considering that the LRO orbit has been refined to an accuracy of 14 m and the LOLA data is used as reference, an accuracy of within 20 m in object space (i.e., the lunar body fixed frame) can be expected for the regional adjustment.

The global adjustment of the multi-mission image and laser altimetry network is based on a combined photogrammetric solution. Three types of tie points (intra-strip tie point linking the stereo images of the same strip, inter-strip tie points linking neighboring images of the same sensor, and inter-sensor tie point linking images from different sensors) will be used in the global adjustment. Weights for different observations (tie point measurements) from multi-mission data will be determined according to their a priori standard deviations, which are related to feature matching accuracy and image resolution. The unknowns for the global adjustment include the 3D ground positions of the tie points and the correction parameters (e.g., affine parameters in image space) for the RPCs. The five LRRRs will be used as absolute control points. The LOLA data will be used as vertical constraints in a way that the ground positions of the image tie points obtained through multi-image triangulation using the image EO parameters should be consistent with a local surface determined by the nearby LOLA points.

It is an interesting issue whether the LOLA points should also be adjusted in the global adjustment, as well as in the regional adjustment. In previous research, adjustment of LOLA point was done by back-projecting the LOLA points onto the images using the sensors models and taking the projected image points as observations; but due to their high accuracy, the LOLA points were only slightly adjusted in local areas (Wu *et al.*, 2014). For a global adjustment, this would significantly increase the computing effort. We will further study the necessity and effectiveness of adjusting LOLA points in the photogrammetric adjustment.

Comparing with the regional adjustment, the global adjustment further refines the 3D coordinates and the image model parameters in a global optimization manner and with absolute control from LRRRs. We expect the accuracy of the resultant tie points can reach 20 m to 30 m.

From a computation point of view, the global adjustment is the process of solving large-scale matrix equations. Sparse matrix technique is necessary to improve the efficiency of the solution. There will be hundreds of thousands of images involved in the global adjustment, which poses a great challenge to the adjustment models, as well as the feature extraction and matching algorithms. Adoption of Big Data technology in the whole process is necessary and may be indispensable.

As part of the outputs of global adjustment, adjusted 3D ground coordinates of the tie points, along with their image coordinates in the related images, will be gathered to form the new generation of LGCN. More matched feature points can be used to densify the LGCN; their ground coordinates are calculated by multi-image triangulation (space intersection) using the adjusted image model parameters, i.e., RPCs and the correction parameters. As a result, the accuracy and point density of the new LGCN should be much higher than those of the ULCN2005.

4 CONCLUDING REMARKS

With the availability of huge volumes of high-resolution images and altimetry data covering the entire lunar surface, and with the development of new photogrammetric techniques, it is both desirable and feasible to construct a new generation lunar global control network using high-resolution data newly acquired by multiple missions in the 21st century. In this initiative, we proposed a technical framework, described the relevant data, elaborated and discussed the key techniques for construction of a new-generation LGCN.

To realize this initiative not only requires enhancement and integration of advanced photogrammetric techniques, but also involves massive data processing work. It requires considerable funding to support multiple teams to work together for the common goal. Big Data technology should be considered and adopted in the implementation of the initiative. International collaboration is particular important and indispensable for construction of the new LGCN using multi-mission data.

ACKNOWLEDGEMENTS

This study was supported in part by National Natural Science Foundation of China under Grants 41671458 and 41590851.

NOTE

* Corresponding author. Email: dikc@radi.ac.cn

REFERENCES

Alcantarilla, P.F., Nuevo, J. & Bartoli, A. (2013) Fast explicit diffusion for accelerated features in nonlinear scale spaces. *Proceedings British Machine Vision Conference (BMVC)*, Bristol, UK. pp. 13.1–13.11.

Archinal, B.A., Rosiek, M.R., Kirk, R.L. & Redding, B.L. (2006) *The Unified Lunar Control Network 2005: USGS Open-File Report*. Available from: http://pubs.usgs.gov/of/2006/1367/ULCN2005-OpenFile.pdf.

Archinal, B.A., Rosiek, M.R., Kirk, R.L., Hare, T.L. & Redding, B.L. (2007) Final completion of the unified lunar control network 2005 and topographic model. *38th Lunar and Planetary Science Conference*, League City, TX, USA. p. 1904.

Barker, M.K., Mazarico, E., Neumann, G.A., Zuber, M.T., Haruyama, J. & Smith, D.E. (2016) A new lunar digital elevation model from the Lunar Orbiter Laser Altimeter and SELENE Terrain Camera. *Icarus*, 273, 346–355.

Bay, H., Ess, A., Tuytelaars, T. & Van Gool, L. (2008) SURF: speeded up robust features. *Computer Vision and Image Understanding (CVIU)*, 110(3), 346–359.

Becker, K.J., Archinal, B.A., Hare, T.M., Kirk, R.L., Howington-Kraus, E., Robinson, M.S. & Rosiek, M.R. (2015) Criteria for automated identification of stereo image pairs. *Lunar and Planetary Science Conference*, The Woodlands, TX, USA. pp. 46, 2703.

Davies, M.E., Colvin, T.R., Meyer, D.L. & Nelson, S. (1994) The unified lunar control network: 1994 version. *Journal of Geophysics Research*, 99(E11), 23211–23214.

Di, K., Ma, R. & Li, R. (2003) Rational functions and potential for rigorous sensor model recovery. *Photogrammetric Engineering and Remote Sensing*, 69(1), 33–41.

Di, K., Liu, Y., Liu, B., Peng, M. & Hu, W. (2014) A self-calibration bundle adjustment method for photogrammetric processing of Chang'E-2 stereo lunar imagery. *IEEE Transaction on Geoscience and Remote Sensing*, 52(9), 5432–5442.

Di, K., Xu, B., Liu, B., Jia, M. & Liu, Z. (2016) Geopositioning precision analysis of multiple image triangulation using LRO NAC lunar images. *Proceedings of 23rd ISPRS Congress, Commission IV*, 12–19 July, Prague, Czech Republic. pp. 369–374.

Edwards, K.E., Colvin, T.R., Becker, T.L., Cook, D., Davies, M.E., Duxbury, T.C., Eliason, E.M., Lee, E.M., McEwen, A.S., Morgan, H., Robinson, M.S. & Sorensen, T. (1996) Global digital mapping of the moon. *27th Lunar and Planetary Science Conference*, Houston, TX, USA. pp. 335–336.

Estes, N.M., Hanger, C.D., Licht, A.A. & Bowman-Cisneros, E. (2013) Lunaserv Web map service: history, implementation details, development, and uses. *44th Lunar and Planetary Science Conference*, The Woodlands, TX, USA. abstract #2069.

Haruyama, J., Hara, S., Hioki, K., *et al.* (2012) Lunar global digital terrain model dataset produced from SELENE (Kaguya) terrain camera stereo observations. *43rd Lunar and Planetary Science Conference*, The Woodlands, TX, USA. abstract #1200.

Kang, Z., Luo, Z., Hu, T. & Gamba, P. (2015) Automatic extraction and identification of lunar impact craters based on optical data and DEMs acquired by the Chang'E satellites. *IEEE Journal of Selected Topics in Applied Earth Observations and Remote Sensing*, 8(10), 4751–4761.

Kirk, R., Archinal, B., Gaddis, L. & Rosiek, M. (2012) Lunar cartography: progress in the 2000s and prospects for the 2010s. *International Archives of the Photogrammetry, Remote sensing and Spatial Information Sciences*, B4. pp. 489–494.

Li, C.L., Ren, X., Liu, J.J., Wang, F.F., Wang, W.R., Yan, W. & Zhang, G.H. (2015) A new global and high resolution topographic map product of the moon from Chang'E-2 image data. *46th Lunar Planetary Science Conference*, The Woodlands, TX, USA. abstract #1638.

Liu, B., Liu, Y., Di, K. & Sun, X. (2014) Block adjustment of Chang'E-1 images based on rational function model. *Remote Sensing of the Environment: 18th National Symposium on Remote Sensing of China*. International Society for Optics and Photonics. Wuhan, China. pp. 91580G–91580G.

Liu, B., Xu, B., Di, K. & Jia, M. (2016) A solution to low RFM fitting precision of planetary orbiter images caused by exposure time changing. *International Archives of the Photogrammetry, Remote Sensing and Spatial Information Sciences*, XLI-B4. pp. 441–448.

Liu, B., Jia, M., Di, K., Oberst, J., Xu, B. & Wan, W. (2017) Geopositioning precision analysis of multiple image triangulation using LROC NAC lunar images. *Planetary and Space Science*. http://dx.doi.org/10.1016/j.pss.2017.07.016.

Liu, B., Di, K., Wang, B., Tang, G., Xu, B., Zhang, L. & Liu, Z. (2015) Positioning and precision validation of Chang'E-3 Lander based on multiple LRO NAC images (in Chinese with English abstract). *Chinese Science Bulletin*, 60, 2750–2757.

Liu, Y. & Di, K. (2011) Evaluation of rational function model for geometric modeling of Chang'E-1 CCD images. *International Achieves of the Photogrammetry, Remote Sensing and Spatial Information Sciences*, Guilin, China, 38, Part 4/W25. pp. 121–125.

Liu, Z., Di, K., Peng, M., Wan, W., Liu, B., Li, L., Yu, T., Wang, B., Zhou, J. & Chen, H. (2015) High precision landing site mapping and rover localization for Chang'e-3 mission. *Science China-physics Mechanics & Astronomy*, 58(1), 1–11.

Lowe, D.G. (2004) Distinctive image features from scale-invariant keypoints. *International Journal of Computer Vision*, 60(2), 91–110.

Mazarico, E., Rowlands, D.D., Neumann, G.A., Smith, D.E., Torrence, M.H., Lemoine, F.G. & Zuber, M.T. (2012) Orbit determination of the Lunar Reconnaissance Orbiter. *Journal of Geodesy*, 86(3), 193–207.

Ping, J.S. (2016) Experiment of lunar radio phase ranging using Chang'E-3 lander. *47th Lunar and Planetary Science Conference*, The Woodlands, TX, USA. p. 1339.

Robinson, M., Brylow, S., Tschimmel, M., *et al.* (2010) Lunar Reconnaissance Orbiter Camera (LROC) instrument overview. *Space Science Reviews*, 150, 81–124.

Schimerman, L.A. (1973) *Lunar Cartographic Dossier*, Volume I. NASA and the Defense Mapping Agency, St. Louis, MO, USA.

Scholten, F., Oberst, J., Matz, K.D., Roatsch, T., Wählisch, M., Speyerer, E. & Robinson, M. (2012) GLD100: the near-global lunar 100 m raster DTM from LROC WAC stereo image data. *Journal of Geophysical Research: Planets*, 117(E12).

Smith, D.E., Zuber, M.T., Jackson, G.B., *et al.* (2010) The Lunar Orbiter Laser Altimeter investigation on the Lunar Reconnaissance Orbiter mission. *Space Science Review*, 150, 209–241.

Smith, D.E., Zuber, M.T., Neumann, G.A., *et al.* (2010) Initial observations from the Lunar Orbiter Laser Altimeter (LOLA). *Geophysical Research Letters*, 37, L18204.

Smith, D.E., Zuber, M.T., Neumann, G.A., *et al.* (2017) Summary of the results from the lunar orbiter laser altimeter after seven years in lunar orbit. *Icarus*, 283, 70–91.

Speyerer, E.J., Wagner, R.V., Robinson, M.S., Licht, A., Thomas, P.C., Becker, K., Anderson, J., Brylow, S.M., Humm, D.C. & Tschimmel, M. (2016) Pre-flight and onorbit geometric calibration of the Lunar Reconnaissance Orbiter Camera. *Space Science Review*, 200, 357–392.

Tran, T., Howingtonkraus, E., Archinal, B., Rosiek, M., Lawrence, S., Gengl, H., Nelson, D., Robinson, M., Beyer, R., Li, R., Oberst, J. & Mattson, S. (2010) Generating digital terrain models from LROC stereo images with SOCET SET. *41th Lunar and Planetary Science Conference*, The Woodlands, TX, USA. p. 2515.

Wagner, R.V., Speyerer, E.J., Burns, K.N., Danton, J. & Robinson, M.S. (2012) Revised coordinates for Apollo hardware. *International Archives of the Photogrammetry, Remote Sensing and Spatial Information Sciences*, XXXIX-B4. pp. 517–521.

Wu, B. & Liu, W.C. (2017) Calibration of boresight offset of LROC NAC imagery for precision lunar topographic mapping. *ISPRS Journal of Photogrammetry and Remote Sensing*, 128, 372–387.

Wu, B., Hu, H. & Guo, J. (2014) Integration of Chang'E-2 imagery and LRO laser altimeter data with a combined block adjustment for precision lunar topographic modeling. *Earth and Planetary Science Letters*, 391, 1–15.

Zhao, B., Yang, J., Wen, D., Gao, W., Chang, L., Song, Z., Xue, B. & Zhao, W. (2011) Overall scheme and on-orbit images of Chang E-2 lunar satellite CCD stereo camera. *Science in China Series E Technological Sciences*, 54(9), 2237–2242.

Zuber, M.T., Smith, D.E., Watkins, M.M., *et al.* (2013) Gravity field of the moon from the Gravity Recovery and Interior Laboratory (GRAIL) mission. *Science*, 339, 668–671.

Zuo, W., Li, C. & Zhang, Z. (2014) Scientific data and their release of Chang'E-1 and Chang'E-2. *Chinese Journal of Geochemistry*, 33, 24–44.

Planetary Remote Sensing and Mapping – Wu et al.
© 2019 Taylor & Francis Group, London, ISBN 978-1-138-58415-0

Chapter 2

Basic geodetic and dynamical parameters of Saturn's moon Enceladus

A key target of future exploration

A. Stark, H. Hussmann, J. Oberst, B. Giese, F. Sohl, D. Shoji, K. Wickhusen and M. Wählisch

ABSTRACT: The small satellite Enceladus is moving near the equatorial plane and deep in the gravity field of the giant planet Saturn. Owing to tidal interaction with its primary, Enceladus has adopted a pronounced three-axial ellipsoidal shape and is tidally locked. Its common rotational and orbital periods are about 1.37 days. As the equator of Saturn is inclined to the planet's orbital plane, Enceladus has pronounced seasons. This paper summarizes our current knowledge regarding the geodetic and dynamic parameters of this unique satellite. With the observed cryo-volcanism at the south pole, Enceladus is thought to harbor a water ocean beneath its icy crust and represents therefore a prime target for future space exploration. We point out that the years 2032/2033 will be very favorable for a lander mission in the south-pole area, which benefits from permanent illumination at that time.

1 INTRODUCTION

Saturn is accompanied by more than 60 satellites of varying size and orbital characteristics. Among these are six major satellites that are larger than 400 km in diameter. Sorted by distance from Saturn and beginning with the innermost satellite, these are: Mimas, Enceladus, Tethys, Dione, Rhea, Titan, and Iapetus. These satellites have generally small orbital eccentricities and have their orbit close to the equatorial plane of Saturn. With a radius of about 252 km, Enceladus is Saturn's sixth-largest moon (Fig. 2.1), intermediate in size between its neighboring satellites Mimas and Tethys with mean radii of 199 and 530 km, respectively.

NASA's Cassini spacecraft was the first to enter orbit about Saturn after Pioneer 11 (1979) and Voyager 1 and 2, which had performed flybys only (1980/1981). Cassini began its tour through the Saturn system in July 2004 including 22 targeted flybys of Enceladus and terminated its mission in 2017 by plunging into the atmosphere of Saturn. The decision to terminate the mission by completely destroying the spacecraft was taken to avoid contamination of the potentially habitable moons Enceladus and Titan. The Cassini spacecraft was equipped with the onboard Imaging Science Subsystem (ISS) consisting of a high-resolution Narrow Angle Camera (NAC) with a focal length of 2000 mm, and a Wide Angle Camera (WAC) with a focal length of 200 mm (Porco *et al.*, 2004). ISS delivered several hundred Enceladus images in different resolutions, from which control point networks and high-quality maps were produced.

Unexpectedly, Cassini's magnetometer showed evidence for a dynamic atmosphere at Enceladus during the first Enceladus flyby (Dougherty *et al.*, 2006). Later cryo-volcanic activity in the south-polar terrain of the satellite was identified by the Cassini ISS as the source of gas and dust particles evaporating and escaping from the surface (Porco *et al.*, 2006). The so-called 'tiger stripes' – almost parallel ridges near the south pole of Enceladus – were identified as cryo-volcanic sites that remained active throughout the 14 years of the Cassini mission (Fig. 2.1). There is no doubt that

Figure 2.1. Mosaic of 21 false-color images acquired by the Cassini ISS NAC on 14 July 2005. The bottom part shows the 'tiger stripes' at Enceladus' south pole.
(Image Credit: NASA/JPL/Space Science Institute)

liquid water is required as a source region of the jets within the tiger-stripe region. However, extension, location, formation, physical and chemical characteristics, and possible maintenance of such a liquid water reservoir are still unclear (for a comprehensive overview, see e.g., Spencer and Nimmo, 2013). After detection of the plumes, there has been a long debate whether Enceladus contains a global ocean or whether the reservoir of liquid water is locally confined to the south-polar region. The global ocean model is supported by the fracture pattern on Enceladus. The distribution and orientation of fractures on the icy surface of Enceladus are consistent with the stress field induced by non-synchronous rotation on long timescales when a global ocean is present (Patthoff and Kattenhorn, 2011). The latter is de-coupling the ice shell from the deep interior. Therefore, the ice shell responds to tidal torques without being tightly connected to the deep interior. Analysis of gravity measurements is also consistent with a global ocean (McKinnon, 2015). Further evidence for a global ocean comes from the large amplitude of physical librations (~0.12°), which requires a global liquid layer (Thomas *et al.*, 2016). Physical librations are small periodic changes of Enceladus' spin rate due to tidal torques exerted by Saturn on the tri-axial figure of Enceladus on the timescale of the satellite's orbital period of 1.37 days. Thomas *et al.* (2016) revealed that a 26-to-31-km thick global ocean under a 21-to-26-km ice layer is consistent with the observed libration amplitude. For a regional ocean, Enceladus' libration amplitude would be significantly reduced to about 0.03° (Thomas *et al.*, 2016), which is not consistent with the observation. However, a global subsurface ocean cannot be maintained for a long time by equilibrium tidal heating (~1.1 GW, Meyer and Wisdom, 2007) with the current orbital eccentricity of 0.0047 (Roberts and Nimmo, 2008). Thus, the eccentricity may have been larger in the past. However, the long-term orbital evolution of Enceladus is not sufficiently constrained to have a complete picture of the satellite's eccentricity as a function of time. Periodic changes in eccentricity along with periodic changes of the tidal heating rate have been studied in thermal-orbital evolution scenarios (e.g., Shoji *et al.*, 2014). In addition, with a global ocean it is difficult to explain the asymmetrical surface activity between the northern and southern hemisphere. In principle, strong tidal heating can be generated not only at the southern hemisphere but also at the northern hemisphere. Enceladus may have large lateral variation of ice thickness (McKinnon, 2015), and the variation may be related to the surface activity. In summary, there is strong evidence for a global subsurface ocean on Enceladus. However, it would be located under an ice shell of at least 21 km thickness. Some global ocean models suggest even 60 to 70 km ice thickness above the ocean (e.g., Běhounková *et al.*, 2015).

Strong evidence in favor of a regional ocean is that the plume emissions and high heat flux are concentrated only around the south-polar terrain (e.g., Porco *et al.*, 2006). If the ocean is localized at the south pole, strong tidal heating is induced only in that area. This can explain the concentration of the plume emission and the heat flux (Běhounková *et al.*, 2012; Tobie *et al.*, 2008). Compared to the global ocean model, a localized ocean can be relatively easily maintained by ~1.1 GW of the equilibrium heating rate. However, the tidal heat generated by the current eccentricity is much smaller than 1.1 GW in the case the ocean is localized (Běhounková *et al.*, 2012). Thus, evolution of the orbital eccentricity should be considered in the case of the regional ocean as well as for the global ocean model. It has been observed that the magnitude of the plume emissions changes with the orbit, which is induced by the dynamic change of Enceladus' tidal stress (Hedman *et al.*, 2013). However, the time change of the brightness of the emissions is delayed from the change of the stress field by about five hours (Nimmo *et al.*, 2014). This delay can be explained if the ocean extends 45° to 60° from the south pole and the depth to the ocean is around 30 km. However, a global ocean model with 60-to-70-km ice thickness is also consistent with the observation (Běhounková *et al.*, 2015).

While the amount and local distribution of water is not entirely clear on Enceladus, there is evidence about the presence of liquid water inside Enceladus. Getting access to these reservoirs, possibly near the tiger stripes with a landed element and assessing the biological potential of Enceladus' subsurface is one of the major goals in future exploration of the outer Solar System. For that purpose accurate maps and precise knowledge on the geodetic and dynamic parameters of Enceladus are required, which will be summarized in this chapter (see also Oberst *et al.* (2017)).

2 GEODETIC AND DYNAMICAL PARAMETERS

In the following we summarize the knowledge with respect to Enceladus' geodetic and dynamical parameters including orbital and rotational characteristics as well as implications from gravity field and shape measurements.

2.1 *Orbit*

Benefitting from Cassini radio tracking and astrometric observations, the ephemerides of the satellites of Saturn could be greatly improved. Enceladus revolves around Saturn in the planet's equatorial plane (inclination ~ 0.01°) on a slightly eccentric orbit at a semi-major axis of 238,020 km corresponding to 3.9 Saturn radii. The present orbit can be approximated by an ellipse with an eccentricity of $e = 0.0049 \pm 0.0011$ (where the given "uncertainty" of e visualizes the deviations of the orbit from a perfect ellipse). The distance to Saturn's center of mass varies between 237,250 and 239,570 km (Jacobson, 2015). Enceladus' orbital parameters are significantly influenced by its neighboring moons, notably Dione. Both satellites, Enceladus and Dione, are locked in a 2:1 mean-motion resonance. (Enceladus is completing two orbits around Saturn every one orbit completed by Dione.) Enceladus' orbit is subject to secular perturbations by its neighboring satellites, in particular by massive Titan, as well as by the resonant perturbations of Dione. The latter is forcing Enceladus' eccentricity to a mean value of $e = 0.0049$. However, due to secular perturbations, the osculating orbital elements including the eccentricity are oscillating around their mean values on various frequencies.

The sidereal (i.e. with respect to stars) orbital period of Enceladus about Saturn is $T_{orb} = 1.370218$ days. However, as Saturn's gravity field has significant non-spherical components mainly due to the planet's oblateness, the orbital ellipse precesses by 0.33796 ± 0.00037 ° d^{-1} with a period of 2.9164 ± 0.0032 years. Consequently, the time between consecutive pericenter passages (the anomalistic period) is 2.54 minutes longer than the sidereal period. The orbital plane of Enceladus is slightly inclined, on average by 5 arc seconds, to the equatorial plane of Saturn (Fig. 2.2). Thereby, with periodicities of 2.36 and 4.98 years the orbit plane precesses about Saturn's spin pole (Giese and Rambaux, 2015).

2.2 *Rotation*

2.2.1 Coordinate system definition

Solar System planets and satellites have common definitions for their coordinate systems, supported by the International Astronomical Union (e.g., Archinal *et al.*, 2018). The origin is located at the center of mass, the z-axis points into the direction of the spin pole (angular momentum vector), while the x- and y-axes define the equatorial plane. The x-axis passes through the prime meridian (0° longitude). For Enceladus, the prime meridian is fixed by the location of crater Salih, which is defined to be at 5° western longitude (Davies and Katayama, 1983). However, with the current definition of the prime meridian constant the prime meridian (and the x-axis) is offset from the direction to Saturn ("long-axis system") by ~4° on average.

2.2.2 Spin pole

Enceladus' spin pole is not fixed in inertial space but is believed to track the motion of the orbit pole as shown in Figure 2.2. Any obliquity (angle between spin pole and orbit pole) is predicted to be smaller than 1.62 arc seconds (Baland *et al.*, 2016). The amplitude of the orbit pole precession is on the order of 0.01° and so is the expected amplitude of spin pole precession. Pre-Cassini spin models for Enceladus recommended by IAU (Seidelmann *et al.*, 2002) had pole solutions as:

$$\begin{pmatrix} \delta(T) \\ \alpha(T) \end{pmatrix} = \begin{pmatrix} 83.52° - 0.004°\,T \\ 40.66° - 0.036°\,T \end{pmatrix} \tag{1}$$

where T is time measured in centuries from J2000 epoch.

However, new control point calculations using high-resolution Cassini images (Giese, 2014) have fixed the mean (over seven years) spin pole orientation at $(\delta,\alpha) = (83.54°, 40.59°)$. The new solution is close to the mean values of the orbit pole orientation (Fig. 2.2). This supports the assumption that the spin pole is tracking the motion of the orbit pole.

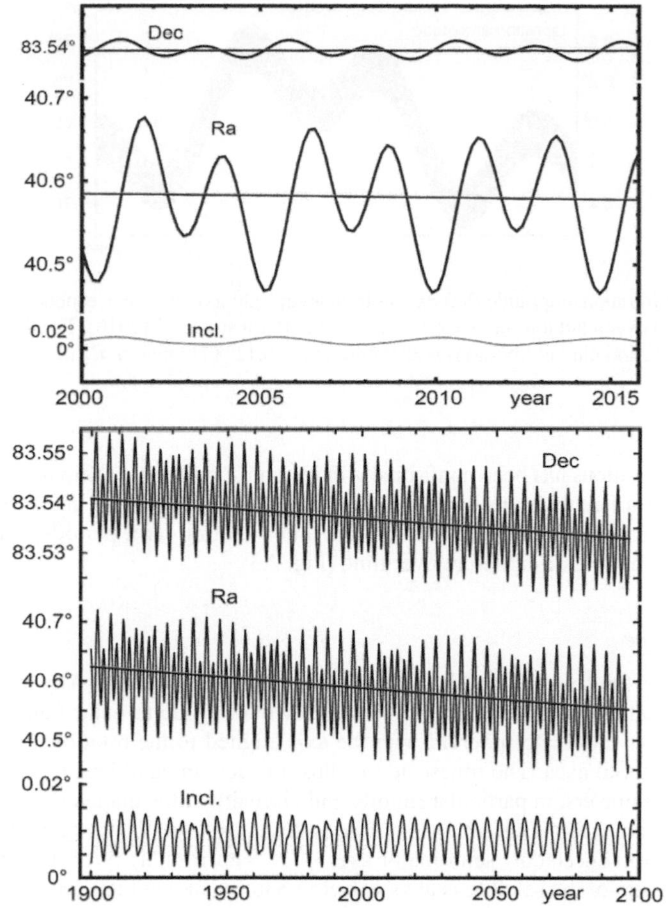

Figure 2.2. ICRF declination (**δ** = Dec) and right ascension (**α** = Ra) of Enceladus' orbit pole determined from JPL ephemeris data (sat365.bsp, Jacobson, 2015). The top and bottom panel depict the short-term (15 years) and long-term (200 years) evolution, respectively. Red lines indicate the orientation of Saturn's spin pole, which is the average of the oscillating orbit pole orientation of Enceladus. The lower curve in the upper panel shows the inclination of the orbit pole to Saturn's spin pole, varying between 0.0032° and 0.0139° around a mean value of 0.0086°. (Figure adopted from Oberst *et al.* [2017].)

2.2.3 Rotation and longitudinal libration

Enceladus is tidally locked in synchronous rotation, i.e. the spin period is always equal to the orbital period T_{orb} At uniform rotation (IAU model) the prime meridian angle is then given by (Archinal *et al.*, 2018):

$$W = 6.32^\circ + \frac{360^\circ}{T_{orb}}\, d = 6.32^\circ + 262.7318996^\circ\, d, \qquad (2)$$

where d is time in days from the J2000 epoch.

However, due to orbital perturbations and a non-spherical shape (see section 2.3), Enceladus experiences forced librations in longitude, which are superimposed on the uniform rotation. There are both diurnal (1.37 days) and long-period librations (3.9 and 11.2 years) (Rambaux *et al.*, 2010).

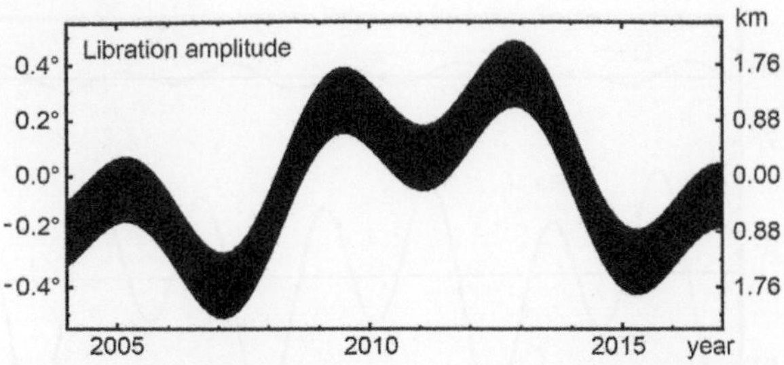

Figure 2.3. Total libration amplitude (left axis unit: degree; right axis unit: corresponding arc lengths along Enceladus' equator) over a full long-term cycle according to Rambaux *et al.* (2010). The thickness of the curve is due to the short-period diurnal librations with amplitude of 0.12° (Thomas *et al.*, 2016).

The presence of librations has been confirmed within control point calculations (Giese *et al.*, 2011) with specifically diurnal librations having amplitude of 0.12° (Thomas *et al.*, 2016). By combination of the three libration periods (one short-term and two long-term periods), the total libration amplitude shows an intricate signature over time (Fig. 2.3).

2.3 *Size and shape*

Size and shape are fundamental geodetic data for any planet or satellite. Any non-rigid spherical body that is constantly rotating around a fixed axis would – due to centrifugal forces – tend to assume the shape of an oblate spheroid with the axis aligned to the rotational axis being shorter than the two equatorial axes. The timescale to adjust to such an equilibrium state will depend on the rheological parameters, in particular rigidity and viscosity of the planetary material, and on the spin rate. In addition to rotational distortion, Enceladus is exposed to strong tidal forces due to its proximity to Saturn (Enceladus' semi-major axis ~ 3.9 Saturn radii). With Enceladus in its tidal lock, one hemisphere of the satellite is always facing Saturn (like in the Earth-Moon system), and one equatorial axis is always pointed to Saturn (neglecting deviations due to Enceladus' orbital eccentricity and librations). This leads on long timescales to a tri-axial ellipsoidal shape with the a-axis pointing to Saturn being the longest, and the rotational axis c being the shortest. Intermediate is the b-axis located in the equatorial plane and pointing in the direction of Enceladus' trailing hemisphere, completing a right-handed system. The satellite's equilibrium shape is a superposition of rotational and tidal effects. With respect to tides, caution has to be taken. We are referring here to tidal effects on an infinite timescale, implying that the satellite responds like a fluid on these long timescales (order of 10^6 or even 10^9 years). These, so-called static tides, are usually parameterized by the fluid Love number k_f, which depends on the internal density distribution but not on rheological parameters. It is the completely relaxed shape a fluid (or layered fluids) would assume after infinite time. However, in reality deviations arise from this ideal case due to radial and/or lateral density variations, non-zero rigidity of the planetary material even on these long timescales, and changes in the orbital and rotational configuration. Further modifications arise from Enceladus' non-negligible orbital eccentricity (see section 2.4).

Early shape models for Enceladus were produced from Voyager flyby observations using limb-fitting techniques (Dermott and Thomas, 1994). Using Cassini data, the shape models were updated by combinations of limb-fitting and control-point analysis (Table 2.1). In contrast to earlier findings, Enceladus was found to have a shape far from equilibrium. In particular, for Enceladus' three-axial ellipsoidal shape, a = 256.2 km, b = 251.4 km, c = 248.6 km, one may find $(a - c)/(b - c) = 2.7$, while for hydrostatic (tidal and rotational) equilibrium this parameter should be 4.2 (McKinnon, 2015).

Table 2.1 Shape parameters of Enceladus. *a*, *b*, and *c* are ellipsoid axes and *R* is the mean radius.

a [km]	*b [km]*	*c [km]*	*R [km]*	*Reference*
256.3 ± 0.3	247.3 ± 0.3	244.6 ± 0.5	249.4 ± 0.3	Dermott and Thomas (1994) (also: IAU: Davies *et al.* (1996))
256.6 ± 0.6	251.4 ± 0.2	248.3 ± 0.2	252.1 ± 0.2	Thomas *et al.* (2007) (also IAU: Archinal *et al.* [2018])
256.6 ± 0.3	251.4 ± 0.2	248.3 ± 0.2	252.1 ± 0.2	Thomas (2010)
256.2 ± 0.3	251.4 ± 0.2	248.6± 0.2	252.24 ± 0.2	Thomas *et al.* (2016)
256.53	251.45	248.66	252.22	Tajeddine *et al.* (2017)

2.4 *Diurnal tides*

If the orbit of Enceladus were circular, its tri-axial shape would be constant over time. However, as the actual orbit of Enceladus is elliptical, additional tidal bulges form in response to the periodic forcing on the orbital period of ~1.37 days. In this case rheological parameters, e.g. rigidity and viscosity, are crucial for the satellites' response on this much shorter forcing frequency (as compared to the infinite timescale discussed in section 2.3).

In case of eccentricity tides, the dynamic displacement of the tidal bulges is caused by two mechanisms (see the details in Murray and Dermott, 1999). One is due to changing distance to Saturn (radial tides) with the tidal bulges growing and decreasing according to changing tidal forces. The other mechanism is due to optical (also called geometrical) libration, which causes a slight shift (<0.6°) of the tidal bulge with respect to the direction to Saturn (librational tides). Both effects are of the same order of magnitude and depend on the orbital eccentricity of the satellite. In case of Enceladus with an eccentricity of 0.0049 forced by the resonance with Dione, the amplitude of the radial displacement is estimated at ~5 m (Hurford *et al.*, 2007, 2009).

By these periodical displacements of tidal bulges, changing stress patterns can be induced (Hurford *et al.*, 2007, 2009, 2012; Smith-Konter and Pappalardo, 2008). Recent analyses show evidence for a correlation of plume activity with the tidal stresses on the diurnal 1.37-day cycle. Due to normal stresses, water-filled cracks may open down to the liquid water reservoir located at least a few kilometers in the subsurface. Whereas the activity for individual jets is not synchronous with the variation of tidal stresses, the overall activity within the south-polar terrain varies in phase with the tidal frequency (Hedman *et al.*, 2013; Porco *et al.*, 2014).

The dynamic displacement of the tidal bulges produces frictional heat by dissipation within the ice. In fact, Enceladus is one of the most dissipative icy satellites in our Solar System. Due to conversion of orbital energy into thermal energy, Enceladus' eccentricity is normally expected to decrease and eventually drop to zero. However, the 2:1 orbital mean-motion resonance with Dione maintains Enceladus' eccentricity at a mean value of 0.0049. As a consequence diurnal tides are ongoing as long as the satellites are locked in resonance in spite of the strong dissipation inside Enceladus. This mechanism is very similar to Io in the Jupiter system being locked in resonance with Europa. Strong dissipation inside Io leads to partial melting of the mantle rock and is responsible for the vigorous volcanism observed on Io's surface. In spite of strong dissipation inside Io, the orbital eccentricity is not damped due to resonant forcing of Europa.

2.5 *Mass and gravity field*

With the Cassini mission it was possible to determine the mass and higher quadrupole moments of Enceladus' gravity field during close flybys. Radio Doppler data acquired by the Deep Space

Network result in $GM = 7.2111 \pm 0.0125$ km^{-3} s^{-2} or mass $M = 1.0805 \pm 0.0019 \times 10^{20}$ kg (Jacobson, 2015).

Hence, spacecraft approaching Enceladus have to cope with strong orbital perturbations by Saturn and are limited to move in so-called "quasi-satellite orbits" (Russell and Lara, 2009). The Hill sphere of Enceladus, in which Enceladus dominates gravitational motion, has a radius of only approximately 950 km, i.e. ~3.8 Enceladus radii (Spahn, Albers, *et al.*, 2006; Spahn, Schmidt, *et al.*, 2006).

In addition to GM, the degree-2 gravity potential has been determined from close flybys (Iess *et al.*, 2014). The only significant non-zero terms are the zonal harmonic $J_2 = 5435.2 \pm 34.9 \times 10^{-6}$ and the sectorial harmonic $C_{22} = 1549.8 \pm 15.6 \times 10^{-6}$ with a ratio $J_2/C_{22} = 3.51 \pm 0.05$. As for the shape parameters, this ratio differs from that expected for a differentiated body in hydrostatic equilibrium (expected value around 3.24). The inferred moment of inertia factor is 0.335 (Iess *et al.*, 2014), suggesting a differentiated body. Enceladus' mean density suggests a rock-to-ice ratio of about 60–40 wt%, which would imply a rock core of about 150–170 km thickness (depending on assumed rock density) and an H_2O layer of about 60-to-80-km thickness assuming full differentiation between the ice and rock component (Schubert *et al.*, 2007). Deviations from hydrostatic state and local variations either in ice thickness or in the distribution of rock inside the core refine the simple picture of spherically symmetric differentiation (Iess *et al.*, 2014). From the degree-2 gravity data McKinnon (2015) concludes that Enceladus has a core with radius and density of 190 km and 2450 kg m^{-3}, respectively, and that a global sub-surface ocean is covered by a thick (50-km average) ice shell that is substantially thinner (~30 km) beneath the south-pole terrain.

3 CONTROL POINT NETWORKS

Control point networks are essentially catalogues of prominent surface features for which body-fixed coordinates are precisely known. The coordinates of the points are typically determined jointly from original measurements of point coordinates in the large numbers (blocks) of overlapping images by so-called "bundle block adjustment techniques". The control point coordinates (when available in 3D) are an important framework for the production of shape models. Also, from the tracking of control points over time, rotational parameters of the planet or satellite may be determined.

An early control point network construction involved 38 Cassini images with resolutions ranging from 190 to 1220 meters per pixel (m/pixel) and covering a time span of six years. This resolution and time interval was appropriate for measuring the long-period librations shown in Figure 2.3 (Giese *et al.*, 2011). In total, 1057 image points corresponding to 186 individual ground points were measured, aiming at a dense and uniform distribution across the surface. These ground points have mean point precisions in (x, y, z) of (±318 m, ±288 m, ±28 m) (Giese *et al.*, 2011). More recently, new control point networks were built. The network by Thomas *et al.* (2016) involved 340 images and positions of 488 control points. Nadezhdina *et al.* (2016) measured 14,121 tie-point positions corresponding to 1128 control points with mean point precisions (reported as 1-sigma) in (x, y, z) of (150 m, 120 m, 90 m). The analysis of Enceladus' shape and topography by Tajeddine *et al.* (2017) contains 54 limb profiles with surface coordinates of 41,780 points as well as 6245 stereogrammetrically derived control points. The authors report a mean uncertainty of 155 m in the control point position, with a variation between 30 to 650 m depending on image resolution and number of observations per point. Another Enceladus global control network was produced at the USGS. A total of 586 images in CLR, GRN, UV3, and IR3 filters with a spatial resolution between 50 and 500 m pix^{-1} and with phase angles less than 120 degrees were selected to establish the control net (Becker *et al.*, 2016).

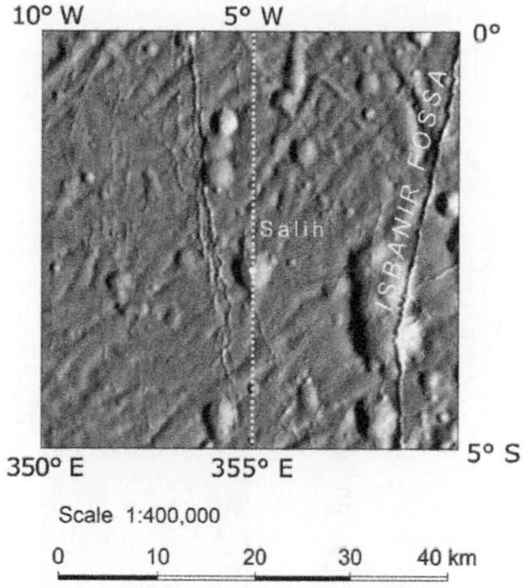

Figure 2.4. The crater Salih (marked by a white dot) located at 5°W (dashed line) defines the prime meridian of Enceladus (Adapted from Roatsch *et al.* [2018]. Credit: DLR)

4 MAPS

Using the images from ISS, Enceladus image mosaics and maps have been prepared by several teams worldwide – including a complete "atlas" (resolution of 1:400,000) (Roatsch *et al.*, 2018). Relevant map products include

- German Aerospace Center (DLR) map and atlas (Roatsch *et al.*, 2008, 2013, 2018)
- Lunar Planetary Institute (LPI) map (Schenk, 2014)
- Moscow State University of Geodesy and Cartography (MIIGAiK) map (Nadezhdina *et al.*, 2016; Zubarev *et al.*, 2014)
- US Geological Survey (USGS) map (Becker *et al.*, 2016; Bland *et al.*, 2015).

New controlled global, south, and north polar maps are published in 2016 and partially updated in 2018 (Becker *et al.*, 2016; Bland *et al.*, 2015). The maps use the surface position of the prime meridian as defined by the IAU cartography working group (Archinal *et al.*, 2018) through the small crater Salih (Fig. 2.4), located at 5°W longitude and 5.16°S latitude.

The Cassini imaging team proposed 64 names for prominent geological features, in addition to the 22 features already named by the Voyager team that are used in the maps (IAU WGPSN, 2018). By international agreement, features on Enceladus are named after people or locations in the medieval Middle Eastern literary epic *The Thousand Nights and a Night*. The quadrangle scheme of the DLR atlas consists of 15 tiles (Fig. 2.5), which conforms to the quadrangle scheme proposed by Greeley and Batson (1990) for larger satellites. Mosaics and maps are archived as standard products in the Planetary Data System (PDS) (https://pds-imaging.jpl.nasa.gov/volumes/carto.html) and are available at the DLR-Europlanet website (http://europlanet.dlr.de/Cassini-Atlases).

5 ILLUMINATION CONDITIONS

With its mean semi-major axis of 9.537 astronomical units (AU) and eccentricity of 0.054 Saturn's distance to the Sun varies between 9 AU and 10.1 AU. Thus, the total solar flux received in

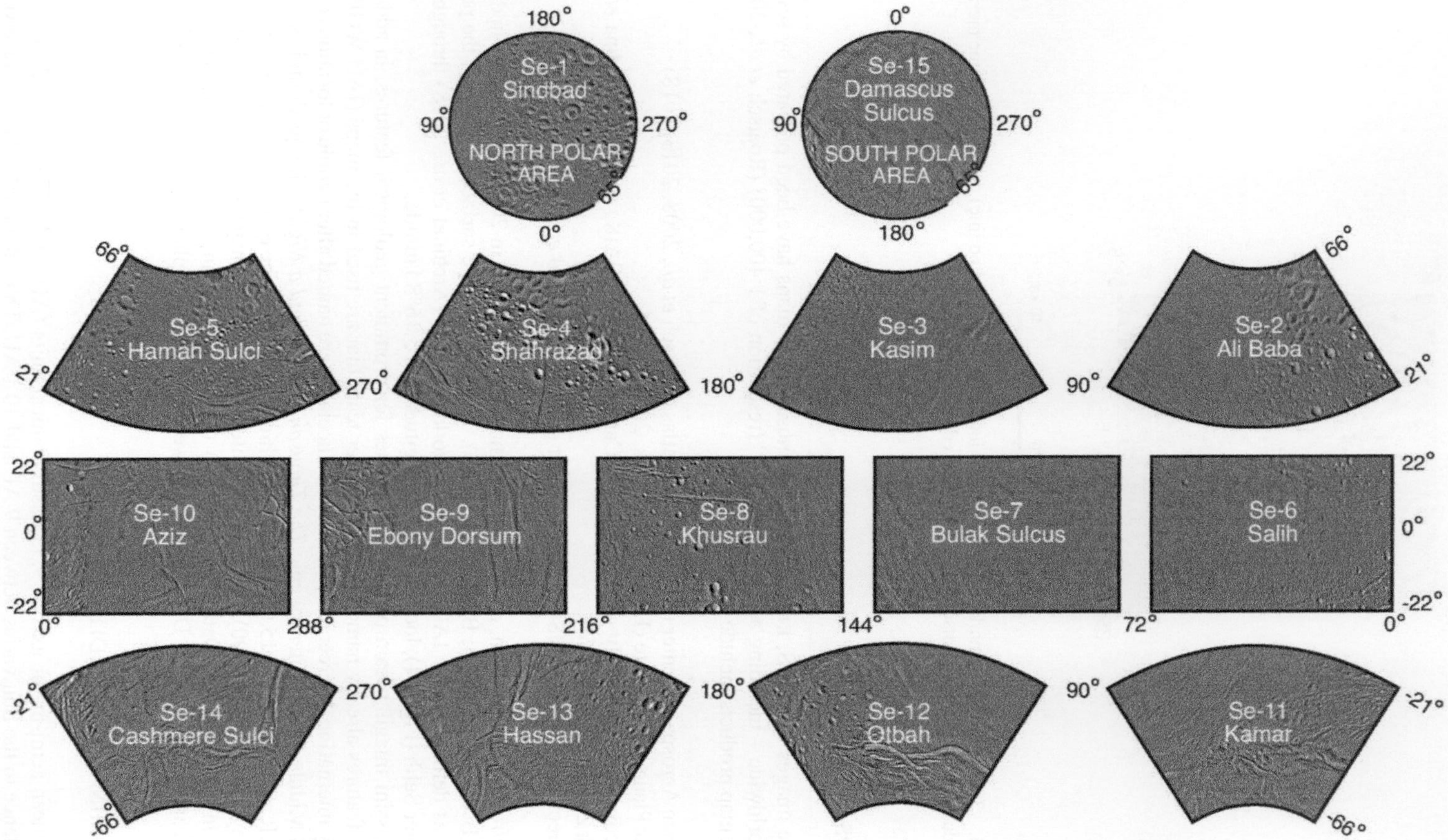

Figure 2.5. Quadrangle mapping scheme for Enceladus. Credit: DLR.

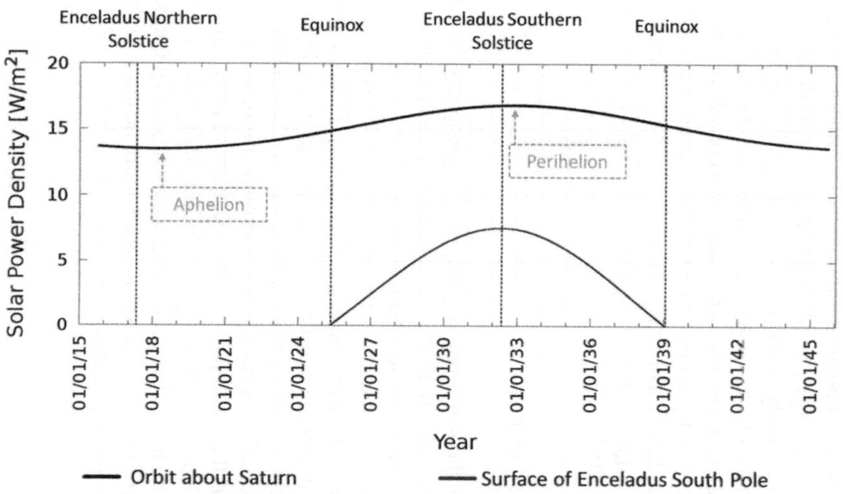

Figure 2.6. Solar flux for areas perpendicular to the incoming solar radiation (black) and for areas on the surface at the south pole (red). Saturn's aphelion and perihelion are also indicated. (Adapted from Oberst *et al.* [2017])

orbit around Saturn varies between 17 W m^{-2} and 13 W m^{-2} (1 m^2 unit area, pointed at the Sun). Saturn's equatorial plane is tilted by 26.73° with respect to its orbit plane, and as Enceladus' orbit inclination (with respect to Saturn's equatorial plane) and obliquity are small, the moon follows the 29.46-year seasonal cycle of its parent planet (Oberst *et al.*, 2017).

Saturn will be at its next equinox position and enter southern summer season in May 2025, which will last until January 2039. Summer peak (southern solstice) will be in May 2032 (Fig. 2.6). Hence, this season will last 13.7 years. Owing to the eccentric orbit of Saturn (and equinox times not being symmetric with respect to the solar orbit period), the northern summer season will be significantly longer (15.7 years). The seasonal patterns in the illumination levels are superposed by Saturn's changing solar distance. In November 2032, Saturn will be at its perihelion position, which will result in maximum solar irradiation on the southern hemisphere (peak solar power density of ~ 17 W m^{-2}). It is interesting to note that southern solstice almost exactly coincides with Saturn's perihelion passage providing maximum possible fluxes at Enceladus' south-pole region. This changes only slowly on the timescale of perihelion precession of Saturn's orbit. Whether this has an effect on the thermal state of the surface layers similar to effects suggested for the surface of Neptune's satellite Triton (Brown *et al.*, 1990) or whether this has an impact of the observed north and south asymmetry on Enceladus activity is not known. From the illumination perspective, the years 2032 and 2033 will be most favorable for a lander mission in the south-pole area, which would benefit from permanent illumination, with the Sun rising up to 26.7° elevation during the day (Oberst *et al.*, 2017).

6 SUMMARY AND CONCLUSION

With its erupting plumes in the south-pole region and its prospects for habitability, Enceladus is one of the most fascinating targets in the outer Solar System for future exploration. Current observations and model interpretations suggest that the small icy satellite contains a global water ocean underneath an ice shell of a few tens of kilometers in thickness. In view of this evidence for a global ocean, the asymmetry with respect to the ongoing thermal and dynamic activity on Enceladus, which seems to be confined to the south-polar region, is difficult to explain. Further investigation, including precise modeling of the geodetic state of Enceladus, is required to resolve this issue. We propose dedicated Enceladus flyby or orbiter missions, involving landed elements preferably to be deployed in the south-polar region (Dachwald *et al.*, 2016; Mitri *et al.*, 2018). Precise maps and exact geodetic characterization, including high-resolution and topographic mapping of the south-polar region, would be required to guarantee safe landings. Benefitting from the favorable illumination conditions at the south pole, the timeframe close to the year 2032 would be best suited for a lander in the south-pole area.

ACKNOWLEDGEMENTS

A. Stark was funded by a research grant from the Helmholtz Association and German Aerospace Center (DLR) (PD-308). The authors thank an anonymous reviewer for helpful comments for the improvement of an earlier version of the manuscript.

REFERENCES

Archinal, B.A., Acton, C.H., A'Hearn, M.F., Conrad, A., Consolmagno, G.J., Duxbury, T., Hestroffer, D., *et al.* (2018) Report of the IAU working group on cartographic coordinates and rotational elements: 2015. *Celestial Mechanics and Dynamical Astronomy*, 130(3), 22. doi:10.1007/s10569-017-9805-5.

Baland, R-M., Yseboodt, M. & Van Hoolst, T. (2016) The obliquity of Enceladus. *Icarus*, 268, 12–31. doi:10.1016/j.icarus.2015.11.039.

Becker, T.L., Bland, M.T., Edmundson, K.L., Soderblom, L.A., Takir, D., Patterson, G.W., Collins, G.C., Pappalardo, R.T., Roatsch, T. & Schenk, P.M. (2016) Completed global control network and basemap of Enceladus. *47th Lunar and Planetary Science Conference*, The Woodlands, TX, USA, abstract #2342.

Běhounková, M., Tobie, G., Choblet, G. & Čadek, O. (2012) Tidally-induced melting events as the origin of south-pole activity on Enceladus. *Icarus*, 219(2), 655–664. doi:10.1016/j.icarus.2012.03.024.

Běhounková, M., Tobie, G., Čadek, O., Choblet, G., Porco, C. & Nimmo, F. (2015) Timing of water plume eruptions on Enceladus explained by interior viscosity structure. *Nature Geoscience*, 8, 601. doi:10.1038/ngeo2475.

Bland, M.T., Becker, T.L., Edmundson, K.L., Patterson, G.W., Collins, G., Pappalardo, R.T., Kattenhorn, S., Roatsch, T. & Schenk, P. (2015) A new Enceladus base map and global control network in support of geological mapping. Paper presented at the *46th Lunar and Planetary Science Conference*, The Woodlands, TX, USA.

Brown, R.H., Kirk, R.L., Johnson, T.V. & Soderblom, L.A. (1990) Energy sources for Triton's geyser-like plumes. *Science*, 250(4979), 431–435. doi:10.1126/science.250.4979.431.

Dachwald, B., Kowalski, J., Baader, F., Espe, C., Feldmann, M., Francke, G. & Plescher, E. (2016) Enceladus explorer: next steps in the development and testing of a steerable subsurface ice probe for autonomous operation. *Enceladus and the Icy Moons of Saturn*, Boulder, CO, USA, abstract #3031.

Davies, M.E., Abalakin, V.K., Bursa, M., Lieske, J.H., Morando, B., Morrison, D., Seidelmann, P.K., Sinclair, A.T., Yallop, B. & Tjuflin, Y.S. (1996) Report of the IAU/IAG/COSPAR working group on cartographic coordinates and rotational elements of the planets and satellites: 1994. *Celestial Mechanics and Dynamical Astronomy*, 63(2), 127–148. doi:10.1007/bf00693410.

Davies, M.E. & Katayama, F.Y. (1983) The control networks of Mimas and Enceladus. *Icarus*, 53(2), 332–340. doi:10.1016/0019-1035(83)90153-7.

Dermott, S.F. & Thomas, P.C. (1994) The determination of the mass and mean density of Enceladus from its observed shape. *Icarus*, 109(2), 241–257. doi:10.1006/icar.1994.1090.

Dougherty, M.K., Khurana, K.K., Neubauer, F.M., Russell, C.T., Saur, J., Leisner, J.S. & Burton, M.E. (2006) Identification of a dynamic atmosphere at Enceladus with the Cassini magnetometer. *Science*, 311(5766), 1406–1409. doi:10.1126/science.1120985.

Giese, B. (2014) An upper limit on Enceladus' obliquity. *European Planetary Science Congress 2014*, Cascais, Portugal, abstract #EPSC2014-419.

Giese, B. & Rambaux, N. (2015) Enceladus' long-period physical librations. *European Planetary Science Congress 2015*, Nantes, France, abstract #EPSC2015-867.

Giese, B., Hussmann, H., Helfenstein, P., Thomas, P.C. & Neukum, G. (2011) Enceladus: evidence for librations forced by Dione. *EPSC-DPS Joint Meeting 2011*, abstract #EPSC-DPS2011-976.

Greeley, R. & Batson, R.M. (1990) *Planetary Mapping*, Volume 6. Cambridge University Press, Cambridge, UK.

Hedman, M.M., Gosmeyer, C.M., Nicholson, P.D., Sotin, C., Brown, R.H., Clark, R.N., Baines, K.H., Buratti, B.J. & Showalter, M.R. (2013) An observed correlation between plume activity and tidal stresses on Enceladus. *Nature*, 500, 182. doi:10.1038/nature12371.

Hurford, T.A., Helfenstein, P. & Spitale, J.N. (2012) Tidal control of jet eruptions on Enceladus as observed by Cassini ISS between 2005 and 2007. *Icarus*, 220(2), 896–903. doi:10.1016/j.icarus.2012.06.022.

Hurford, T.A., Helfenstein, P., Hoppa, G.V., Greenberg, R. & Bills, B.G. (2007) Eruptions arising from tidally controlled periodic openings of rifts on Enceladus. *Nature*, 447, 292. doi:10.1038/nature05821.

Hurford, T.A., Bills, B.G., Helfenstein, P., Greenberg, R., Hoppa, G.V. & Hamilton, D.P. (2009) Geological implications of a physical libration on Enceladus. *Icarus*, 203(2), 541–552. doi:10.1016/j.icarus.2009.04.025.

Iess, L., Stevenson, D.J., Parisi, M., Hemingway, D., Jacobson, R.A., Lunine, J.I., Nimmo, F., *et al.* (2014) The gravity field and interior structure of Enceladus. *Science*, 344(6179), 78–80. doi:10.1126/science.1250551.

Jacobson, R.A. (2015) SAT375. *JPL Satellite Ephemeris*. Available from: https://naif.jpl.nasa.gov/pub/naif/generic_kernels/spk/satellites/sat375.cmt

McKinnon, W.B. (2015) Effect of Enceladus's rapid synchronous spin on interpretation of Cassini gravity. *Geophysical Research Letters*, 42(7), 2137–2143. doi:10.1002/2015GL063384.

Meyer, J. & Wisdom, J. (2007) Tidal heating in Enceladus. *Icarus*, 188(2), 535–539. doi:10.1016/j.icarus.2007.03.001.

Mitri, G., Postberg, F., Soderblom, J.M., Wurz, P., Tortora, P., Abel, B., Barnes, J.W., *et al.* (2018) Explorer of Enceladus and Titan (E2T): investigating ocean worlds' evolution and habitability in the solar system. *Planetary and Space Science*, 155, 73–90. doi:10.1016/j.pss.2017.11.001.

Murray, C.D. & Dermott, S.F. (1999) *Solar System Dynamics*. Cambridge University Press, Cambridge, UK.

Nadezhdina, I.E., Zubarev, A.E., Brusnikin, E.S. & Oberst, J. (2016) A libration model for Enceladus based on geodetic control point network Analysis. Paper presented at *the International Archives of the Photogrammetry, Remote Sensing & Spatial Information Sciences*.

Nimmo, F., Porco, C. & Mitchell, C. (2014) Tidally modulated eruptions on Enceladus: Cassini ISS observations and models. *The Astronomical Journal*, 148(3), 46. doi:10.1088/0004-6256/148/3/46.

Oberst, J., Hussmann, H., Giese, B., Sohl, F., Shoji, D., Stark, A., Wickhusen, K. & Wählisch, M. (2017) Enceladus geodetic framework. Paper presented at the *2017 International Symposium on Planetary Remote Sensing and Mapping*, Hong Kong.

Patthoff, D.A. & Kattenhorn, S.A. (2011) A fracture history on Enceladus provides evidence for a global ocean. *Geophysical Research Letters*, 38(18). doi:10.1029/2011GL048387.

Porco, C.C., DiNino, D. & Nimmo, F. (2014) How the geysers, tidal stresses, and thermal emission across the south polar terrain of Enceladus are related. *The Astronomical Journal*, 148(3), 45. doi:10.1088/0004-6256/148/3/45.

Porco, C.C., Helfenstein, P., Thomas, P.C., Ingersoll, A.P., Wisdom, J., West, R., Neukum, G., *et al.* (2006) Cassini observes the active south pole of Enceladus. *Science*, 311(5766), 1393–1401. doi:10.1126/science.1123013.

Porco, C.C., West, R.A., Squyres, S., Mcewen, A., Thomas, P., Murray, C.D., Delgenio, A., *et al.* (2004) Cassini imaging science: instrument characteristics and anticipated scientific investigations at Saturn. *Space Science Reviews*, 115(1), 363–497. doi:10.1007/s11214-004-1456-7.

Rambaux, N., Castillo-Rogez, J.C., Williams, J.G. & Karatekin, O. (2010) Librational response of Enceladus. *Geophysical Research Letters*, 37(4), A118. doi:10.1029/2009gl041465.

Roatsch, T., Kersten, E., Hoffmeister, A., Wählisch, M., Matz, K.D. & Porco, C.C. (2013) Recent improvements of the Saturnian satellites atlases: Mimas, Enceladus, and Dione. *Planetary and Space Science*, 77, 118–125. doi:10.1016/j.pss.2012.02.016.

Roatsch, T., Kersten, E., Matz, K.D., Bland, M.T., Becker, T.L., Patterson, G.W. & Porco, C.C. (2018) Final Mimas and Enceladus atlases derived from Cassini-ISS images. *Planetary and Space Science*. doi:10.1016/j.pss.2018.05.021.

Roatsch, T., Wählisch, M., Giese, B., Hoffmeister, A., Matz, K.D., Scholten, F., Kuhn, A., *et al.* (2008) High-resolution Enceladus atlas derived from Cassini-ISS images. *Planetary and Space Science*, 56(1), 109–116. doi:10.1016/j.pss.2007.03.014.

Roberts, J.H. & Nimmo, F. (2008) Tidal heating and the long-term stability of a subsurface ocean on Enceladus. *Icarus*, 194(2), 675–689. doi:10.1016/j.icarus.2007.11.010.

Russell, R.P. & Lara, M. (2009) On the design of an Enceladus science orbit. *Acta Astronautica*, 65(1), 27–39. doi:10.1016/j.actaastro.2009.01.021.

Schenk, P. (2014) Blue pearls for Rhea: color-mapping Saturn's icy moons. *Planetary Report*, 34, 8–13.

Schubert, G., Anderson, J.D., Travis, B.J. & Palguta, J. (2007) Enceladus: present internal structure and differentiation by early and long-term radiogenic heating. *Icarus*, 188(2), 345–355. doi:10.1016/j.icarus.2006.12.012.

Seidelmann, P.K., Abalakin, V.K., Bursa, M., Davies, M.E., de Bergh, C., Lieske, J.H., Oberst, J., *et al.* (2002) Report of the IAU/IAG working group on cartographic coordinates and rotational elements of the planets and satellites: 2000. *Celestial Mechanics and Dynamical Astronomy*, 82(1), 83–111. doi:10.1023/a:1013939327465.

Shoji, D., Hussmann, H., Sohl, F. & Kurita, K. (2014) Non-steady state tidal heating of Enceladus. *Icarus*, 235, 75–85. doi:10.1016/j.icarus.2014.03.006.

Smith-Konter, B. & Pappalardo, R.T. (2008) Tidally driven stress accumulation and shear failure of Enceladus's tiger stripes. *Icarus*, 198(2), 435–451. doi:10.1016/j.icarus.2008.07.005.

Spahn, F., Albers, N., Hörning, M., Kempf, S., Krivov, A.V., Makuch, M., Schmidt, J., Seiß, M. & Miodrag, S. (2006) E ring dust sources: implications from Cassini's dust measurements. *Planetary and Space Science*, 54(9), 1024–1032. doi:10.1016/j.pss.2006.05.022.

Spahn, F., Schmidt, J., Albers, N., Hörning, M., Makuch, M., Seiß, M., Kempf, S., *et al.* (2006) Cassini dust measurements at Enceladus and implications for the origin of the E ring. *Science*, 311(5766), 1416–1418. doi:10.1126/science.1121375.

Spencer, J.R. & Nimmo, F. (2013) Enceladus: an active ice world in the Saturn system. *Annual Review of Earth and Planetary Sciences*, 41(1), 693–717. doi:10.1146/annurev-earth-050212-124025.

Tajeddine, R., Soderlund, K.M., Thomas, P.C., Helfenstein, P., Hedman, M.M., Burns, J.A. & Schenk, P.M. (2017) True polar wander of Enceladus from topographic data. *Icarus*, 295, 46–60. doi:10.1016/j.icarus.2017.04.019.

Thomas, P.C. (2010) Sizes, shapes, and derived properties of the saturnian satellites after the Cassini nominal mission. *Icarus*, 208(1), 395–401. doi:10.1016/j.icarus.2010.01.025.

Thomas, P.C., Tajeddine, R., Tiscareno, M.S., Burns, J.A., Joseph, J., Loredo, T.J., Helfenstein, P. & Porco, C. (2016) Enceladus's measured physical libration requires a global subsurface ocean. *Icarus*, 264, 37–47. doi:10.1016/j.icarus.2015.08.037.

Thomas, P.C., Burns, J.A., Helfenstein, R., Squyres, S., Veverka, J., Porco, C., Turtle, E.P., *et al.* (2007) Shapes of the saturnian icy satellites and their significance. *Icarus*, 190(2), 573–584. doi:10.1016/j.icarus.2007.03.012.

Tobie, G., Čadek, O. & Sotin, C. (2008) Solid tidal friction above a liquid water reservoir as the origin of the south pole hotspot on Enceladus. *Icarus*, 196(2), 642–652. doi:10.1016/j.icarus.2008.03.008.

WGPSN, IAU (2018) *Gazetteer of planetary nomenclature by the International Astronomical Union (IAU) Working Group for Planetary System Nomenclature (WGPSN)*. Available from: https://planetarynames.wr.usgs.gov/.

Zubarev, A., Kozlova, N., Kokhanov, A., Oberst, J., Nadezhdina, I., Patraty, V. & Karachevtseva, I. (2014) Geodesy and cartography methods of exploration of the outer planetary systems: Galilean satellites and Enceladus. *40th COSPAR Scientific Assembly*, Moscow, Russia.

Planetary Remote Sensing and Mapping – Wu et al.
© 2019 Taylor & Francis Group, London, ISBN 978-1-138-58415-0

Chapter 3

On the applicability of physically defined height systems for telluric planets and moons

R. Tenzer and I. Foroughi

ABSTRACT: Except for some specific purposes, the geometric heights on telluric planets (Mercury, Venus, and Mars) and moons have almost exclusively been defined with respect to the geometric reference surface (sphere or ellipsoid) that mathematically approximates their size and shape, or with respect to the center of the planet/moon. On Earth, this type of heights represents the geodetic heights that are taken with respect to the reference ellipsoid and are measured by the Global Navigation Satellite Systems (GNSSs). For most of the practical and scientific applications, however, the physical (orthometric) heights are used. These heights are defined with respect to the geoid, which is an equipotential surface that best approximates the mean sea surface. The realization of a similar concept of physical heights for telluric planets and moons is discussed in this study, with the emphasis on their practical use. We inspect this aspect on selected study areas and discuss possible benefits of using physical heights, mainly in the context of gravity-driven mass transport as well as the choice of potential landing sites of future satellite missions.

1 INTRODUCTION

In space-science applications, the height information is determined from processing satellite orbital positions and elevations above the planetary surface obtained from the satellite radar altimetry (or other remote-sensing methods). As a consequence, the geometric heights (of the planetary/lunar surface) are typically defined with respect to the geometric reference surface (either the sphere or the ellipsoid in presence of the polar flattening). Nevertheless, certain applications, such as studies of gravity-driven mass movements (lava, ice, aquifers), require the physical heights to be defined with respect to the equipotential surface. By analogy with terrestrial applications, the definition of physical heights could be done with respect to the geoid surface, which represents the physical approximation of the size and shape of a particular planet or moon. Since, the satellite orbital parameters have been used to determine global gravity models for Mercury, Venus, Mars, and Moon, the definition of physical heights becomes possible. The need for physically defined heights has been already to some extent acknowledged. In Mars explorations, for instance, the physical heights that govern atmospheric pressure during entry, descent, and landing are of paramount importance. This aspect might be especially important when planning future missions landing at locations characterized by extreme topography, such as Tharsis region or Valles Marineris on Mars. The importance of applying the physical heights is also closely related to gravity-driven mass transport in order to better interpret and understand some geological processes forming the planetary and lunar surface, such as sediment or lava infill of impact craters and surface depressions, or the assessment of paleo-shorelines and stream channel-flow directions. The basic theoretical and some practical aspects of defining the physical heights on telluric planets and moons were addressed by Tenzer *et al.* (2018). They proposed and applied a refined method for defining the physical heights from gravity and topographic models and demonstrated that differences between the geometric and physical heights are at some places significant. Following their concept, we apply this method here to investigate differences between the geometric and physical heights in

Tharsis region including Valles Marineris on Mars. We also examine lunar topographic features of South Pole-Aitken, Imbrium, and Serenitatis impact craters.

2 METHOD

The difference between the geometric (geodetic) height h and the physical (orthometric) height H represents the geoidal height N, so that:

$$N = h - H \tag{1}$$

The geometric heights are defined with respect to the adopted geometric reference surface (either the ellipsoid or the sphere). As follows from Equation (1), the physical heights are taken with respect to the geoid surface, which is defined by the geoidal heights measured with respect to the geometric reference surface.

The geoidal heights have been often calculated only approximately using the following expression (e.g., Wieczorek, 2015):

$$N^* \approx \frac{\mathrm{GM}}{\mathrm{R}\gamma} \sum_{n=0}^{\bar{n}} \sum_{m=-n}^{n} \mathrm{T}_{n,m}\, \mathrm{Y}_{n,m} \tag{2}$$

where GM is the centric gravitational constant (i.e. the product of Newton's gravitational constant and the total mass of a planet/moon), R is the mean radius of the sphere which approximates the geoid surface, $\mathrm{Y}_{n,m}$ are the (fully normalized) surface spherical functions of degree n and order m, $\mathrm{T}_{n,m}$ are the (fully normalized) coefficients of the disturbing potential T, γ is the normal gravity at the (geometric) reference surface, and $\bar{n}$ is the upper summation index of spherical harmonics.

The computation of the geoidal heights according to Equation (2) is correct only over regions with zero (or negative) heights, because the coefficients $\mathrm{T}_{n,m}$ describe the external gravity field (corrected for the normal gravity component). In the case of positive heights $H > 0$, the topographic mass density distribution has to be taken into consideration to evaluate the disturbing potential on the geoid. Tenzer *et al.* (2018) presented the method for more rigorous computing of the geoidal heights that takes into consideration also the topography. They evaluated the geoidal heights according to the expression that treats the topographic and non-topographic contributions individually. It reads:

$$\begin{aligned} N = {} & \frac{\mathrm{GM}}{\mathrm{R}\gamma} \sum_{n=0}^{\bar{n}} \sum_{m=-n}^{n} \left(\frac{\mathrm{R}}{\mathrm{R}+H} \right)^{n+1} \mathrm{T}_{n,m}\, \mathrm{Y}_{n,m}(\Omega) \\ & + \frac{\mathrm{GM}}{\mathrm{R}\gamma} \sum_{n=0}^{\bar{n}} \left[1 - \left(1 + \frac{H}{\mathrm{R}} \right)^{-n-1} \right] \sum_{m=-n}^{n} \mathrm{V}^{\mathrm{T}}_{n,m}\, \mathrm{Y}_{n,m}(\Omega) \\ & - \frac{\mathrm{GM}}{\mathrm{R}} \sum_{n=0}^{\bar{n}} \sum_{m=-n}^{n} \mathrm{V}^{\mathrm{bias}}_{n,m}\, \mathrm{Y}_{n,m}(\Omega) \\ & + \frac{\mathrm{GM}}{\mathrm{R}\gamma} \sum_{n=0}^{\bar{n}} \sum_{m=-n}^{n} \left[1 - \left(1 + \frac{H}{\mathrm{R}} \right)^{-n-1} \right] \mathrm{T}^{\mathrm{NT}}_{n,m}\, \mathrm{Y}_{n,m}(\Omega) \end{aligned} \tag{3}$$

The topographic contribution is computed from the topographic-potential and topographic-bias coefficients $\mathrm{V}^{\mathrm{T}}_{n,m}$ and $\mathrm{V}^{\mathrm{bias}}_{n,m}$ respectively. These coefficients are defined by:

$$V^{T}_{n,m} = \frac{3}{2n+1}\frac{\rho^{T}}{\rho}\sum_{k=0}^{n+2}\binom{n+2}{k}\frac{1}{k+1}\frac{H^{(k+1)}_{n,m}}{R^{k+1}}, \quad V^{bias}_{n,m} = 3\frac{\rho^{T}}{\rho}\sum_{k=1}^{2}\frac{1}{k+1}\frac{H^{(k+1)}_{n,m}}{R^{k+1}} \tag{4}$$

where ρ denotes the mean density of the whole planet/moon, and ρ^T is the mean topographic density of a planet/moon. The coefficients $V^{T}_{n,m}$ and $V^{bias}_{n,m}$ are computed from the topographic coefficients $H^{(k)}_{n,m}$:

$$H^{(k)}_{n} = \frac{2n+1}{4\pi}\iint_{\sigma} H^{k}\, P_{n}(t)\, d\sigma = \sum_{m=-n}^{n} H^{(k)}_{n,m}\, Y_{n,m} \quad (H>0) \tag{5}$$

where P_n is the Legendre polynomial of degree n defined for the argument t of cosine of the spherical angle ψ, i.e. $t = \cos\psi$; $d\sigma$ is the infinitesimal surface element; and σ is the full spatial angle.

The non-topographic contribution (i.e., the third constituent on the right-hand side of Equation 3) is evaluated from the coefficients $T^{NT}_{n,m} = T_{n,m} - V^{T}_{n,m}$ of the no-topography disturbing potential T^{NT} (Vaníček *et al.*, 2005).

3 NUMERICAL EXAMPLES

To demonstrate the relevance of using the physical heights, we investigated the geoidal heights (i.e. differences between the geometric and physical heights) at locations characterized by extreme topography. For this purpose, we selected the study area of Tharsis region and Valles Marineris on Mars. Tharsis region is a vast and complex topographic rise comprising the largest shield volcanoes of Ascraeus, Arsia, Pavonis, and Olympus Mons. Olympus Mons is the highest topographic feature in the solar system. In contrast, Valles Marineris rift valleys represent the deepest places on Mars. We further selected South Pole-Aitken impact crater and compared the results within the study area of Imbrium and Serenitatis impact craters on the Moon. South Pole-Aitken basin is the largest and deepest crater on the Moon. The highest elevations are found to the northeast of this crater, and it has been suggested that this area might represent thick ejecta deposits that were emplaced during an oblique South Pole-Aitken basin impact event. Other large impact basins, among them Imbrium and Serenitatis, are also characterized by low elevations and elevated rims.

3.1 *Martian topographic features*

To date, most of the Martian topography is known from the NASA's MOLA (Mars Orbiter Laser Altimeter) mission with kilometer resolution (Smith *et al.*, 2001). We used these data to generate the topographic coefficients and then computed the geometric heights on a 1×1 arc-degree grid within the study area comprising Tharsis region and Valles Marineris with a spectral resolution complete to the spherical harmonic degree 120. We used the same spectral resolution to compute the geoidal heights from the latest Martian gravity model MRO120D. The MRO120D gravity model was derived with a spectral resolution complete to the spherical harmonic degree 120 based on analysis of tracking data, particularly of the Mars Reconnaissance Orbiter (MRO) spacecraft (Konopliv *et al.*, 2016). The geometric heights are shown in Figure 3.1a. The maximum geometric heights of Olympus Mons reach 21.0 km, while the lowest places along Valles Marineris have depths up to −5.3 km. In Figure 3.1b, we plotted differences between the geometric and physical heights. We could see that the topographic mass surplus of the whole Tharsis region significantly modifies the geoidal geometry. Additional large localized modifications of the geoidal geometry are due to volcanic rises of Ascraeus, Arsia, and Pavonis Mons, with maximum differences between the geometric and physical heights up to about 1.7 km at these locations. The most pronounced

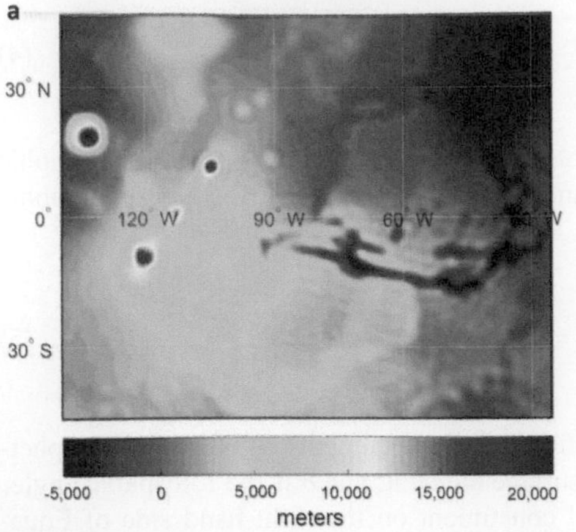

Figure 3.1a. Regional maps of (**a**) geometric heights

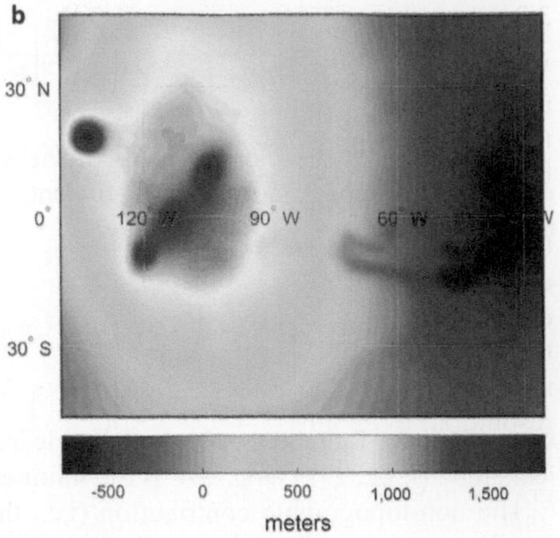

Figure 3.1b. Regional maps of (**b**) geoidal heights at the study area of Tharsis region and Valles Marineris on Mars

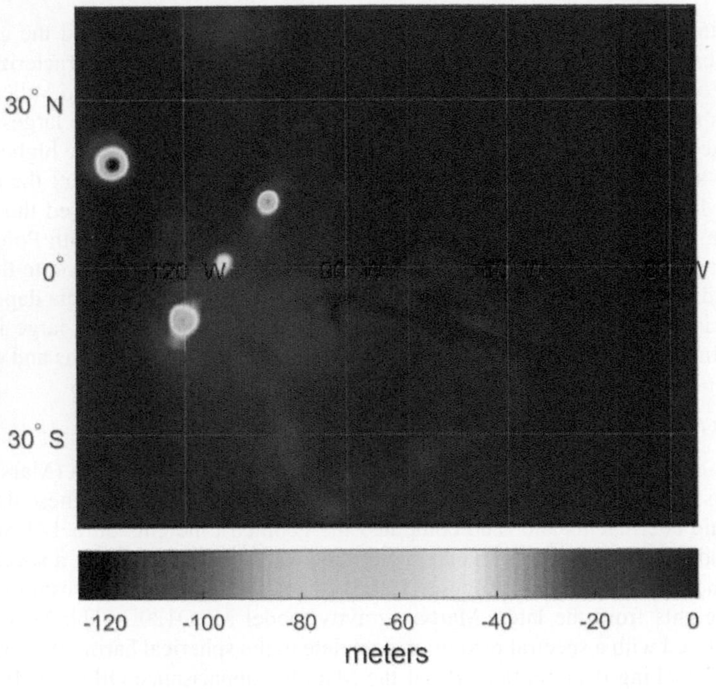

Figure 3.2. Differences between values of the geoidal heights computed accurately (Equation 3) and approximately (Equation 2) at the study area of Tharsis region and Valles Marineris on Mars

differences are seen at the location of Olympus Mons. The maximum geoidal heights there reach 1.8 km. In contrast, we could see large negative differences along Valles Marineris rift valleys up to about −0.7 km. We also checked the differences between values of the geoidal heights computed accurately (Equation 3) and approximately (Equation 2). As seen in Figure 3.2, these differences vary within a large interval from −125.6 m to 0.9 m.

3.2 *Lunar topographic features*

We further investigated differences between the geometric and physical heights at two study areas, namely for South Pole-Aitken impact crater and for Imbrium and Serenitatis impact craters. For the Moon, topographic and gravity models were determined with the extreme precision based on processing data measured over the last decade by Lunar Reconnaissance Orbiter (LRO) and Gravity and Interior Laboratory (GRAIL) missions. Based on the LRO laser altimetry, lunar topography models with a resolution of about 500 m or much better have been produced (Smith *et al.*, 2010), and refined through combination with data from other lunar missions, e.g. by Barker *et al.* (2016) to about 60 m resolution or better. Tracking data and inter-satellite distances from the GRAIL mission (Zuber *et al.*, 2013) are the basis for very high-resolution spherical harmonic models of the lunar gravity field, e.g. to degree and order 900, equivalent to about 6 km detail resolution, (Konopliv *et al.*, 2014; Lemoine *et al.*, 2014) and finer (e.g., Goossens *et al.*, 2014). In this study we used the SLDEM2015 lunar elevation model (Barker *et al.*, 2016) and the lunar gravity model GRGM900C (Lemoine *et al.*, 2014) to compute the geometric and geoidal heights with the spectral resolution complete do the spherical harmonic degree of 900. We note here that the latest released lunar gravity model GRGM1200A has a spectral resolution up to degree 1500. The results within the study area of South Pole-Aitken impact crater are presented in Figure 3.3. We could see (Fig. 3.3a) that the geometric heights reach maxima up to about 8.5 km at its rim, while the surface inside the crater deepens to about −8.9 km. This topography significantly modifies the equipotential geometry (see Fig. 3.3b), showing that the geoidal heights inside the crater are mostly negative with minima of about −295 m, while maxima under the crater rim are up to about 395 m.

We repeated the computation at the study area covering Imbrium and Serenitatis impact craters. Compared to South Pole-Atkins impact crater, the topographic manifestation of these two craters is much less prominent (Fig. 3.4a). Moreover, the crater rim in both cases in not exhibited significantly. Nevertheless, we see large modifications of the geoidal geometry up to about 0.4 km (Fig. 3.4b).

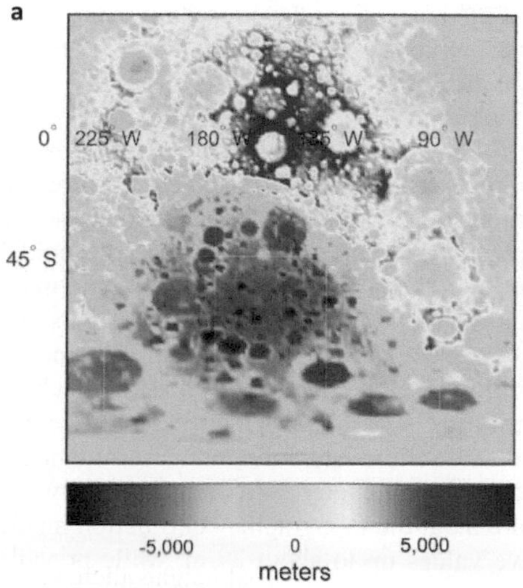

Figure 3.3a. Regional maps of (**a**) geometric heights

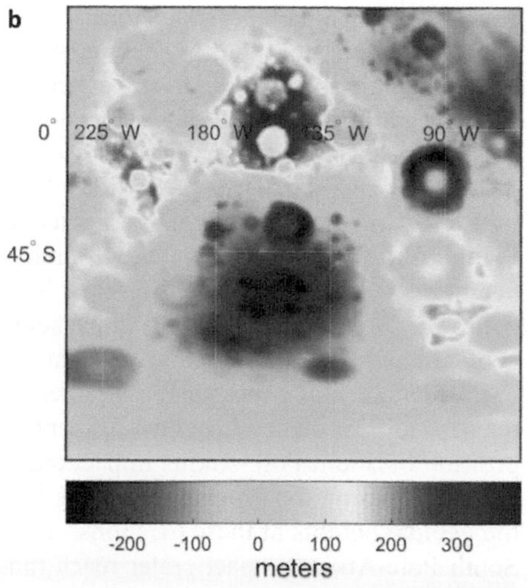

Figure 3.3b. Regional maps of (**b**) geoidal heights at the study area of South Pole-Aitken impact crater on the Moon

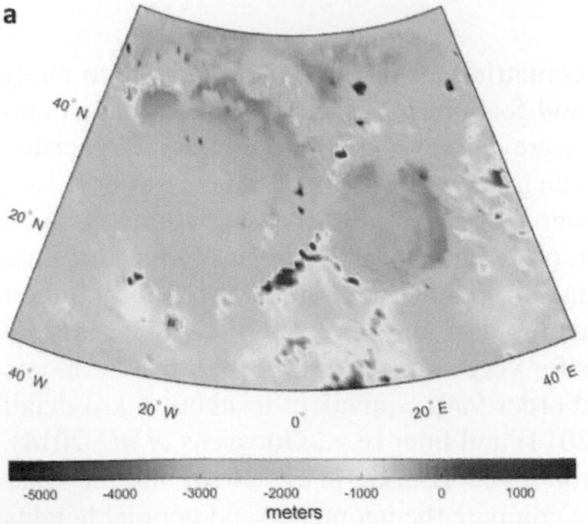

Figure 3.4a. Regional maps of (**a**) geometric heights

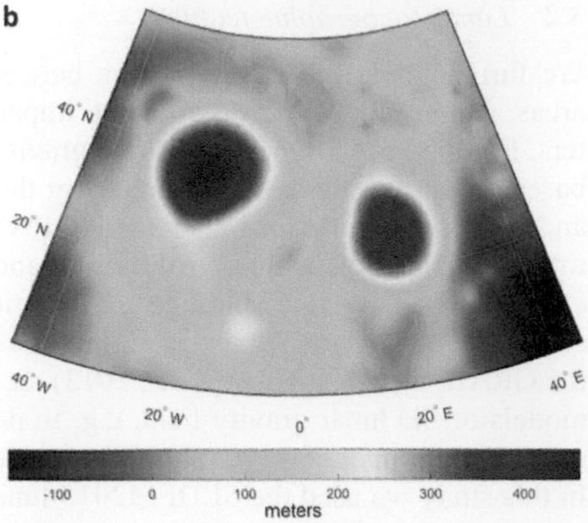

Figure 3.4b. Regional maps of (**b**) geoidal heights at the study area of Imbrium and Serenitatis impact craters on the Moon

4 DISCUSSION

We observed large modifications of the geoidal equipotential geometry by Tharsis volcanic accumulations with additional, localized modifications by Ascraeus, Arsia, Pavonis, and Olympus Mons shield volcanoes. This finding has implications, for instance, on the gravimetric interpretation of this volcanic region, particularly when computing the topographic gravity correction that is required in studies of the crust and mantle structure (geometry of magmatic chambers and hotspots). Similarly, large modifications of the geoidal geometry by Valles Marineris rift walleyes should be taken into consideration when interpreting the gravity-driven mass transport, especially ancient or more recent lava-flow channels.

We also confirmed large modifications of the geoidal geometry by crater topography. In this case, the physical heights should be used when interpreting the geological composition of the basin floor and its possible infill by sediments or lava flows. Moreover, implications are related to the understanding of the origin of impact craters. In the absence of seismic data, the crustal thickness could be estimated based on only gravity and topographic models. A realistic estimation of the crustal thickness in this particular case is important to know if the crater was formed by high- or low-velocity impact. It is speculated, for instance, that South Pole-Atkins crater was created by a low-velocity and low-angle impact. The most interesting finding is that when compared to South Pole-Atkins impact crater, we could see a completely different manifestation of the geoidal geometry at locations of Imbrium and Serenitatis impact craters. Whereas the topographic mass deficiency of South Pole-Atkins impact crater propagates to the large negative geoidal heights, the geoidal heights at Imbrium and Serenitatis impact craters reach large positive values. This finding has a fundamental consequence on interpreting the origin of these craters as well as their subsequent modifications by lava flow or mantle uplift. A striking contrast between the geological composition of South Pole-Atkins impact crater compared to Imbrium and Serenitatis is also evident when comparing the gravitational contribution of topographic geometry (of a uniform density) on the geoidal heights at these locations. As seen in Figure 3.5a, the topographic contribution inside South Pole-Atkins impact crater reach mostly positive values up to about 29 m, while prevails negative over elevated crater rim. In Imbrium and Serenitatis this pattern is opposite, with negative values inside these two impact craters.

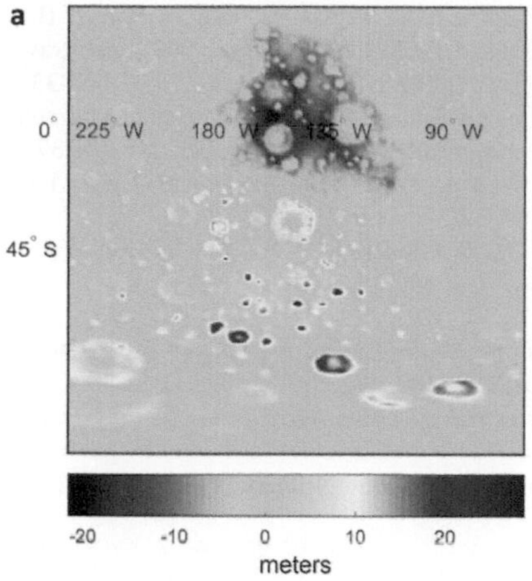

Figure 3.5a. Topographic contribution to the geoidal heights at the study area of: (**a**) South Pole-Aitken

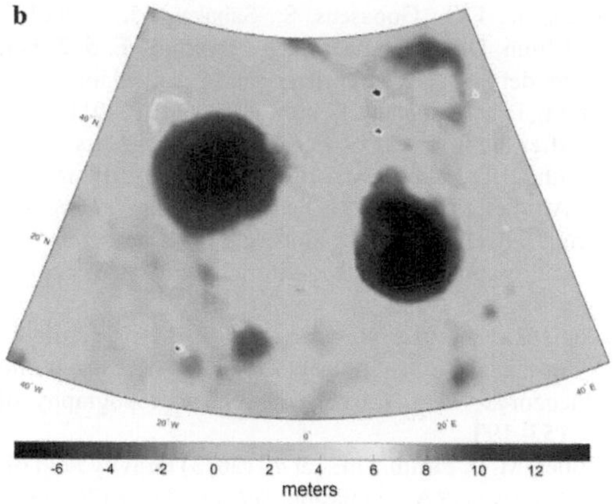

Figure 3.5b. Topographic contribution to the geoidal heights at the study area of: (**b**) Imbrium and Serenitatis impact craters on the Moon

5 SUMMARY

We have demonstrated on selected examples the application of physical heights. As shown, the differences between the geometric and physical heights could be significant. For Tharsis region we estimated that these differences reach about 1.8 km in contrast to Valles Marineris where these differences have opposite signature and reach −0.7 km. Similarly, large modifications up to about 0.5 km were confirmed on the Moon. Such large differences have also impact on a gravimetric interpretation of planetary/lunar surface formations. The evidence was given by different modifications of the geoidal geometry at the location of South Pole-Atkins impact crater, compared to Imbrium and Serenitatis impact craters. In the former case the geoidal heights inside the crater reach large negative values, while in the latter case we detected large positive values. This finding indicates that the origin or further evolution of these craters differ. Another aspect of using the physical heights defined with respect to the geoidal equipotential geometry is a more accurate interpretation of gravity-driven mass transport, for instance, in the case of Valles Marineris rift walleyes formed by lava flows.

REFERENCES

Barker, M.K., Mazarico, E., Neumann, G.A., Zuber, M.T., Haruyama, J. & Smith, D.E. (2016) A new lunar digital elevation model from the Lunar Orbiter Laser Altimeter and SELENE Terrain Camera. *Icarus*, 2723(15), 346–355.

Goossens, S., Sabaka, T.J., Nicholas, J.B., Lemoine, F.G., Rowlands, D.D., Mazarico, E., Neumann, G.A., Smith, D.E. & Zuber, M.T. (2014) High-resolution local gravity model of the south pole of the Moon from GRAIL extended mission data. *Geophysical Research Letters*, 41(10), 3367–3374.

Konopliv, A.S., Park, R.S. & Folkner, W.M. (2016) An improved JPL Mars gravity field and orientation from Mars orbiter and lander tracking data. *Icarus*, 274, 253–260.

Konopliv, A.S., Park, R.S., Yuan, D.-Y., *et al.* (2014) High-resolution lunar gravity fields from the GRAIL primary and extended missions. *Geophysical Research Letters*, 41, 1452–1458.

Lemoine, F.G., Goossens, S., Sabaka, T.J., Nicholas, J.B., Mazarico, E., Rowlands, D.D., Loomis, B.D., Chinn, D.S., Neumann, G.A., Smith, D.E. & Zuber, M.T. (2014) GRGM900C: a degree 900 lunar gravity model from GRAIL primary and extended mission data. *Geophysical Research Letters*, 41(10), 3382–3389.

Smith, D.E., Zuber, M.T., Frey, H.V., *et al.* (2001) Mars Orbiter Laser Altimeter (MOLA): experiment summary after the first year of global mapping of Mars. *Journal of Geophysical Research*, 106(E10), 23689–23722.

Smith, D.E., Zuber, M.T., Neumann, G.A., *et al.* (2010) Initial observations from the Lunar Orbiter Laser Altimeter (LOLA). *Geophysical Research Letters*, 37, L18204.

Tenzer, R., Foroughi, I., Sjöberg, L.E., Bagherbandi, M., Hirt, Ch. & Pitoňák, M. (2018) Definition of physical height systems for telluric planets and moons. *Surveys in Geophysics*, 39(3), 313–335. doi:10.1007/s10712-017-9457-8.

Vaníček, P., Tenzer, R., Sjöberg, L.E., Martinec, Z. & Featherstone, W.E. (2005) New views of the spherical Bouguer gravity anomaly. *Geophysical Journal International*, 159, 460–472.

Wieczorek, M.A. (2015) Gravity and topography of the terrestrial planets. *Treatise on Geophysics*, 10, 153–193.

Zuber, M.T., Smith, D.E., *et al.* (2013) Gravity field of the moon from the Gravity Recovery and Interior Laboratory (GRAIL) mission. *Science*, 339(6120), 668–671.

Section II

Planetary exploration missions and sensors

Planetary Remote Sensing and Mapping – Wu et al.
© 2019 Taylor & Francis Group, London, ISBN 978-1-138-58415-0

Chapter 4

Comet 67P/Churyumov-Gerasimenko through the eyes of the Rosetta/OSIRIS cameras

C. Güttler and the OSIRIS Team

ABSTRACT: The ESA Rosetta Spacecraft followed comet 67P/Churyumov-Gerasimenko for more than two years (August 2014 to September 2016) on its way around the Sun. The bi-lobed comet with its peculiar shape shows landscapes and surface features never seen before. When the comet became active, approaching its perihelion in August 2015, spectacular dust jets and strong, short-lived outbursts were observed. These left their footprints on the comet's surface, which was therefore constantly evolving. The comet's morphology, its activity, and resulting changes on the surface are now being interpreted to better understand how comets work and what the driving mechanisms behind their activity are. A comet's evolution (through activity) and its origin are the two driving questions in cometary science, and one cannot be understood without the other. The Rosetta mission and – with an emphasis in this overview article – images of the OSIRIS cameras provided the most comprehensive dataset to shape our understanding of comets for decades.

1 INTRODUCTION

The Rosetta Spacecraft was launched in March 2004 on its journey to a comet named 67P/Churyumov-Gerasimenko. On its 10-year-long journey it flew by the two asteroids Steins and Lutetia in 2008 and 2010 before reaching comet 67P in August 2014. As a cornerstone mission of the European Space Agency, the Rosetta spacecraft carried a plethora of 11 complementary scientific instruments and the lander Philae to land on the comet in November 2014. One of the instruments is the scientific camera system OSIRIS (Optical, Spectroscopic, and Infrared Remote Imaging System; Keller *et al.*, 2007), which results are in the spotlight of this article. It is a system of two cameras, the Narrow Angle Camera (NAC) with a field of view of 2 x 2 degrees and the Wide Angle Camera (WAC) with a field of view of 12 x 12 degrees. These are designed to be complementary in that they can study small-scale surface features as well as the large cometary dust coma at the same time.

When Rosetta arrived at comet 67P on 5 August 2014, we discovered a nucleus with a particular bi-lobed shape and a rich variety of surface morphologies such as steep cliffs, terraces, circular pits, as well as a variation between smooth and rugged terrains (Fig. 4.1, left). The nucleus morphology as well as the first characterization of physical parameters were described by Sierks *et al.* (2015) and Thomas *et al.* (2015). Among these are the comet's rotation rate (12.4 hours in 2014), its volume (18.56 km^3) and mass ($1 \cdot 10^{13}$ kg), the consequential very low bulk density (537.8 kg m^{-3}), and the geometric albedo (5.9% at 550 nm). More details on the characterization and mapping are provided in Section 2.

With an unprecedented image resolution (approx. 1 m $pixel^{-1}$ at the time of arrival and down to 1 mm $pixel^{-1}$ on 30 Sep 2016), the images showed the importance of air fall, surface dust transport, mass wasting, and insolation weathering, and offer some support for subsurface fluidization and mass loss through the ejection of large chunks of material (Thomas *et al.*, 2015). The variety of surface textures points at a variety also in processes that create these. The uppermost surface is not fully "pristine" in large parts but shaped by evolutionary processes and air fall in the northern hemisphere. Fractures are present wherever the surface is not covered with air fall material; textures in areas associated with fractures typically appear more brittle than others.

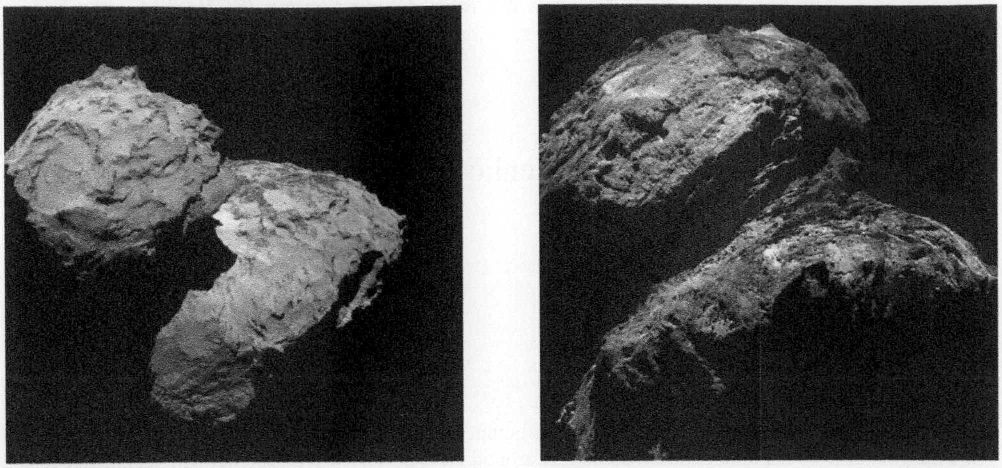

Figure 4.1. The two hemispheres of comet 67P. The northern hemisphere (left) is covered with dust mantles, while the southern hemisphere (right) is not covered and appears rugged. The scale is 4.1 km x 3.3 km x 1.8 km for the large lobe and 2.6 km x 2.3 km x 1.8 km for the small lobe (Sierks *et al.*, 2015).

As the comet approached the Sun, illumination conditions were gradually changing. While in August 2014 only the northern regions were illuminated (Fig. 4.1, left), the subsolar latitude was changing towards perihelion such that in August 2015 the southern hemisphere was fully illuminated (Fig. 4.1, right). The comet showed a different face in the south with surface morphologies that show hardly any dust blankets and less cliffs and pits. Textures on the southern hemisphere appear rugged and more consolidated. Since the minimum of the subsolar latitude (southernmost) per orbit coincides with the time of maximum insolation, it is now commonly accepted that the south is more representative of the comet's interior while the north is cloaked under blankets of air fall dust (see section 3 for details).

2 CHARACTERIZATION AND MAPPING

A requirement for the success of the Rosetta Mission was an early characterization of the comet's shape, rotational state, and coordinate system. Early digital terrain models were developed in the OSIRIS team, relying on different techniques for complementarity and redundancy. One model is generated by Stereo Photoclinometry (SPC), presented by Jorda *et al.* (2016; and updated since then), the other is based on Stereo Photogrammetry (SPG), presented by Preusker *et al.* (2015, 2017). These shape models are the basis for many surface related studies such as global thermal modelling, strength measurement from imagery, or composition and photometry maps to name just a few. Furthermore, the shape models were tied to a commonly agreed coordinate system, the "Cheops System" (Fig. 4.2, left), where the z-axis is determined by the comet's rotation axis, and the x/y-axes are constrained by the comet's largest boulder Cheops fixed to a longitude of 142 degrees (see Preusker *et al.* [2015] for the rigorous definition).

A morphology-driven cartography scheme was presented by Sierks *et al.* and Thomas *et al.* and later refined by El-Maarry *et al.* (2016) to include the southern hemisphere. To enable communication between scientists within and among Rosetta teams, 26 morphologically distinct regions were defined and named after ancient Egyptian deities. A cartographic map of these regions is displayed in Figure 4.2 (right). Due to the differences in textural appearance among regions, the idea is that different regions represent different physical properties and/or evolutionary processes. This is however not proven to this point.

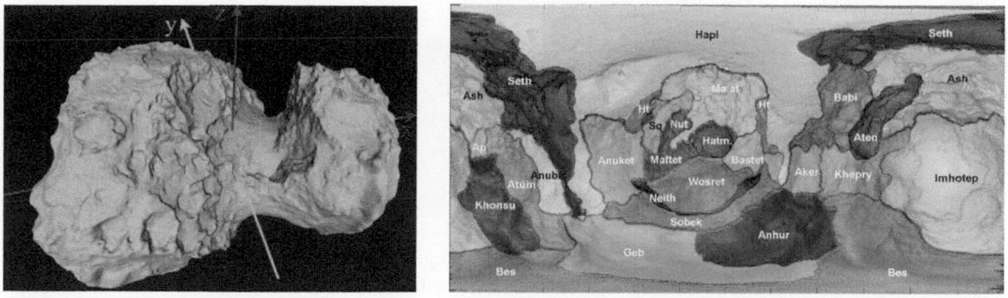

Figure 4.2. The coordinate system referred to as "Cheops" (left; adapted from Sierks *et al.*, 2015) and the definition of morphologically distinct regions (right; adapted from El-Maarry *et al.*, 2016). The longitudes on the horizontal axis range from −180 to +180 degrees (left to right) and the latitudes in the vertical from −90 to 90 degrees (bottom to top).

3 SELECTED SCIENCE CASES

This section provides a few examples of interesting scientific findings from Rosetta. This is scratching at the surface and by no means complete. For deeper reading, the interested reader is referred to the several Rosetta special issues, hosting the papers references below.

3.1 *Changes and fallback*

While the comet's static surface morphology is already interesting, one has to keep in mind that activity is constantly changing its upper surface layers. Observed changes on 67P are summarized by El-Maarry *et al.* (2017), of which the most striking are cliff collapses, extension of fractures, the mobilization of a 13,000 metric-ton boulder, the appearance and/or receding of shallow scarps, disappearance and formation of aeolian-like ripples, and other erosional transport mechanisms. Most of these are directly associated with solar irradiation and resulting gas and dust activity, as they appear around times when the associated region receives its maximum irradiation. The strongest changes occur around perihelion, at smallest heliocentric distances.

Two examples for transport mechanisms of unconsolidated material are shown in Figure 4.3. The left image pair displays changes in the Ma'at region, colloquially referred to as Honeycomb structures (Hu *et al.*, 2017). When Rosetta arrived at comet 67P in autumn 2014, the surface textures in the white circles were not particularly distinguishable from their surroundings. Then, in March 2015, when the sun was close to zenith over this region, the surface changed its roughness into a texture that named these features into honeycombs. Later that year, after the comet's perihelion passage, the textures reverted into a similar appearance as in 2014. A detailed analysis by Hu *et al.* (including topography analysis and thermal modelling) indicates that surface material is *removed* towards March 2015, uncovering the observed textures, and *re-deposited* later.

The right pair of images in Figure 4.3 shows dune-like ripples in the Hapi region, in the smooth neck between the two lobes. The two observations here are well before and well after perihelion and it can be noticed that the ripples change their appearance. The striking observation however is that the ripples *disappeared* in the perihelion timeframe (not shown in Fig. 4.3) and *re-appeared* at the same location. The full process to form these ripples is not well understood but it is striking that, in spite of being globally rare, they form again at the exact same location. Details are described by El-Maarry *et al.* (2017).

These two examples among many can be explained by the repeated liftoff and deposition of material. This is explained on a global scale by Keller *et al.* (2017): It was noticed early on that the northern and southern hemispheres of the comet show a very different morphology in that

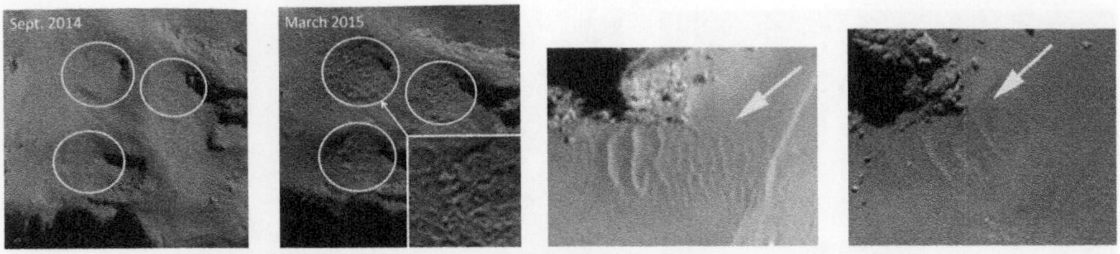

Figure 4.3. Left pair: "Honeycomb features" in the Ma'at region in September 2014 (left) and March 2015 (right). (Adapted from Hu *et al.* (2017.) Right pair: Dunes in the Hapi region in September 2014 (left) and June 2016 (right). (Adapted from El-Maarry *et al.* (2017).)

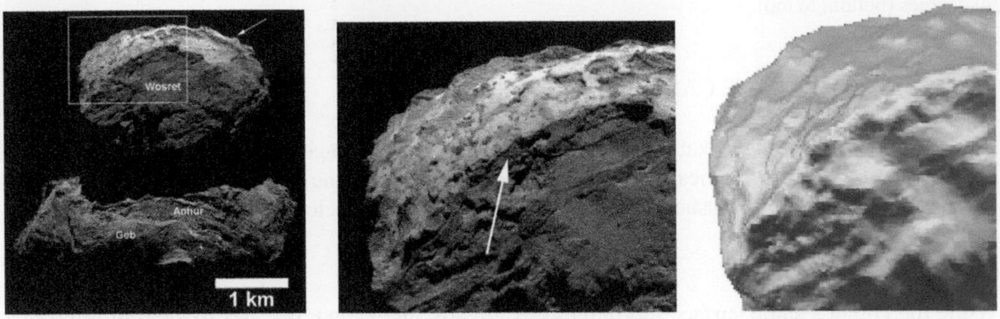

Figure 4.4. Fallback in the transition between regions Maftet and Wosret (cf. Fig. 4.2, right). Overview of the comet (left), detail at the boundary region (center, white arrow) and thermal model of the same (right). (Adapted from Keller *et al.* (2017).)

the northern hemisphere is smooth with dust deposits and the southern hemisphere, lacking these deposits, is rugged with an appearance of consolidated material (Fig. 4.1). Keller *et al.* pointed out that this is particularly notable in the near-equatorial transition between the Maftet and the Wosret regions (Fig. 4.4, left and middle). A thermal model, integrated over the full orbit of 67P, shows a similarly striking difference in the overall heat input at this transition (Fig. 4.4, right; see Keller *et al.* for details). The overall picture is that the south is much more active and blows off large amounts of material around perihelion. Some of this is re-deposited (see Lai *et al.*, 2016), and deposits in the north cannot be lifted until northern summer, at significantly larger heliocentric distances, thus with less heat input and activity, respectively. A net mass deposition in the north and mass loss in the south explains many observations.

3.2 *Activity*

Changes on the surface are driven by the activity of the comet. As a comet approaches the sun, its surface warms up and ices (H_2O, CO_2, CO, and many more volatile species) under the desiccated, dusty surface sublimate. The expanding gases form a sub-surface pressure, which escapes into space and drags dust particles along. While the gas can be observed from ground and by several Rosetta instruments, the dust dominates the light emission (scattering in that case) in the visible wavelength range. An example of OSIRIS dust observations is shown in Figure 4.5.

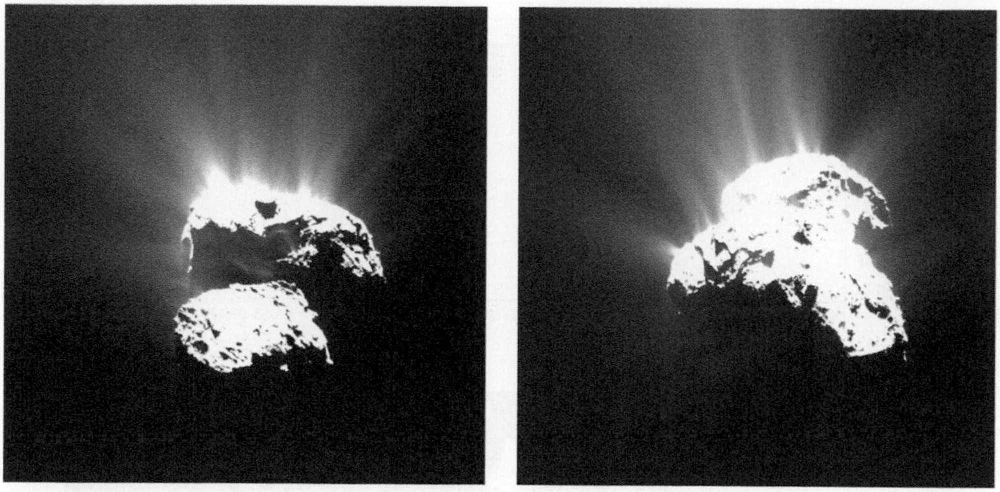

Figure 4.5. Activity around perihelion in August 2015, showing multiple jet features. Most of the features in this example are stable for hours and recurring over multiple rotations. The images were taken on 26 and 27 August from a distance of 400 km to the nucleus.

Dust activity was observed early on in the mission (Tubiana *et al.*, 2015) and intensified towards perihelion. Activity can be grouped into (a) a diffuse dust background, (b) regular, re-occurring jet activity, (c) transient jet or plume formation, and (d) strong, transient outbursts [note that this is not a rigorous definition of activity types but rather meant for illustrative purposes]. The examples in Figure 4.5 show dust jets, which are typically stable over a few hours (as long as their foot point is illuminated), undetectable at local night, and often re-appear in the next local morning.

The sources and mechanisms of activity are being intensely studied (Shi *et al.*, 2016; Vincent *et al.*, 2016a, 2016b). While the lifting of dust particles is not easy due to cohesion (Blum *et al.*, 2017), the sources of jets can often be associated with sub-surface temperatures (Shi *et al.*, 2016) or topography (Vincent *et al.*, 2016a). In contrast, strong outbursts – mostly during perihelion but also at surprisingly large heliocentric distances – cannot be explained by the thermal equilibrium from solar irradiation. Sub-surface energy storage is required to explain these eruptions (Agarwal *et al.*, 2017).

3.3 *Rotation rate*

The last example connects surface changes and activity. When Rosetta approached comet 67P, Mottola *et al.* (2014) measured the rotation rate to be 12.40 hours, around 21 minutes shorter than measurements from the comet's last apparition (Lowry *et al.*, 2012). It was already then assumed that the spin-rate change was due to non-symmetric activity patterns during the last apparition. This was confirmed as the comet approached the sun towards its 2015 perihelion passage when Rosetta could constantly monitor the spin state in detail. The spin period during the Rosetta mission phase is shown in Figure 4.6 (top) and a clear spin-up from 12.4 to 12.05 hours can be seen around summer 2015.

This spin-up is also associated with a >500-m-long crack in the northern Hapi and Anuket regions, shown in Figure 4.6 (bottom). It was proposed by Sierks *et al.* (2015) that this crack could be the result of a flexure between the comet's head and the body lobe. This flexure would be the result of a torque from differently acting forces on the two lobes. The change in spin rate is the result of such a global torque, induced by non-symmetric activity patterns (Keller *et al.*, 2015). Follow-up observations in June 2016 confirmed the expectation that the crack is extending and new, parallel fractures are opening nearby (El-Maarry *et al.*, 2017).

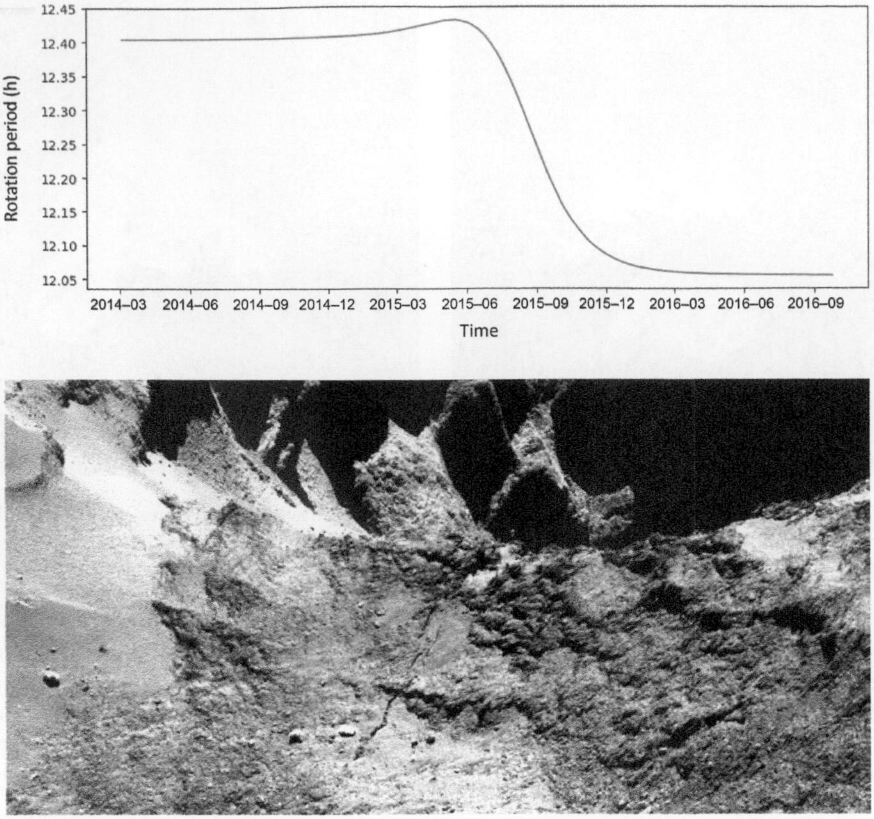

Figure 4.6. Changing rotation rate around perihelion in summer 2015 (top) and a global crack in the Anuket region – the comet's neck – in the center of the image (bottom)

4 SUMMARY

The driving question arising from Rosetta is of the origin of comet 67P. How primordial is the body and how did it evolve since its formation 4.5 billion years ago? These questions are being addressed and discussed in the community (Blum *et al.*, 2017; Davidsson *et al.*, 2016). The problem with comets in general is that their today's appearance changed from its primordial one though the effects exemplified above. Only a deep understanding of the evolutionary processes (activity and changes) allows us to draw conclusions about its origin. It was the intention of this article to provide a glimpse into the dynamic field of cometary science and the exciting challenges ahead to form a consistent picture on comet formation and evolution.

ACKNOWLEDGEMENT

The article is the summary of a keynote presentation at the International Symposium on Planetary Remote Sensing and Mapping in Hong Kong, given on 14 August 2017. I hereby acknowledge the contribution of the whole Rosetta/OSIRIS Team, whose scientific papers and discussions form the basis of this article.

OSIRIS was built by a consortium led by the Max-Planck-Institut für Sonnensystemforschung, Göttingen, Germany, in collaboration with CISAS, University of Padova, Italy, the Laboratoire d'Astrophysique de Marseille, France, the Instituto de Astrofísica de Andalucia, CSIC, Granada, Spain, the Scientific Support Office of the European Space Agency, Noordwijk, The Netherlands, the Instituto Nacional de Técnica Aeroespacial, Madrid, Spain, the Universidad Politéchnica de Madrid, Spain, the Department of Physics and Astronomy of Uppsala University, Sweden, and the Institut für Datentechnik und Kommunikationsnetze der Technischen Universität Braunschweig, Germany.

The support of the national funding agencies of Germany (DLR), France (CNES), Italy (ASI), Spain (MEC), Sweden (SNSB), and the ESA Technical Directorate is gratefully acknowledged. We thank the Rosetta Science Ground Segment at ESAC, the Rosetta Mission Operations Centre at ESOC and the Rosetta Project at ESTEC for their outstanding work enabling the science return of the Rosetta Mission.

REFERENCES

Agarwal, J., *et al.* (2017) Evidence of sub-surface energy storage in comet 67P from the outburst of 2016 July 03. *The Monthly Notices of the Royal Astronomical Society*, 469, S606–S625.

Blum, J., *et al.* (2017) Evidence for the formation of comet 67P/Churyumov-Gerasimenko through gravitational collapse of a bound clump of pebbles. *The Monthly Notices of the Royal Astronomical Society*, 469, S755–S773.

Davidsson, B.J.R., *et al.* (2016) The primordial nucleus of comet 67P/Churyumov-Gerasimenko. *Astronomy and Astrophysics*, 592, A63.

El-Maarry, M.R., *et al.* (2016) Regional surface morphology of comet 67P/Churyumov-Gerasimenko from Rosetta/OSIRIS images: the southern hemisphere. *Astronomy and Astrophysics*, 593, A110.

El-Maarry, M.R., *et al.* (2017) Surface changes on comet 67P/Churyumov-Gerasimenko suggest a more active past. *Science*, 355, 1392–1395.

Hu, X., *et al.* (2017) Seasonal erosion and restoration of the dust cover on comet 67P/Churyumov-Gerasimenko as observed by OSIRIS onboard Rosetta. *Astronomy and Astrophysics*, 604, A114.

Jorda, L., *et al.* (2016) The global shape, density and rotation of Comet 67P/Churyumov-Gerasimenko from preperihelion Rosetta/OSIRIS observations. *Icarus*, 277, 257–278.

Keller, H.U., *et al.* (2007) OSIRIS – the scientific camera system onboard Rosetta. *Space Science Reviews*, 128, 433–506.

Keller, H.U., *et al.* (2015) The changing rotation period of comet 67P/Churyumov-Gerasimenko controlled by its activity. *Astronomy and Astrophysics*, 579, L5.

Keller, H.U., *et al.* (2017) Seasonal mass transfer on the nucleus of comet 67P/Churyumov-Gerasimenko. *The Monthly Notices of the Royal Astronomical Society*, 469, S357–S371.

Lai, I.-L., *et al.* (2016) Gas outflow and dust transport of comet 67P/Churyumov-Gerasimenko. *The Monthly Notices of the Royal Astronomical Society*, 462, S533–S546.

Lowry, S., *et al.* (2012) The nucleus of Comet 67P/Churyumov-Gerasimenko: a new shape model and thermophysical analysis. *Astronomy and Astrophysics*, 548, A12.

Mottola, S., *et al.* (2014) The rotation state of 67P/Churyumov-Gerasimenko from approach observations with the OSIRIS cameras on Rosetta. *Astronomy and Astrophysics*, 569, L2.

Preusker, F., *et al.* (2015) Shape model, reference system definition, and cartographic mapping standards for comet 67P/Churyumov-Gerasimenko – stereo photogrammetric analysis of Rosetta/OSIRIS image data. *Astronomy and Astrophysics*, 583, A33.

Preusker, F., *et al.* (2017) The global meter-level shape model of comet 67P/Churyumov-Gerasimenko. *Astronomy and Astrophysics*, 607, L1.

Shi, X., *et al.* (2016) Sunset jets observed on comet 67P/Churyumov-Gerasimenko sustained by subsurface thermal lag. *Astronomy and Astrophysics*, 586, A7.

Sierks, H., *et al.* (2015) On the nucleus structure and activity of comet 67P/Churyumov-Gerasimenko. *Science*, 347(6220), 1044.

Thomas, N., *et al.* (2015) The morphological diversity of comet 67P/Churyumov-Gerasimenko. *Science*, 347(6220), 44.

Tubiana, C., *et al.* (2015) 67P/Churyumov-Gerasimenko: activity between March and June 2014 as observed from Rosetta/OSIRIS. *Astronomy and Astrophysics*, 573, A62.
Vincent, J.-B., *et al.* (2016a) Summer fireworks on comet 67P. *The Monthly Notices of the Royal Astronomical Society*, 462, S184–S194.
Vincent, J.-B., *et al.* (2016b) Are fractured cliffs the source of cometary dust jets? Insights from OSIRIS/Rosetta at 67P/Churyumov-Gerasimenko. *Astronomy and Astrophysics*, 587, A14.

Planetary Remote Sensing and Mapping – Wu et al.
© 2019 Taylor & Francis Group, London, ISBN 978-1-138-58415-0

Chapter 5

The BepiColombo Laser Altimeter (BELA)

An instrument for geodetic investigations of Mercury

H. Hussmann, J. Oberst, A. Stark, G. Steinbrügge

ABSTRACT: BELA, the Laser Altimeter on ESA's and JAXA's joint BepiColombo mission, will be launched to Mercury in October 2018. The instrument is equipped with a Nd:YAG-Laser, operating typically at 10 Hz. The instrument is designed to acquire range measurements up to a distance of 1050 km, enabling global coverage of the planet. A unique full digitization of the returned pulse is foreseen supporting albedo and surface roughness measurements on the scale of the laser footprint. The goal of BELA is to obtain a low-order shape and global topographic model of the planet, and to measure Mercury's rotation, including long-period librations. Also, assuming a two-year mission in orbit, we aim at measurements of Mercury tides. Results from BELA will impose strong constraints on interior structure models and will improve our understanding of formation and evolution of the planet.

1 INTRODUCTION

Size, shape and rotation constitute basic geodetic data for any planet. This is particularly true for Mercury, where the precise knowledge on the global shape and the rotational state reveals important information on the inner workings of this planet. Owing to its proximity to the Sun, its elliptical orbit and its moments of inertia, the planet is captured in a 3:2 spin-orbit coupling, a rotational resonance unique in the Solar System. On its eccentric orbit, Mercury is subjected to periodic tidal deformation.

Laser altimeter experiments are particularly suited to determine geodetic parameters of the planets from orbit and produce precise maps of topography and surface morphologies. This paper discusses the BepiColombo Laser Altimeter (BELA), which will be launched to Mercury in 2018 to provide unique information about the innermost planet of the Solar System.

1.1 *Basic work principles of an altimeter*

Altimetry provides the distance from a spacecraft to a planet's surface by precisely measuring the round-trip travel time of a short pulse of laser light (Fig. 5.1). The travel time is measured from the time the laser pulse is fired to the time the laser light is reflected back from the surface and received by the instrument's telescope. This telescope is carefully aligned with the laser emitter and focuses the returning laser light onto a detector.

The range R from the spacecraft to the laser pulse footprint on the surface can be obtained through the simple relation $R = c\,\Delta t/2$, where c is the speed of light in vacuum and Δt is the time delay between transmission and reception of the signal (travel time). In addition to range, the amplitude and shape of the returning pulse reveal characteristics of the target, such as surface reflectance and roughness. Typically, the laser transmitter produces very short (few nanoseconds) near-infrared laser pulses (1064 nm wavelength). State-of-the-art laser transmitters are capable to fire tens of pulses per second.

The range measurements are converted to map coordinates and elevations for each laser pulse by combining the range data with information on the position and orientation of the spacecraft at the time the laser pulse was fired. While the spacecraft position along its flight path is typically

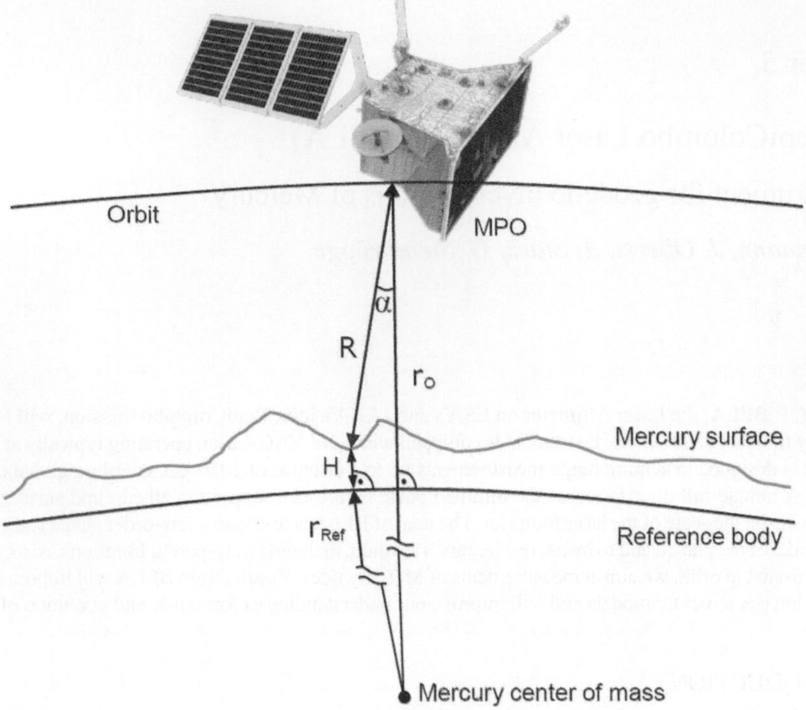

Figure 5.1. Working principle of a laser altimeter. Laser pulses are sent to surface (red arrow). With knowledge of the position and attitude of the spacecraft the topographic height *H*, i.e. the elevation of the surface with respect to a reference shape, can be derived. Nominally, the laser altimeter is nadir-pointed, i.e. alpha is a very small angle.

modeled from radio tracking data, the orientation of the spacecraft, i.e. the direction in which the laser pulse was fired, is established by the attitude control system of the spacecraft.

1.2 *Altimeters in planetary science*

Laser altimeters are powerful tools for many aspects in planetary geodesy and dynamics. While the early Apollo capsules orbiting near the equatorial plane of the Moon carried laser altimeters already (Sjogren and Wollenhaupt, 1973), the first global topographic maps of the Moon were derived by the LIght Detection And Ranging (LIDAR) experiment (Zuber *et al.*, 1994) on the Clementine spacecraft. More recent exploration efforts of the Moon included laser altimeters on the Japanese mission Kaguya (Araki *et al.*, 2009), on the Chinese Chang'E missions (Ouyang *et al.*, 2010), and on the Indian Chandrayaan-1 mission. Most complete and most accurate datasets were obtained with the Lunar Orbiter Laser Altimeter (LOLA) on board NASA's Lunar Reconnaissance Orbiter. With its high shot rate (28 Hz), capability of making multiple (five) measurements per shot, and its outstanding measurement resolution (laser spot size of only ~5 m), LOLA provided global maps of high-resolution topography, slopes, roughness and albedo of the Moon (Smith *et al.*, 2017). LOLA also made successful measurements of tidal deformation (Mazarico, Barker, *et al.*, 2014). The instrument provided critical support to spacecraft navigation, by using tracking data from ground-based laser ranging stations (Bauer *et al.*, 2016, 2017) as well as height differences at crossing points of the laser altimeter profiles as observables (Mazarico *et al.*, 2011).

The first laser altimeter flown on an asteroid mission was the NEAR-Shoemaker Laser Rangefinder (NLR) onboard the NEAR-Shoemaker spacecraft that went in orbit around asteroid 433

Eros in 2000 (Zuber *et al.*, 2000). The Light Detection and Ranging instrument (LIDAR) onboard the Japanese Hayabusa-1 spacecraft was operational for three months from September to November 2005 in orbit around asteroid 25143 Itokawa (Mukai *et al.*, 2007).

The terrestrial planets Mars and Mercury have been successfully investigated by laser altimeters flown on NASA's Mars Global Surveyor (Smith *et al.*, 2001) and MESSENGER missions (Zuber *et al.*, 2012). The Mars Orbiter Laser Altimeter (MOLA) on the Mars Global Surveyor (MGS) spacecraft acquired altimeter data in orbit around Mars from 1999 to 2001. The instrument operated at a frequency of 10 Hz and collected more than 600 million range measurements, which were basis for a precise global topographic reference map, still in use today (Smith *et al.*, 2001). The Mercury Laser Altimeter (MLA) was operational in orbit around Mercury from March 2011 until April 2015. Due to its highly elliptic polar orbit and a maximum measurement distance of up to 1500 km, only the northern hemisphere of Mercury could be covered by MLA data (Zuber *et al.*, 2012).

2 MERCURY

Mercury is only slightly larger than Earth's moon, but, e.g. smaller than the Galilean satellite Ganymede. Owing to its proximity to the Sun, observations from Earth are very challenging, and until recently, the planet has been poorly explored. NASA's Mariner 10 spacecraft performed three flybys of Mercury in 1974/1975 and returned the first close-up images, covering about 45% of the planet's surface. The MESSENGER (MErcury Surface, Space ENvironment, GEochemistry and Ranging) spacecraft was launched in 2004, performed three flybys (two in 2008 and one in 2009) before entering orbit around the planet in March 2011. It began a comprehensive remote sensing campaign (Solomon *et al.*, 2001, 2008) until the mission was terminated by the spacecraft's crash on the planet in April 2015. The phase immediately before the intended crash was used to get high-resolution observations of Mercury's surface.

2.1 *Mercury rotation*

Mercury moves around the Sun in 88 Earth days. It exhibits an unusual dynamic state. With its rotation locked in a 3:2 spin-orbit resonance, the planet rotates exactly three times, as it orbits the Sun twice (Fig. 5.2). Both the spin axis and the orbit-plane normal precess around the instantaneous Laplace plane normal with a period near 320,000 years. In addition, because of the planet's slightly eccentric orbit, the tidal torque of the Sun on the asymmetric mass distribution of the planet forces a physical libration in longitude, i.e., a small oscillation about the mean rotation rate.

The first accurate observations of Mercury's rotation (Pettengill and Dyce, 1965) demonstrated the distinct resonance of the planet (Colombo, 1965). On the basis of Mariner 10 images the rotation period was determined as 58.6461 ± 0.005 days (Klaasen, 1976). More recently, Earth-based radar observations revealed Mercury's small obliquity (~2.0 arc minutes) and the amplitude of its forced libration (Margot *et al.*, 2007, 2012).

Mercury's locked rotation can constrain models of the interior structure of the planet through observations from an orbiting spacecraft alone. Laser altimeter tracks from three years of near-continuous MESSENGER orbital observations were co-registered with geometrically rigid stereo terrain models to study rotation parameters, in particular librational motion, spin axis orientation and mean rotation (Stark *et al.*, 2015). A large libration amplitude of 38.9 ± 1.3 arc seconds was found in agreement with the earlier radar observations, confirming that Mercury possesses a liquid outer core, decoupled from the mantle. The mean rotation rate was observed to be $6.13851804 \pm 9.4 \times 10^{-7\circ}\ d^{-1}$, significantly higher than the expected resonant rotation rate. It was suggested that Mercury is undergoing long-period librational motion, resulting from planetary perturbations of its orbit. On the basis of the new estimates of Mercury's rotation, updates for the interior structure of Mercury (Knibbe and van Westrenen, 2018) and an update of the planet's reference frame could be obtained (Stark *et al.*, 2018).

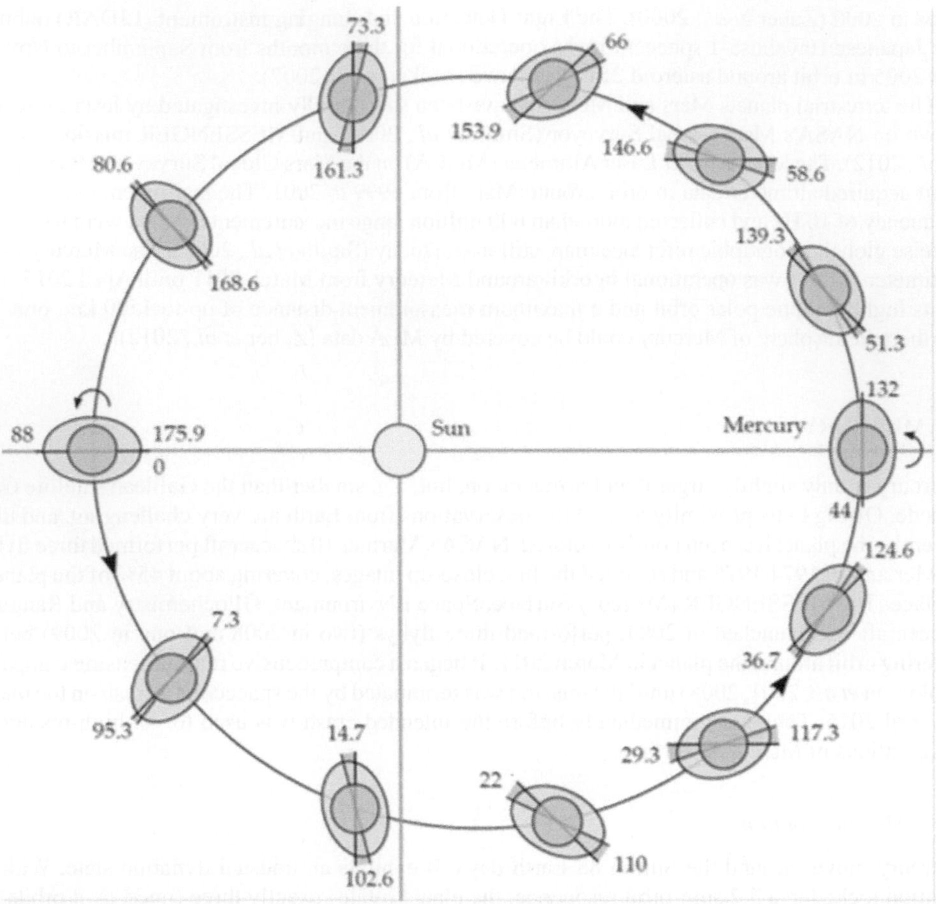

Figure 5.2. Rotation of Mercury within two revolutions about the Sun. The red line indicates the uniform rotation of Mercury's core and the black line shows the orientation of Mercury's mantle subjected to longitudinal librations. The libration angle in the first and second orbit is indicated by the red and blue shaded areas, respectively. For visualization purposes, the libration amplitude was scaled by a factor of 1800. The spacing between the rotation states represents Mercury at equal time intervals (7.33 days). The numbers indicate the time in days since the beginning of the first orbit at perihelion (Stark, 2015).

Recent interior structure models suggest a core radius of about 2000 km surrounded by a 400-km thick mantle and 35-km thick crust (Hauck *et al.*, 2013; Padovan *et al.*, 2015). Unfortunately, the existence of a solid inner core could not be unambiguously confirmed by MESSENGER data. Measurements of the tidal deformations by BepiColombo, however, could provide constraints on the size of the solid inner core (Steinbrügge, Padovan, *et al.*, 2018).

2.2 *Mercury's surface and shape*

Doppler and ranging measurements yielded estimates of Mercury's gravity field, particularly the coefficients of terms up to degree and order 50 in a spherical harmonic expansion (Mazarico, Genova, *et al.*, 2014). However, owing to the eccentric spacecraft orbit, a good model resolution was achieved in the northern hemisphere only. Several complementary techniques have been used to study Mercury's topography with MESSENGER data, including laser altimetry by the Mercury

Laser Altimeter (MLA) (Zuber *et al.*, 2012), measurements of radio occultation times (Perry *et al.*, 2015), limb profiling (Elgner *et al.*, 2014) and stereo imaging (Oberst *et al.*, 2010; Preusker *et al.*, 2011, 2017). Topographic models on regional and local scales were derived with stereo photogrammetry using images from the early Mercury flybys and in MESSENGER's orbit phase. Currently, gridded models are being produced, involving > 6000 images and 6 billion ground points, which have resolutions as low as 222 meters per grid element (Preusker *et al.*, this book). In contrast, radio occultation and limb data suffer from their limited resolution.

The traces of the thermal evolution of Mercury are well recorded in the geological features on its surface. The contraction process created the ubiquitous lobe-shaped scarps or cliffs, some hundreds of kilometers long and soaring up to a kilometer high (Byrne *et al.*, 2014). Mercury's surface has been constantly reshaped by volcanic activity, some of which possibly quite recently (Ernst *et al.*, 2015; Head *et al.*, 2011).

Like in the case of the Moon, Mercury's small obliquity and its rough topography prevents direct sunlight from reaching the floor of some polar craters. Such crates may represent "cold traps", favorable to the accumulation of volatiles (Neumann *et al.*, 2013). Indeed, Earth-based radar identified radar-bright spots in the north (Slade *et al.*, 1992) and in Mercury's south-polar region (Harmon *et al.*, 1994), which have been interpreted as evidence of water ice deposits. Thermal modeling of heat propagation within the regolith indicated that – even with Mercury's massive solar input – water ice potentially could be stable for billions of years (Paige *et al.*, 2013). However, with the lack of in-situ observations, the geological settings of these radar-bright spots could not be established. Images from MESSENGER's orbital mission provided the first nearly complete view of Mercury's north-polar region under a range of illumination conditions. It was quickly established that radar-bright features near Mercury's north pole are associated with locations persistently shadowed in the images. Within 10° of the pole, almost all craters larger than 10 km in diameter host radar-bright deposits (Chabot *et al.*, 2012, 2018). Measurements of surface reflectance of permanently shadowed areas by MLA revealed regions concentrated on poleward-facing slopes of anomalously dark and bright deposits at the laser wavelength of 1064 nm (Neuman *et al.*, 2013). The observations of optically bright regions are consistent with surface water ice, whereas dark regions are consistent with a surface layer of complex organic material that likely overlies buried ice and provides thermal insulation. Impacts of comets or volatile-rich asteroids could have provided both the dark and bright deposits. Similar studies for the south pole are awaiting new data from future missions.

3 THE BEPICOLOMBO MISSION

BepiColombo is a joint mission to Mercury under responsibility of the European Space Agency (ESA) and the Japan Aerospace Exploration Agency (JAXA), scheduled for launch in October 2018 and for arrival in late 2025. A nominal mission of one year is foreseen, with a possible one-year extension. ESA is responsible for the overall mission design, the integration and the launch. BepiColombo consists of two spacecraft, the Mercury Planetary Orbiter (MPO) provided by ESA and the Mercury Magnetospheric Orbiter (MMO) provided by JAXA. The two spacecraft will be inserted in different orbits optimized for their respective measurements (Benkhoff *et al.*, 2010).

3.1 *The BepiColombo Laser Altimeter (BELA)*

In 2004 ESA confirmed 11 selected payload instruments for the MPO, one of which is the BepiColombo Laser Altimeter (BELA). The instrument is designed to acquire range measurements up to a distance of 1050 km from Mercury's surface, implying that global altimetry measurements will be possible, allowing coverage of the southern hemisphere for the first time.

For BELA full digitization of the transmitted and returned pulse is foreseen (Thomas *et al.*, 2007). The onboard software is capable of analyzing the return pulse by using polynomial fits to approximate the pulse shapes. If requested, the fully digitized pulse can be returned to ground. This will greatly improve the instrument performance when the signal-to-noise is small (Gunderson and Thomas, 2010). Furthermore, the shape of the return pulse bears information on the albedo and the roughness of the surface on the scale of the laser footprint (~ 40 to 80 m).

BELA is equipped with two redundant Nd:YAG-lasers, capable of generating 50 mJ laser pulses at 1064 nm wavelength. The laser can be operated from 1 to 10 Hz. The receiver is a Cassegrain-type telescope with an aperture of 20 cm and for detection of the reflected laser pulses an APD ("Avalanche Photo Diode") is used. As is the case for other instruments on board the MPO, precautions against the intense solar radiation have to be taken. Figure 5.3 shows the proto-flight model of BELA after integration on the MPO. During orbital operation phases BELA needs an open nadir–orientated aperture. Therefore, the instrument is designed to protect the laser and the telescope from high solar radiation and planetary infrared fluxes in order to keep the temperatures within the acceptable range. The purpose of the Transmitter Baffle Unit (TBU) is to protect the inner side of the transmitter from direct sunlight, to reflect the incoming sunlight and to block

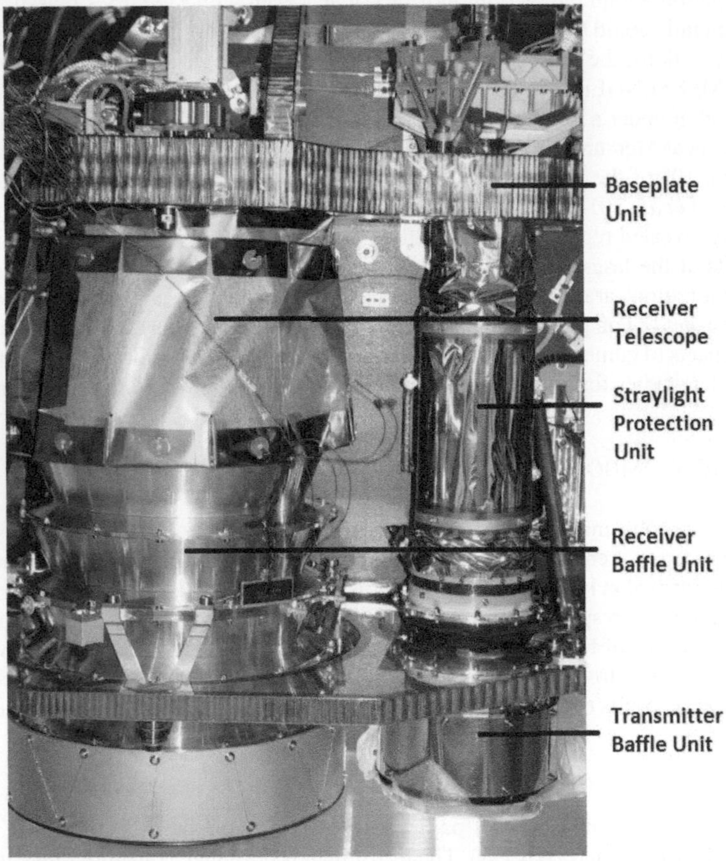

Figure 5.3. BELA proto-flight model mounted on the MPO. The receiver is shown on the left and the transmitter on the right. The receiver telescope which is located under the MLI has a diameter of 20 cm. Here the instrument is shown including protection covers at the bottom (to be removed before flight) in front of the transmitter and receiver baffles, respectively.

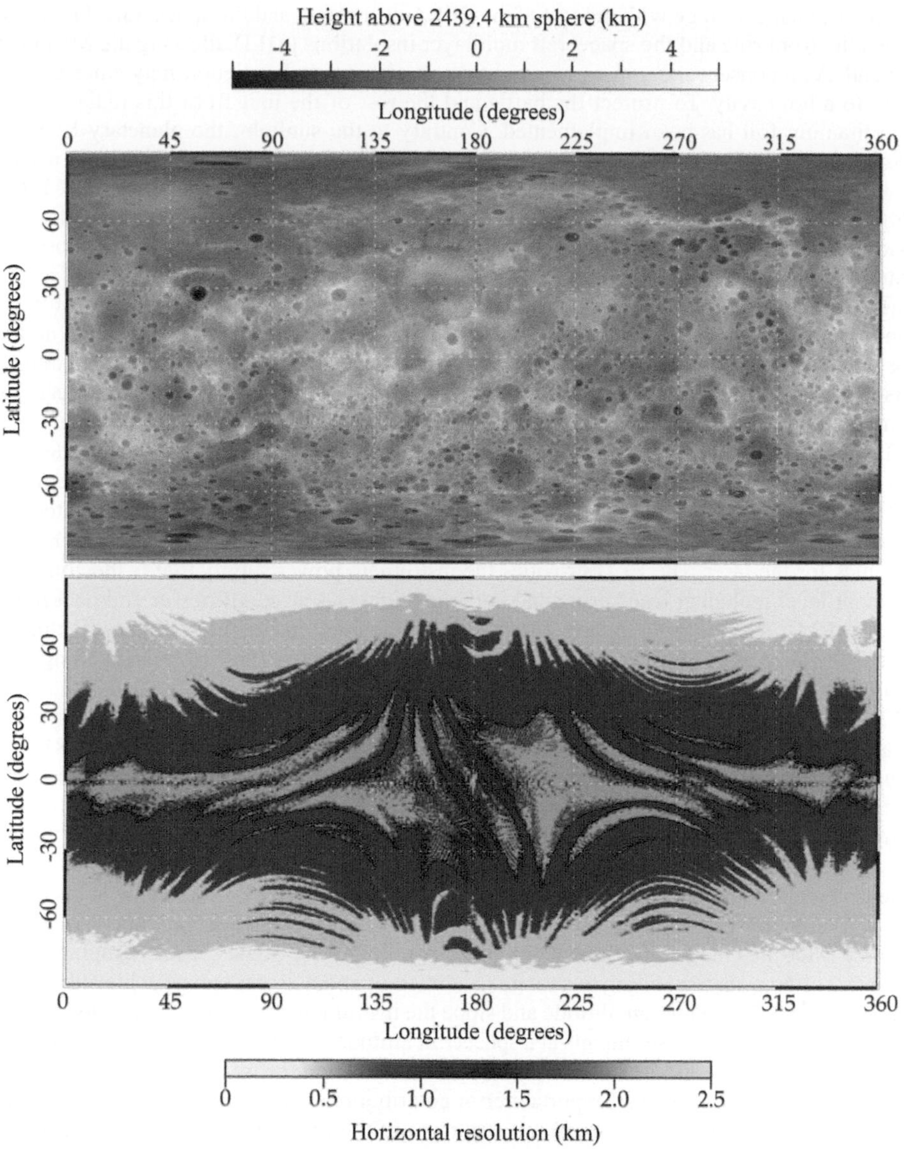

Figure 5.4. Top: Topographic map of Mercury obtained from stereo images of the MESSENGER spacecraft (Becker *et al.*, 2016). Bottom: Best horizontal resolution of a topographic map based on BELA profiles obtained after two years of operation in Mercury orbit (Steinbrügge *et al.*, 2018).

direct infrared (IR) flux from the planet. It is composed of a front ring located on the outer side of the spacecraft, therefore exposed to the highest radiation. The front ring is well insulated from the baffle and prevents the front section of this component from being struck directly by solar radiation, keeping the temperature of the baffle sufficiently low. The baffle is a Stavroudis type and consists of nine vanes. It is made of aluminum with a diamond turned surface on the inside which provides a good specular reflectivity. The outside of the baffle surface is polished aluminum reducing the heat exchange with the rest of the BELA instrument and the spacecraft. There is a gap between the front ring and the spacecraft multilayer insulations (MLI) allowing the MLI blankets to expand. As a consequence, in some conditions environmental radiation may enter the system leading to a hot cavity. To protect the baffle and the rest of the unit from this radiation a gold-coated titanium foil has been implemented. Contrary to the sunlight, the planetary IR radiation cannot be reflected by the baffle, as the aperture is always nadir-orientated. Therefore a sapphire filter has been implemented which provides a narrow bandpass for the laser beam at 1064 nm, while blocking all other wavelength mainly by reflection. The filter also resists to the thermal stresses, especially to the expected high temperature. Remaining straylight shall be absorbed by the Straylight and anti-contamination Protection Unit (SPU) so that the heat flux to the Laser Head Box (LHB) is minimized. The second purpose of the SPU is to protect the front lens of the beam expander as well as the filter against contamination. The SPU does not dissipate heat and is not exposed to the environment. Therefore its temperature is determined only by the surrounding elements, mainly TBU, LHB and the optical bench on which the SPU is mounted. The "heart" of the transmitter side is the LHB. It contains an Nd:YAG-laser emitting pulses at a wavelength of 1064 nm. The laser is pumped longitudinally having the advantage over conventional side-pumped lasers that the absorption path length can be longer making the laser less sensitive to temperature variations of the pump diodes. Two electronic boxes (not shown in Fig. 5.3) are located behind the laser: the Laser Electronics Unit (LEU) and the Electronics Unit (ELU). The LEU houses the electronics for the laser. The ELU provides the instrument power supply that is the controller on instrument level including laser firing and is the instrument-spacecraft power and data interface. Inside the ELU, the electronic boards (or PCBs) are mounted inside metal frames, including the Power Control module (PCM), the Digital Processing Modules (DPM) and the Range Finder Module (RFM), which hosts the digital pulse fitting algorithm.

BELA is a bistatic instrument, with the receiver side having a similar setup as the transmitter. It is also equipped with a larger front ring followed by the Receiver Baffle Unit (RBU). The telescope is then mounted together with the laser on the Baseplate Unit (BPU). Behind the telescope is the focal plane assembly which hosts the APD. The signal is transmitted via the Analog Electronics Unit to the RFM, where the laser pulses are processed and transmitted as science data to the DPM.

The transmitted and returned pulses are sampled with a bin-size of 12.5 ns which would correspond to a range resolution of 1.875 m in this so-called "coarse-detection" analysis. However, due to the filter-matching algorithms within the range finder electronics, a sub-sampling accuracy smaller than 1.5 ns corresponding to a range resolution of about 20 cm can be achieved under optimum conditions. Depending on altitude and slope the instrument error stays below 80 cm even for steep slopes of 40° and measurements at a spacecraft altitude of 1050 km (Steinbrügge, Stark *et al.*, 2018). However, due to additional error sources the range accuracy increases to several meters for typical measurements. The most important error contributions are (a) the pointing uncertainties of the instrument due to remaining small misalignment of the transmitter with respect to the spacecraft reference frame (corresponding to an estimated range error of ~ 9.5 m on Mercury's surface taking slope statistics on Mercury into account), (b) radial orbit errors (~1.8 m), (c) guidance errors due to the uncertainty in spacecraft position that translates into an additional angle with respect to the nadir axis (included in (a)) and (d) uncertainties in the rotational state of Mercury (~2 m). Mainly due to uncertainties in the pointing, the along-track and cross-track errors at the surface are about 40 m at 400 km altitude and 120 m at 1200 km altitude. Concepts to calibrate the pointing of the transmitter with respect to the camera pointing, which will reduce the pointing errors significantly, are currently under study. Other error sources are related to electronics and clock drifts. Since both signals the transmitter as well as the receiver signal undergo the same electronic signal chain, the electronic delays cancel out. Further, clock drifts on long timescales (much longer than a few milliseconds, the typical duration of a range measurement) of the range finder clock will be calibrated using an onboard pulse per second signal over the mission duration. Assuming uncorrelated errors, the overall range measurement error is less than 10 m assuming specific roughness values on different spatial scales (12.1 m at 200-m baseline and 6.4 m at 50-m baseline) and an albedo of 0.19 (Steinbrügge, Stark *et al.*, 2018).

3.2 *BELA science objectives*

The main measurement objective for BELA is to derive a global network of laser tracks which allows for studies of topography from global to local scales. BELA will improve on Mercury's global shape and global topography. As we also expect globally complete gravity data from the mission, joint inversions of gravity and topography will be possible. This will allow us to develop global crustal thickness models and, for example, to study the interior structures and compensation states of large impact basins.

Benefitting from global coverage, BELA will also improve on the rotation parameters. Using techniques that were developed for MESSENGER, the estimates of Mercury's rotational rate, obliquity, as well as amplitude and phase of the physical librations may improve. In particular, it will be interesting to verify the rotation rate measured from MESSENGER data and to identify a possible long-period libration term.

Another key objective will be the measurements of the tidal Love numbers, i.e., the deformation of the planet due to tides raised by the strong gravity field of the Sun. This can be accomplished with the help of laser profile cross-over points, where the tidal deformation can be retrieved by a differential height measurement. The distribution of cross-over points, which will finally depend on the actual signal-to-noise ratio and on the given operation scenario, is shown in Figure 5.5. Due to MPO's polar orbit, most of the cross-over points are located at high latitudes for which tidal amplitudes are only moderate, not exceeding a few tens of centimeters. At equatorial regions, where tidal amplitudes can exceed 1 m the number of cross-over points is rather limited. In a realistic simulation of BELA measurements the cross-over points were analyzed simultaneously in an inversion process solving for the tidal signal. With a known tidal potential at the time of the passes it is possible to solve for the tidal Love number h_2 with good accuracy, assuming a sufficient number of cross-over points from two years in orbit (Steinbrügge, Stark, *et al.*, 2018). This analysis

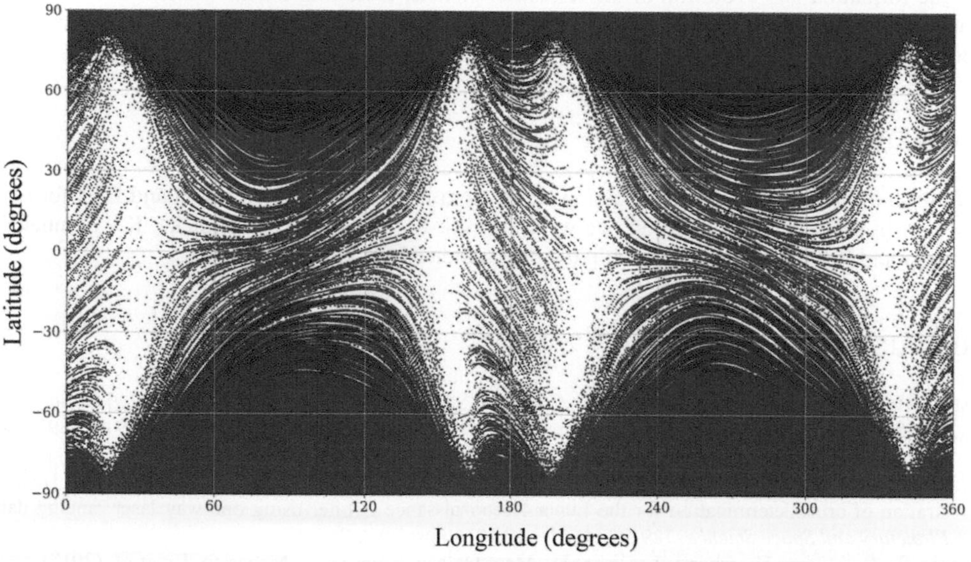

Figure 5.5. Distribution of BELA cross-over points considering an operation time of two years. Points marked in red are within the instrument's detection range of about 1200 km. Points marked in blue denote possible ground-track crossings based on the orbit geometry.

will allow constraining models of Mercury's interior. In particular, this may resolve the question of the size and physical state of an inner solid core.

BELA will be mapping geologic formations to obtain a complete inventory and 3D structures of Mercury's impact basins and large craters, as well as volcanic and tectonic features. The global mapping of tectonic fault systems will allow a complete reconstruction of the thermo-elastic history and global contraction of the planet. In addition, BELA's measurements of surface roughness and albedo will clearly establish the abundance and the characteristics of volatiles in the polar areas, including the south-polar region.

4 SUMMARY AND CONCLUSION

Mercury is certainly a unique planet, raising key questions to our understanding of the Solar System formation and evolution: Why is Mercury's density markedly higher than that of all other terrestrial planets? Is the core of Mercury liquid or solid? Is Mercury tectonically and volcanically active today? Why does such a small planet possess an intrinsic magnetic field, while Venus, Mars and the Moon do not have any? What is the origin and nature of the deposits in the permanently shadowed craters of the polar regions? It is these questions that will be tackled by the BepiColombo mission including its 16 scientific instruments on both orbiters, MPO and MMO. BELA will focus mainly on the geodetic characterization of the innermost planet.

Laser altimeters of the next generation will continue to be successful in planetary exploration. In the near future laser altimeters will be also used to study the Galilean satellites, further asteroids and possibly comets. Beginning next year, the Osiris-Rex Laser Altimeter (OLA) will map the small asteroid 101955 Bennu, equipped with a novel 3D laser imaging device. The Ganymede Laser Altimeter (GALA) on ESA's JUICE (Jupiter Icy Satellites Explorer) will investigate the Galilean satellites. Coping with challenging environment of high radiation levels, the instrument aims at the detection of sub-surface oceans and habitable worlds underneath the icy crust of the moons (Hussmann, 2015).

The data obtained by BELA will greatly improve our understanding of planet Mercury as well as the formation and evolution of the terrestrial planets, including Earth. While BELA has been mounted on the spacecraft in October 2016 (Fig. 5.3), the BELA team is now anxiously awaiting BepiColombo's launch this year (2018).

ACKNOWLEDGEMENTS

We wish to thank the BELA engineering teams from DLR, University of Bern and ESA for their support in the project. In addition, we wish to thank K. Wickhusen, F. Lüdicke, K. Gwinner, E. Hauber and F. Sohl for their support in the preparation of this manuscript.

REFERENCES

Araki, H., Tazawa, S., Noda, H., Ishihara, Y., Goossens, S., Sasaki, S., Kawano, N., *et al.* (2009) Lunar global shape and polar topography derived from kaguya-LALT laser altimetry. *Science*, 323(5916), 897–900. doi:10.1126/science.1164146.

Bauer, S., Hussmann, H., Oberst, J., Dirkx, D., Mao, D., Neumann, G.A., Mazarico, E., *et al.* (2016) Demonstration of orbit determination for the Lunar Reconnaissance Orbiter using one-way laser ranging data. *Planetary and Space Science*, 129, 32–46. doi:10.1016/j.pss.2016.06.005.

Bauer, S., Hussmann, H., Oberst, J., Dirkx, D., Mao, D., Neumann, G.A., Mazarico, E., *et al.* (2017) Analysis of one-way laser ranging data to LRO, time transfer and clock characterization. *Icarus*, 283, 38–54. doi:10.1016/j.icarus.2016.09.026.

Becker, K.J., Robinson, M.S., Becker, T.L., Weller, L.A., Edmundson, K.L., Neumann, G.A., Perry, M.E. & Solomon, S.C. (2016) First global digital elevation model of Mercury. *47th Lunar and Planetary Science Conference*, Houston, TX, USA.

Benkhoff, J., van Casteren, J., Hayakawa, H., Fujimoto, M., Laakso, H., Novara, M., Ferri, P., Middleton, H.R. & Ziethe, R. (2010) BepiColombo – comprehensive exploration of Mercury: mission overview and science goals. *Planetary and Space Science*, 58(1–2), 2–20. doi:10.1016/j.pss.2009.09.020.

Byrne, P.K., Klimczak, C., Sengor, A.M.C., Solomon, S.C., Watters, T.R. & Hauck, II, S.A. (2014) Mercury's global contraction much greater than earlier estimates. *Nature Geoscience*, 7(4), 301–317. doi:10.1038/ngeo2097.

Chabot, N.L., Shread, E.E. & Harmon, J.K. (2018) Investigating Mercury's south polar deposits: Arecibo radar observations and high-resolution determination of illumination conditions. *Journal of Geophysical Research: Planets*, n/a–n/a. doi:10.1002/2017JE005500.

Chabot, N.L., Ernst, C.M., Denevi, B.W., Harmon, J.K., Murchie, S.L., Blewett, D.T., Solomon, S.C. & Zhong, E.D. (2012) Areas of permanent shadow in Mercury's south polar region ascertained by MESSENGER orbital imaging. *Geophysical Research Letters*, 39(9), L09204. doi:10.1029/2012gl051526.

Colombo, G. (1965) Rotational period of the planet Mercury. *Nature*, 208(5010), 575. doi:10.1038/208575a0.

Elgner, S., Stark, A., Oberst, J., Perry, M.E., Zuber, M.T., Robinson, M.S. & Solomon, S.C. (2014) Mercury's global shape and topography from MESSENGER limb images. *Planetary and Space Science*, 103, 299–308. doi:10.1016/j.pss.2014.07.019.

Ernst, C.M., Denevi, B.W., Barnouin, O.S., Klimczak, C., Chabot, N.L., Head, J.W., Murchie, S.L., *et al.* (2015) Stratigraphy of the Caloris basin, Mercury: implications for volcanic history and basin impact melt. *Icarus*, 250, 413–429. doi:10.1016/j.icarus.2014.11.003.

Gunderson, K. & Thomas, N. (2010) BELA receiver performance modeling over the BepiColombo mission lifetime. *Planetary and Space Science*, 58(1–2), 309–318. doi:10.1016/j.pss.2009.08.006.

Harmon, J.K., Slade, M.A., Velez, R.A., Crespo, A., Dryer, M.J. & Johnson, J.M. (1994) Radar mapping of Mercury's polar anomalies. *Nature*, 369(6477), 213–215.

Hauck, II, S.A., Margot, J.L., Solomon, S.C., Phillips, R.J., Johnson, C.L., Lemoine, F.G., Mazarico, E., *et al.* (2013) The curious case of Mercury's internal structure. *Journal of Geophysical Research: Planets*, 118(6), 1204–1220. doi:10.1002/jgre.20091.

Head, J.W., Chapman, C.R., Strom, R.G., Fassett, C.I., Denevi, B.W., Blewett, D.T., Ernst, C.M., *et al.* (2011) Flood volcanism in the northern high latitudes of Mercury revealed by MESSENGER. *Science*, 333(6051), 1853–1856. doi:10.1126/science.1211997.

Hussmann, H. (2015) The Ganymede Laser Altimeter (GALA). *AGU Fall Meeting Abstracts*, San Francisco, CA, USA.

Klaasen, K.P. (1976) Mercury's rotation axis and period. *Icarus*, 28(4), 469–478. doi:10.1016/0019-1035(76)90120-2.

Knibbe, J.S. & van Westrenen, W. (2018) The thermal evolution of Mercury's Fe – Si core. *Earth and Planetary Science Letters*, 482(Supplement C), 147–159. https://doi.org/10.1016/j.epsl.2017.11.006.

Margot, J.L., Peale, S.J., Jurgens, R.F., Slade, M.A. & Holin, I.V. (2007) Large longitude libration of Mercury reveals a molten core. *Science*, 316(5825), 710–714. doi:10.1126/science.1140514.

Margot, J.L., Peale, S.J., Solomon, S.C., Hauck, S.A., Ghigo, F.D., Jurgens, R.F., Yseboodt, M., Giorgini, J.D., Padovan, S. & Campbell, D.B. (2012) Mercury's moment of inertia from spin and gravity data. *Journal of Geophysical Research-Planets*, 117(E12), E00L9. doi:10.1029/2012je004161.

Mazarico, E., Barker, M.K., Neumann, G.A., Zuber, M.T. & Smith, D.E. (2014) Detection of the lunar body tide by the Lunar Orbiter Laser Altimeter. *Geophysical Research Letters*, 41(7), 2282–2288. doi:10.1002/2013GL059085.

Mazarico, E., Rowlands, D.D., Neumann, G.A., Smith, D.E., Torrence, M.H., Lemoine, F.G. & Zuber, M.T. (2011) Orbit determination of the Lunar Reconnaissance Orbiter. *Journal of Geodesy*, 86(3),193–207. doi:10.1007/s00190-011-0509-4.

Mazarico, E., Genova, A., Goossens, S., Lemoine, F.G., Neumann, G.A., Zuber, M.T., Smith, D.E. & Solomon, S.C. (2014) The gravity field, orientation, and ephemeris of Mercury from MESSENGER observations after three years in orbit. *Journal of Geophysical Research-Planets*, 119(12), 2417–2436. doi:10.1002/2014JE004675.

Mukai, T., Abe, S., Hirata, N., Nakamura, R., Barnouin-Jha, O.S., Cheng, A.F., Mizuno, T., *et al.* (2007) An overview of the LIDAR observations of asteroid 25143 Itokawa. *Advances in Space Research*, 40(2), 187–192. doi:https://doi.org/10.1016/j.asr.2007.04.075.

Neumann, G.A., Cavanaugh, J.F., Sun, X., Mazarico, E.M., Smith, D.E., Zuber, M.T., Mao, D., *et al.* (2013) Bright and dark polar deposits on Mercury: evidence for surface volatiles. *Science*, 339(6117), 296–300. doi:10.1126/science.1229764.

Oberst, J., Preusker, F., Phillips, R.J., Watters, T.R., Head, J.W., Zuber, M.T. & Solomon, S.C. (2010) The morphology of Mercury's Caloris basin as seen in MESSENGER stereo topographic models. *Icarus*, 209(1), 230–238. doi:10.1016/j.icarus.2010.03.009.

Ouyang, Z.Y., Li, C., Zou, Y.L., Zhang, H.B., Lü, C., Liu, J.Z., Liu, J.J., *et al.* (2010) Primary scientific results of Chang'E-1 lunar mission. *Science China Earth Sciences*, 53(11), 1565–1581. doi:10.1007/s11430-010-4056-2.

Padovan, S., Wieczorek, M.A., Margot, J.L., Tosi, N. & Solomon, S.C. (2015) Thickness of the crust of Mercury from geoid-to-topography ratios. *Geophysical Research Letters*, 42(4), 1029–1038. doi:10.1002/2014gl062487.

Paige, D.A., Siegler, M.A., Harmon, J.K., Neumann, G.A., Mazarico, E.M., Smith, D.E., Zuber, M.T., Harju, E., Delitsky, M.L. & Solomon, S.C. (2013) Thermal stability of volatiles in the north polar region of Mercury. *Science*, 339(6117), 300–303. doi:10.1126/science.1231106.

Perry, M.E., Neumann, G.A., Phillips, R.J., Barnouin, O.S., Ernst, C.M., Kahan, D.S., Solomon, S.C., *et al.* (2015) The low-degree shape of Mercury. *Geophysical Research Letters*, 42(17), 6951–6958. doi:10.1002/2015GL065101.

Pettengill, G.H. & Dyce, R.B. (1965) A radar determination of the rotation of the planet Mercury. *Nature*, 206, 1240.

Preusker, F., Oberst, J., Head, J.W., Watters, T.R., Robinson, M.S., Zuber, M.T. & Solomon, S.C. (2011) Stereo topographic models of Mercury after three MESSENGER flybys. *Planetary and Space Science*, 59(15), 1910–1917. doi:10.1016/j.pss.2011.07.005.

Preusker, F., Stark, A., Oberst, J., Matz, K.-D., Gwinner, K., Roatsch, T. & Watters, T.R. (2017) Toward high-resolution global topography of Mercury from MESSENGER orbital stereo imaging: a prototype model for the H6 (Kuiper) quadrangle. *Planetary and Space Science*, 142, 26–37. doi:10.1016/j.pss.2017.04.012.

Sjogren, W.L. & Wollenhaupt, W.R. (1973) Lunar shape via the Apollo laser altimeter. *Science*, 179(4070), 275–278. doi:10.1126/science.179.4070.275.

Slade, M.A., Butler, B.J. & Muhleman, D.O. (1992) Mercury radar imaging: evidence for polar ice. *Science*, 258(5082), 635–640. doi:10.1126/science.258.5082.635.

Smith, D.E., Zuber, M.T., Frey, H.V., Garvin, J.B., Head, J.W., Muhleman, D.O., Pettengill, G.H., *et al.* (2001) Mars Orbiter Laser Altimeter: experiment summary after the first year of global mapping of Mars. *Journal of Geophysical Research-Planets*, 106(E10), 23689–23722. doi:10.1029/2000je001364.

Smith, D.E., Zuber, M.T., Neumann, G.A., Mazarico, E., Lemoine, III, F.G., Head, J.W., Lucey, P.G., *et al.* (2017) Summary of the results from the lunar orbiter laser altimeter after seven years in lunar orbit. *Icarus*, 283, 70–91. doi:10.1016/j.icarus.2016.06.006.

Solomon, S.C., McNutt, Jr., R.L., Gold, R.E., Acuna, M.H., Baker, D.N., Boynton, W.V., Chapman, C.R., *et al.* (2001) The MESSENGER mission to Mercury: scientific objectives and implementation. *Planetary and Space Science*, 49(14–15), 1445–1465. doi:10.1016/S0032-0633(01)00085-X.

Solomon, S.C., McNutt, Jr., R.L., Watters, T.R., Lawrence, D.J., Feldman, W.C., Head, J.W., Krimigis, S.M., *et al.* (2008) Return to Mercury: a global perspective on MESSENGER's first Mercury flyby. *Science*, 321(5885), 59–62. doi:10.1126/science.1159706.

Stark, A. (2015) *Observations of Mercury's Rotational State From Combined MESSENGER Laser Altimeter and Imaging Data*. PhD thesis, Technische Universität Berlin, Berlin, Germany.

Stark, A., Oberst, J., Preusker, F., Burmeister, S., Steinbrügge, G. & Hussmann, H. (2018) The reference frames of Mercury after MESSENGER. *Journal of Geodesy*, 92(9), 949–961.

Stark, A., Oberst, J., Preusker, F., Peale, S.J., Margot, J.-L., Phillips, R.J., Neumann, G.A., Smith, D.E., Zuber, M.T. & Solomon, S.C. (2015) First MESSENGER orbital observations of Mercury's librations. *Geophysical Research Letters*, 42(19), 7881–7889. doi:10.1002/2015gl065152.

Steinbrügge, G., Stark, A., Hussmann, H., Wickhusen, K. & Oberst, J. (2018) The performance of the BepiColombo Laser Altimeter (BELA) prior launch and prospects for Mercury orbit operations. *Planetary and Space Science*, 159, 84–92. doi:10.1016/j.pss.201804.017.

Steinbrügge, G., Padovan, S., Hussmann, H., Steinke, T., Stark, A. & Oberst, J. (2018) Viscoelastic tides of Mercury and the determination of its inner core size. *Journal of Geophysical Research: Planets*. In press.

Thomas, N., Spohn, T., Barriot, J.P., Benz, W., Beutler, G., Christensen, U., Dehant, V., *et al.* (2007) The BepiColombo Laser Altimeter (BELA): concept and baseline design. *Planetary and Space Science*, 55(10), 1398–1413. doi:10.1016/j.pss.2007.03.003.

Zuber, M.T., Smith, D.E., Lemoine, F.G. & Neumann, G.A. (1994) The shape and internal structure of the moon from the clementine mission. *Science*, 266(5192), 1839–1843. doi:10.1126/science.266.5192.1839.

Zuber, M.T., Smith, D.E., Cheng, A.F., Garvin, J.B., Aharonson, O., Cole, T.D., Dunn, P.J., *et al.* (2000) The shape of 433 eros from the NEAR-Shoemaker laser rangefinder. *Science*, 289(5487), 2097–2101. doi:10.1126/science.289.5487.2097.

Zuber, M.T., Smith, D.E., Phillips, R.J., Solomon, S.C., Neumann, II, G.A., Hauck, S.A., Peale, S.J., *et al.* (2012) Topography of the Northern hemisphere of Mercury from MESSENGER laser altimetry. *Science*, 336(6078), 217–220. doi:10.1126/science.1218805.

Planetary Remote Sensing and Mapping – Wu et al.

Chapter 6

Mars orbit optical remote sensor

High- and medium-resolution integrated stereo camera

Q. Meng, D. Wang and J. Dong

ABSTRACT: The high- and medium-resolution integrated stereo camera (HiMeRISC) is a two-component system consisting of a high-resolution camera (HRC) and a medium-resolution stereo camera (MRSC) designed for stereo mapping of the Martian surface and high-spatial-resolution imaging of selected regions. An off-axis three-mirror system is used in the HRC, with a focal length of 5000 mm, a field-of-view of 2° and an F-number of 12.5. Over 10-km-wide areas can be imaged in a resolution of 0.5 m pix-1 (at 300-km orbit altitude by the HRC). The MRSC has a focal length of 32 mm, a field-of-view of 66°, an F-number of 6.8 and a working wavelength range of 450–750 nm. Areas with widths of 250 km can be imaged in a resolution of 50 m pix-1 (at 300-km orbit altitude). MRSC was aligned, and the test results showed good image quality. HiMeRISC can provide various imaging modes with the combination of the two cameras, and it will be particularly useful in the geological research in Mars-exploring experiments.

1 INTRODUCTION

In 1958, the United States launched the first human deep-space probe, 'Pioneer 0', which played a significant role in human deep space exploration history. Over the past 60 years, various countries have accomplished more than 200 deep-space exploration missions (Ye and Peng, 2006). Among them, the majority are lunar exploration missions. Countries that have conducted lunar exploration missions include the United States, the former Soviet Union/Russia, Europe, Japan, China and India. These countries have conducted about 126 lunar exploration missions, including 56 from the United States (including 10 manned lunar exploration missions, which succeeded 37 times and failed 19 times, with a success rate of 66%) and 64 from Russia (succeeded 21 times, failed 43 times). Europe, Japan and India have each conducted one exploration mission. China has launched four lunar exploration missions, including the Chang'e-1, Chang'e-2, Chang'e-3 and the Chang'e reentry test devices (Ouyang *et al.*, 2010; Sun *et al.*, 2013; Zhao *et al.*, 2011).

Mars is the closest planet to the Earth in the solar system, and its orbital parameters, volume, density and other important parameters are very similar to those of Earth (Cattermole, 2012). Therefore, Mars is another key target in the deep space exploration. Every 26 months, there is one-time Mars opposition, where Mars is just 60 million kilometres from Earth. It is a particularly close approach for the Red Planet, and it also means that Mars exploration missions at those times will cost less. Even so, because of the long distance, complex environment and other factors, the Mars exploration success rate is only about 50%. There are higher risks for deep space orbit missions than for Earth orbit missions, and failure will result in waiting for a longer time for the next launch window. The mission specificity requires the selected payload to have multiple functions to better obtain abundant information during a one-time mission.

Optical remote sensing is a key approach to obtaining target information remotely. With the Mars 1M No. 1 spacecraft, which was developed by the Soviet Union and identified as the first Mars spacecraft in human history, the optical remote sensors were frequently carried by the orbiters. This setup is of great significance for researching and developing Mars orbit optical remote sensors.

1.1 *Mars orbit optical sensors in recent decades*

In recent decades, some important Mars exploration missions have been accomplished successfully. Among them, Mars Global Surveyor (MGS) (Malin and Edgett, 2001), Mars Express (Chicarro *et al.*, 2004) and Mars Reconnaissance Orbiter (MRO) (Zurek and Smrekar, 2007) are the outstanding representatives.

MGS is a robotic spacecraft developed by NASA's Jet Propulsion Laboratory (JPL) and launched in November 1996. Mars Express is a space exploration mission conducted by the European Space Agency (ESA). The spacecraft was launched on 2 June 2003. The Mars Express mission was the first planetary mission attempted by ESA. MRO was built by Lockheed Martin under the supervision of JPL. MRO was launched on 12 August 2005 and attained Martian orbit on 10 March 2006.

Without exception, optical sensors were selected as the primary payload for these Mars exploration missions. Among these optical sensors, some were panchromatic, some were chromatic, some were push-broom cameras and some were stereo cameras. However, they all had the same characteristic: high resolution. The representative Mars orbit optical remote sensors with high resolution developed in recent decades are summarised as follows:

1.1.1 High-resolution Mars Orbiter Camera

The Mars Orbiter Camera (MOC) narrow-angle system is a 3.5-m-focal-length camera that was selected as the payload in the MGS mission. The camera uses a Ritchey – Chretien optical system, which can achieve a 1.4 m pix^{-1} ground sampling distance (GSD) (Albee *et al.*, 2001).

1.1.2 High-resolution stereo camera

The high-resolution stereo camera (HRSC) was selected as the primary payload for the Mars Express. HRSC is a push-broom scanning camera with nine charge-coupled device (CCD) line detectors mounted in parallel on a focal plane. Its GSD is 10 m pix^{-1} at an altitude of 250 km, with an image swath of 53 km and 2.3 m pix^{-1} for an additional framing CCD device, called a super-resolution channel (SRC). The SRC provides frame images imbedded in the basic HRSC swath at a five-times-higher resolution. The SRC comprises a 1024 x 1024 CCD array and lightweight mirror optics, with its optical axis parallel to the HRSC camera head optical axis. The HRSC design permits stereo imaging with triple to quintuple panchromatic along-track stereo, including a nadir, forward- and afterward-looking (±18.91°) and two inner (±12.81°) stereo line sensors (Jaumann *et al.*, 2007; Neukum and Jaumann, 2004).

1.1.3 High-resolution imaging science experiment

The high-resolution imaging science experiment (HiRISE) camera was launched in August 2005 onboard NASA's MRO spacecraft. The camera uses a 50 cm, f/24 all-reflective optical system and a time delay and integration (TDI) detector assembly to map the surface of Mars from an orbital altitude of ~300 km. The HiRISE focal plane comprises a staggered array of fourteen charge-coupled devices (CCDs) with a pixel instantaneous field of view (IFOV) of 1 μrad. The nominal spatial resolution of HiRISE images from a 250 x 320 km orbit is 0.25–0.32 m pix^{-1} (Gallagher *et al.*, 2005).

1.2 *Development trend of Mars orbit optical sensors*

12.1 Sub-metre resolution fine optical remote observation

In the past 20 years, with the gradual improvement of the key technologies of optical remote sensors, such as large-aperture primary mirror support, thermal control and onboard image processing, sensors can observe Mars in high resolution (GSD less than 10 m pix^{-1}). The emphasis on Mars visible imaging remote sensing also gradually evolved from the early mid-resolution (hundreds of metres) global remote sensing to high-resolution fine observation of local areas.

Currently, the Mars Express has completed GSD of 100-m observations on 100% of the Martian surface and GSD of 10-m observations on 50% of the Martian surface. Based on these data, we have been able to generate a more detailed Mars map. For high-resolution observation below GSD of 10 m, the current global coverage of all projects does not exceed 1%. Therefore, carrying out fine observation, especially high-resolution observation below 1 m, will be a main development trend of Mars remote sensing observations in the future. Accurate data on the surface rocks and geomorphology of Mars can be obtained by sub-metre resolution imaging, which is an important basis for judging the evolution of Mars. However, because of technical limitations, only the HiRISE camera has reached resolutions under a GSD of 1 m. Therefore, sub-metre-resolution fine observation will also be a trend in the future for Mars exploration.

1.2.2 Visible spectrum observation and near-infrared observation

The work spectrum is one of the important parameters of remote sensing cameras. The appropriate choice of work spectrum can obtain a higher signal-to-noise ratio (SNR), and the images can better reflect the actual situation of the object. From the available data analysis, iron ore is one of the most widely distributed minerals detected on Mars, followed by layered silicate, sulfate, olivine and other minerals. From the spectral characteristics of these typical minerals, the main high-reflectivity intervals of the mineral components are mainly concentrated in the 500–800 nm spectral range (Murchie *et al.*, 2007). Therefore, near-infrared spectral imaging is of great significance for Mars observation.

1.2.3 High-resolution three-dimensional imaging of the Martian surface

Topographic features are the main targets of high-resolution remote sensing observations on Mars. Most geologic features, such as volcanoes, gullies and craters, are three-dimensionally distributed in space, and thus it is difficult to obtain accurate data of these features in the vertical direction only by two-dimensional images. HRSC stereoscopic imaging function is achieved through the use of multi-line CCD detectors. Because the focal plane panchromatic channel for HiRISE is a single line, its platform attitude must be adjusted to achieve three-dimensional imaging.

Based on previous analysis, sub-metre high-resolution stereo optical observation is a developing trend of Mars orbit optical sensors (Wang *et al.*, 2014).

2 PLANNING AND DESIGN OF THE MARS ORBIT OPTICAL SENSOR

Reconnaissance cameras and mapping cameras are the two major types of orbit optical sensors. Reconnaissance cameras are mainly used to confirm whether a target exists or not, and mapping cameras are mainly used to obtain mapping data. Furthermore, with the development of optical sensors, stereo cameras are widely used in stereo mapping. According to the deep-space exploration

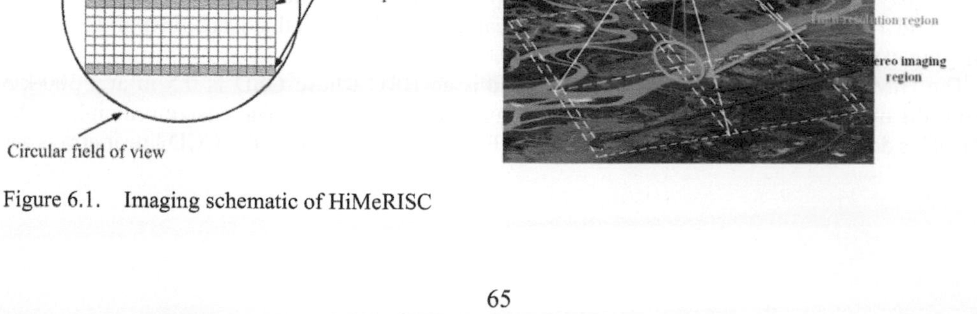

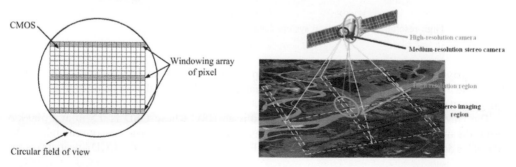

Figure 6.1. Imaging schematic of HiMeRISC

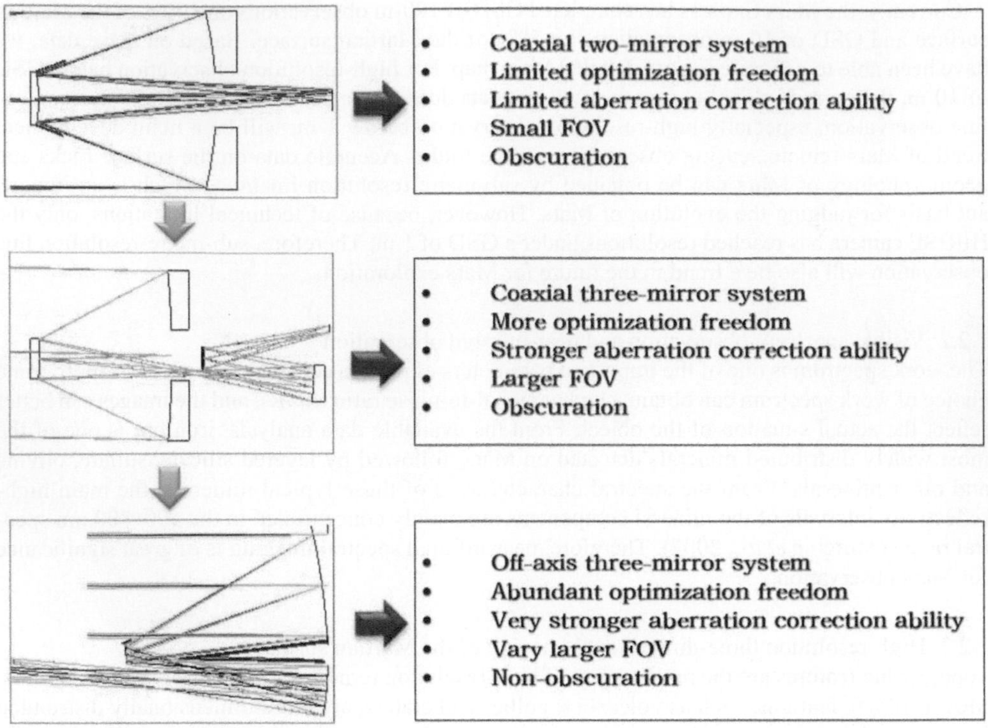

Coaxial two-mirror system
Limited optimization freedom
Limited aberration correction ability
Small FOV
Obscuration
Coaxial three-mirror system
More optimization freedom
Stronger aberration correction ability
Larger FOV
Obscuration
Off-axis three-mirror system
Abundant optimization freedom
Very stronger aberration correction ability
Vary larger FOV
Non-obscuration

Figure 6.2. Optical system comparison

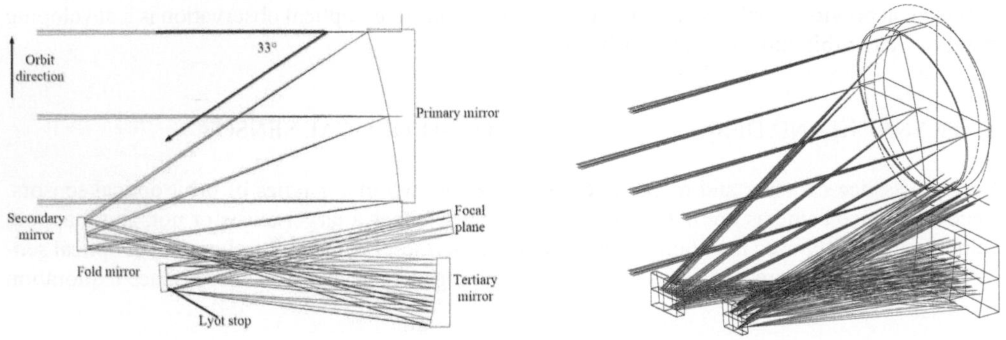

Figure 6.3. High-resolution TMA optical system design

optical payload progress, we propose to develop a Mars orbit optical sensor comprising both a high- and a medium-resolution integrated stereo camera (HiMeRISC) that has both high-resolution imaging capability and stereo imaging capability.

The HiMeRISC consists of two cameras. One is an HRC whose GSD is 0.5 m at a 300-km orbit; the other is an MRSC whose GSD is 50 m at the 300-km orbit. The design principle of the MRSC is derived and improved from the HRSC. For the HRSC, three TDI – CCD linear arrays are mounted in the instrument focal plane. The TDI – CCD arrays have panchromatic, chromatic and near-infrared imaging functions. In the MRSC, a CMOS is used in its focal plane, and the camera has an area array imaging function. Furthermore, the CMOS can be windowed at some arrays in the cross-track direction. The windowed pixels in the array can achieve a line array function to perform stereo imaging (see Fig. 6.4).

The HiMeRISC has multiple imaging modes. Besides the stereo imaging mode, the medium-resolution camera can image a wide region, a result of its wide FOV, and the HRC can image a key region in high resolution for specific requirements, such as searching for ground targets, finding the best landing region and so on.

2.1 *Optical system design of the high-resolution camera*

The HRC adopts an off-axis three-mirror anastigmatic (TMA) optical system (Meng *et al.*, 2018). This type of optical system has been getting more attention since the discovery of the advantages of much wider fields of view and non-obscuration, which makes a better spot diagram energy concentration (Meng *et al.*, 2014). With the progress of computing ability and optical manufacture and alignment, off-axis TMA optical systems have been applied in various optical remote sensors successfully. Figure 6.2 shows a comparison of typical optical systems. Figure 6.3 shows the TMA optical system design of the HRC.

The optical system has a focal length of 5000 mm, an FOV of 2° and an F-number of 12.5. An area 10-km wide can be imaged in a resolution of 0.5 m pix^{-1} (at 300-km orbit altitude). The optical system has a compact structure, and the distance between the primary mirror and the secondary mirror is less than 800 mm. The telephoto ratio value reaches 6. The camera focal plane subsystem uses three TDI – CCD arrays (see in Fig. 6.4). Each TDI – CCD array has five working spectrum bands: the panchromatic band, red band, green band, blue band and infrared band. Furthermore, two pieces of CMOS detectors are laid on the edge of the focal plane. The CMOS detectors are chromatic detectors, and thus they can achieve not only area array imaging but also video imaging.

The optical system design results of the modulation transfer function (MTF) in all FOVs approach the diffractive limit. The wavefront average value (RMS value) is 0.025λ (λ = 0.6328 μm), and the maximum grid distortion value is 0.1% (Figure 6.5). The detailed design parameters of the HRC optical system are listed in Table 6.1.

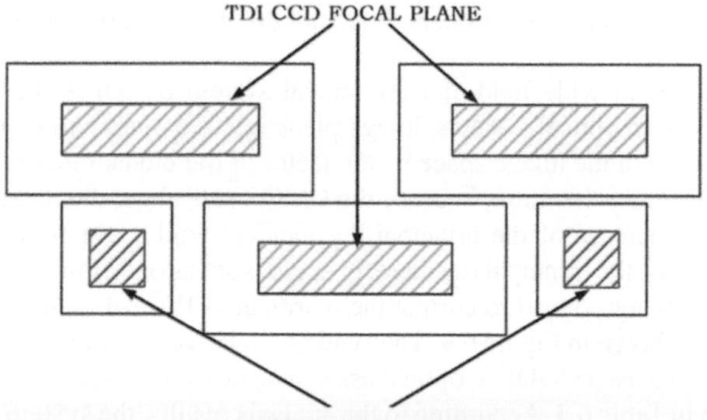

Figure 6.4. Focal plane layout

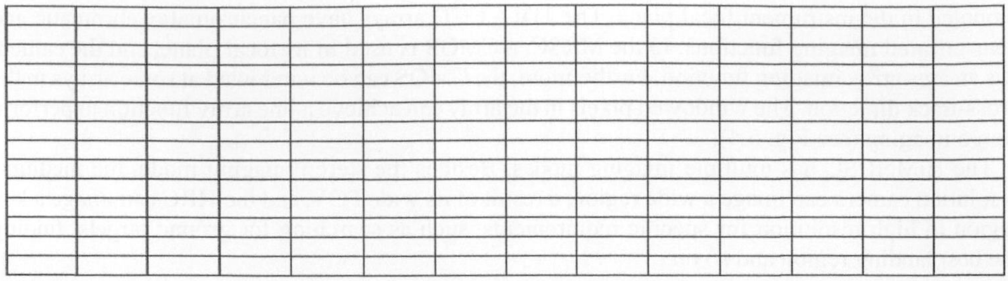

Figure 6.5. Grid distortion

Table 6.1 HRC optical system design parameters

Focal length	5000 mm				
Ground sampling distance (GSD)	≯50 cm pix^{-1} (at 300-km altitude)				
Field of view (FOV)	2°				
Swath width	≮10 km (at 300-km altitude)				
Obscuration	0				
MTF	>50%@50 cycles/mm@632.8 nm				
Wavefront error	Avg. 0.025λ RMS (λ = 0.6328 μm)				
Optical distortion	0.1%				
Spectral bands (nm)	Panchromatic	Red	Green	Blue	Infrared
	400–900	640–760	520–640	520–640	760–1000
Telephoto ratio	6				

2.2 *Optical system design of the medium-resolution stereo camera*

In the MRSC optical system, two kinds of optical configurations are combined to achieve the design goal. In the optical system shown in Figure 6.6, one part is a retrofocus lens, which can decrease the ray incidence angle, and the other part is an achromatism microscope. The two parts combine to correct aberrations. As described in the previous section, to achieve stereo imaging, we strictly control the possible lens distortion, and the maximum distortion allowed is 0.1%, as shown in Figure 6.7.

One of the difficulties in wide field-of-view optical system design is the relative illumination value of marginal FOVs. In optical systems, image-plane relative illumination will decrease with the increase of the field angle in the image space by the factor of the biquadrate of the cosine of the field angle in the image space. Furthermore, because the CMOS self-align, the detector relative response will decrease with the increase of the principal ray incident angle, which is shown in Figure 6.8. During the design process, the principal ray incident angle is set as one of the optimisation merit functions, and a stop coma is introduced to correct the marginal FOV relative to the illumination value. The correction result is shown in Figure 6.9. The synthetic relative illumination is shown in Table 6.2.

After optimisation, we used CODE V optical assistance design software to analyse the tolerance; the results are shown in Table 6.3. According to the analysis results, the system adopts the mounting centring alignment method, and the lens is shown in Figure 6.10. The optical system was tested by using the MTF measuring equipment shown in Figure 6.11, and the testing results after alignment are shown in Figure 6.12. The figure shows that the MTF of each FOV all reached 0.4@80/mm, which meets the requirement of the static MTF in the laboratory.

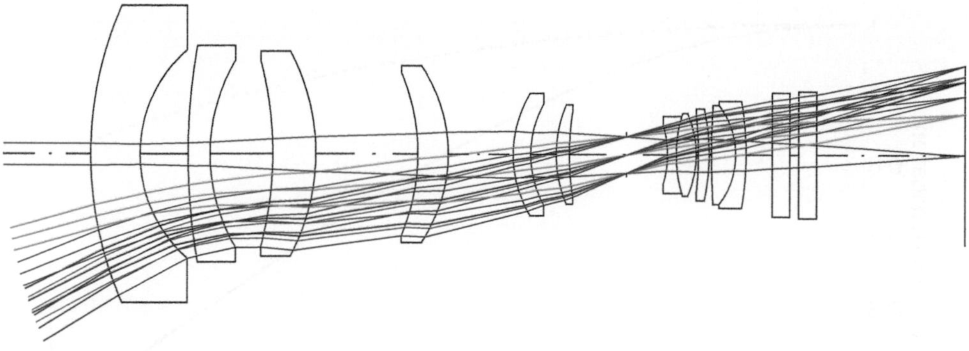

Figure 6.6. Optical system configuration

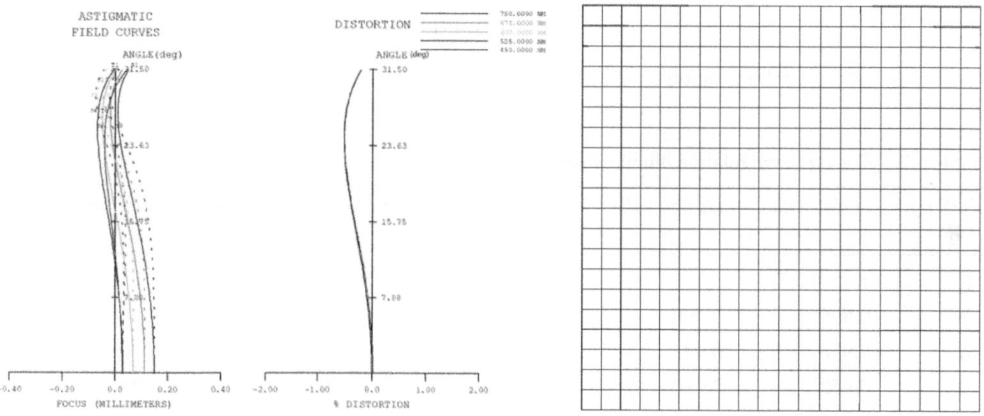

Figure 6.7. Astigmatic field curves and distortion

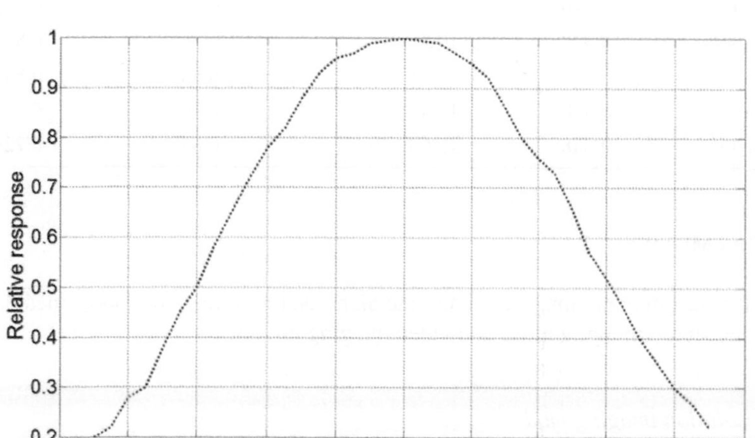

Figure 6.8. Angular response

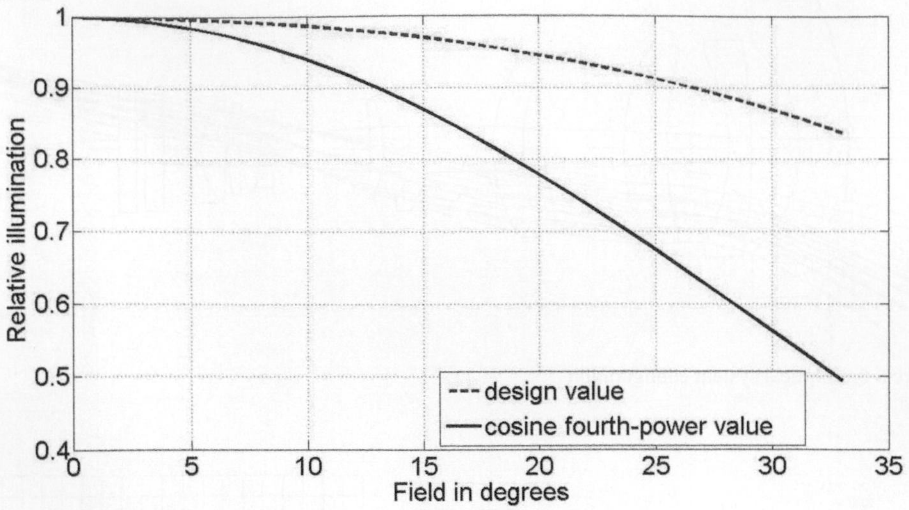

Figure 6.9. Relative illumination of optical system image plane

Table 6.2 The synthetic relative illumination

FOV (°)	*cos⁴ω value*	*Design value*	*Principal ray incident angle*	*CMOS angular response*	*Multiple response*
0	1	1	0	1	1
3	0.995	0.999	1.27	1	0.999
6	0.978	0.995	2.54	0.99	0.985
9	0.952	0.990	3.82	0.99	0.980
12	0.915	0.981	5.12	0.98	0.961
15	0.871	0.970	6.42	0.98	0.951
18	0.818	0.956	7.75	0.97	0.927
21	0.760	0.940	9.11	0.96	0.902
24	0.696	0.920	10.5	0.96	0.883
27	0.630	0.897	11.94	0.94	0.843
30	0.563	0.870	13.43	0.90	0.783
33	0.495	0.838	15.00	0.87	0.729

3 IMAGING MODES

The HiMeRISC has multiple imaging modes: the high-resolution imaging mode, medium-resolution imaging mode, stereo imaging mode and video imaging model.

3.1 *High-resolution imaging mode*

HRC can image the Martian surface in a resolution of 0.5 m pix^{-1} (at 300-km orbit altitude). The high-resolution imaging model has the following advantages. First, this mode can be used to observe some definitive key regions, e.g., the landing site or other regions of scientific interests.

Table 6.3 Tolerance of the MRSC

Lens Number	*Surface Number*	*Manufacture tolerance*				*Glass tolerance*		*Alignment tolerance*			
		N	*RMS*	*Δd (mm)*	*$\chi('')$*	*Δnd*	*Δvd*	*Lens distance (mm)*	*Lens tilt ('')*	*Group decentre (mm)*	*Bonding decentre ('')*
1	1	1	1/60	0.02	30	0.0005	0.5%	0.02	20	0.01	–
	2	1	1/60								
–	–	–	–	–	–	–	–	–	–	–	–
10	23	1	1/60	0.02	30	0.0005	0.5%	0.02	20	0.01	–
	24	1	1/60								

Figure 6.10. Alignment completed

Figure 6.11. MTF testing

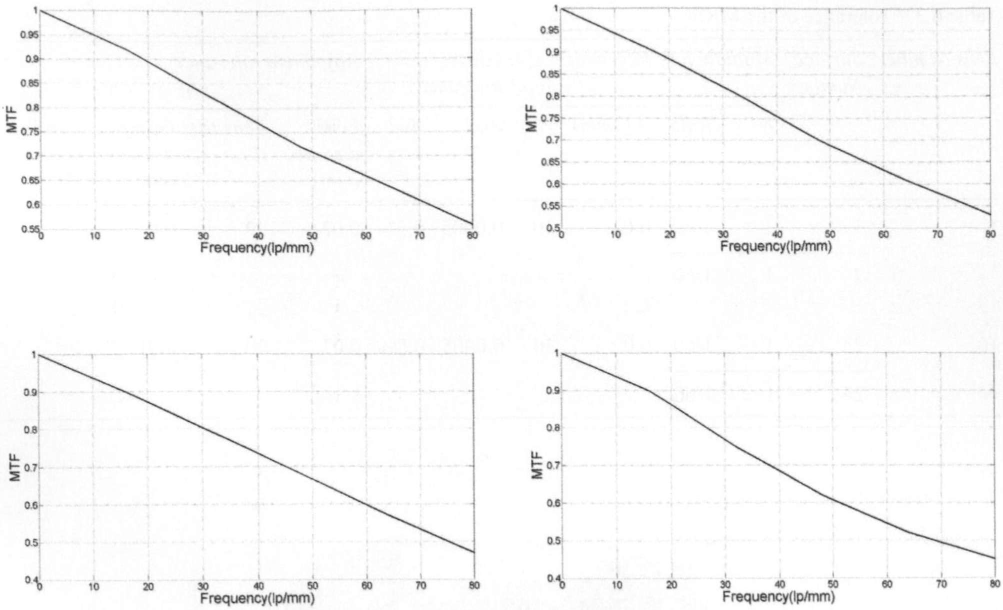

Figure 6.12. MTF-tested curves: (a) 0, (b) 0.5, (c) 0.7 and (d) 1 FOV

Table 6.5 Relationship between spatial resolution and binning

Orbit altitude (km)	*Binning*	*GSD (m)*
300	5 x 5	2.50
300–350	4 x 4	$\ngtr$2.30
350–450	3 x 3	$\ngtr$2.25
450–750	2 x 2	$\ngtr$2.50
750–1500	none	$\ngtr$2.50

Second, it can be used to observe the rover and its trajectory on the ground. Third, it can be used to image certain regions with flexibility according to the MRSC's feedback.

However, the GSD will decrease with the increase of the orbit altitude, so we formulated a 2.5 m GSD invariable-resolution imaging mode, as shown in Table 6.5. In this imaging mode, the camera will use 5 x 5, 4 x 4, 3 x 3, 2 x 2 and 1 x 1 (none) binning for different orbit altitudes to maintain the 2.5-m GSD.

3.2 *Medium-resolution imaging mode*

The medium-resolution imaging mode, which is operated by the MRSC, can achieve a resolution of 50 m pix^{-1} (at 300-km orbit altitude), with a wide imaging swath of 250 km x 190 km. Because of the advantage of large imaging area ability, the MRSC can be used for general remote sensing surveys. Images acquired by the MRSC can be used to produce maps of the Martian surface on a global scale.

3.3 *Stereo imaging mode*

We can use the MRSC to achieve stereo imaging. According to windowing of the CMOS at some arrays in the cross-track direction, the windowed pixels in the array achieve a line array function to perform stereo imaging. The MRSC has a large FOV, so it is conducive to large base/height ratio imaging.

3.4 *Video imaging mode*

The video imaging mode can achieve on both the HRC and the MRSC. Two pieces of CMOS are assembled on the HRC focal plane, as described earlier, and the CMOS combined with the long-focal-length optical system can take high-resolution videos. Furthermore, the MRSC's CMOS also can achieve video imaging mode in a medium resolution. Video imaging mode is a useful and novel mode. Using this mode, we can take, for example, Martian dust storm videos, Martian vehicle-moving videos and so on.

4 SUMMARY

A high- and medium-resolution integrated stereo camera (HiMeRISC), which consists of one high-resolution and one medium-resolution stereo camera, is designed and manufactured. The HRC is compact and lightweight, and it can photograph in a resolution of 0.5 m pix^{-1} (at 300-km orbit altitude). Besides the TDI – CCD arrays, two pieces of CMOS are assembled on the focal plane to achieve video capabilities. The medium-resolution stereo camera (MRSC) is a wide field-of-view optical remote sensor that can photograph at a resolution of 50 m pix^{-1} with an imaging width of 250 km x 190 km (at 300-km orbit altitude). Furthermore, the MRSC can achieve not only large-area-array imaging but also stereo imaging by combining with the CMOS.

The HiMeRISC has abundant imaging modes, including a high-resolution imaging mode, medium-resolution imaging mode, video imaging mode, stereo imaging mode and so on. Based on these imaging modes, more scientific data can be obtained.

REFERENCES

Albee, A.L., Arvidson, R.E., Palluconi, F. & Thorpe, T. (2001) Overview of the Mars global surveyor mission. *Journal of Geophysical Research: Planets*, 106(E10), 23291–23316.

Cattermole, P. (2012) *Mars: The Story of the Red Planet*. Springer Science & Business Media, New York, NY, USA.

Chicarro, A., Martin, P. & Trautner, R. (2004) The Mars express mission: an overview. In: *Mars Express: The Scientific Payload*, Volume 1240. European Space Agency. pp. 3–13,

Gallagher, D., Bergstrom, J., Day, J., Martin, B., Reed, T., Spuhler, P. & Tommeraasen, M. (2005) Overview of the optical design and performance of the High-Resolution Science Imaging Experiment (HiRISE). In: *Current Developments in Lens Design and Optical Engineering VI*, Volume 5874. International Society for Optics and Photonics. p. 58740K.

Jaumann, R., Neukum, G., Behnke, T., Duxbury, T.C., Eichentopf, K., Flohrer, J. & Hoffmann, H. (2007) The High-Resolution Stereo Camera (HRSC) experiment on Mars Express: instrument aspects and experiment conduct from interplanetary cruise through the nominal mission. *Planetary and Space Science*, 55(7–8), 928–952.

Malin, M.C. & Edgett, K.S. (2001) Mars global surveyor Mars orbiter camera: interplanetary cruise through primary mission. *Journal of Geophysical Research: Planets*, 106(E10), 23429–23570.

Meng, Q., Wang, W., Ma, H. & Dong, J. (2014) Easy-aligned off-axis three-mirror system with wide field of view using freeform surface based on integration of primary and tertiary mirror. *Applied Optics*, 53(14), 3028–3034.

Meng, Q., Wang, H., Wang, W. & Yan, Z. (2018) Desensitization design method of unobscured three-mirror anastigmatic optical systems with an adjustment-optimization-evaluation process. *Applied Optics*, 57(6), 1472–1481.

Murchie, S., Arvidson, R., Bedini, P., Beisser, K., Bibring, J.P., Bishop, J. & Darlington, E.H. (2007) Compact Reconnaissance Imaging Spectrometer for Mars (CRISM) on Mars Reconnaissance Orbiter (MRO). *Journal of Geophysical Research: Planets*, 112(E5).

Neukum, G. & Jaumann, R. (2004) HRSC: The high-resolution stereo camera of Mars Express. In: *Mars Express: The Scientific Payload*, Volume 1240. European Space Agency. pp. 17–35.

Ouyang, Z., Li, C., Zou, Y., Zhang, H., Lü, C., Liu, J. & Bian, W. (2010) Primary scientific results of Chang'E-1 lunar mission. *Science China Earth Sciences*, 53(11), 1565–1581.

Sun, Z., Jia, Y. & Zhang, H. (2013) Technological advancements and promotion roles of Chang'e-3 lunar probe mission. *Science China Technological Sciences*, 56(11), 2702–2708.

Wang, W., Dong, J. & Meng, Q. (2014) Current status and developing tendency of visible spectral remote sensing camera for mars observation. *Chinese Optics*, 7(2), 208–214.

Ye, P.J. & Peng, J. (2006) Deep space exploration and its prospect in China. *Engineering Science*, 10, 2.

Zhao, B., Yang, J., Wen, D., Gao, W., Chang, L., Song, Z. & Zhao, W. (2011) Overall scheme and on-orbit images of Chang'E-2 lunar satellite CCD stereo camera. *Science China Technological Sciences*, 54(9), 2237.

Zurek, R.W. & Smrekar, S.E. (2007) An overview of the Mars Reconnaissance Orbiter (MRO) science mission. *Journal of Geophysical Research: Planets*, 112(E5).

Planetary Remote Sensing and Mapping – Wu et al.
© 2019 Taylor & Francis Group, London, ISBN 978-1-138-58415-0

Chapter 7

Mission profile and design challenges of Mars landing exploration

J. Dong, Z. Sun, W. Rao, Y. Jia, C. Wang, B. Chen and Y. Chu

ABSTRACT: Mars is the planet that most closely resembles Earth. Exploring Mars benefits the identification of the origin of life on the planet and potential habitats for human beings. This chapter describes the highlights of the recent Mars landing exploration missions by NASA and the European Space Agency and introduces the first Mars exploration mission from China. Referring to the previous and planned Mars exploration missions, this chapter also presents a preliminary study of the profile design of the first Mars exploration mission from China, including the landing site selection, main engineering constraints and the entry, descent and landing process. Finally, the key challenges and verification of the mission are discussed. The studies presented in this chapter can provide a useful reference for future Mars exploration missions.

ACRONYMS

Entry, Descent and Landing (EDL)
Flight Path Angle (FPA)
Entry Interface Point (EIP)
Mars Orbit Insertion (MOI)
Inertial Measurement Unit (IMU)
Disk-Gap-Band (DGB)
Centre of Gravity (CoG)
Guidance, Navigation and Control (GNC)
Phenolic Impregnated Carbon Ablator (PICA)
Thermal Protection System (TPS)
Terminal Descent Radar (TDR)
Terrain Relative Navigation (TRN)
Computational Fluid Dynamics (CFD)
Ultra High Frequency (UHF)
Reaction Control System (RCS)
Failure Detection, Isolation and Recovery (FDIR)

1 INTRODUCTION

Of all of the planets in the solar system, Mars is the one that most closely resembles Earth, which is a complex and vast world with a long history. Mars has become the focus of future deep space exploration, which will benefit the identification of the origin of life on the planet and discovery of potential habitats for human beings. China has successfully landed on the Moon, and our next step is to explore Mars.

This chapter introduces the recently planned Mars landing exploration missions, including Insight and Mars 2020 rover missions (NASA), ExoMars 2020 mission (European Space Agency, ESA) and the first Mars exploration mission from China. It also presents the considerations for the profile design of a Mars landing exploration mission and key challenges faced by the mission.

2 RECENT MARS LANDING EXPLORATION MISSIONS

2.1 *InSight*

The Mars InSight (Interior Exploration using Seismic Investigations, Geodesy and Heat Transport) mission has two scientific objectives: (1) to understand the formation and evolution of the terrestrial planets through investigation of the interior structure and processes of Mars and (2) to determine the present level of tectonic activity and the meteorite impact rate on Mars. The payload of the InSight lander includes: (1) a seismometer to measure the pulse of Mars and study the waves created by marsquakes, thumps of meteorite impacts and hot molten magma churning at great depths; (2) the heat flow and physical properties probe (HP^3) to measure the temperature of Mars, determine how much heat is still flowing through the interior of the planet and shed light on how Mars formed and evolved; and (3) a radio science instrument (RISE) to measure the reflexes of Mars as the sun pushes and pulls it in its orbit, providing clues to the size and composition of the deep inner core of the planet (Smrekar and Banerdt, 2014).

InSight is scheduled to launch in May 2018 followed by a six-month cruise to Mars. The planned landing site is Elysium Planitia. The mission is expected to last one Martian year.

Although the design of InSight was inherited from the Phoenix lander, its landing mission presents four additional challenges: (1) InSight will enter the atmosphere at a higher velocity (6.3 km/s versus 5.6 km/s for Phoenix). (2) InSight will have more mass entering the atmosphere compared to Phoenix (608 kg versus 573 kg). (3) InSight will land at an elevation approximately 1.5 km higher than the landing site of Phoenix and will, therefore, have less time for deceleration. The elevation around the landing site of Phoenix was almost -4 km. (4) InSight will land during the northern hemisphere autumn on Mars when the lander may encounter global dust. Some of the changes to the entry, descent and landing (EDL) system of InSight compared to that of Phoenix were: (1) a thicker heat shield to, in part, handle dust storms; (2) a parachute that opens at a higher speed; and (3) parachute suspension lines of a stronger material (NASA, 2018).

2.2 *Mars 2020 rover*

The Mars 2020 rover mission is another planned Mars robotic landing exploration with four scientific objectives: (1) to identify environments capable of supporting microbial life; (2) to look for signs of possible past microbial life in those habitable environments, particularly in special rocks known to preserve signs of life over time; (3) to collect core rock and 'soil' samples and store them in a 'cache' on the surface of Mars that a future mission could potentially return to Earth; and (4) to test oxygen production from the Martian atmosphere for humans. The payload of the Mars 2020 rover will include: (1) Mastcam-Z, an advanced camera system with panoramic and stereoscopic imaging capability with the ability to zoom, which will also determine the mineralogy of the Martian surface and assist with rover operations; (2) MEDA (Mars Environmental Dynamics Analyser), a set of sensors that will provide measurements of temperature, wind speed and direction, pressure, relative humidity and dust size and shape; (3) MOXIE (Mars Oxygen Isru Experiment), an exploration technology investigation that will produce oxygen from the Martian atmospheric carbon dioxide; (4) PIXL (Planetary Instrument for X-Ray Lithochemistry), an X-ray fluorescence spectrometer that will also contain an imager of high resolution to determine the fine-scale elemental composition of the Martian surface materials; (5) RIMFAX (Radar Imager for Mars' Subsurface Experiment), a ground-penetrating radar that will provide centimetre-scale resolution of the geologic structure of the subsurface; (6) SHERLOC (Scanning Habitable Environments with Raman and Luminescence for Organics and Chemicals), a spectrometer that will provide fine-scale imaging and use an ultraviolet (UV) laser to determine the fine-scale mineralogy and detect organic compounds; and (7) SuperCam, an instrument that will provide imaging, chemical composition analysis and mineralogy, and detect the presence of organic compounds in rocks and regolith from a distance (Farley and Schulte, 2014).

The launch of the mission will be in July or August 2020. Its design is based on the successful Mars Science Laboratory (MSL) mission architecture of NASA, including the Curiosity rover and landing system. The Mars 2020 rover mission will have the following major new technologies for EDL: a range trigger for mortaring the parachute, terrain-relative navigation, MEDLI2 (MSL Entry, Descent and Landing Instrumentation) and its EDL cameras and microphone.

Terrain-relative navigation will use onboard analysis of downward-looking images taken during descent and match them to the map that will indicate the zones designated unsafe for landing. As it is descending, the spacecraft will determine whether it is headed for one of the unsafe zones (e.g., towards dangerous ground up to about 300 m in diameter) and divert itself towards safer ground. MEDLI2 is a next-generation sensor suite that collects temperature and pressure measurements on the heat shield and afterbody during EDL. Atmospheric data from MEDLI2 and MEDA will be used to obtain the atmospheric density and winds, which will be helpful for future missions to Mars. The Mars 2020 rover mission will also add multiple descent cameras, including parachute monitoring cameras, a descent-stage 'down-look' camera, a rover 'up-look' camera and a rover 'down-look' camera. The EDL system will also include a microphone to capture the sounds during EDL (Farley and Schulte, 2014).

2.3 *ExoMars 2020*

The ExoMars programme will demonstrate key flight and in situ enabling technologies for future Mars exploration missions. The scientific goals are: (1) to search for signs of past and present life on Mars; (2) to investigate the water and geochemical environments as a function of depth in the shallow subsurface; (3) to investigate Martian atmospheric trace gases and their sources; (4) to characterise the surface environment.

The spacecraft of the ExoMars 2020 mission consists of a carrier module, a descent module and a rover module. The descent module consists of a landing platform, front shield, backshell and a parachute system. The parachute system is a two-stage system with supersonic and subsonic canopies. The rover payload is comprised of two different instrument sets that can conduct exhaustive scientific research of the Martian exobiology conditions: (1) a survey instrument set, including a panoramic camera, ground-penetrating radar, infrared spectrometer, Dynamic Albedo of Neutrons detector, infrared spectrometer for subsurface studies and close-up imager; (2) an analytical instrument set, including a hyperspectral imager, organic molecule analyser and Raman laser spectrometer (Musetti *et al.*, 2014).

The ExoMars 2020 mission will deliver a European rover and a Russian surface platform to the surface of Mars. During the launch and cruise phase, the carrier module will transport the surface platform and the rover within a single aeroshell. The descent module will separate from the carrier shortly before reaching the Martian atmosphere. Similar to past missions, during the descent phase, a heat shield will protect the payload from severe heat flux. Parachutes, thrusters and damping systems will reduce the speed and allow a controlled landing on the surface of Mars. After the landing, the rover will egress from the platform to start its science mission. The rover will analyse the physical and chemical properties of Martian samples by drilling. The drill is designed to extract samples from various depths, down to a maximum of two metres. Once collected, a sample will be delivered to the analytical laboratory of the rover, which will perform mineralogical and chemistry testing, including the identification of organic substances. The rover is expected to travel several kilometres during its mission (Rodionov, 2015).

2.4 *First Mars exploration mission of China*

China plans to launch its first Mars exploration mission in 2020. The scientific objectives of the mission are: (1) to study the appearance and geological structure of the Martian surface; (2) to analyse Martian soil characteristics and water distribution; (3) to detect the Martian physical field and internal structure; (4) to investigate the atmosphere-ionosphere and climate conditions. The

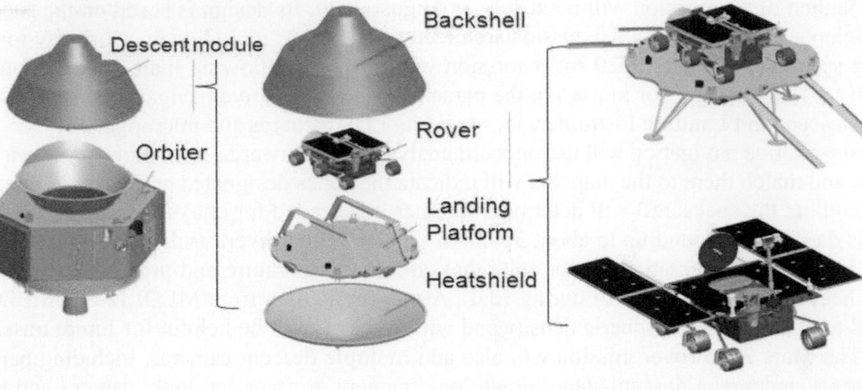

Figure 7.1. Vehicle configuration

spacecraft is composed of an orbiter and a descent module, which will complete orbiting, landing and roving in one launch. The descent module is comprised of a heat shield, backshell, landing platform and rover. These major spacecraft components are shown in Figure 7.1.

The orbiter will carry the descent module into the Martian orbit. Periodical checking of the onboard instruments will be performed during the Earth-Mars cruise phase, including instrument calibrations, battery charges and other maintenance. When the landing time is reached, the descent module will be separated from the orbiter after the altitude of the perigee is decreased. When the top of the atmosphere is reached, the descent module will begin the EDL process. During this process, images of the Martian surface will be taken by the cameras after the heat shield is separated from the entry module. The key parameters will be measured by the sensors installed in the heat shield and backshell similar to past Mars exploration missions.

After landing on the Martian surface, the rover will be released from the landing platform to begin the main science mission. The payload of the rover will include a terrain camera, sub-surface radar, multispectral camera, magnetometer and anemometer. The terrain camera will acquire images of the surrounding environment. The sub-surface radar will obtain information on the internal structure of Mars. The multispectral camera will be used to analyse the surface components. The magnetometer will gauge the weak magnetic field on Mars. The anemometer will measure the atmospheric parameters.

The orbiter will complete the relay mission for the rover in the relay orbit. When the lifetime of the rover ends, the orbiter will enter into a remote sensing orbit to image the Martian surface for more than one year.

3 SELECTION OF THE LANDING SITE

The selection of an accessible landing site is the primary step in surface exploration. The landing site is determined by both scientific goals and engineering constraints, of which the two factors should be well balanced and carefully decided. The engineering constraints are the key concerns of the mission. The Viking and Mars Pathfinder (MPF) missions took the engineering constraints into consideration at the primary design phase and selected probable landing sites.

There are various types of landforms on Mars, such as plains, craters, mountains, canyons and riverbeds. The characteristics of the elevation, slope and relief are obvious, and the rocks are widely distributed. In addition, some areas are covered with thick dust arising from Mars storms. The atmosphere density of Mars is approximately 1% that of the Earth. The engineering constraints are mainly related to the atmospheric environment, surface characteristics of the landing sites and mission trajectory. The surface characteristics include sunlight, thermal constraints, local elevation, rock distribution, dust thickness, slopes and relief that can affect the EDL safety and rover mobility. Data acquired from previous Mars missions can help select possible landing sites.

The primary landing site is selected according to the constraints mentioned above. Of most importance, the epoch of arrival should not be in the season of the Mars global dust storms. The landing site selection follows the mission schedule (i.e., Phase I – EDL; Phase II – surface survey by a rover) (Dong *et al.*, 2015).

3.1 *Phase I – EDL*

During the EDL phase, the atmospheric density, elevation and slopes are the main factors to be considered. The atmospheric density is altitude-dependent because it is associated with sensitive events in the EDL sequence, such as peak deceleration, deployment of parachutes, terminal descent velocity and initiation of powered descent. The elevation of the landing site must be low enough to have sufficient atmosphere above the site for a safe landing because the spacecraft relies on the atmosphere for deceleration during the descent. A low elevation will provide thicker atmosphere and longer duration for deceleration.

Therefore, the main consideration in the EDL phase is to select a region with a low elevation, such as the plain areas of the Martian surface that are typically smooth. The available landing region with a latitude higher than 5°N and an elevation lower than −3 km (referenced to the Mars Orbiter Laser Altimeter (MOLA) data) satisfies the EDL mission.

3.2 *Phase II – Rover survey*

The environmental factors on Mars depend on the latitude. According to the local wind speed, lighting and temperature, the plain within the region of 5°N to 30°N can be selected as primary, which ensures that the solar array of the lander will provide adequate power and warmth at all times.

Three plains were selected according to the elevation, slope and latitude as shown in Figure 7.2. According to Golombek *et al.* (2012), Region 3 (box, left) belongs to the A-level that has less thermal inertias, high reflectivity and thick dust (Fig. 7.3). Region 1 (polygon, middle) is mainly in the B-level, and Region 2 (polygon, right) is in the C-level. Thus, Regions 1 and 2 were selected. Region 1 has an abundance of riverbeds and craters. Therefore, the terrain is more complicated than in Region 2. In the next step, we will select the candidate landing sites from these two regions according to the scientific objectives and analyse if the terrain situation is satisfied. Several landing ellipses will be discussed and selected before the launch. The conditions of the slopes and rock abundance should be focused on when selecting the landing ellipses.

3.2.1 Slopes

Terrain relief features and slopes may influence radar measurements and affect the stability of the landing platform. The radar uses multiple beams to measure velocity and altitude. The initial measurements are acquired after jettisoning the front shield while the vehicle is still hanging under the parachute. Continuous measurements are performed throughout the descent phase. Over the entire descent trajectory, slopes at various length scales can alter the knowledge of the 'distance to ground at landing' with potentially serious consequences on fuel consumption, control authority and landing conditions.

The maximally permissible slope for touching the Martian surface safely is normally between 8° and 15° according to past missions. The slopes can be analysed based on digital elevation models (DEMs) of the surface. There are topographic datasets of the Martian surface with different resolutions collected from past missions. The main data resources include MOLA measurements, images from the High-Resolution Imaging Science Experiment (HiRISE), the Context Camera (CTX) and the High-Resolution Stereo Camera (HRSC). The spatial resolution of the MOLA DEM is 463 m $pixel^{-1}$. HiRISE images are usually 0.25 to 0.5 m $pixel^{-1}$, which results

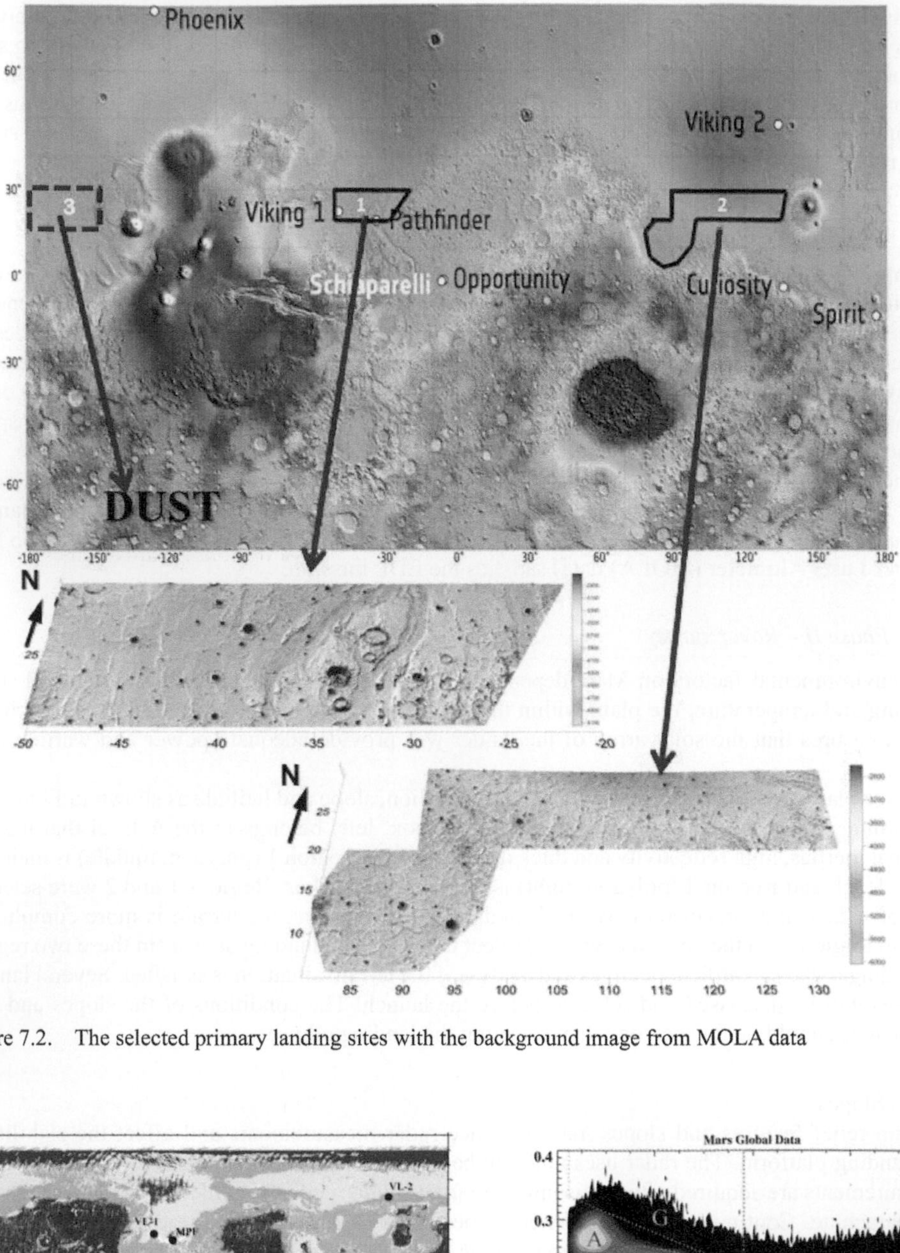

Figure 7.2. The selected primary landing sites with the background image from MOLA data

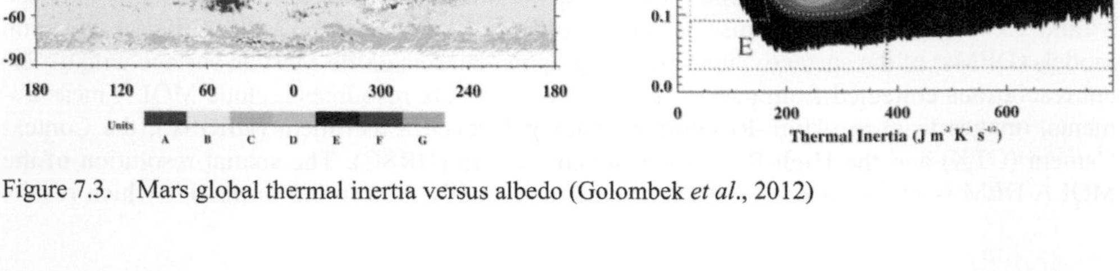

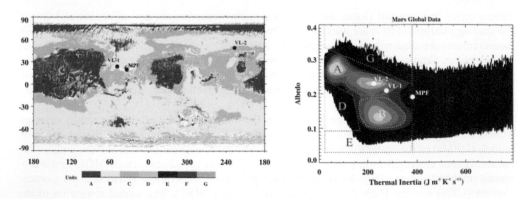

Figure 7.3. Mars global thermal inertia versus albedo (Golombek *et al.*, 2012)

in a DEM spacing of 1 to 2 m. The resolution of CTX images is 6 m pixel^{-1} with the generated DEMs normally having a resolution of 20 m. The resolution of the HRSC DEMs is about 50 m. However, the coverage areas of the HiRISE, CTX and HRSC are regional. Only the MOLA data cover the global surface.

Low-resolution DEMs have large baselines when calculating the slopes, which lead to the smoothness of the terrain, and the details of the terrain surface might be neglected. For DEMs covering the same area, if the resolution of the DEM decreases, the derived average slope will also decrease. An amplification function can be used to calibrate the slopes from low-resolution DEMs with large baselines to slopes with smaller baselines (Wang and Wu, 2017). Our future analysis of slopes will use the available data sources and functions to analyse the slope conditions for different candidate landing sites.

3.2.2 Rock abundance

A general consideration for selecting candidate landing sites is the rock abundance. Large quantities of rocks bigger than 0.5 m are not considered safe for landing or rover mobility. The thermal measurements from the Viking Infrared Thermal Mapper (IRTM) will be used to study the surface rock abundance of the candidate landing sites. Although the resolution of the IRTM data is relatively low (60 km pixel^{-1}), the rock abundances for five landing sites (Viking 1 and 2, Pathfinder, Spirit and Opportunity), which were analysed based on the IRTM data, are highly consistent with the surface measurements based on the images acquired by the landers and rovers (Golombek *et al.*, 2012).

4 THE EDL PROCESS

There are typically three phases in the EDL: the entry, parachute descent and powered descent phases (Prakash *et al.*, 2008). The descent module will perform a lifting entry into the Martian atmosphere followed by a descent phase under a Disk-Gap-Band (DGB) parachute. A retro-engine will slow down the landing platform up to the point where a short free fall (or touchdown) will be initiated. Finally, a crushable structure will absorb the remaining energy at landing.

4.1 *Initial condition*

The entry corridor is important for EDL. The selection of the Flight Path Angle (FPA) at the Entry Interface Point (EIP) depends on many constraints, such as the maximum heat flux, altitude of the parachute deployment and landing accuracy. The FPA is affected by many errors, such as the initial navigation, orbital manoeuvre and separation velocity errors. An angle of attack is commanded that is similar to the trim angle before EDL. The commanded bank angle is set according to the initial entry orbital error.

Several minutes prior to atmospheric entry, the spacecraft will finish slewing to the entry attitude. The error of the attitude determination before entry is also critical for the Mach number estimation and the velocity of touching the Martian surface (Kipp *et al.*, 2006). A final health and status check of the instruments will be conducted before Mars entry. The command sequences must be loaded for EDL. The loading of surface sequences and communication windows are needed for the first several Martian days.

The orbit determination solution should be accurate enough to meet entry knowledge requirements before EDL. From this solution, an entry state file will be generated and fed into the final EDL parameters update process. The EDL configuration files that result from this process will be uplinked to the spacecraft at E-3 hours, giving enough time to resend the file if necessary.

During the EDL operations, weather observations will be performed by the orbiter. The prediction of the landing site will use the data and atmosphere calculation model to support the final orbital manoeuvre and the update of parameters before EDL (Prince *et al.*, 2011).

4.2 *Entry phase*

4.2.1 Entry trajectory design

Both the ballistic and lifting entries have been used for the entry phase. The lifting entry is better to guarantee the parachute deployment altitude, parachute deployment attack angle and landing accuracy for the descent module with a high ballistic coefficient. It is necessary to maximise lift until the last possible moment before parachute deployment.

Based on the lifting entry, once the magnitude of the filtered drag acceleration climbs 0.2 g (Martin *et al.*, 2015), the entry guidance is used to predict the downrange flown and command a bank angle to correct for any range errors. Simultaneously, the guidance is employed to monitor the cross-range to the target and command a bank reversal whenever the cross-range crosses a dead-band threshold.

4.2.2 Parachute deployment constraints

Constraints on the parachute deployment conditions affect the guidance design to ensure adequate margins for the dispersed trajectories to meet the requirements.

1 Deployment altitude: As a propulsive descent system will be used after parachute deceleration, there is a timeline margin requirement that allows sufficient time to be spent on the parachute, radar and powered descent to land safely. This timeline is often roughly translated into a desired minimum chute deployment altitude relative to the surface. Normally, a minimum altitude of 6 to 8 km (Gavin and Craig, 2011) above the ground is necessary.

2 Mach number: The Mach number at chute deployment has two effects on the chute: aeroheating and inflation dynamics. If the Mach number is too high, the chute may fail due to excessive heating at the stagnation point or experience a violent inflation that excessively loads the chute. The acceptable deploy Mach number ranged from 1.1 to 2.3 for the MSL and Viking missions. In addition, the non-linear and non-stationary parachute forces increase exponentially above Mach 1.4 and become severe at Mach 2, which will cause IMU saturation according to the ExoMars 2016 landing mishap (Nielsen, 2017).

3 Dynamic pressure: Sufficient dynamic pressure at chute deployment is critical to ensuring inflation. If the dynamic pressure is too low, the resulting peak inflation loads may cause the chute to fail.

4 Chute opening loads: The chute loads that the entry vehicle and payload structure are designed for are another constraint on deployment conditions. The magnitude of the design chute opening load is a function of the chute drag that varies with Mach number, the inflation time of the chute and the dynamic pressure at the time of inflation.

5 Total angle of attack: The parachute should be deployed at a low total angle of attack to avoid the line sail that will affect the asymmetric canopy inflation and cause the high dynamic oscillatory motion. Therefore, the trim angle of attack should be zeroed out prior to parachute deployment. There are two ways to implement this task: (i) jettison of the internal entry balance masses to null the CoG-offset used for entry and (ii) trim-tab deployment to change the flow field (Kipp *et al.*, 2006).

6 Mach number error: Parachute deployment is triggered based on the Mach number. The probe estimates the Mach number via correlation with navigated velocity and knowledge of the Martian atmosphere. It is necessary to constrain the conditions that can be encountered on any given day within the landing window while also considering all potential landing locations and parachute deployment altitudes. Sound speed, which dictates the conversion from wind relative velocity to Mach number, can vary as much as 1% to 2% of the inertial velocity (Kipp *et al.*, 2006). More importantly, winds, whether gusty or steady state, create a significant error between inertial and wind relative velocity that is not estimable during entry. Furthermore, the error estimates on navigated velocity are nearly 2% of the inertial velocity (Kipp *et al.*, 2006). For MSL, the uncertainty in these parameters is determined statistically by sampling multiple Martian days and local solar times predicted by computer simulations of the Martian

atmosphere (Way, 2011). The value of the mean wind velocity on the landing date will be uploaded to the probe to decrease the Mach bias.

4.3 *Parachute descent phase*

For the previous successful Mars landing missions, the descent phase was based on a single stage supersonic DGB parachute. Following parachute deployment, a vehicle can quickly decelerate to subsonic conditions, which will bring the descent module to vertical velocities of about 80 to 120 m s^{-1}. Thus, the parachute system acts to burn over 75% of the remaining kinetic energy in just one to two minutes. During this phase, the front shield is separated, and the radar altimeter is activated to work with the IMU and trigger the back-shield separation based on relative navigation with respect to the Martian surface.

Heat-shield separation must satisfy two requirements: positive separation from the flight system with no re-contact and satisfactory distance to ensure no beam of the radar is obscured after activation. The separation mechanism needs to create initial separation velocity to avoid short-term re-contact. A sufficient ballistic coefficient difference should exist between the heat shield and the entry vehicle to result in continuous positive separation. Both Mach number and time delay triggers were used to separate the heat shield in the past (Raiszadeh *et al.*, 2011).

For the Mars 2020 rover mission, some new technologies will be used to increase the landing precision and the probability of a safe landing. Its range trigger will deploy the parachute based on the position of the spacecraft relative to the desired landing target. The range trigger could deliver the Mars 2020 rover to the exact spot in the planned landing area, and reduce the size of the landing ellipse (an oval-shaped landing area target) by more than 50%. The smaller ellipse size will allow the mission team to land at some sites where a larger ellipse would be too risky given they would include more hazards on the surface. Another potential advantage of testing the range trigger is that it would reduce the risk of any future Mars Sample Return mission because it would help that mission land closer to samples cached on the surface (Farley and Schulte, 2014).

4.4 *Powered descent phase*

Taking MSL as an example, the Guidance, Navigation and Control (GNC) software computes the appropriate altitude to separate from the backshell and start the powered descent based on the estimated vertical velocity. Backshell separation is followed by a short time of free fall to avoid short-term re-contact with the backshell. A backshell avoidance manoeuvre was added to mitigate the risk of backshell/parachute re-contact of the lander during terminal descent and at touchdown. MSL adopted the out of plane divert manoeuvre to move 30 m apart (Martin *et al.*, 2015). Phoenix performed a slight attitude adjustment in the plane (Grover, 2007).

When the altitude is reached before landing or the touching sensors work, the engine will be shut down. The crushable structure will absorb the remaining energy at landing. Following these events, the motors for the solar arrays will warm up and prepare to unfurl its solar panels, which is an important activity to ensure the lander has enough power for surface operations. This task together with others will take place autonomously on the landing day without human intervention. Next, the high-gain antenna will start to establish contact with the Earth station, and the Ultra High Frequency (UHF) link to the orbiter will be ready for transmission. Afterwards, the panoramic camera will start to look around, and the scientific instruments will begin the surface exploration mission.

5 KEY CHALLENGES AND VERIFICATION OF THE MARS LANDING MISSION

5.1 *Structure of the descent module*

The structure of the descent module will protect the spacecraft from the intense heating, pressure, shear stresses and parachute and deployment forces experienced during entry. The chute opening

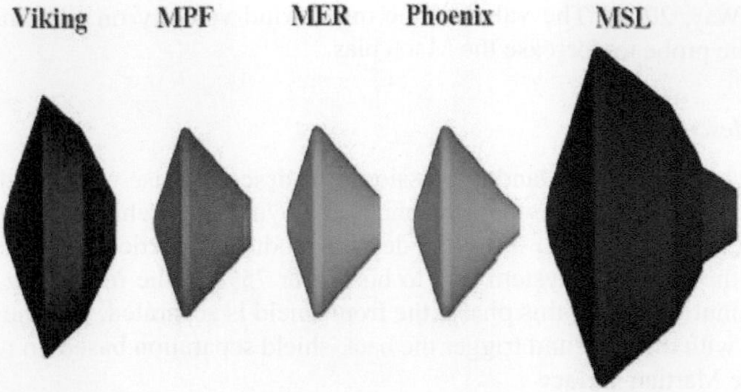

Figure 7.4. Different descent module configurations of NASA Mars missions

loads are the main constraints for the structure design. In addition, the parachute pull angle (i.e., the angle between the parachute total force vector and the negative axis of symmetry of the aeroshell) may affect both aeroshell dynamics and structural loads (Cruz *et al.*, 2014).

The front shield is generally based on the typical 70° sphere-cone geometry, the heritage of the Viking design and successfully adopted in subsequent NASA missions to Mars (Fig. 7.4). The Phenolic Impregnated Carbon Ablator (PICA) was selected as the forebody Thermal Protection System (TPS). The TPS must be validated using the arcjet test.

5.2 *Disk-gap-band parachute*

For the Mars atmospheric conditions, design trades in the NASA Mars landing missions, including options with two-stage parachutes, resulted in a baseline of a large supersonic DGB parachute (Cruz *et al.*, 2013; 2014). The diameter is a balance between the chute opening loads and deceleration requirements.

The parachute qualification program will be divided into four phases representing key functional risks: (i) mortar deployment, (ii) canopy inflation, (iii) supersonic performance and (iv) subsonic performance. Each phase will contain specific components that are evaluated across various test venues. The drag coefficients as a function of Mach number for the DGB parachute can be tested via a wind tunnel.

The rocket flight test (Fig. 7.5) will supply the real Mach number and dynamic pressure environment to verify every phase, but it cannot cover the extreme conditions due to the limited test number. This test is used to modify the core parameters of the parachute model. Therefore, the Computational Fluid Dynamics (CFD) flow field simulation and dynamic calculation are vital to establishing the parachute performance parameters.

The parachute models are extremely important for simulation and identify the effect of the parachute inflation. The area oscillation is the typical phenomenon in the parachute descent phase that will affect the design of the Inertial Measurement Units (IMU) and structure of the backshell. The rocket flight can also verify the dynamic characteristics of the parachute.

5.3 *Descent inertial measurement unit*

The EDL system contains the descent IMU to propagate the spacecraft state. The IMU saturation, which arises from the parachute deployment, will affect the navigation error and may cause a mishap (e.g., the ExoMars 2016 mission). The IMU design must balance the measurement range and accuracy. The navigation test will be performed under the vibration condition of the parachute deployment, separation trigger and powered descent to confirm the index. Before attitude determination and navigation in flight, the IMU should also be carefully calibrated.

Figure 7.5. An example of a parachute flight test via rocket

5.4 *Terminal descent radar*

Terminal descent radar consists of four radar beams canted at various angles, each of which independently measures the slant range and velocity. The radar is designed to operate over a wide range of conditions from shortly after heat shield separation until the landing platform touches down on the ground. The surface materials presented at the landing site should be radar reflective, which can provide sufficient backscatter signals to enable measurement of the altitude and velocity with respect to the ground during the descent. The radar design will be based on the ranges of the backscattering at nadir and off-nadir.

The radar field test can be separated into two parts to cover the high to medium altitude and velocity ranges of operation. The test on a transport plane can cover the high altitude and velocity range that the radar will operate in after the heat shield separation. The test in a powered flight experiment will verify the low altitude and velocity envelope. A simulation testbed is also developed for product verification and end-to-end simulations.

5.5 *EDL and science mission communication*

A suite of X-Band and UHF antennas were utilised to maintain communication both directly to Earth and to the orbiter during the EDL and science mission phases in the NASA Mars landing missions. During EDL, the descent module can only use the UHF link with the orbiter to return data back to Earth. After landing, the rover or the lander can communicate to Earth directly via the X-Band antenna and establish the communication with the orbiter periodically.

The test examines not only the link from the spacecraft to the orbiter testbeds but involves the respective ground data systems of the orbiter and the descent module and represents a high-fidelity test of the landing communication flow. The UHF and X-band functional and compatibility tests ensure that the communication system works and communicates properly with the orbiter and the ground station, respectively (Kornfeld *et al.*, 2014).

5.6 *Engines of the platform*

The main engine for reducing the final velocity needs to generate high and adjustable thrust. The hot-fire, vibration and thermal vacuum tests are necessary to verify the requirements. The potential adverse interaction between the supersonic flow and RCS firings can result in a reversal of the control action with possibly catastrophic results.

The hot-fire test demonstrates that the engines satisfy the performance requirements (e.g., throttle range, specific impulse and thrust) after the vibration test. The thruster valve assembly is qualified separately and undergoes functional tests between the environmental testing. In addition to the development program, the model is constructed in the flight dynamic simulations to support the system test (Kornfeld *et al.*, 2014).

5.7 *Flight system simulation and test*

The hardware and software of the spacecraft should be tested to validate the timeline and GNC behaviours, considering the specific environment on Mars. The tests normally include the following two types: (i) Real-time closed-loop execution of EDL hardware and software and (ii) functional testing.

In addition to the flight mode test for most hardware, the dynamic simulation is also necessary for the design of the key parameters and FDIR (Failure Detection, Isolation and Recovery). The model of the parachute deployment dynamics is very critical for navigation because the large canopy motion due to unsteady wake dynamics may cause large riser angle variations, including bridle slacking and asymmetric canopy inflation, which cause IMU saturation (e.g., ExoMars 2016 mission). In addition, mathematical models of the critical hardware, such as the IMU, the TDR, the thrusters and the main engine, should also be established accurately to support the end-to-end simulations.

6 SUMMARY

This chapter describes the highlights of the recent Mars landing exploration missions by NASA and ESA and introduces the first Mars exploration mission from China. It also presents a preliminary study for the profile design of this first mission, including the landing site selection, main engineering constraints and the EDL process. Key challenges and verification of the mission are discussed. The studies presented in this chapter can provide a useful reference for future Mars exploration missions.

REFERENCES

Cruz, J.R., Way, D.W., Shidner, J.D., *et al.* (2013) Parachute models used in the Mars Science Laboratory entry, descent, and landing simulation. *The 22nd AIAA Aerodynamic Decelerator Systems Technology Conference*, 25–28 March 2013, Daytona Beach, CA, USA.

Cruz, J.R., Way, D.W., Shidner, J.D., *et al.* (2014) Reconstruction of the Mars Science Laboratory parachute performance. *Journal of Spacecraft and Rockets*, 51, 1185–1196.

Dong, J., Meng, L.Z., Zhao, Y., *et al.* (2015) Selection of the martian landing site based on the engineering constraints. *The 66th International Astronautical Congress*, Jerusalem, Israel.

Farley, K. & Schulte, M. (2014) Mars 2020 mission: science rover. *The 8th Mars Conference*, 18 July 2014, Pasadena, CA, USA.

Golombek, M., Grant, J., Kipp, D., *et al.* (2012) Selection of the Mars Science Laboratory landing site. *Space Science Reviews*, 170, 641–737.

Grover, R. (2007) Evolution of the Phoenix EDL system architecture. *International Planetary Probe Workshop*, Bordeaux, France.

Kipp, D., Martin, M.S., Essmiller, J., *et al.* (2006) Mars Science Laboratory entry, descent, and landing triggers. *The 2006 IEEE Aerospace Conference*, Big Sky, Montana, March 4–11, Paper 1445.

Kornfeld, R.P., Prakash, R. & Devereaux, A.S. (2014) Verification and validation of the Mars Science Laboratory/curiosity rover entry, descent, and landing system. *Journal of Spacecraft and Rockets*, 51, 1251–1269.

Martin, M.S., Mendeck, G.F., Brugarolas, P.B., *et al.* (2015) In-flight experience of the Mars Science Laboratory guidance, navigation and control system for entry, descent, and landing. *CEAS Space Journal*, 7, 119–142.

Mendeck, G.F. & Craig, L.E. (2011) Entry guidance for the 2011 Mars Science Laboratory mission. *Atmospheric Flight Mechanics Conference*, 8–11 August 2011, Portland, OR, USA, AIAA-2011-6639.

Musetti, B., Vinai, B., Allasio, A., *et al.* (2014) The ExoMars 2018 mission. *The 65th International Astronautical Congress*, 29 September to 03 October, Toronto, Canada, Paper 25027.

NASA (2018) *InSight mission overview*. Available from: https://mars.nasa.gov/insight/mission/timeline/edl.

Nielsen, T.T. (2017) EXOMARS 2016-Schiaparelli anomaly inquiry. Available from: http://exploration.esa.int/mars/59176-exomars-2016-schiaparelli-anomaly-inquiry/

Prakash, R., Burkhart, P.D., Chen, A., *et al.* (2008) Mars Science Laboratory entry, descent, and landing system overview. *The 2008 IEEE Aerospace Conference*, 1–8 March, Big Sky, Montana, Paper 1531.

Prince, J., Desai, P., Queen, E. & Grover, M. (2011) Mars Phoenix Entry, Descent, and Landing Simulation Design and Modeling Analysis. Journal of Spacecraft and Rockets, 48. 756-764.

Raiszadeh, B., Desai, P. & Michelltree, R. (2011) Mars exploration rover heat shield recontact analysis. *The 21st Aerodynamic Decelerator Systems Technology Conference and Seminar*, 23–26 May 2011, Dublin, Ireland, AIAA-2011-2584.

Rodionov, D. (2015) ExoMars 2018 surface platform experiment proposal information package. *EXM-SP-EPIP-IKI-0001*. Available from: http://exploration.esa.int/mars/55699-exomars-2018-surface-platform-experiment-proposal-information-package/

Smrekar, S. & Banerdt, B. (2014) The InSight mission to Mars. *The 8th Mars Conference*, 18 July 2014, Pasadena, CA, USA.

Wang, Y. & Wu, B. (2017) Improved large-scale slope analysis on Mars based on correlation of slope derived with different baselines. *International Archives of the Photogrammetry, Remote Sensing & Spatial Information Sciences*, XLII-3/W1. pp. 155–161.

Way, D. (2011) On the use of a range trigger for the Mars Science Laboratory entry descent and landing. *The 2011 IEEE Aerospace Conference*, Big Sky, Montana, Paper 1142.

Section III

Geometric information extraction from planetary remote sensing data

Planetary Remote Sensing and Mapping – Wu et al.
© 2019 Taylor & Francis Group, London, ISBN 978-1-138-58415-0

Chapter 8

Correcting spacecraft jitter in HiRISE images

S. S. Sutton, A. K. Boyd, R. L. Kirk, D. Cook, J. W. Backer, A. Fennema, R. Heyd, A. S. McEwen and S. D. Mirchandani

ABSTRACT: Mechanical oscillations or vibrations on spacecraft, also called pointing jitter, cause geometric distortions and/or smear in high-resolution digital images acquired from orbit. Geometric distortion is especially a problem with pushbroom sensors, such as the High Resolution Imaging Science Experiment (HiRISE) instrument on-board the Mars Reconnaissance Orbiter (MRO). Geometric distortions occur at a range of frequencies that may not be obvious in the image products, but can cause problems with stereo image correlation in the production of digital elevation models, and in measuring surface changes in time series with orthorectified images. The HiRISE focal plane comprises a staggered array of fourteen charge-coupled devices (CCDs) with pixel instantaneous field of view (IFOV) of 1 microradian. The high spatial resolution of HiRISE makes it both sensitive to, and an excellent recorder of jitter. We present an algorithm using Fourier analysis to resolve the jitter function for a HiRISE image that is then used to update instrument pointing information to remove geometric distortions from the image. Implementation of the jitter analysis and image correction is performed on selected HiRISE images made available to the public. Results show marked reduction of geometric distortions. This work has applications to similar cameras operating now (such as the Lunar Reconnaissance Orbiter Camera Narrow Angle Camera (LROC NAC) on-board the Lunar Reconnaissance Orbiter) and to the design of future instruments (such as the Europa Imaging System, planned for the Europa Clipper mission).

1 INTRODUCTION

High-resolution imaging from low orbits or low-altitude flybys is achievable with pushbroom imaging. This mode of digital imaging builds an image line by line as the camera is flown over the ground target. Multiple detectors can be arranged to increase swath width, accommodate different color filters or to modify imaging modes within an observation. A significant advantage of pushbroom imaging is the use of time delay and integration (TDI) (McGraw *et al.*, 1980, 1986) to achieve a useful signal-to-noise ratio (SNR) in spite of very short line times, without causing excessive smear. The pushbroom method also allows for an arbitrary image length to be acquired up to the limits of on-board memory capacity or thermal limits. Along with the increasing use of higher resolution imaging comes the issue of sensitivity to platform stability during imaging, especially when using TDI. This paper describes the general problem of spacecraft jitter and a specific correction algorithm applied to the High Resolution Imaging Science Experiment (HiRISE) (McEwen *et al.*, 2007) operating on board the Mars Reconnaissance Orbiter (MRO) (Zurek and Smrekar, 2007). As more high-resolution instruments are flown, the issue of spacecraft jitter will need to be addressed in the design stages as well as during operation, especially for stereo mapping, change detection, high-resolution multispectral imaging or multi-sensor data fusion. The approach described here has general applicability to similar instruments and image data.

1.1 *Background*

The small instantaneous field of view (IFOV) of a high-resolution camera operating on an orbiting spacecraft leads to the requirement for a high degree of stability during imaging. Vibrations and mechanical motions that are of a frequency comparable to or shorter than the integration time (including TDI) lead to image smear, so the spacecraft must have acceptable stability on these short timescales (10 ms for MRO/HiRISE). Lower-frequency motions with an amplitude comparable to or larger than the IFOV of the detector can cause geometric distortions in the image. Spacecraft jitter is defined here as high-frequency periodic motion (Fig. 8.1), which might not be measured accurately by the spacecraft attitude control system. Distortions from jitter in the images are not usually visible to the human eye. However, image processing techniques that rely on accurate image correlation, such as change detection and stereo matching, are sensitive to even slight geometric distortions. Jitter also complicates band-to-band color registration, if the color acquisition is not simultaneous.

The presence of spacecraft jitter in orbiting single and multi-detector pushbroom imaging systems is widely acknowledged, especially as more systems obtain higher spatial resolution (Ayoub *et al.*, 2008; Eastman *et al.*, 2007; Hochman *et al.*, 2004; Kirk *et al.*, 2008; Li *et al.*, 2008; Teshima and Iwasaki, 2008; Theiler *et al.*, 1997). Even sub-pixel distortions can be problematic, as shown in Teshima and Iwasaki (2008). The need for correction of spacecraft jitter increases with increasing numbers of overlapping orbital high-resolution pushbroom images (Ayoub *et al.*, 2008; Eastman *et al.*, 2007).

Jitter measurement methods employed by other groups include multi-temporal image correlation (Ayoub *et al.*, 2008; Kirk *et al.*, 2008; Teshima and Iwasaki, 2008), charge-coupled device (CCD) to CCD or band-to-band correlation (Hochman *et al.*, 2004; Teshima and Iwasaki, 2008; Theiler *et al.*, 1997) and comparing to a reference image (Ayoub *et al.*, 2008; Teshima and Iwasaki, 2008). Multi-temporal image correlation, or comparison to a reference image assumed to be stable, is not always feasible, and is more likely to be a useful technique for Earth-orbiting sensors due to the abundance of repeat imaging and established ground control. Planetary image data do not have the benefit of abundant ground control, and finding an appropriate reference image is not possible most of the time due to the uncertainties in other spacecraft pointing and mapping or vastly differing spatial resolution of the different image datasets.

The approach described here uses Fourier analysis to solve for the absolute spacecraft pointing motion using knowledge of the focal plane layout. All of the jitter information is derived from measuring pixel offsets in the images. The jitter signal measured from the images is analyzed in frequency space. The derived function is compared to the input data to optimize the solution and minimize aliasing. The derived jitter function is ultimately combined with spacecraft pointing data to transform the image pixels to remove distortions in the images. The primary reason this

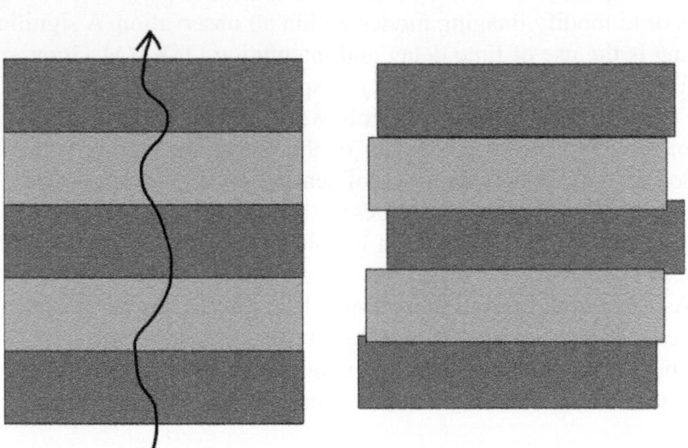

Figure 8.1. Cartoon illustrating the effects of jitter in pushbroom imaging. (Left) Camera motion with exaggerated jitter over the ground target. (Right) Reconstructed image with jitter distortions.

approach to modeling jitter is effective for HiRISE is the very narrow angular separation between overlapping detectors. This results in minimal stereo parallax between detectors.

1.2 *HiRISE instrument characteristics*

The HiRISE instrument consists of a 0.5 m primary mirror along with two smaller mirrors and two fold mirrors (one of which is attached to a focus mechanism), focusing light from the surface of Mars onto a staggered array of 14 CCDs, each 2048 pixels wide by 128 TDI stages (McEwen *et al.*, 2007). The CCDs are arranged on the focal plane assembly with vertical (along-track) offsets and horizontal (cross-track) overlaps (Fig. 8.2). The width of an image comprises 10 CCDs spanning the full-swath width, collecting visible light in the 550–850 nm range (RED0–9). The center 20% of the swath also has two detectors in the near-infrared, 800–1000 nm, (IR10–11) and two detectors in the 400–600 nm or blue-green wavelengths (BG12–13). Each RED CCD on the HiRISE focal plane overlaps the coverage in the cross-track direction of the adjacent CCD by approximately 48 pixels.

HiRISE has different imaging modes made up of combinations of TDI with 8, 32, 64 or 128 lines and pixel binning of 1 (unbinned), 2, 4, 8 or 16 for improved SNR (McEwen *et al.*, 2007). Binning and TDI can be set for each CCD. Detector pixels have an IFOV of 1 μrad. The nominal spatial scale from the 250 x 320 km orbit is 0.25–0.32 m pixel^{-1}, or slightly lower when pointing off-nadir by up to 30°. A complete HiRISE observation is made up of the image strips from the 10 RED CCDs stitched together, incorporating the overlapping areas to produce a final full-resolution (unbinned) product that is up to 20,000 pixels wide by an image length, determined by imaging modes and camera memory capacity, typically no more than 120,000 lines (Bergstrom *et al.*, 2004).

1.3 *Jitter calibration and mitigation on MRO*

HiRISE pointing stability requirements are driven by the line time and number of TDI stages used. Smeared pixels occur when the rate of along-track motion does not match the integration time over the TDI stages, or there is cross-track motion or alignment error preventing summation down TDI columns. To avoid smeared pixels, it is essential that the spacecraft remain stable in both the along- and cross-track directions over the TDI integration timescale. Frequency response functions for MRO were modeled and tested on the pre-launch spacecraft configuration for a variety of operational scenarios (Gasparinni, 2005). Although pre-launch modeling of spacecraft jitter on MRO was within the requirement of 2.5 pixels within 3-sigma (three standard deviations from the mean) (Gasparinni, 2005), this would have permitted substantial distortion and/or blur. Better performance was (mostly) obtained in flight. Smear of >1/4 pixel in HiRISE images due to jitter occurs if

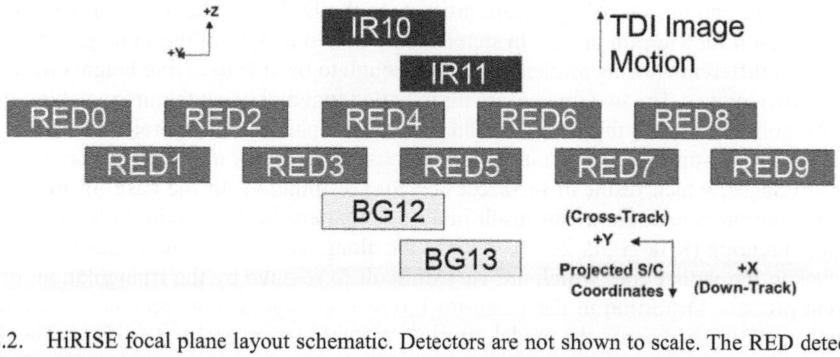

Figure 8.2. HiRISE focal plane layout schematic. Detectors are not shown to scale. The RED detectors are not perfectly aligned in each of two rows; instead, some are slightly offset to mitigate aliasing at higher frequencies. The IR and BG detectors provide information at lower frequencies.

the amplitude of motion is greater than 0.25 μrad over ~10 milliseconds. Fewer TDI stages can be used to minimize potential jitter distortions at the cost of SNR (Bergstrom *et al.*, 2004). Mitigating strategies in-flight, such as high-stability mode, greatly reduce but do not eliminate, jitter affecting HiRISE images.

MRO's on-board pointing information is measured with two star trackers that sample at 10 Hz, and three-laser gyro Inertial Measurement Units (IMUs) that sample at 200 Hz (Lee *et al.*, 2003). Spacecraft attitude data from the star trackers and the IMUs are used to create the reconstructed spacecraft pointing kernel (Acton, 1996). There is thought to be some amount of noise or drift in the IMU data, which may even introduce other errors into the pointing reconstruction. Therefore the IMU data are not a reliable source of information for correcting high-frequency jitter.

The approach to reducing jitter distortions in HiRISE images presented here is to use the image data, rather than to try to completely eliminate the sources of the motion (which is not possible). The likelihood of inflight jitter was anticipated by the designers of the HiRISE focal plane (McEwen *et al.*, 2007). To maximize the jitter information obtainable from the image data, several of the CCDs in one row of the array were shifted off their nominal baseline by a small amount to vary the time separation between adjacent detectors. Although this variation is small, it was expected to be able to prevent aliasing of jitter frequencies to a given time separation, which would prevent detection of those frequencies in the images. The longer time separation between the color (IR and BG) and the RED CCDs expands the dataset, and also has the advantage of wider cross-track coverage, rather than the ~48 pixels in the RED – RED pairs. The disadvantage of the IR and BG detectors for measuring jitter is that they usually need to be binned at least 2 x 2 for adequate SNR.

1.4 *Effects of jitter in HiRISE images*

Although the jitter in HiRISE images is for the most part within mission specifications, it can be significant enough to negatively affect color registration and Digital Terrain Model (DTM) production. Jitter also causes discontinuities in the image along RED – RED CCD seams in the full reconstructed observation. Accurate color registration is necessary to stack the IR and BG data with the RED data to make a three-band image. To register the color in HiRISE the distortions in the BG and IR data are measured relative to RED using a grid network of points. The output is a control network that is used to perform a splined interpolation line by line (Becker *et al.*, 2007). This produces satisfactory color registration, but does not remove the geometric distortions in an absolute sense.

Another process affected negatively by jitter is DTM production. HiRISE DTMs, at 1 m horizontal resolution and vertical precision of <1 m, are a high-value derived product. They are used for science as well as mission planning and landing site assessment (Golombek, Bellutta *et al.*, 2012; Golombek, Grant *et al.*, 2012; Kirk *et al.*, 2008, 2011). Since the jitter varies from image to image, producing a DTM with a HiRISE stereo pair is complicated by geometric distortions. These distortions can introduce significant artifacts to the DTM, or altogether prevent derivation of a good solution for a terrain model. In stereo analysis, two images of the same ground target are acquired from different viewing angles, differing enough to be able to derive heights from the parallax in the two images. The two images are initially tied together using feature matching tie points and control points. They are then transformed to epipolar space, so that parallax is ideally only in the cross-track, or x-direction. Cross-track distortions cause errors in the x-parallax between the two stereo images, which result in incorrect elevation estimation. In the case of line scan cameras, if this motion is periodic it can result in a ripple pattern in the terrain model parallel to the cross-track direction (Kirk *et al.*, 2003). Jitter in the along-track direction creates high-frequency mismatches in the y-direction, which are very difficult to remove by the triangulation, or bundle adjustment process. Distortion in the along-track direction degrades the results in area matching algorithms, resulting in noise in the model, spurious matches (blunders) and artifacts. The discontinuities along the seams of the RED CCDs also result in linear elevation artifacts of typically >1 m in HiRISE DTMs. It is important to remember that artifacts that appear in DTMs due to jitter are a result of the combined effects of jitter, if present, in either or both of the stereo images. Rather than trying to remove these distortions from the DTM post-production (the only option available to Kirk *et al.* (2003) with Mars Orbital Camera (MOC) images), we use the procedure described here to minimize jitter distortions in the source image data. The negative effects of jitter on DTM production and change detection (using orthorectified images) motivate the need for an absolute jitter correction algorithm that is general enough to be largely automated. The algorithm to derive the jitter in an absolute sense, followed by image correction, is described in the following sections.

2 METHODS

Observations of the measured jitter signal in HiRISE images reveal that there is no predictable, regular or repeatable pattern. The motions appear to be periodic, which leads us to take the approach of solving for the frequencies of absolute motions using Fourier analysis. This novel approach allows the pointing history of HiRISE to be modeled more accurately than what is provided by the spacecraft ephemeris. The updated pointing history is then used to project the images, minimizing distortions caused by jitter.

The three stages of our algorithm are detailed in the following sections. We rely on the freely available Integrated Software for Imagers and Spectrometers, v.3.x (ISIS3) (Anderson *et al.*, 2004) to gather these data, as well as for many other steps in the process. Each of these stages is part of a data processing subsystem called HiPrecision implemented at the HiRISE Operations Center (HiROC). HiPrecision is a two-branch software processing subsystem (Fig. 8.3). The HiRISE Jitter-Analyzed CK (HiJACK) branch, applies the methods described here to correct geometric distortions in HiRISE images. The NoProj branch performs geometric correction only, without modeling and correcting for jitter, as not all HiRISE images require jitter correction. The term 'kernels' refers to data files in the formats developed by NASA's Navigation and Ancillary Information Facility (NAIF). NAIF archives and provides tools for working with spacecraft and planetary ephemerides, navigation geometry and instrument orientation data, called SPICE (http://naif.jpl.nasa.gov/naif/). In particular, kernel files containing instrument position and pointing information are called camera kernels (CK). These products are freely available through NASA's Planetary Data System (PDS) Geosciences, Imaging and Navigation Nodes.

2.1 *Jitter measurement*

We measure jitter in the along-track (image line or row) and cross-track (image sample or column) directions. Jitter in a third direction, twist (yaw), is conceivable but has not been observed to be significant on MRO. To solve for jitter in an absolute sense, the along-track spacing (time difference) between the CCDs on the focal plane is taken into account when considering the pixel offsets. The image data analyzed are a set of CCD image strips that all overlap a common CCD (e.g., RED3–4, RED4–5 and BG12–RED4, or RED4–5, RED5–6 and IR11–RED5). The image data from each CCD have been geometrically and radiometrically calibrated in the standard HiRISE image processing pipeline (Becker *et al.*, 2007; Eliason *et al.*, 2007) but have not been map projected. If any of the CCDs have been binned, those image strips are enlarged to match the spatial scale of the lowest binning of the set of CCDs. Color CCDs (IR and BG) are typically binned by a factor of 2 or 4 to increase SNR.

In the ideal case, where the spacecraft is entirely stable during imaging, a surface feature imaged in one detector should appear at a predictable corresponding location in the overlapping region of the adjacent detector. Deviations from expected object locations in the overlapping areas of each CCD image strip are measured using the ISIS3 application *hijitreg*, which was created specifically to measure offsets in HiRISE images to aid in color registration. Briefly, *hijitreg* creates a grid or column of control points in one image, and performs a search for the corresponding feature in the overlapping image using an area-based pattern-matching algorithm. Points are measured every 20

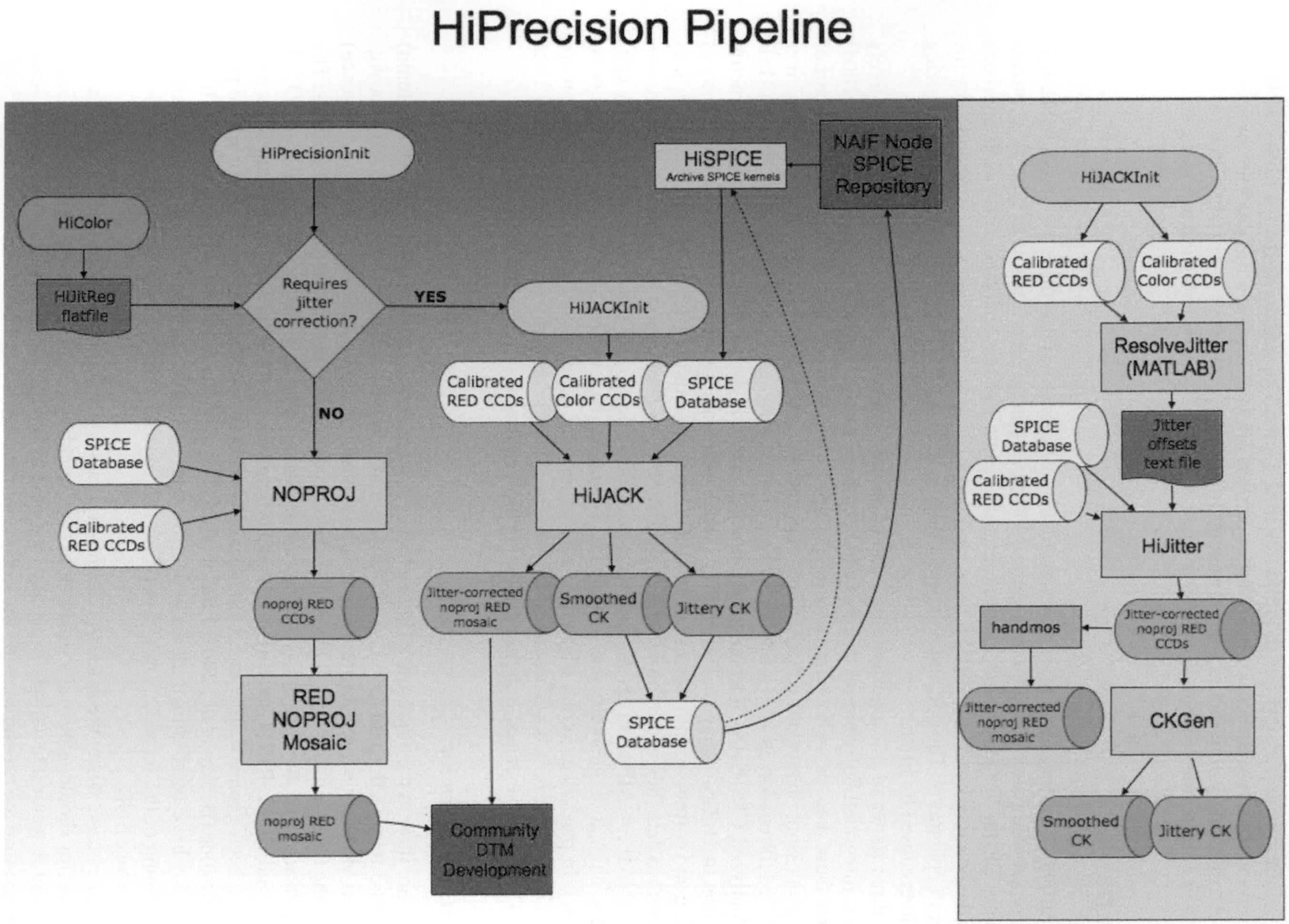

Figure 8.3. HiPrecision processing subsystem flowchart showing the two branches that produce the NoProj, geometrically corrected products and the HiJACK branch that additionally performs jitter correction. The HiJACK branch is detailed in the right panel. Calibrated products are pulled from intermediate steps in the HiRISE image processing pipeline.

lines, an increment chosen by testing for the best balance between resolving the jitter signature and processing time. Four columns of control points across the color image strips are measured to improve statistics, taking advantage of the fact that the IR and BG CCDs completely overlap the RED CCDs in the cross-track direction. The area-based matching algorithm finds a maximum correlation at subpixel precision within user-specified search parameters. This best-fit match location is described as a pixel offset in sample (cross-track) and line (along-track) from the expected location of the feature. When describing jitter in image space, we use the terms sample and line, which correspond to the orbital directions cross-track and along-track, respectively.

The output of *hijitreg* is a text file listing the program input parameters and the measured matches, ephemeris times and match quality statistics for each point successfully measured. The output from *hijitreg* contains three text files – one for each pair of CCDs, which are the input for the following step.

2.2 *Fourier analysis*

The pixel offsets measured by *hijitreg* must be of high-enough density and quality to ensure that the derived solution of the jitter motion is reliable. Datasets that contain large gaps or sparse matches will not yield a reliable solution, and are rejected. The coregistration data are somewhat noisy, and can contain spurious matches, especially in areas of the image that are low contrast or contain fine-scale repeating patterns. Multiple columns of points across the IR/BG–RED image pairs are averaged together across each row. Point offset measurements are filtered with a boxcar median function that rejects any point for which the magnitude of the offset is larger than a threshold value from its neighbors. For the boxcar filter, the window width is 11 points and the tolerance threshold is 2 pixels. From here onward, sample and line datasets are treated separately. A further filtering step is done to minimize noise. In the frequency domain, a low-pass Gaussian filter is used to reduce some of the noise in the data (Fig. 8.4). A bicubic spline interpolation is performed on each dataset to create a uniformly spaced series. This step is necessary because although the data points are sampled at evenly spaced intervals, not all sampled points return a valid offset value.

We wish to model the pointing offset of the camera, $j(t)$, as a function of time, but what is observed is the relative offset, $F(t)$, between two observations of the same feature in different CCDs, separated by an interval, Δt, such that:

$$F(t) = j(t + \Delta t) - j(t) \tag{1}$$

The functions $F(t)$ and $j(t)$ can each be represented over the time duration of measured offsets, L, by the Fourier series

$$F(t) = \frac{1}{N}\sum_{i=0}^{N-1} a_i \sin\left(\frac{2\pi i}{L}t\right) + b_i \cos\left(\frac{2\pi i}{L}t\right) \tag{2}$$

$$j(t) = \frac{1}{N}\sum_{i=0}^{N-1} A_i \sin\left(\frac{2\pi i}{L}t\right) + B_i \cos\left(\frac{2\pi i}{L}t\right) \tag{3}$$

where N is the number of samples obtained. Because we have resampled the relative jitter measurements $F(t)$ to the uniform spacing $t_k = kL/N$ where N is a power of 2, the coefficients a_i and b_i can be obtained efficiently by use of the Fast Fourier Transform (FFT). The FFT algorithm is used as implemented in MATLAB, which uses the FFTW library (www.fftw.org/) (Frigo and Johnson, 1998) to compute the Discrete Fourier Transform (DFT). The FFT (Cooley and Tukey, 1965) is appropriate because the data we are working with are sets of discrete samples in the spatial domain. Substituting (2) and (3) into (1), using the angle-sum formulae of trigonometry, and identifying terms results in a set of algebraic relations between the known coefficients a_i and b_i and the desired coefficients A_i and B_i:

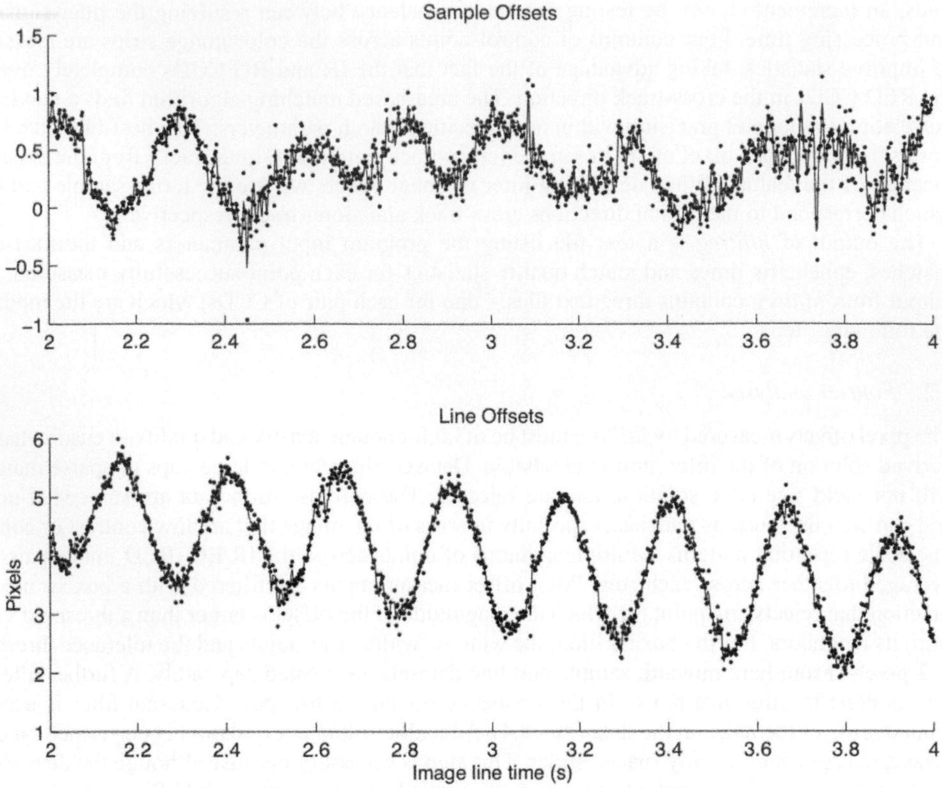

Figure 8.4. Detail of plot of original pixel offsets (dots) and smoothed data (solid line) from ESP_019988_1750 RED4-RED5

$$a_i = A_i(\cos(\alpha_i) - 1) + B_i\sin(\alpha_i) \tag{4}$$

$$b_i = B_i(\cos(\alpha_i) - 1) - A_i\sin(\alpha_i) \tag{5}$$

where $\alpha_i = 2\pi i\Delta t/L$. Solving for A_i and B_i, we then have

$$A_i = -\frac{1}{2}\left(\frac{a_i\sin(\alpha_i) + b_i\cos(\alpha_i) + b_i}{\sin(\alpha_i)}\right) \tag{6}$$

$$B_i = \frac{1}{2}\left(\frac{a_i\cos(\alpha_i) - b_i\sin(\alpha_i) + a_i}{\sin(\alpha_i)}\right) \tag{7}$$

Equation (2) can then be evaluated by an inverse FFT, yielding the values of $F(t)$ at the discrete times t_k. Note that when $\alpha_i = 2\pi n$ for any integer n, in other words, when there is an integer number of cycles of jitter between measurements, the motion repeats exactly. In this case $a_i = b_i = 0$ and the coefficients in Equations (6) and (7) are unbounded so that A_i and B_i cannot be reconstructed. The matching process is thus blind to such frequencies. At frequencies close to these singularities the reconstructed motion will be subject to large round-off errors. An important special case of this phenomenon is $n = 0$. Comparison of relative jitter can never constrain the absolute pointing

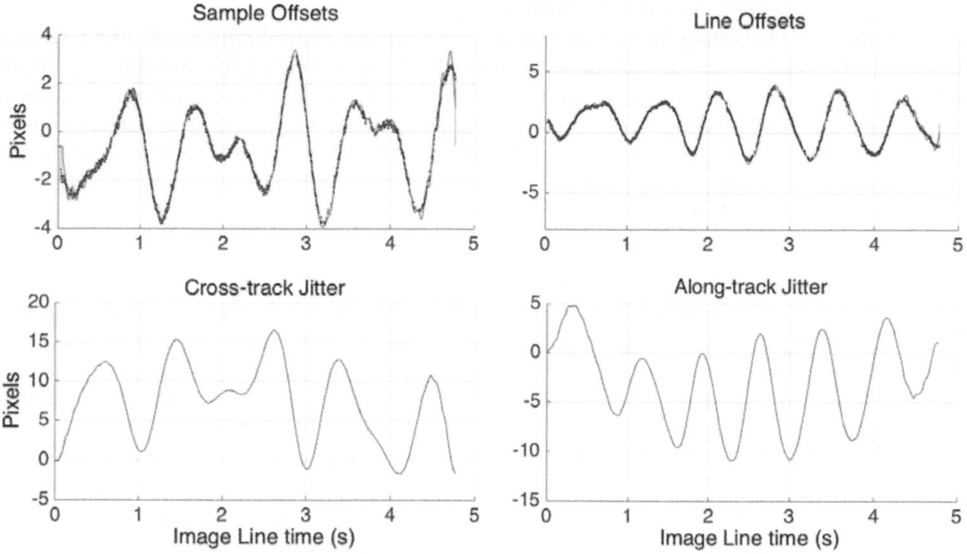

Figure 8.5. Output of ResolveJitter for PSP_007556_2010. Top row shows sample and line offset data for the RED4–5 pair (red) with predicted jitter from solution overplotted in black. Bottom panels show the modeled jitter function.

error averaged over the whole image, so that this zero frequency term in the series for j must be supplied by the *a priori* pointing history provided in the form of the NAIF SPICE C-kernel (CK). To avoid inaccurate reconstruction of the jitter at the other blind or near-blind frequencies, those components that are within a given tolerance range of Δt of a particular CCD pair are set to zero, and the overall solution is based on the solutions for the other two CCD pairs rather than all three. To optimize this masking process we perform a grid search on the width of the frequency window near each blind frequency that is to be excluded, and pick the width that generates the solution best matching the input jitter-difference observations. Finally, the solution is smoothed with a Gaussian low-pass filter applied in the frequency (Fourier transform) domain. The degree of smoothing (i.e., the bandwidth of the filter) is selected by a second grid search to minimize the error in reproducing the jitter differences (Fig. 8.5).

The output from the above algorithm is a text file that describes the derived jitter function in terms of pixel offsets and the corresponding ephemeris line time, at evenly spaced line intervals. The sample and line jitter functions are described separately. This text file is read into the subsequent step that uses the derived function to transform the images. Additionally, the jitter function can be used to estimate the minimum pixel smear by taking the derivative of the sample jitter function (interpolated) from line to line. Very high-frequency jitter that is not measured could add to this minimum smear.

2.3 *Image correction*

The ISIS3 application *hijitter* performs the image transformation. First, the sample and line pixel offsets are converted to rotation angles and combined with the reconstructed MRO pointing information, as described in the spacecraft observation geometry data, or SPICE (Acton, 1996). The jitter corrections are applied and used to update the SPICE camera pointing information – stored as a binary large object (blob) in the image labels, but ultimately derived from a Camera pointing Kernel (CK) file – before finally projecting all the RED cubes with the ISIS3 application *noproj* to remove camera distortions.

2.3.1 Convert pixel offsets to angular rotations

Hijitter begins by converting the line and sample pixel offsets from the jitter text file to rotation angles and later combines them with the reconstructed MRO pointing kernel quaternions. First, the model of feature offsets, $j(t)$, is converted to a model of spacecraft and camera rotation. Because the rotations are small (tens of μrad at most), the order in which they occur is not significant. We use the Navigation and Ancillary Information Facility (NAIF) (Acton, 1996) routine *eul2m* to construct a jitter rotation matrix that transforms a vector from the nominal camera coordinates to the true, jittery camera coordinates.

2.3.2 Combine jitter with camera pointing

The next step is to combine the jitter matrix with the rotation information recorded in the C-kernel, or CK SPICE blob, to produce an updated CK. This is done as a function of time, either at the sampling times of the original CK, or if necessary, to incorporate all information in the jitter function, at a higher sampling rate. The CK can be interpolated by using NAIF routines. The jitter functions $j(t)$ are represented by Fourier series and can thus be evaluated at any time of interest.

Ideally, we would obtain the camera matrix obtained from the CK, which would relate the nominal camera pointing to the J2000 inertial coordinate system (Müller and Jappel, 1977; Seidelmann *et al.*, 1980). In reality, there are several problems with simply multiplying the rotations together. First, there are indications that the camera matrix contains high-frequency noise, such as from the IMU data, which does not represent the motions of the camera (nominal, jitter, or otherwise). This noise is excluded from the final rotation matrix by smoothing. Second, the jitter model will be relatively accurate for higher frequencies but may drift over longer periods and depart from the true motion of the camera. Thus, we have a motivation to high-pass filter the jitter model in some way. Third, the portions of the frequency domain for which the camera matrix and the jitter vector are valid may overlap, so that the same (non-nominal, but real) motions of the camera may be recorded in both datasets. We filter the two datasets in a complementary fashion, so that any given frequency of motion is represented by one source, the other, or by a weighted combination that gives the correct amplitude for any frequency that is represented by both.

2.3.3 Project to ideal camera space

The jitter corrections are applied and used to update the SPICE before finally projecting all the RED image strips with *noproj*. This transformation is done by projecting pixels from the real camera down to a nominal surface and back up into the ideal and smoothly moving camera. This is one of the steps at which the small viewing angle variation within an observation becomes important, in that the nominal surface need not be topographically accurate. The HiRISE *ideal* camera is a single 20,000-sample line scan camera with no optical distortion and centered in the HiRISE focal plane between RED4 and RED5. When the *noproj*'ed image strips are mosaicked together, they form a single ideal camera observation. The jitter corrections from the corrected CK are applied to the input pixels to map them to correct positions on the ground. These corrected positions are then mapped to the ideal HiRISE camera with a smoothed CK to produce an image with minimal jitter distortions.

3 RESULTS

There are several ways to assess the quality of the image correction. The ultimate goal of this work is to remove geometric distortions caused by spacecraft motion in all of the HiRISE CCD image strips. If this was completely successful, then there would be (a) no surface feature mismatches along CCD seams, and (b) perfect color registration. Another measure of the quality of the correction can be seen in improvements to the quality of DTMs.

To make a first quality check on the solution, we run *hijitreg* again on the RED4-RED5 CCDs and compare plots to see if the apparent jitter has been removed. Typical results show a substantial reduction in amplitude of the jitter motions, as measured in the CCD-to-CCD pixel offsets with *hijitreg* (Fig. 8.6). The solution spans the time common to all three CCDs used in the analysis. Imaging in the IR CCD starts the latest, so the correction does not affect the first ~200 lines of the RED CCD image. This is seen as a sharp drop at the beginning of the plotted data after correction (blue points) in Figure 8.6. Processing in the HiPrecision subsystem will account for this by excluding lines in the final output image data that are outside of the corrected time span. In most cases, the average amplitude is reduced to <1 pixel. Comparison of the maximum, mean and standard deviation of the jitter magnitude, before and after correction, shows that for all cases where the mean is larger than 0.5 pixel, the correction shows a definite improvement, with the mean error being reduced to <0.5 pixels (Fig. 8.7).

Color registration between the IR, BG and RED bands shows improvement over no correction at all, but in many cases is still not as accurate as the relative correction performed in the HiRISE color processing (Fig. 8.8). As with the measurements of the jitter plots, the color fringing can be seen to be reduced from >1 pixels to ~1 pixel, or better.

Qualitatively, improvements can also be seen as a lessening of discontinuities along image strip seams (Fig. 8.9). Both the reduction in jitter amplitude and the reduced discontinuities along CCD seams allows for improved stereo analysis resulting in better DTM quality. In most cases where jitter correction is applied to stereo pairs, the DTM has fewer artifacts and therefore requires less

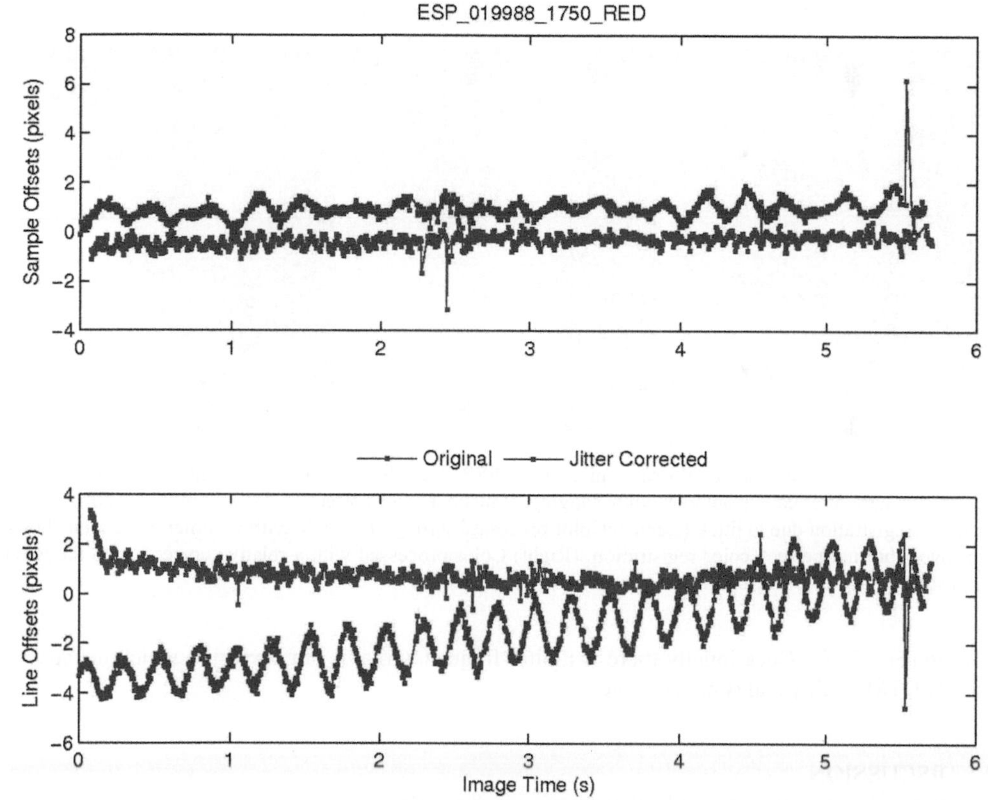

Figure 8.6. Plot of *hijitreg* results from ESP_019988_1750 RED4-RED5 before (red) and after (blue) jitter correction. The spikes in the after-correction (blue) plot fall below the outlier detection threshold, and do not indicate a significant effect on the jitter-corrected image

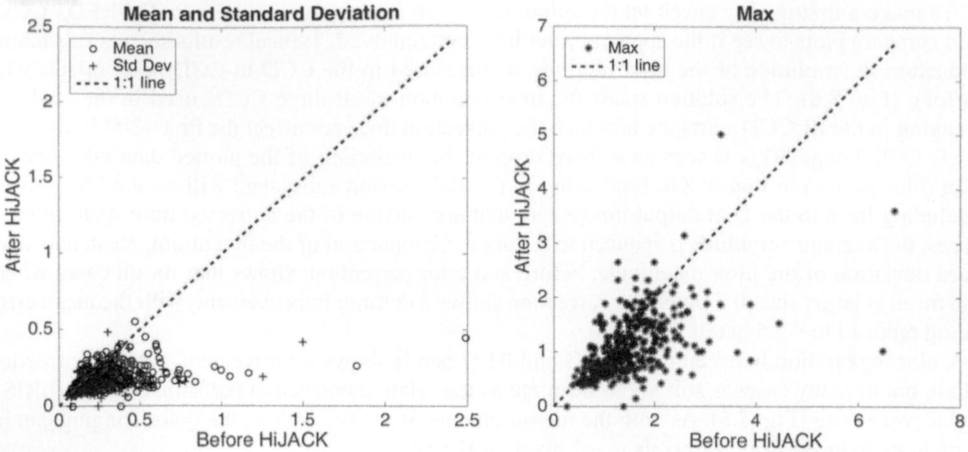

Figure 8.7. Comparison of the mean and standard deviation (left) of the jitter magnitude (pixels) before and after correction. Best correction points lie below the 1:1 red diagonal line. For the mean and standard deviation tests, there is improvement for all cases where the jitter magnitude is >0.5 pixels. The maximum magnitude is not as effective of a measure of the correction due to outliers that are not identified as such by the measurement routine.

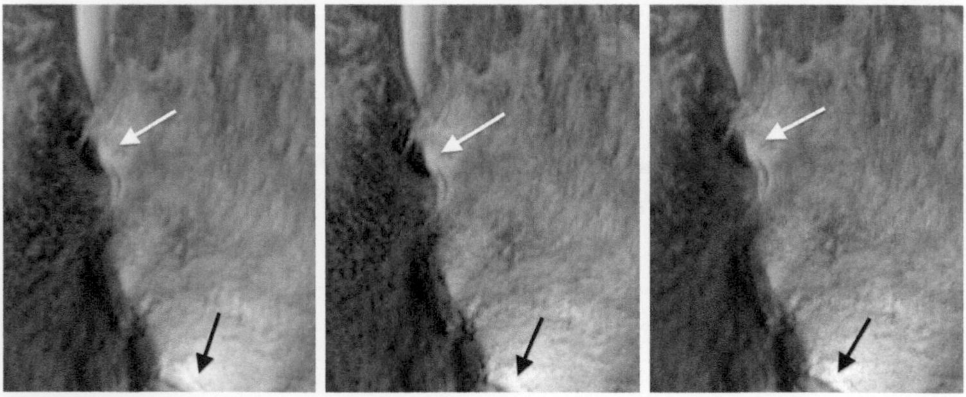

Figure 8.8. Color registration comparison in detail of ESP_012039_2010. The black and white arrows point out areas where jitter causes obvious color fringing. (Left) No color correction shows obvious color fringing from misregistration due to jitter. (Center) Color processed through HiJACK with no other correction shows improved, but not perfect, color registration. (Right) Color processed with a relative correction to the RED band shows good color registration.

editing (Fig. 8.10). Occasionally there is a jitter frequency that is not corrected which negatively affects DTM quality and is not editable.

4 DISCUSSION

It is necessary to combine the jitter reconstruction results with a smoothed version of the *a priori* pointing. This has to be done at some level because the jitter modeling amounts to integration of a signal that is given in differential form. Thus it very definitely cannot be used to estimate absolute

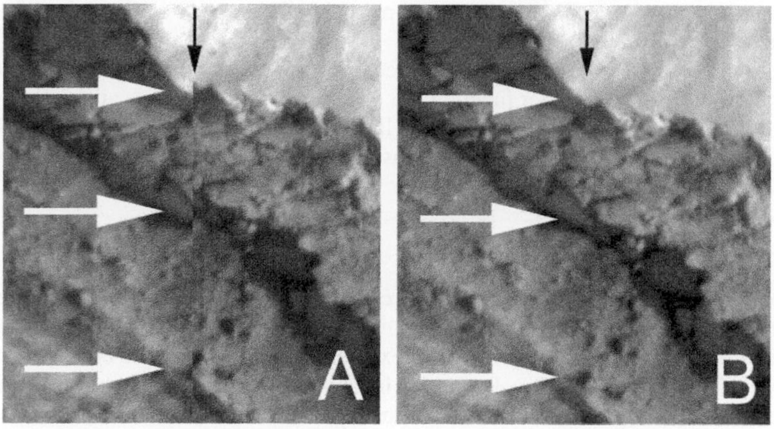

Figure 8.9. Seam (black arrows) between CCD strips in ESP_019988_1750. White arrows indicate places where features are obviously misaligned in A, and where those features are now well aligned across the seam in B, which has been processed through HiJACK.

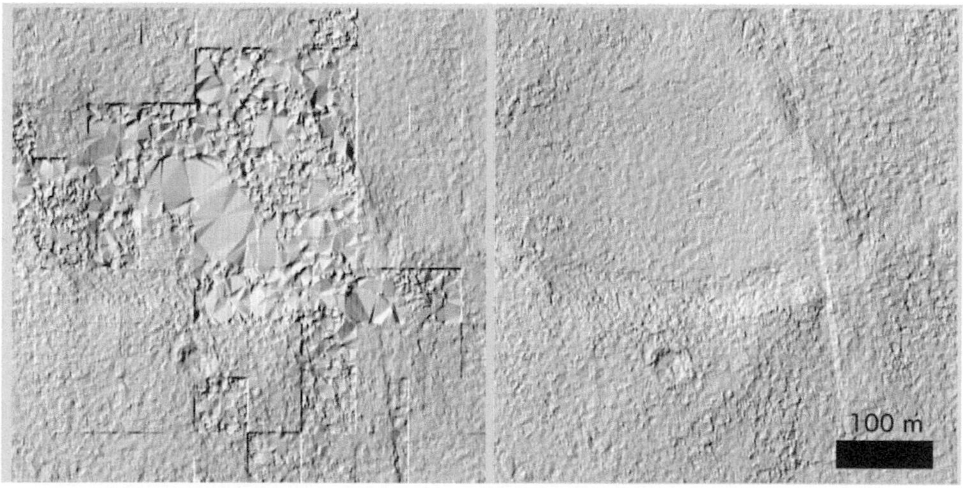

Figure 8.10. Detail of shaded relief map of a DTM made from ESP_035895_1965 and ESP_036172_1965, which had mean jitter magnitudes of 0.8717 and 0.41184 pixels, respectively. (Left) Shaded relief image of the DTM made with non-jitter-corrected images. Artifacts are failures of the stereo correlator due to excessive mismatch between images. (Right) The same area of the DTM produced from jitter-corrected images. The seam between detectors is quantitatively reduced but does not disappear. Examination of elevation profiles across the seam confirms that the offset is reduced.

pointing (the fundamental frequency) and it is destined to be weak at some range of low frequencies. Our internal analysis of repeat observation coordinates finds better agreement with smoothed pointing kernels (CKs) than with the unsmoothed CKs. However, an analysis of what fraction of information in the standard CKs is useful has not been undertaken. Our results show that combining the jitter reconstruction with the smoothed CK succeeds in adding the higher frequency jitter signal (presumably not captured in the CK) to the lower-frequency spacecraft oscillations without duplicating information. We can clearly make the assumption that the jitter signal is not captured in the CKs because otherwise the jitter would be corrected in the image reconstruction upon applying SPICE data, specifically the reconstructed CKs.

Color registration after HiJACK is not perfect (Fig. 8.8), because the transformation of each band is based on the estimated jitter function, for which not all frequencies are perfectly resolved. The relative correction achieved by shifting the lines in the color bands to match the RED image strips produces better results at this time, because the color image lines are shifted to match the measured jitter offsets exactly.

4.1 *Improvements to DTM production*

Jitter-corrected HiRISE stereo pairs improve stereo correlation in DTM production. The main benefit of jitter correction with HiJACK is the reduction of jitter amplitude, which improves results in stereo matching algorithms. In the worst cases, excessive y-parallax creates blunders in the terrain model that require interactive editing. Reduction in jitter also reduces the persistent effect along the CCD strip seams in DTMs. Jitter correction of stereo pairs can reduce the size of these seams from ~5 m to <2 m. In some cases, the jitter correction does not completely eliminate jitter issues from stereo images. Based on the results shown in the above section, the color registration of the jitter-corrected color is not perfect, but it is often an improvement over the uncorrected data. Further work needs to be done to develop a robust process to orthorectify color data with jitter-corrected images.

4.2 *Binning and TDI*

Jitter amplitude is affected by binning, because the binned data (usually the IR/BG image strips) are rescaled to match the bin mode of the RED image strips during the color registration process. This rescaling causes the amplitude of the offsets to also rescale. It does not change the frequency information. For example, reducing the RED data by a factor of two to match 2×2 binned color data showed the amplitude of jitter measurements to be approximately half as large (measured in pixels) as that of the offsets measured from enlarging the color data to match the scale of the RED. The amplitude of the offsets in these tests was reduced by approximately the same amount as the reduction scale. This result implies that binning must be taken into account when interpreting the amplitude of the measured jitter offsets.

The implementation of the algorithm presented here assumes each line in the image is integrated over the programmed line time, without consideration of TDI mode. The effects of TDI should be incorporated into future improvements to the algorithm. For example, the centroid of the point-spread function varies in a TDI sensor, in the presence of jitter (Hochman *et al.*, 2004).

4.3 *Applying lessons learned to future instrument design*

The jitter correction method described here had not been developed when HiRISE was designed. Without this knowledge, the offsets and overlaps between CCDs are not optimal. Variation of the along-track CCD offsets for some triplets are enough to provide differing frequency information, but timing differences are actually very small compared to the large time difference between the IR/BG and RED CCDs. This creates a range of frequencies that are not well resolved between those two extremes. Future instruments can benefit from this work. In particular there is increasing use of larger detector arrays, which could provide greater flexibility to design for jitter measurement over a broad range of frequencies. Future instruments could use complementary metal-oxide semiconductor (CMOS) detectors (also called active pixel sensors) that allow for customizing image line spacing to optimize the ability to resolve non-aliased jitter frequencies. For example the Europa Imaging System (EIS) (Turtle *et al.*, 2016) has a wide-angle and a narrow-angle camera, each with 2048 (along-track) x 4096 (cross-track) pixels. With 2048 along-track lines, we can read out multiple line (or digital TDI) sections over a range of pixel separations, to best detect all jitter frequencies. Active pixel sensors now provide CCD-level image quality (Janesick *et al.*, 2014), and the independent readout of each pixel allows additional capabilities. For example, jitter corrections could be applied in the digital processing unit prior to summing lines for TDI, thus minimizing smear. Diagonal TDI is possible if the camera cannot be oriented correctly for column-aligned TDI (McEwen *et al.*, 2012). The approach presented here can be generalized to other sensors if the parallax angle between overlapping readouts is small. Active pixel sensors have the advantage that the spacing between readout areas can be selected rather than being built in. This would allow for flexibility in planning imaging modes to measure and solve for ranges of frequencies.

5 CONCLUSION

Results of the jitter correction algorithm and image transformation presented here show a significant reduction of geometric distortions in HiRISE images. Despite planning for the effects of spacecraft jitter by slightly offsetting detectors on the focal plane, some insensitivity to certain frequencies of motion is apparent. Therefore, the overall geometric distortions are minimized, but not completely removed.

Use of jitter-corrected HiRISE images shows marked improvements in DTM quality, specifically in terrain models that were previously rendered unacceptable by jitter-induced distortions. Implementation of the derived jitter solution allows the HiRISE team to create and release precision geometry products to NASA's PDS Imaging Node. Complementary to the release of such image products will be the creation and archiving of the jitter-corrected pointing kernels (SPICE CK) to the NAIF node. These image and ancillary data products are freely provided to the science community.

REFERENCES

Acton, C. (1996) Ancillary data services of NASA's navigation and ancillary information facility. *Planetary and Space Science*, 44(1), 65–70.

Anderson, J., Sides, S.C., Soltesz, D.L., Sucharski, T.L. & Becker, K.J. (2004) Modernization of the integrated software for imagers and spectrometers. *Lunar and Planetary Institute Science Conference*, 35, 2039.

Ayoub, F., Leprince, S., Binet, R., Lewis, K., Aharonson, O. & Avouac, J.-P. (2008) Influence of camera distortions on satellite image restoration and change detection applications. *Geoscience and Remote Sensing Symposium, IGARSS, IEEE International*, Boston, MA, USA, 1072–1075.

Becker, K., Anderson, J., Sides, S., Miller, E., Eliason, E. & Keszthelyi, L. (2007) Processing HiRISE images using ISIS3. *Lunar and Planetary Institute Science Conference*, League City, TX, USA, abstract #1779.

Bergstrom, J.W., Delamere, W.A. & McEwen, A.S. (2004) MRO High Resolution Imaging Science Experiment (HiRISE): instrument test, calibration, and operating constraints. *55th Internationall Astronautical Congress*, Vancouver, Canada.

Cooley, J.W. & Tukey, J.W. (1965) An algorithm for the machine calculation of complex Fourier series. *Mathematics of Computation*, 19, 297–301.

Eastman, R., Le-Moigne, J. & Netanyahu, N. (2007) Research issues in image registration for remote sensing. *IEEE Conference on Computer Vision and Pattern Recognition*, Minneapolis, MN, USA. pp. 1–8.

Eliason, E., Castalia, B., Espinoza, Y., Fennema, A., Heyd, R., Leis, R., McArthur, G., McEwen, A.S., Milazzo, M., Motazedian, T., Schaller, C. & Spitale, I. (2007) HiRISE data processing and standard data products. *Lunar and Planetary Institute Science Conference*, League City, TX, USA, abstract #2037.

Frigo, M. & Johnson, S.G. (1998) FFTW: An adaptive software architecture for the FFT. *Proceedings of the International Conference on Acoustics, Speech, and Signal Processing*, Volume 3. pp. 1381–1384.

Gasparinni, T. (2005) Jitter analysis summary. *Technical Report*, Lockheed Martin, MRO-FAR-008: Revision 1.

Golombek, M., Bellutta, P., Calef, F., Fergason, R., Hoover, R., Huertas, A., Kipp, D., Kirk, R., Parker, T., Sun, Y. & Sladek, H. (2012) Surface characteristics and traversability of the Gale crater Mars Science Laboratory landing site. *Lunar and Planetary Institute Science Conference*, The Woodlands, TX, USA, Volume 43. Abstract #1608.

Golombek, M., Grant, J., Kipp, D., Vasavada, A., Kirk, R., Fergason, R., *et al.* (2012) Selection of the Mars Science Laboratory landing site. *Space Science Reviews*, 170, 641–737.

Hochman, G., Yitzhaky, Y., Kopeika, N., Lauber, Y., Citroen, M. & Stern, A. (2004) Restoration of images captured by a staggered time delay and integration camera in the presence of mechanical vibrations. *Applied Optics*, 43, 4345–4354.

Janesick, J., Elliott, T., Andrews, J., Tower, J., Bell, P., Teruya, A., Kimbrough, J. & Bishop, J. (2014) Mk x Nk gated CMOS imager. *Proceedings of the SPIE*, San Diego, CA, USA, 9211, 921106–921113.

Kirk, R.L., Howington-Kraus, E., Redding, B., Galuszka, D., Hare, T., Archinal, B., Soderblom, L. & Barrett, J. (2003) High-resolution topomapping of candidate MER landing sites with Mars Orbiter Camera narrow-angle images. *Journal of Geophyscial Research*, 108(E12), 8088.

Kirk, R.L., Howington-Kraus, E., Galuszka, D., Redding, B., Antonsen, J., Coker, K., Foster, E., Hopkins, M., Licht, A., Fennema, A., Calef, F., Nuti, S., Parker, T. & Golombek, M. (2011) Wall-to-wall 1-m topographic coverage of the Mars Science Laboratory candidate landing sites. *Lunar and Planetary Institute Science Conference*, The Woodlands, TX, USA, Volume 42. Abstract #2407.

Kirk, R.L., Howington-Kraus, E., Rosiek, M.R., Anderson, J.A., Archinal, B.A., Becker, K.J., Cook, D.A., Galuszka, D.M., Geissler, P.E., Hare, T.M., Holmberg, I.M., Keszthelyi, L.P., Redding, B.L., Delamere, W.A., Gallagher, D., Chapel, J.D., Eliason, E.M., King, R. & McEwen, A.S. (2008) Ultra-high resolution topographic mapping of Mars with MRO HiRISE stereo images: meter-scale slopes of candidate Phoenix landing sites. *Journal of Geophysical Research*, 113, E00A24.

Lee, S., Skulsky, J., Chapel, D., Cwynar, R., Gehling, R. & Delamere, A. (2003) Mars Reconnaissance Orbiter design approach for high-resolution surface imaging. *26th Annual AAS Guidance and Control Conference*, Breckenridge, CO, USA.

Li, R., Di, K., Hwangbo, J. & Chen, Y. (2008) Rigorous photogrammetric processing of HiRISE stereo images and topographic mapping at Mars Exploration Rover landing sites. *Lunar and Planetary Institute Science Conference*, League City, TX, USA, abstract #1864.

McEwen, A.S., Janesick, J., Elliot, S.T., Turtle, E.P., Strohbehn, K. & Adams, E. (2012) Radiation-hard camera for Jupiter system science. *International Workshop on Instrumentation for Planetary Missions*, 10–12 October 2012, Greenbelt, MD, USA.

McEwen, A.S., Eliason, E., Bergstrom, J., Bridges, N.T., Hansen, C.J., Delamere, W.A., Grant, J.A., Gulick, V.C., Herkenhoff, K.E., Keszthelyi, L., Kirk, R.L., Mellon, M.T., Squyres, S.W., Thomas, N. & Weitz, C.M. (2007) Mars Reconnaissance Orbiter's High Resolution Imaging Science Experiment (HiRISE). *Journal of Geophysical Research*, 112, E05S02.

McGraw, J.T., Angel, J.R.P. & Sargent, T.A. (1980) A Charge-Coupled Device (CCD) transit-telescope survey for galactic and extragalactic variability and polarization. *SPIE Applications of Digital Image Processing to Astronomy*, Pasadena, CA, USA, Volume 264. pp. 20–28.

McGraw, J.T., Cawson, M. & Keane, M. (1986) Operation of the CCD/Transit instrument (CTI). *Proceedings of the SPIE, Instrumentation in Astronomy*, Tucson, AZ, USA.

Müller, E.A. & Jappel, A. (eds) (1977) *Trans. Int. Astron. Union, Vol. XVI B, Proc. 16th General Assembly, Grenoble, 1976*. Reidel, Dordrecht, The Netherlands. p. 60.

Seidelmann, P.K., Kaplan, G.H. & Van Flandern, T.C. (1980) New celestial reference system. *International Astronomical Union Colloquium*, 56, 305–316.

Teshima, Y. & Iwasaki, A. (2008) Correction of attitude fluctuation of Terra spacecraft using ASTER/swir imagery with parallax observation. *IEEE Transactions on Geoscience and Remote Sensing*, 46, 222–227.

Theiler, J., Henderson, B. & Smith, G. (1997) Algorithms using inter-band cross-correlation for pixel registration and jitter reconstruction in multi-channel push-broom imagers. *Proceedings of the SPIE*, Bellingham, WA, USA, 3163, 22–32.

Turtle, E.P., McEwen, A.S., Osterman, S.N., Boldt, J.D., Strohbehn, K. & EIS Science Team (2016) The Europa Imaging System (EIS), a camera suite to investigate Europa's geology, ice shell, and potential for current activity. *3rd International Workshop on Instrumentation for Planetary Mission,* Pasadena, CA, USA, *LPI Contributions*, Volume 1980. p. 4091.

Zurek, R.W. & Smrekar, S.E. (2007) An overview of the Mars Reconnaissance Orbiter (MRO) science mission. *Journal of Geophysical Research*, 112(E5), E05S01.

Planetary Remote Sensing and Mapping – Wu et al.
© 2019 Taylor & Francis Group, London, ISBN 978-1-138-58415-0

Chapter 9

Community tools for cartographic and photogrammetric processing of Mars Express HRSC images

R. L. Kirk, E. Howington-Kraus, K. Edmundson, B. Redding, D. Galuszka, T. Hare and K. Gwinner

ABSTRACT: In this chapter we describe the software we have developed for photogrammetric processing of images from the Mars Express High Resolution Stereo Camera (MEX HRSC) to produce digital topographic models (DTMs) and orthoimages, as well as testing we have performed. HRSC has returned images, including stereo and color coverage of most of Mars at decameter scales. The instrument team has developed an extremely powerful processing pipeline and delivered a large number of high-level data products, but our independent software is nevertheless of interest because it provides a check on the standard products, sheds light on the capabilities of software elements we use for multiple missions besides HRSC, and is publicly available, giving users the opportunity to make products that may not (yet) be released by the team and custom products such as local mosaics. We have tested our software on images of three areas: Candor Chasma and Nanedi Valles (both the subject of past DTM comparisons reported by Heipke *et al.*, 2007) and Gale crater, which was extensively mapped at pixel scales 50 times finer than HRSC before its selection as the landing site of the Curiosity rover. We find the vertical precision and mean deviation from the altimetry data used as a control reference for our DTMs to be comparable to the nadir image pixel size. The horizontal resolution of the DTMs appears to be an order of magnitude coarser than the lower limit of 3–5 image pixels that is commonly stated.

1 INTRODUCTION

The High Resolution Stereo Camera (HRSC) on the Mars Express orbiter (Neukum *et al.*, 2004) is a multi-line pushbroom scanner that can obtain stereo and color coverage of targets in a single overpass. Since commencing operations in 2004 HRSC has imaged more than 98% of Mars with ground sample distances of 10 m (nadir channel at periapse) and greater (76% at 20 m or better). The instrument team uses the Video Image Communication And Retrieval (VICAR) software to produce and archive a range of data products from uncalibrated and radiometrically calibrated images to controlled digital topographic models (DTMs) and orthoimages and regional mosaics of DTM and orthophoto data (Gwinner *et al.*, 2009, 2016; Gwinner, Scholten *et al.*, 2010b). DTMs have been produced for about 44% of the planet to date. Alternatives to this highly effective standard processing pipeline are nevertheless of interest to researchers who do not have access to the full VICAR suite and may wish to make topographic products or perform other (e. g., spectrophotometric) analyses prior to the release of the highest level products. We have therefore developed software to ingest HRSC images and model their geometry in the US Geological Survey's Integrated Software for Imagers and Spectrometers (ISIS3; Sides *et al.*, 2017), which can be used for data preparation, geodetic control, and analysis, and the commercial photogrammetric software SOCET SET and SOCET GXP (® BAE Systems; Miller and Walker, 1993, 1995) which can be used for independent production of DTMs and orthoimages.

The initial implementation of this capability utilized the then-current ISIS2 system and the generic pushbroom sensor model of SOCET SET, and was described in the DTM comparison of independent photogrammetric processing by different elements of the HRSC team (Heipke *et al.*, 2007). A major drawback of this prototype was that neither software system then allowed for push-broom images in which the exposure time changes from line to line. Except at periapsis, HRSC makes such timing changes every few hundred lines to accommodate changes of altitude and velocity in its elliptical orbit. As a result, it was necessary to split observations into blocks of constant exposure time, greatly increasing the effort needed to control the images and collect DTMs.

Here, we describe a substantially improved HRSC processing capability that incorporates sensor models with varying line timing in the current ISIS3 system (Sides *et al.*, 2017) and SOCET SET. This enormously reduces the work effort for processing most images and eliminates the artifacts that arose from segmenting them. In addition, the software takes advantage of the continuously evolving capabilities of ISIS3 and the improved image matching module NGATE (Next Generation Automatic Terrain Extraction, incorporating area- and feature-based algorithms, multi-image and multi-direction matching) of SOCET SET, thus greatly reducing the need for manual editing of DTM errors. We have also developed a procedure for geodetically controlling the images to Mars Orbiter Laser Altimeter (MOLA) data by registering a preliminary stereo topographic model to MOLA by using the point cloud alignment (*pc_align*) function of the NASA Ames Stereo Pipeline (ASP; Moratto *et al.*, 2010). This effectively converts inter-image tiepoints into ground control points in the MOLA coordinate system. The result is improved absolute accuracy and a significant reduction in work effort relative to manual measurement of ground control. *The ISIS3 and ASP software used are freely available; SOCET SET is a commercial product.* We have recently ported our SOCET SET HRSC sensor model to the Community Sensor Model (CSM; Community Sensor Model Working Group, 2010; Hare and Kirk, 2017) standard utilized by the successor photogrammetric system SOCET GXP that is currently offered by BAE. This CSM source code has now been released under a BSD open source license and is available from USGS's GitHub site (https://github.com/USGS-Astrogeology/CSM-CameraModel).

We illustrate current HRSC processing capabilities with three examples, of which the first two come from the DTM comparison of 2007. Candor Chasma (h1235_0001) was a near-periapse observation with constant exposure time that could be processed relatively easily at that time. We show qualitative and quantitative improvements in DTM resolution and precision as well as greatly reduced need for manual editing, and we illustrate some of the photometric applications possible in ISIS. At the Nanedi Valles site we can now process all three long-arc orbits (h0894_0000, h0905_0000 and h0927_0000) without segmenting the images. Finally, processing image set h4235_0001, which covers the landing site of the Mars Science Laboratory (MSL) rover and its rugged science target of Aeolis Mons in Gale crater, provides a rare opportunity to evaluate DTM resolution and precision because extensive High Resolution Imaging Science Experiment (HiRISE) DTMs are available (Golombek *et al.*, 2012). The HiRISE products have ~50x smaller pixel scale so that discrepancies can mostly be attributed to HRSC. We use the HiRISE DTMs to compare the resolution and precision of our HRSC DTMs with the (evolving) standard products.

We find that the vertical precision of HRSC DTMs is comparable to the pixel scale but the horizontal resolution may be 15–30 image pixels, depending on processing. This is significantly coarser than the lower limit of 3–5 pixels based on the minimum size for image patches to be matched. Stereo DTMs registered to MOLA altimetry by surface fitting typically deviate by 10 m or less in mean elevation. Estimates of the RMS deviation are strongly influenced by the sparse sampling of the altimetry, but range from <50 m in flat areas of our test sites to ~100 m in rugged areas.

1.1 *Motivation*

We had several goals in undertaking this development:

- To provide an independent verification of the results of the stereo pipeline used to produce archival products by the mission team

- To assess the quality of DTMs we could produce (using software and techniques we apply to many other missions) in relation to other approaches and especially those tailored specifically for HRSC
- To enable members of the planetary community who do not have access to the specialized VICAR software used by the HRSC team to produce their own DTMs and orthorectified (map-projected) image products, particularly in the interval between the release of the images and the delivery of higher-level derived products by the team
- To provide researchers the ability to make different tradeoffs between artifacts in the topographic products, smoothing to reduce these artifacts at the expense of lost resolution, or effort spent on manual editing than the standard products offer
- To make ISIS2/3 processing capabilities that are unique or particularly strong, in particular photometric modeling and correction (Kirk *et al.*, 2000) and photoclinometry (shape-from-shading; Kirk *et al.*, 2003), available for use with HRSC data

The capabilities described below are now available to the planetary science community in the latest releases of ISIS3 and through the NASA-USGS Planetary Photogrammetry Guest Facility (Kirk *et al.*, 2009), which provides access to SOCET SET and soon SOCET GXP.

1.2 *Technical approach*

For HRSC, as for a wide variety of other planetary imagers, we currently utilize BAE Systems' SOCET SET for stereo processing, including controlling images by bundle adjustment, producing initial DTMs by automated image matching, interactive quality control and editing of DTMs, and projection of images onto the DTMs to form orthoimages. We use ISIS3 to ingest the images and metadata in standard formats used by the mission and translate them into formats readable by SOCET SET. ISIS3 can also be used to orthorectify images (using an already existing DTM), and to re-ingest the SOCET products. It provides a host of standard functions such as image display and measurement, map transformations, mosaicking, and formatting of products for use with other (e. g., GIS) software or for PDS archiving. To avoid the need to develop HRSC-specific radiometric calibration software, we make use of the "Level 2" image products, which are already calibrated but still in native camera geometry (Scholten *et al.*, 2005). (Note that these images would be called "Level 1" in the system of Batson (1990) commonly used in descriptions of ISIS2/3 processing.)

1.3 *Relation to past work*

At the start of HRSC operations in 2004, the USGS was developing a new software system, ISIS3 (Anderson *et al.*, 2004) to replace its earlier ISIS2 (Eliason, 1997; Gaddis *et al.*, 1997; Torson and Becker, 1997) software. Because the new system was not yet fully operational, we opted to use ISIS2 and implemented programs to ingest HRSC Level 2 images in VICAR and Planetary Data System (PDS) formats and to translate the images from ISIS2 to SOCET SET format. We also created sensor model software to enable geometric calculations including orthorectification and photometric modeling with existing ISIS2 programs. The generic pushbroom scanner sensor model was used in SOCET SET.

A major shortcoming of both the ISIS2 and SOCET SET sensor models at the time was that they assumed a constant exposure time per line. HRSC typically changes its exposure time within an image (as often as every few hundred lines), so it was necessary to split observations into multiple files and handle them separately. Neither sensor model handled images reduced by averaging blocks of pixels into "macropixels," so such images had to be enlarged to full size before use. Shortcomings of SOCET SET made it impossible to constrain the various channels (fore and aft stereo, nadir, etc.) of the HRSC to move together during the control calculation, and to perform stereomatching between more than two images at a time (though the situation was still better than ISIS2, which had no software to control pushbroom images until 2005 and has no automated

DTM production to date). None of these problems were insurmountable, but they had two general consequences: (a) much of the strength of HRSC as a multi-line stereo scanner was lost because the images had to be controlled separately and matched in pairs rather than multiples; and (b) the labor required when mapping with HRSC increased enormously because large numbers of image segments had to be controlled independently and matched in many different pairwise combinations (as well as at different grid spacings to produce best results on both steep and bland areas), and then the results combined to produce a single DTM. These difficulties were directly reflected in the conclusions of the HRSC team's DTM comparison project (Heipke *et al.*, 2007): that the quality of the SOCET DTMs was reasonable but not as good as those produced by algorithms that made use of multiple images in matching, and that the human work effort greatly exceeded that for other approaches. On the positive side, we were able to demonstrate unique ISIS capabilities for photometric modeling, "sharpening" of the DTMs by photoclinometry, and photometric processing (Kirk, Howington-Kraus, Galuszka, Redding, Hare, Heipke *et al.*, 2006; Kirk *et al.*, 2006a, 2006b).

In 2009 we began developing ISIS3 software for HRSC, including a sensor model that handled changing exposure times, but competing priorities prevented us from completing this work. Incidental progress was steady in both ISIS3 and SOCET SET over the next few years. ISIS3 has matured rapidly and includes both interactive and manual tools for collecting the tiepoints needed to control images. Its control adjustment program, *jigsaw* (Edmundson *et al.*, 2012), now adjusts trajectory as well as pointing, handles pushbroom scanners, and can impose the constraint that the channels of a multi-line scanner like HRSC must adjust together. BAE has fixed the issues that limited our earlier HRSC processing and has developed a new image matching module (Next Generation Automatic Terrain Extraction, or NGATE) that performs dense matching with feature-based as well as area-based methods (Zhang, 2006; Zhang *et al.*, 2006). Subsequent development of NGATE has greatly improved its performance on "desert" (bare-ground) surfaces; though developed for the Earth, this has proved extremely helpful for extraterrestrial mapping.

In 2012–2013 we returned to the problem of improving pushbroom sensor models. The end result (Kirk *et al.*, 2014) was an improved set of core routines for the ISIS3 pushbroom sensors and a new "USGS pushbroom sensor model" for SOCET SET. These developments share a common code base and the following features:

- Faster and more robust solution algorithm to determine the image line on which a given ground point appears
- Handling of constant or varying line exposure times in the same base model
- Handling of pixel-averaging modes and detectors at arbitrary locations in the focal plane
- Handling of images obtained by spacecraft rotation as well as translation, allowing (for example) mapping Phobos

With these software developments, we produced DTMs based on a "conventional" approach to control that was based on the manual collection of ground control points that are identifiable in both the images and the MOLA global altimetry dataset (Smith *et al.*, 2001). The quality of initial DTM products from NGATE without interactive editing or merging of the results from multiple image combinations is similar to or better than that of the highly edited products submitted for the 2006–7 DTM comparison (Kirk *et al.*, 2014). In particular, it was no longer necessary to merge multiple DTM segments or to edit the almost featureless plateau areas surrounding Candor Chasma; the new NGATE algorithm interpolates such terrain with far fewer artifacts than the older method.

In 2011 we began to experiment with controlling stereopairs (initially HiRISE, for which the images are acquired on separate orbits; Kirk *et al.*, 2008) by applying surface fitting techniques (cf. Lin *et al.*, 2010). This approach greatly reduces interactive effort and can be more robust than the conventional approach to stereo control. Rather than searching interactively for identifiable ground control points in the MOLA dataset, we now perform an initial, strictly relative control adjustment, make a coarse initial DTM in arbitrary coordinates, and then determine the transformation that fits this free-floating DTM to the MOLA surface. We initially used a commercial package, Geomagic Control, to do the fitting, but the point-cloud alignment routine *pc_align* of Ames Stereo Pipeline (Moratto *et al.*, 2010), provides the same functionality and is open source. By applying the same transformation to the image-to-image tiepoints in the free-floating coordinate system, we effectively convert them to ground control points that can be used in a final, absolute control calculation, which also includes tiepoints between adjacent orbit strips for multi-orbit projects.

This paper describes our complete process for control of HRSC images and production of DTMs and orthoimages. In addition to implementing the needed sensor models and other software, we have developed and documented procedures for processing single-orbit and multi-orbit datasets. The efficacy of this system is demonstrated on three test datasets.

2 TEST DATASETS

The two datasets used for the HRSC DTM comparison (Heipke *et al.*, 2007) have been analyzed and documented in detail and provide an ideal benchmark for both DTM quality and work effort. The first of these was derived from a single observation (h01235_0001) over western Candor Chasma. The images cover a range of approximately −8.4° to 0.25° latitude and 282.0° to 284.5° East longitude (we use planetocentric coordinates throughout). The area mapped was somewhat smaller, extending to only −3.4° latitude, because the northern part of the coverage consisted mainly of very bland plateaux not especially interesting for topographic mapping. The total area mapped by us is about 18.7 thousand km^2.

This dataset was relatively challenging in the sense of containing both topographic relief of many kilometers and bland plateaus that challenge image matching algorithms. The signal-to-noise ratio was also somewhat less than optimal because of atmospheric haze. On the other hand, the full area was covered with a constant line exposure time, which greatly facilitated our initial mapping (Kirk *et al.*, 2006a, 2006b, 2006c).

The second area chosen for the DTM comparison was a set of three adjacent observations (h0894_0000, h0905_0000, h0927_0000) covering Nanedi Valles. The full dataset covers −0.3° to 14.3° latitude and −50.2° to −45.2° longitude, covering an area of about 198 thousand km^2. Compared to the Candor images, these orbits had higher image quality but the area has substantially less local relief. In 2006–7 we processed parts of orbits h0905 and h0894 covering a limited latitude range 2.2° to 8.1° by dividing the images into blocks of uniform exposure time. This was, needless to say, extremely time consuming both in controlling the image segments and organizing the collection of elevation data from the many overlapping images. Here we present the results of mapping the three full orbits, totaling about 198 thousand km^2, or 3.5 times the area of our earlier Nanedi DTM and more than 10 times the area of the Candor DTM. In order to compensate (in part) for the lower contrast of these images, we applied digital filters as part of the ISIS3 preprocessing. First a 15-×-15-pixel highpass filter was added to the original images to enhance local detail and then the result was lowpass filtered at 3 x 3 pixels to suppress noise.

The third study area presented in this paper is Gale crater, which was selected as the landing site of the Mars Science Laboratory (MSL) Curiosity (Golombek *et al.*, 2012). This site provides an opportunity rare in planetary science to evaluate the precision and resolution of topographic data by comparing them to more precise and higher resolution dataset. The majority of the MSL landing ellipse and a substantial area of Aeolis Mons (also known informally as "Mount Sharp") were mapped with HiRISE stereo images at about 25 cm $pixel^{-1}$, which is roughly a factor of 50 smaller than the HRSC pixel scale. Kirk *et al.* (2011) compared the HiRISE DTMs to an HRSC DTM produced from data acquired on multiple orbits (Gwinner *et al.*, 2010a). Here, we use the same HiRISE DTM mosaic to evaluate the quality of DTMs produced from a more recent HRSC observation, h4235_0001. This observation was selected because it has full resolution and superior image quality, based on subjective evaluation of contrast, sharpness, and compression artifacts, to many of the image sets available at the time of site selection. We evaluate both the Level 4 DTM released by DLR to the NASA Planetary Data System in 2010 and one produced with our own software. The images from orbit h4235_0001 cover latitudes −7.1° to −3.1° and longitudes 136.5° to 137.9°, but we mapped a smaller latitude range −4.9° to −4.25° that overlaps the HiRISE coverage. Within this area, we focused on comparing DTM results in much smaller area of Aeolis Mons (latitude −4.9° to −4.67°, longitude 137.35° to 137.45°), which is topographically rugged and thus presents a dense set of surface features conducive to image matching. Our HRSC DTM was controlled independently to MOLA as described in the next section, and had to be translated 100 m to the east to bring it into visual alignment with the HiRISE DTM. No shift was required to improve the alignment of the Level 4 HRSC DTM, and no consistent shifts on the order of the post spacing between either HRSC DTM and HiRISE could be identified visually.

3 WORKFLOW

This section is intended as an overview of the procedures involved DTM production, with sufficient detail to permit the reader to follow the logic of our processing approach. A standard operating procedure (SOP) document giving step-by-step instructions and details such as statistical weightings for the adjustments is available on request (see section 5).

3.1 *Data preparation*

All HRSC images used are obtained from the NASA Planetary Data System Geosciences Node and first ingested into ISIS3 with the program *hrsc2isis*. We rely on the instrument team's radiometric calibration of the images and use the Level 2 data. The images are then translated for use in SOCET SET by using the script *hrsc4socet_2013_11_04.pl*. This script initializes the images with SPICE orientation data (Acton, 1999), normalizes the images to 8-bit dynamic range for use in SOCET SET, calls *socetlinescankeywords* to prepare the metadata needed, then *isis2raw* to transform the image into a raw format readable by SOCET SET. These files are then imported into SOCET SET as linescan images and constraints are established between the nadir and stereo channels of an observation. MOLA altimetry data (both point clouds and interpolated DTMs) for the project area are also prepared in ISIS3 and transferred to SOCET SET as needed.

3.2 *Single-orbit control*

Geodetic control of a single orbit image set proceeds in three stages. First, tens of tiepoints between the nadir and stereo images are measured manually with the Interactive Point Measurement (IPM) tool and a relative adjustment is performed with the Multi-Sensor Triangulation (MST) tool to remove internal parallax. In this adjustment, omega (across track) and phi (along-track) orientation angles are adjusted along with their first and second time derivatives, weighted very loosely to allow changes of about 1 km but the trajectory is not adjusted. Second, the elevation of a single ground control point (chosen in a relatively smooth area) is measured with IPM and used to adjust constant biases to the spacecraft trajectory to bring the stereo model into moderately close alignment with the Martian surface. The weighting is again on the order of 1 km. Several hundred tiepoints between the images are then collected using Automatic Point Measurement (APM), followed by an adjustment to validate the tiepoint measures and obtain consistent ground coordinate measurements.

Adaptive Automatic Terrain Extraction (AATE) is next used to generate a slightly denser set of matched points in the form of a coarse DTM, which is merged with the tiepoints and control points from the adjustment. Surface fitting is then used to find the optimal alignment of this "free floating" set of points to the surface defined by MOLA altimetry. We currently use the *pc_align* module of the Ames Stereo Pipeline (Moratto *et al.*, 2010) for this function. Once the transformation that aligns the tiepoints with the MOLA surface is determined, they are treated as ground control points at their transformed locations, and any control points measured previously are converted to image-to-image tiepoints with no constraints to ground coordinates.

A final adjustment is performed with the above points (we refer to the converted tiepoints as "pseudo" ground control points because their locations are based on a model-wide fit rather than direct measurement of individual features, and typically weight them at about 100 m horizontally and 10 m vertically, which is less than the precision of true ground points identified in the altimetry data.). In this adjustment, both trajectory (bias and drift) and pointing angles (bias, drift, and accelerations) are adjusted with weights corresponding to a few km. The radial component of the trajectory, which is generally better determined than the horizontal components, is weighted at the level of a few hundred meters. At each stage, trial adjustments may be performed in order to identify inaccurately measured points, which can then be excluded from future adjustment steps, and the images may be filtered to improve matching performance.

3.3 *Multiple orbit control*

Our workflow for multiple orbits begins with obtaining a sparse set of tiepoints along orbits and between orbits where they overlap, using IPM. A relative adjustment of all orbits is performed to remove any image parallax. APM is then run to densify tiepoints both along and between orbits. Using these tiepoints and starting from *a priori* orientation parameters, the workflow proceeds similarly to the single-orbit case, starting with the angular (omega and phi) adjustment, and continuing to the conversion of tiepoints from a coarse DTM into pseudo ground control points. These steps are performed on each orbit individually. The pseudo ground control from the single-orbit adjustments are then combined with the inter-orbit ties for the final adjustment of trajectory and pointing of all orbits simultaneously. Both intra- and inter-orbit tiepoints are weighted at the kilometer level as in the single-orbit case, but those that are converted to pseudo ground control are then weighted more tightly at their surface-fitted locations. To minimize the residuals in this final step we adjust the orientation for the three images of each orbit independently rather than constraining them to be consistent. It is worth noting that this sequence of steps, which presupposes that all images are available at the start of the project so they can be tied together at the outset, is used for historical reasons. Alternative approaches are conceptually possible, such as controlling each orbit individually as described in the previous section, then collecting inter-orbit tiepoints and performing a final adjustment of all orbits simultaneously. We have not tested this approach but it is simpler and could be useful, especially when there is a need to add coverage to an existing block of orbits.

3.4 *DTM extraction*

We use NGATE for primary DTM extraction and enable multi-way matching between the nadir and two stereo images. The low contrast strategy is effective for planetary mapping because it compensates for major brightness variations across the images, allowing local textural features to be detected more readily. NGATE works on a "pyramid" of reduced resolution images and uses both area- and feature-based methods to estimate elevations on a very dense grid, then filters the results from multiple algorithms, points, and image combinations. To reduce mismatches, we generally use the MOLA DTM as a "seed" for the NGATE solution at the coarsest level. NGATE usually produces an extremely "blocky" DTM with relatively flat areas separated by steep slopes. Because one of our primary interests is in estimating surface slopes that might affect the safety of landing a spacecraft, we generally follow DTM extraction in NGATE with a single pass of the older, area-based ATE algorithm applied to the full resolution images and constrained not to change elevations by more than a pixel. This serves as a form of "smart smoothing" of the NGATE DTM.

It is worth noting that in 2006 the quality of DTMs produced by ATE without multi-image matching was such that we performed extensive manual editing and, in particular, replaced very noisy stereo-matched data on the plateau around Candor Chasma with MOLA data. The DTMs presented in this paper have not been edited in this way.

4 RESULTS

4.1 *Candor Chasma*

Figure 9.1 compares our Candor Chasma DTMs from Kirk *et al.* (2006a, 2006b, 2006c) and Kirk *et al.* (2014) at which point the software and procedures described here were mature though not yet released, with MOLA altimetry. The use of NGATE with multi-way matching in the 2014 model greatly reduces the discrepancies between the stereo DTM and altimetry on the steep walls of the canyon. The difference between the 2014 stereo DTM and the interpolated MOLA DTM has a mean ± standard deviation of 30 ± 118 m excluding the areas of obvious edge effects visible in Figure 9.1f. These differences partly reflect the poorer sampling of the MOLA dataset. A similar comparison to the point cloud of MOLA measurements gives a difference of 36 ± 98 m, which compares favorably to the statistics of 11 ± 84 m for the standard team product and 50 ± 122 m for the USGS DTM reported by Heipke *et al.* (2007). It should also be noted that the 2007 comparison was based on only two MOLA tracks whereas the statistics reported here are derived from about 120 tracks.

The ATE algorithm employed in 2006 performed poorly on the bland plateau areas surrounding the canyon, producing artifacts that we describe as "snow angels" with amplitudes of roughly a hundred meters. In fact, the DTM we submitted to the comparison by Heipke *et al.* (2007) was edited to replace these areas with MOLA data. NGATE interpolates only confidently matched features, so the plateau areas are sparsely sampled and appear faceted (Fig. 9.1e) but the RMS deviation from MOLA altimetry is much less, about 50 m.

Figure 9.2 illustrates the improvement in topographic detail from MOLA to the standard HRSC team DTM and our 2006 and 2014 DTMs for an area of the canyon floor with relatively good image texture. The improved resolution and geologic plausibility of the topographic model after refinement by shape from shading (Kirk, Howington-Kraus, Galuszka, Redding, Hare, Heipke *et al.*, 2006; Kirk *et al.*, 2006a, 2006b) is also illustrated.

Nanedi Valles

Figure 9.3 shows color-coded shaded reliefs of our 2006 DTM of part of the Nanedi image set and our current full DTM, along with differences with respect to MOLA altimetry. The local-scale artifacts resulting from the need to divide the images into constant-exposure segments in 2006 (a process that also increased the human workload for control and DTM collection enormously) are clearly apparent in Figure 9.3b. These artifacts are eliminated in the current solution but some end effects are visible in Figure 9.3d. The deviation of the current DTM from MOLA is −1.4 ± 57 m for the DTM in Figure 9.3, −0.5 ± 44 m for the individual points, excluding these end effects. Once again, the limited resolution of MOLA contributes significantly to this difference. This statistic includes the area toward the bottom of the westernmost orbit (h0927_0000) for which the S2 image was unavailable between latitudes 4° and 7°. The stereo DTM deviates from MOLA by about 50 m here. The increasing discrepancy between adjacent orbits toward the south, where their overlap decreases and eventually vanishes, is also noteworthy.

Because the SOCET SET adjustment tool MST only performs linear adjustments to trajectory and quadratic adjustments to camera pointing for pushbroom images, we were concerned that long-arc images could not be controlled at satisfactory precision without breaking them into smaller sections. The stereo-MOLA discrepancies in Figure 9.3d, however, plausibly relate more to the image coverage and overlap than to unmodeled along-orbit orientation variations.

Figure 9.4 shows closeups of part of the Nanedi Valles channel itself. Our current DTM shows improved inter-orbit consistency over our 2006 product and resolution compared to the DLR product from Heipke *et al.* (2007) though it should be emphasized that the standard processing pipeline has been continuously improved since then (e.g., Gwinner *et al.*, 2009, 2010b). Orbit h0894 yielded a

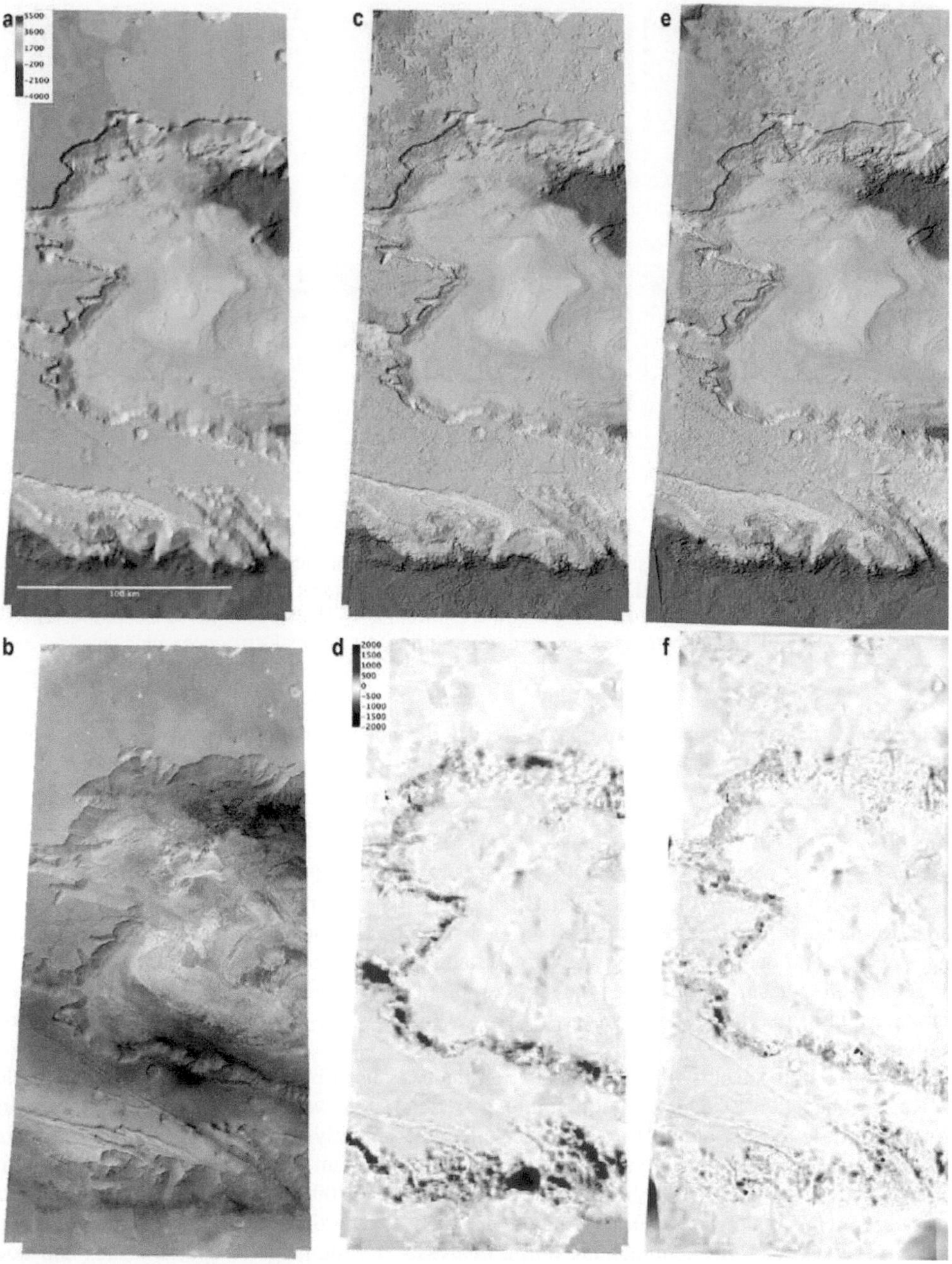

Figure 9.1. Results for Candor comparison area. All images are in Sinusoidal projection with north at top. (a) Color-coded shaded relief from MOLA data. (b) Quasi-natural color composite of HRSC orthoimages. (c) As (a), for USGS 2006 DTM. (d) Difference between 2006 DTM and MOLA. Note the large discrepancies at the canyon edges, where the stereo model "floats" above the true topography by hundreds of meters. (e) As (a), for 2014 DTM. (f) Difference between 2014 DTM and MOLA. Systematic departures from true topography are greatly reduced to ~120 m RMS. Many of the remaining discrepancies are related to inadequate sampling of the MOLA dataset.

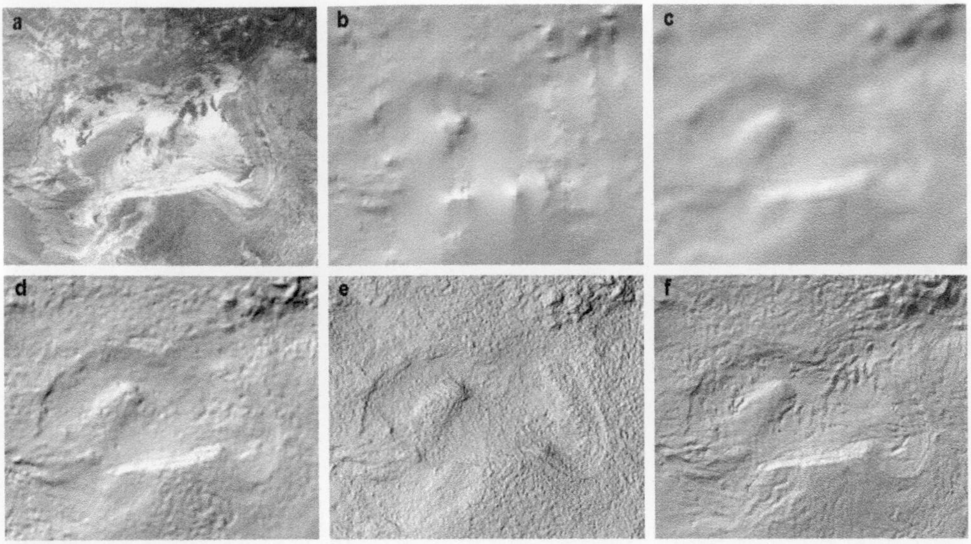

Figure 9.2. Details of Candor Chasma floor from Figure 9.1 (a) Nadir image. (b) Shaded relief from MOLA DTM. (c) HRSC preliminary 200 m DTM from 2007. Note that subsequent archived DTMs have significantly higher quality. (d) USGS DTM from 2006. (e) USGS DTM from 2014 showing improved resolution. (f) 2006 DTM refined by photoclinometry (shape from shading) resulting in more realistic topographic details.

noisier DTM than its neighbors, and this is directly traceable to greater noise in the images, probably due to poorer atmospheric conditions.

4.2 *Gale crater*

Kirk *et al.* (2011) used the extensive HiRISE DTM coverage of the candidate MSL landing site in Gale crater to evaluate the multi-orbit HRSC DTM (Gwinner, Oberst *et al.*, 2010). Here, we make similar comparisons for the current Level 4 DTM produced from a single orbit of high image quality and our own DTM made from the same image set. Such relative comparisons cannot address absolute accuracy, but given the large ratio in pixel scales (0.25 m vs. 12.5 m) comparing to increasingly smoothed versions of the higher-resolution DTM provides information about both resolution and precision.

One approach to quantifying DTM resolution that was used by Heipke *et al.* (2007) is to plot the RMS slope against the horizontal baseline over which it is evaluated. Slopes at every baseline from one post to the size of the DTM can be calculated efficiently by using Fast Fourier Transform methods, and such baseline-slope curves have been an important tool in landing site selection for numerous missions (e.g., Golombek *et al.*, 2012; Kirk *et al.*, 2008). As shown in Figure 9.5a, the slope curves can contain information about the data collection process as well as the surface topography. Advantages of this approach are that it addresses a wide range of spatial scales and does not require a "truth" DTM. Resolution can often be inferred from a break in the derivative of the baseline-slope curve, but having a "truth" DTM is nonetheless useful as an indication of the true surface slope behavior, so that only deviations from this are interpreted as resolution limitations or other data effects. The main drawback of looking at baseline-slope curves is that the results can be ambiguous. For example, an upturn of the curve at short baselines due to localized artifacts (noise) in the DTM can potentially mask the leveling out that is due to limited resolution. Figure 9.5b shows the curves for slopes along north-south baselines on Aeolis Mons, averaged over a 3.75-×-12.8-km area. The Level 4 DTM is clearly deficient in slopes at baselines shorter than about

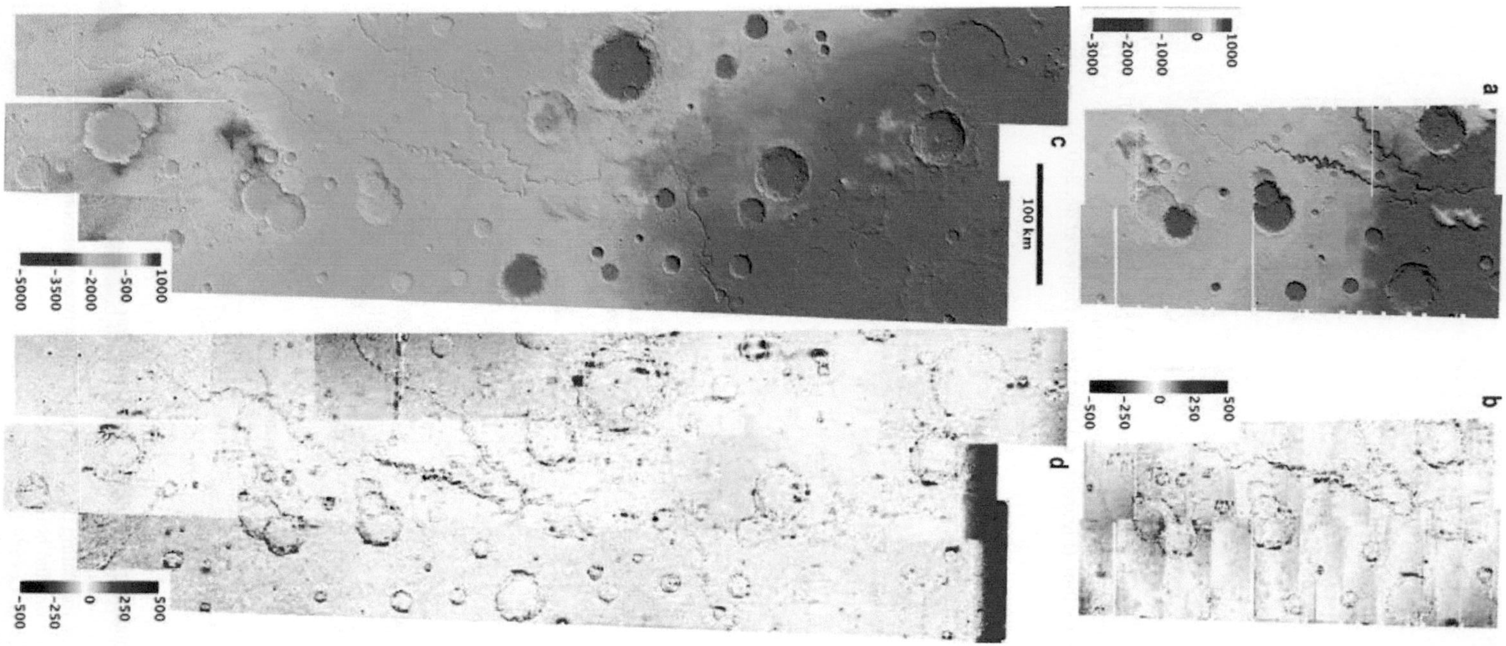

Figure 9.3. Nanedi comparison area. (a) Color-shaded relief of area mapped in 2006. (b) Difference of 2006 DTM from MOLA, showing artifacts related to the segmentation of the images by exposure time. Note different color scale from Figures 9.1d and 9.1f. (c) DTM covering the full longitude extent of three orbits, presented for the first time in this work. (d) Difference between current DTM and MOLA.

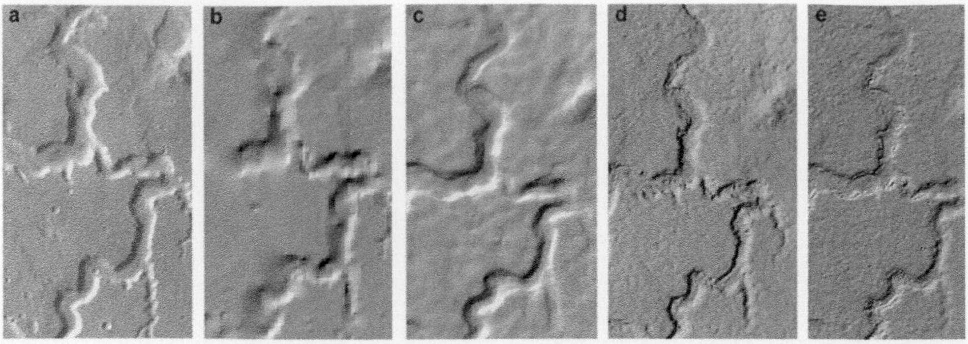

Figure 9.4. Details of Nanedi Valles. (a) Nadir image. (b) Shaded relief from MOLA data. (c) DLR DTM from 2007. (d) USGS DTM using ATE matching algorithm from 2006. (e) USGS DTM using multi-way matching with NGATE algorithm. (Figures a–c adapted from Heipke *et al.*, 2007)

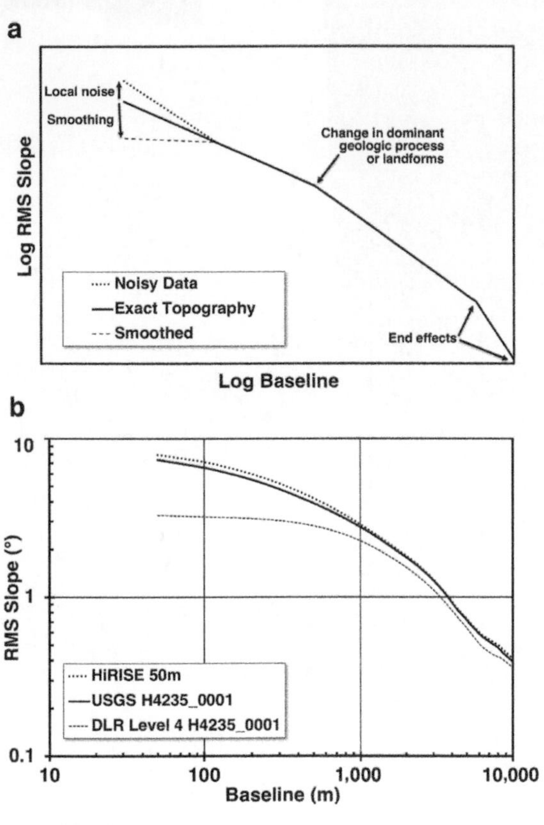

Figure 9.5. Plots of RMS bidirectional slope as a function of the horizontal baseline over which it is measured. A log-log representation is normally used to make clear the fractal (power law) behavior of topography. (a) Schematic illustration showing that the slope curve contains both information about real surface effects (e.g., breaks in derivatives, indicating possible transitions between the dominant geologic processes) and data effects (end effects, resolution, and noise). (b) Actual data for north-south slopes on the flank of Aeolis Mons. Flattening of the curve for the Level 4 DTM indicates limited resolution. The curve for the USGS DTM tracks the full-resolution HiRISE curve more closely but must be interpreted with care. See text for discussion.

1500 m, and the flat curve suggests that there are *no* features contributing slopes over distances smaller than about 500 m. The slope curve from the USGS DTM agrees much more closely with that from HiRISE but there is a noticeable widening of the gap between the curves at 700-m-to-1000-m baselines. It is thus difficult to draw a conclusion about the resolution of the USGS HRSC DTM from its baseline-slope curve.

Because we have a registered high-resolution DTM of Aeolis Mons as a reference, we can assess the quality of the HRSC DTMs much more directly by making post-by-post comparisons. Figure 9.6 shows the standard deviation of the difference between our h4235_0001 Gale DTM and the HiRISE reference, and also the released Level 4 DTM, as a function of smoothing applied to the HiRISE model. The best fit for our DTM occurs with a (boxcar) lowpass filter size of 350 m and the residual deviation, a measure of vertical precision, is 13.25 m. The Level 4 DTM matches this RMS deviation at the same smoothing but most closely resembles the HiRISE DTM smoothed with a 700-m filter, with a RMS deviation of 11.3 m. This is very similar to the result of Kirk *et al.* (2011) for the multi-orbit DTM, 12.5-m RMS for a 700-m in-filter width. We conclude that (a) our processing and that of the HRSC team are achieving generally comparable results but making slightly different tradeoffs between DTM resolution and precision (SOCET SET offers the capability to control this tradeoff, and manual editing provides a means to eliminate the most severe errors while preserving resolution elsewhere); and (b) a precision comparable to the nadir pixel scale is achieved but the horizontal resolution is roughly an order of magnitude coarser than the theoretical minimum DTM resolution of 3–5 image pixels that is frequently quoted in the literature (e.g., Kirk *et al.*, 2008) and somewhat poorer than the empirical results presented by Heipke *et al.* (2007).

This discrepancy deserves further investigation. One possible explanation is that the DTMs being differenced are not precisely registered. If so, the optimal level of smoothing would likely be increased, because the smoothing would help hide the misregistration as well as the differences in local details between the datasets. As noted, however, such misalignments were not seen at the pixel level (in the case of the USGS DTM, after making an overall shift). To quantify the effect that this level of misalignment would have on the best-fitting smoothing, we performed simulations starting with the HiRISE elevation profile shown in Figure 9.7. HRSC-like profiles were generated by smoothing this one-dimensional "truth" dataset with a boxcar filter of specified width and displacing it from its original position by increments of 50 m. The degree of smoothing of the true profile giving the best agreement with each smoothed-and-shifted profile was then estimated by

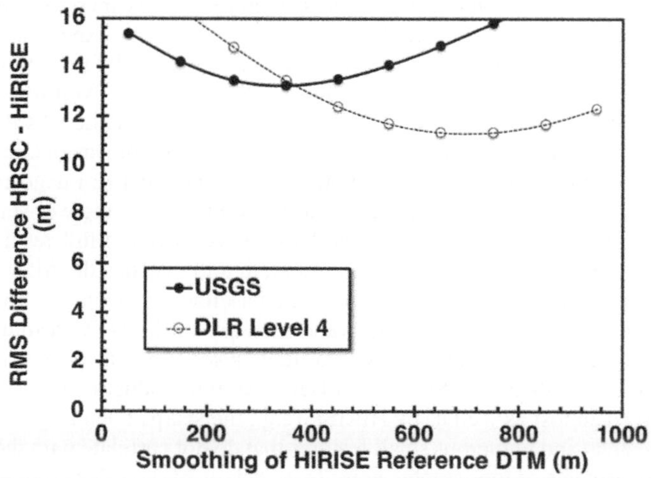

Figure 9.6. Standard deviation of the difference between USGS and HRSC Team Level 4 DTMs for rugged terrain on Aeolis Mons. Error level for the products is similar when evaluated at 350-m smoothing, suggesting they simply make different trades between resolution and precision.

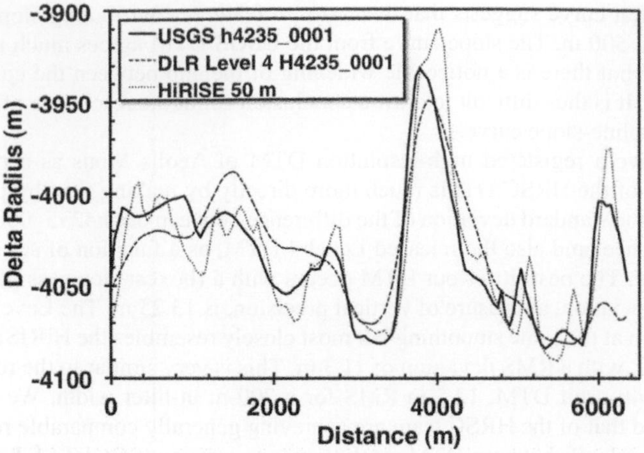

Figure 9.7. Sample profiles in Gale crater illustrating the differences in resolution between USGS and standard HRSC Team DTMs and the HiRISE DTM (downsampled from 1 m to 50 m per post) as revealed by the presence or absence of small (100–200 m wide) features. Note also that there are no consistent shifts between the profiles at the scale of 1–2 posts (50–100 m).

the same process used for the actual HRSC DTMs. Interpolating these estimates to estimate the optimal filter size to a fraction of a post, we find that for shifts of one or more posts it obeys the approximate relation:

$$\text{Optimal filter width} = \tfrac{1}{2}\,(\text{True filter width} + 1\text{ post}) + 3\,(\text{Shift})$$

with a precision of ~0.1 post. Thus, a small shift can emulate a larger amount of blurring. The best-fit filter widths for the USGS and Level 4 DTMs could be explained by misalignments on the order of 2 and 4.6 posts (100 and 230 m) even if they had the full resolution of the 50 m post^{-1} HiRISE DTM.

Though it would account for the statistics, this interpretation is inconsistent with the baseline-slope results for the Level 4 DTM. It is also inconsistent with examination of the DTMs directly, in the form of profiles (e.g., Fig. 9.7) and as terrain-shaded relief images. Figure 9.8 shows shaded relief images of the three DTMs, as well as of the HiRISE DTM after smoothing for its optimal match to each of the HRSC DTMs. (The filter used to match the Level 4 dataset was 650 m across, which is an odd number of posts. The optimal width of 700 m was obtained by interpolating the results for odd filter dimensions.) We note first that the HiRISE image contains coherent features only 2 posts wide (one bright, one dark), corresponding to a single post in the DTM with opposite slopes to either side. Thus, this dataset does indeed have the full resolution possible at 50 m post^{-1}. Second, the Level 4 product appears smooth and matches the filtered HiRISE DTM qualitatively as well as quantitatively. Along with the baseline-slope curves, this result suggests that the 700-m scale of the optimal filter is a valid measure of the DTM resolution. The results for the USGS DTM are more complex. A few areas (e.g., a set of parallel ridges in the lower right corner) resemble the smoothed HiRISE DTM. Elsewhere, the product contains real features that appear to be resolved almost as well as by HiRISE (e.g., the scarps bounding the largest channel, in the center of the area) and numerous small features that do not correlate with those in the "truth" DTM, and thus are artifacts or topographic noise (e.g., details of the ridges running across the top of the area). Thus, the "resolution" estimated for this DTM must be interpreted with caution. It may truly apply to some features, but the DTM also contains a mixture of real features resolved, real features missed, and noise at small baselines, leading to better or worse resolution locally.

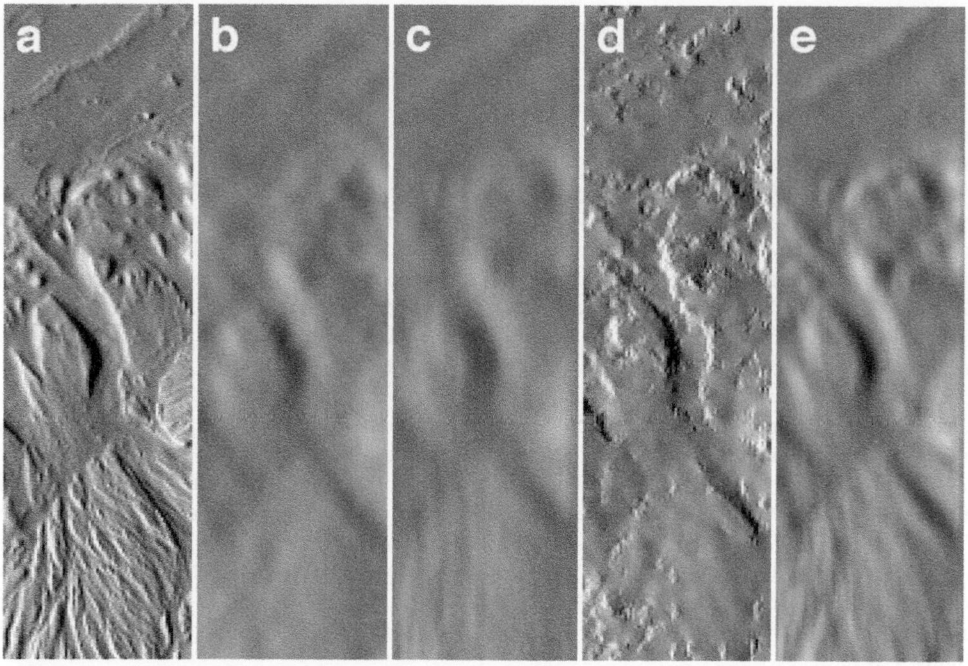

Figure 9.8. Terrain shaded relief images of part of Aeolis Mons, showing the same 3.75-×-12.8-km area in Equirectangular projection at 50 m post^{-1}, north at top, illumination from the left. (a) HiRISE DTM resampled from 1 m to 50 m post^{-1}. (b) DLR Level 4 DTM from HRSC images h4235_h0001. (c) HiRISE DTM smoothed with 650-m boxcar, giving best fit (for a filter of odd dimensions) to Level 4 elevations. (d) USGS DTM from h4235_h0001 contains smooth areas, resolved small geologic features, and small artifacts unrelated to real topography in (a). (e) HiRISE DTM smoothed with 350-m boxcar filter.

Furthermore, the relatively good agreement of slopes over short baselines with the reference is at least partly fortuitous as a result. A final observation is that smoothing the USGS DTM with a 350-m boxcar filter yields a result (not shown) that closely resembles both the Level 4 product and the HiRISE DTM with 650-m smoothing. This adds support to the idea that the DLR pipeline and USGS SOCET SET processing have intrinsically similar performance for these images but choose different amounts of smoothing to suppress local artifacts.

5 SOFTWARE AVAILABILITY

The software described here is, for the most part, already available to the planetary community. The HRSC sensor model and translation programs will be included in an upcoming ISIS3 release. The USGS pushbroom sensor model for SOCET SET and documentation of our Standard Operating Procedures (SOPs) for HRSC processing are available on request from PlanetaryPhotogrammetry@usgs.gov. The updated CSM version of the BAE/USGS pushbroom sensor model for SOCET GXP has now been open sourced and is available from the USGS GitHub site: https://github.com/USGS-Astrogeology/CSM-CameraModel. The Ames Stereo Pipeline, including the surface fitting program *pc_align* used in our workflow, is available at https://ti.arc.nasa.gov/tech/asr/intelligent-robotics/ngt/stereo/. For those who do not have their own SOCET workstations, access to workstations at the USGS in Flagstaff is available through the Planetary Photogrammetry Guest Facility (Kirk *et al.*, 2009).

6 FUTURE WORK

The main challenge currently facing us is that BAE Systems introduced a new processing system called the Geospatial eXploitation Package (GXP) as the successor to SOCET SET in 2006. It has taken the intervening decade for BAE to incorporate full photogrammetric capabilities in GXP and for us to work with them to eliminate Earth-specific restrictions originally built into the new system. GXP allows for user-defined "plug-in" sensor models but uses a new standard for them, the Community Sensor Model (CSMWG, 2010). We have worked closely with the CSM Working Group to generalize the standard so it can be used for planetary work (Hare and Kirk, 2017) and BAE has recently delivered the source code for a CSM version 3.0.3 of the "generic" (but not multi-line) pushbroom sensor model that we had developed for SOCET SET. USGS has just updated this version to allowing for HRSC or variable exposure rate sensor models. Several teams (at the USGS along with Arizona State University and the University of Arizona) are working together to develop new operating procedures for GXP. Our progress can be followed online, and released documentation and code can be obtained from: https://github.com/USGS-Astrogeology/socet_gxp_dev. We anticipate releasing a compiled and ready-to-use GXP-compatible CSM 3.0.3 version of the HRSC sensor model for the planetary community once SOCET GXP version 4.4 is released by BAE (~April 2018). Training on SOCET SET at the Planetary Photogrammetry Guest Facility in Flagstaff has been placed on hold for the past year, but we envision resuming training for DTM production with GXP in the near future. Finally, we have begun to convert the ISIS2 photoclinometry program *pc2d* (Kirk *et al.*, 2003) to an ISIS3-compaitble Python application, and completion of this project will make it straightforward to produce enhanced DTMs similar to Figure 9.3f.

It is also of interest to explore further the question of DTM resolution. First, because the rugged area on Aeolis Mons is quite small (the width of a single HiRISE swath, ~5 km) there is an obvious need to perform similar studies in other (ideally, larger) areas to confirm our results. Second, it would be useful to quantify the precision with which the HRSC and reference DTMs are registered more carefully and explore the effect of introducing offsets with a full DTM rather than a single profile. Third, there are many ways of estimating "resolution" and these should be compared. Other approaches are possible beyond the three used here (slope behavior, comparison to a smoothed reference DTM, and qualitative evaluation of the smallest geologically real features). For example, Heipke *et al.* (2007) assessed small features more quantitatively by counting craters in the DTM and in the (much higher-resolution) image; the diameter at which the majority of craters seen in the image are also identified in the DTM is another estimate of resolution. How do these various estimates relate to one another? Are they highly correlated, as one would hope? If so, what are the proportions between them and how do they compare to the classic definition of resolution as the separation at which two features can be distinguished? HRSC and HiRISE data are nearly ideal for a comparative study of DTM resolution estimates, but the results would be broadly applicable to the evaluation of topographic models from any stereo images.

7 CONCLUSION

We believe that the HRSC stereo processing pipeline that we have developed in ISIS3 and SOCET SET (and soon in GXP) provides a significant added capability to the planetary community even given the outstanding efforts of the instrument team to produce systematic products. With our software, users can make their own DTMs, orthoimages, and mosaics in advance of the release of high-level data, make their own tradeoffs between DTM precision, resolution, and editing time, and leverage the photometric and other analysis capabilities of ISIS3.

ACKNOWLEDGEMENTS

We gratefully acknowledge the support of the NASA Mars Express Project, the Planetary Geology and Geophysics Cartography program (2005–2015) and the NASA-USGS Interagency Agreement for planetary mapping (2016 on) for the work described here.

REFERENCES

Acton, C. (1999) SPICE products available to the planetary science community. *Lunar and Planetary Science Conference*, Houston, TX, USA, Volume 30, abstract #1233.

Anderson, J.A. et al. (2004). Modernization of the Integrated Software for Imagers and Spectrometers. *Lunar and Planetary Science Conference*, League City, TX, USA, Volume 35, abstract #2039.

Batson, R.M. (1990) Cartography. In: Greeley, R. & Batson, R.M. (eds) *Planetary Mapping*. Cambridge University Press, Cambridge, UK, pp. 86–94.

Community Sensor Model Working Group (CSMWG) (2010) *Community Sensor Model Technical Requirements Document*, v. 3. 0, NGA.STND.0017_3.

Edmundson, K.L., Cook, D.A., Thomas, O.H., Archinal, B.A. & Kirk, R.L. (2012) Jigsaw: the ISIS3 bundle adjustment for extraterrestrial photogrammetry. *International Annals of Photogrammetry, Remote Sensing, and Spatial Information Sciences*, Volume I-4. pp. 203–208.

Eliason, E.M. (1997) Production of digital image models using the ISIS system. *Lunar and Planetary Science*, Conference, Houston, TX, USA, Volume 28, abstract #331.

Gaddis, L., Anderson, J., Becker, K., Becker, T., Cook, D., Edwards, K., Eliason, E., Hare, T., Kieffer, H., Lee, E.M., Mathews, J., Soderblom, L., Sucharski, T., Torson, J., McEwen, A. & Robinson, M. (1997) An overview of the Integrated Software for Imaging Spectrometers (ISIS). *Lunar and Planetary Science*, Conference, Houston, TX, USA, Volume 28, abstract #387.

Golombek, M., *et al.* (2012) Selection of the Mars Science Laboratory landing site. *Space Science Reviews*, Volume 170, pp. 641–737. doi:10.1007/s11214-012-9916-y.

Gwinner, K., *et al.* (2016) The High Resolution Stereo Camera (HRSC) of Mars Express and its approach to science analysis and mapping for Mars and its satellites. *Planetary and Space Science*, Volume 126, pp. 93–138.

Gwinner, K., Oberst, J., Jaumann, R. & Neukum, G. (2010a) Regional HRSC multi-orbit digital terrain models for the Mars Science Laboratory candidate landing sites. *Lunar and Planetary Science Conference*, The Woodlands, TX, USA, Volume 41, abstract #2727.

Gwinner, K., Scholten, F., Preusker, F., Elgner, S., Roatsch, T., Spiegel, M., Schmidt, R., Oberst, J., Jaumann, R. & Heipke, C. (2010b) Topography of Mars from global mapping by HRSC high-resolution digital terrain models and orthoimages: characteristics and performance. *Earth and Planetary Science Letters*, 294, 506–519.

Gwinner, K., Scholten, F., Spiegel, M., Schmidt, R., Giese, B., Oberst, J., Heipke, C., Jaumann, R. & Neukum, G. (2009) Derivation and validation of high-resolution digital terrain models from Mars Express HRSC-data. *Photogrammetric Engineering & Remote Sensing*, Volume 75(9), pp. 1127–1142.

Hare, T.M. & Kirk, R.L. (2017) Community Sensor Model standard for the planetary domain. *Lunar and Planetary Science Conference*, The Woodlands, TX, USA, Volume 48, abstract #1111.

Heipke, C., Oberst, J., Albertz, J., Attwenger, M., Dorninger, P., Dorrer, E., Ewe, M., Gehrke, S., Gwinner, K., Hirschmüller, H, Kim, J.R., Kirk, R.L., Mayer, H., Muller, J.-P., Rengarajan, R., Rentsch, M., Schmidt, R., Scholten, F., Shan, J., Spiegel, M., Wählisch, M., Neukum, G. & the HRSC Co-Investigator Team (2007) Evaluating planetary digital terrain models: the HRSC DTM test. *Planetary and Space Science*, Volume 55, pp. 2173–2191. doi:10.1016/j.pss.2007.07.006.

Kirk, R.L., Barrett, J.M. & Soderblom, L.A. (2003) Photoclinometry made simple. . .? *ISPRS Working Group IV/9 Workshop "Advances in Planetary Mapping 2003"*, March 2003, Houston, TX, USA. Available from: http://astrogeology.usgs.gov/Projects/ISPRS/Meetings/Houston2003/abstracts/Kirk_isprs_mar03.pdf.

Kirk, R.L., Edmundson, K.L., Howington-Kraus, E., Redding, B., Thomas, O., Jaumann, R. & the HRSC Co-Investigator Team (2014) Practical processing of Mars Express HRSC images in ISIS and SOCET SET. *8th International Conference on Mars*, Pasadena, CA, USA, 14–18 July 2014, abstract #1161.

Kirk, R.L., Howington-Kraus, E., Galuszka, D., Redding, B., Antonsen, J., Coker, K., Foster, E., Hopkins, M., Licht, A., Fennema, A., Calef, F., Nuti, S., Parker, T.J. & Golombek, M.P. (2011) Near-complete 1-m topographic models of the MSL candidate landing sites: site safety and quality evaluation. *European Planetary Science Conference*, Nantes, France, Volume 6, abstract EPSC2011-1465.

Kirk, R.L., Howington-Kraus, E., Galuszka, D., Redding, B. & Hare, T.M. (2006a) Topomapping of Mars with HRSC images, ISIS, and a commercial stereo workstation. *European Planetary Science Conference*, Berlin, Germany, EPSC2006-A-00487.

Kirk, R.L., Howington-Kraus, E., Galuszka, D., Redding, B. & Hare, T.M. (2006b) Topomapping of Mars with HRSC Images, ISIS, and a commercial stereo workstation. *International Archives of Photogrammetry, Remote Sensing, and Spatial Information Sciences*, Volume 36(4), "Geospatial Databases for Sustainable Development", Goa, India.

Kirk, R.L., Howington-Kraus, E., Galuszka, D., Redding, B., Hare, T.M., Heipke, C., Oberst, J., Neukum, G. & the HRSC Co-Investigator Team (2006c) Mapping Mars with HRSC, ISIS, and SOCET SET. *Lunar and Planetary Science Conference*, League City, TX, USA, Volume 37, abstract #2050.

Kirk, R.L., Howington-Kraus, E. & Rosiek, M.R. (2009) Build your own topographic model: a photogrammetry guest facility for planetary researchers. *Lunar and Planetary Science Conference*, The Woodlands, TX, USA, Volume 40, abstract #1414.

Kirk, R.L., Howington-Kraus, E., Rosiek, M.R., Anderson, J.A., Archinal, B.A., Becker, K.J., Cook, D.A., Galuszka, D.M., Geissler, P.E., Hare, T.M., Holmberg, I.M., Keszthelyi, L.P., Redding, B.L., Delamere, A.W., Gallagher, D., Chapel, J.D., Eliason, E.M., King, R., McEwen, A.S. & the HiRISE Team (2008) Ultrahigh resolution topographic mapping of Mars with MRO HiRISE stereo images: meter-scale slopes of candidate Phoenix landing sites. *Journal of Geophysical Research*, Volume 113, abstract E00A24. doi:10.1029/2007JE003000.

Kirk, R.L., Thompson, K.T., Becker, T.L. & Lee, E.M. (2000) Photometric modelling for planetary cartography. *Lunar and Planetary Science Conference*, Houston, TX, USA, Volume 31, abstract #2025.

Lin, S.-Y., Muller, J.-P., Mills, J.P. & Miller, J.E. (2010) An assessment of surface matching for the automated co-registration of MOLA, HRSC and HiRISE DTMs. *Earth and Planetary Science Letters*, Volume 294, pp. 520–533.

Miller, S.B. & Walker, A.S. (1993) Further developments of Leica digital photogrammetric systems by Helava. *ACSM/ASPRS Annual Convention and Exposition Technical Papers*, New Orleans, LA, USA, Volume 3. pp. 256–263.

Miller, S.B. & Walker, A.S. (1995) Die Entwicklung der digitalen photogrammetrischen Systeme von Leica und Helava. *Z. Photogramm. Fernerkundung*, Volume 63(1), pp. 4–16.

Moratto, Z.M., Broxton, M.J., Beyer, R.A. & Lundy, M. (2010) Ames Stereo Pipeline, NASA's open source Automated stereogrammetry software. *Lunar and Planetary Science Conference*, The Woodlands, TX, USA, Volume 41, abstract #2364.

Neukum, G., Jaumann, R., & the HRSC Co-Investigator Team (2004) *HRSC: The High Resolution Stereo Camera of Mars Express*. ESA Special Publications SP-1240, Noordwijk, The Netherlands.

Scholten, F., Gwinner, K., Roatsch, T., Matz, K., Wahlisch, M., Giese, B., Oberst, J., Jaumann, R. & Neukum, G. (2005) Mars Express HRSC data processing-methods and operational aspects. *Photogrammetric Engineering and Remote Sensing*, Volume 71(10), pp. 1143–1152.

Sides, S.C., Becker, T.L., Becker, K.J., Edmundson, K.L., Backer, J.W., Wilson1, T.J., Weller, L.A., Humphrey, I.R., Berry, K.L., Shepherd, M.R., Hahn, M.A., Rose, C.C., Rodriguez, R., Paquette, A.C., Mapel, J.A., Shinaman, J.R. & Richie, J.O. (2017) The USGS Integrated Software for Imagers and Spectrometers (ISIS 3) instrument support, new capabilities, and releases. *Lunar and Planetary Science Conference*, The Woodlands, TX, USA, Volume 48, abstract #2739.

Smith, D.E., *et al.* (2001) Mars Orbiter Laser Altimeter: experiment summary after the first year of global mapping of Mars. *Journal of Geophysical Reserch*, Volume 107, pp. 23689–23722.

Torson, J. & Becker, K. (1997) ISIS – a software architecture for processing planetary images. *Lunar and Planetary Science Conference*, Houston, TX, USA, Volume 28, abstract #1443.

Zhang, B. (2006) Towards a higher level of automation in softcopy photogrammetry: NGATE and LIDAR processing in SOCET SET. Paper presented at *Geocue Corporation 2nd Annual Technical Exchange Conference*, Nashville, TN, USA, 26–27 September 2006.

Zhang, B., Miller, S., DeVenecia, K. & Walker, S. (2006) Automatic terrain extraction using multiple image pair and back matching. Paper presented at *ASPRS 2006 Annual Conference, American Society for Photogrammetry and Remote Sensing*, Reno, Nevada, 1–5 May 2006.

Planetary Remote Sensing and Mapping – Wu et al.
© 2019 Taylor & Francis Group, London, ISBN 978-1-138-58415-0

Chapter 10

Photogrammetric processing of LROC NAC images for precision lunar topographic mapping

B. Wu, H. Hu and W. C. Liu

ABSTRACT: The narrow-angle camera (NAC) of the lunar reconnaissance orbiter camera (LROC) uses a pair of closely attached pushbroom sensors, NAC Left (NAC-L) and NAC Right (NAC-R), to obtain a large swath of coverage while providing high-resolution imaging. However, the two image sensors do not share the same lens and cannot be modeled geometrically with a single physical model. This leads to problems in the generation of precision lunar topographic models from NAC images, e.g., the boresight misalignment between NAC-L and NAC-R, and the geometrical inconsistencies for the NACs in different tracks and multiple stereo models due to the irregular overlapping areas. Aiming at solving these problems, this chapter presents a rigorous photogrammetric processing method for NAC images, which consists of three major steps. First, using triple-matching tie points between the NAC images, the boresight offsets are improved through a least-squares adjustment. Then, with the initial estimation provided by the calibrated boresight angles, the orientation parameters of a stereo pair of NAC images are further refined through a combined block adjustment. In addition, in order to reduce the number of stereo models, we propose a coupled epipolar rectification method, which merges the NAC-L and NAC-R in the disparity space. Finally, dense matching by a semi-global matching method produces a pixelwise disparity map, from which the high-resolution lunar topographic model can be derived. Experimental results reveal that the presented approach is able to reduce the gaps and inconsistencies caused by the inaccurate boresight offsets between the two NAC cameras and the irregular overlapping regions, and to finally generate precise and consistent 3D topographic models from the NAC stereo images.

1 INTRODUCTION

Three-dimensional topographic models of planetary surfaces are essential for planetary exploration and scientific research (Garvin *et al.*, 1999; Robinson *et al.*, 2010; Wu *et al.*, 2013). Although large-scale topographic models can be obtained from laser altimetric data, such as the lunar orbiter laser altimeter (LOLA) (Smith *et al.*, 2010) and the Mars orbiter laser altimeter (MOLA) (Garvin *et al.*, 1999), these datasets generally have a large spatial resolution (e.g., dozens or hundreds of meters). However, operational and research procedures, such as precision landing, maneuvering rovers, astronaut navigation (Wu, Li *et al.*, 2014), and crater morphometry (Watters *et al.*, 2015), require topographic models with meter-level resolution, which can be obtained from high-resolution cameras such as the narrow-angle camera (NAC) onboard the lunar reconnaissance orbiter (LRO) (Robinson *et al.*, 2010).

Although the NAC provides unprecedented image resolution for the lunar surface (up to 0.5 m pixel^{-1}), the advantages of higher resolution come at the cost of smaller field-of-view (FOV), which is 2.85° using a 700-mm focal length telescope. In this configuration, a swath coverage of 2.5 km is provided by the sensor size of 5000 pixels. To increase the swath angle, two sensors are bundled together, denoted NAC-L and NAC-R, so that the swath coverage can be doubled using the two NACs (Robinson *et al.*, 2010). Stereographic NACs are made possible by pointing the satellite at the same region in another orbiter track to form a sufficient convergent angle.

The special configuration of the image network of the NAC stereo pairs leads to three problems in the photogrammetric processing of NAC images, including: (1) The intra-track inconsistencies between NAC-L and NAC-R, if not corrected, will cause obvious gaps in the overlapping area; (2) The inter-track inconsistencies will cause large vertical disparities and reduce the performance of the successive dense image matching (DIM); and (3) The four images of a stereo pair will generally produce four stereo models, in which two pairs consist of thin and irregular areas as shown in Figure 10.1, and state-of-the-art DIM methods may fail to recover the disparities in these areas. In practice, the commercial software SOCET SET and the Integrated Software for Imagers and Spectrometers (ISIS) system developed by the United States Geological Survey (USGS) have been used to process the LROC NAC images for DEM generation (Burns *et al.*, 2012; Robinson *et al.*, 2012; Tran *et al.*, 2010). Other in-house photogrammetric processing systems have also been used to derive DEMs from the NAC images (Henriksen *et al.*, 2017; Oberst *et al.*, 2010; Li *et al.*, 2011). However, these software systems usually require lots of interactions on the selection of multiple folds tie points for block adjustment and editing of noises after DIM.

In order to alleviate the drudgery of human intervention, this chapter presents an automatic and rigorous photogrammetric processing pipeline of NAC images for precision and high-resolution lunar topographic mapping. The remainder of the chapter is organized as follows. In the next section, we discuss the related works on lunar topographic mapping. Section 3 presents methods to estimate the boresight offsets of the NAC images. Section 4 presents block adjustment and coupled rectification for NAC images, which reduces the stereo models by merging NAC-L and NAC-R in the epipolar space. Section 5 details the DIM methods using the texture-aware semi-global matching (SGM) (Hirschmuller, 2008; Hu, Chen *et al.*, 2016). In section 6, we qualitatively and quantitatively evaluate the proposed method using two stereo pairs of NAC images. We provide concluding remarks in the last section.

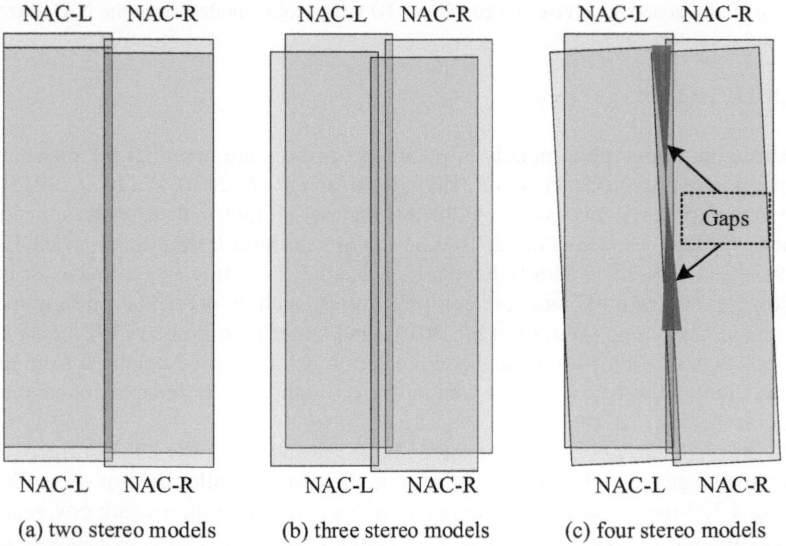

Figure 10.1. Different theoretical model configurations for a stereo pair of NAC images. The last case of four stereo models is the most common. Please note the irregular overlapping region, which may cause gaps because of the insufficient matches obtained.

2 RELATED WORKS

The data for planetary topographic mapping is mainly sourced from two types of sensors: laser altimeters, such as LOLA (Robinson *et al.*, 2010) and MOLA (Smith *et al.*, 2001), and orbiter cameras. The first type of sensor measures the range from the spacecraft to the planet surface and is able to directly calculate the point position with the help of the ephemeris and attitude information of the spacecraft. Although the point spacing along the track is quite dense, the distances between the tracks are quite large, and the final DEM generated from the laser altimeter data suffers from the anisotropic density distribution. On the other hand, topographic mapping using images acquired by the orbiter cameras through photogrammetric processing offers 3D measurements distributed more evenly and generates DEMs of better spatial resolution (Li *et al.*, 2011; Watters *et al.*, 2015; Wu, Hu *et al.*, 2014).

The first step in using satellite images for precision topographic mapping is to conduct the block adjustment of all the involved images to remove any possible geometric inconsistencies among them. Although this issue has been well studied for regular image blocks (Di *et al.*, 2014; Wu, Hu *et al.*, 2014), the block adjustment of a stereo pair of NACs is still a non-trivial problem due to the irregular geometry of the image network and the absence of ground controls. The literature focuses mainly on the properties of the boresight offsets between NAC-L and NAC-R. For example, the temperature-dependent model (Speyerer *et al.*, 2016) improves the orientation parameters of NAC images to generate DEMs of better precision. In addition, for the inter-track inconsistencies between the four NAC images, currently, a combined bundle adjustment is generally used, through automatically detected feature correspondences and the connectivity of the image blocks can be strengthened by manually selected tie points, in a commercial photogrammetry system, SOCET SET from BAE Systems (Henriksen *et al.*, 2017; Mattson *et al.*, 2012; Tran *et al.*, 2010). Only a subset of all the exterior orientation (EO) parameters are adjusted to avoid correlation of the unknowns and make the optimization better posed. However, the above procedure involves a lot of interactions in tie point selection, which is the major bottleneck of massive topographic modeling using NAC images. In this study, we robustly handle the internal inconsistencies through two steps: firstly the boresight offsets are calibrated for each stereo pair to remove the intra-track inconsistencies and then an independent block adjustment for the stereo pairs is conducted in a single free network to remove the inter-track inconsistencies, with the boresight offsets as initial estimation and supported by the automatically matched pairwise, three- and four-folded tie points.

After removing the inconsistencies, the next step is to conduct DIM to generate the image correspondences and create a topographic model from the stereo image pairs. Two main DIM strategies, the local and global methods, can be categorized with the use of smoothness constraints and the inference of disparities, respectively (Hu, Chen *et al.*, 2016; Scharstein and Szeliski, 2002). DIM can be interpreted as the procedure for finding the nearest neighbor field of the two-dimensional (2D) image (Barnes *et al.*, 2009) in the similarity space. To reduce the search complexity, epipolar geometry is generally used to constrain the search to one dimension. Unlike epipolar rectification for a frame camera (Fusiello *et al.*, 2000; Loop and Zhengyou, 1999), the epipolar geometry of pushbroom sensors for most satellite images is a hyperbola rather than a straight line on the stereo pairs (Hirschmuller, 2008; Kim, 2000). Fortunately, because the attitude changes during the collection of a single scene are small, the complicated epipolar geometry can be approximated using some simple functions, such as the affine transformation (Wang *et al.*, 2011), second-order polynomial (Oh *et al.*, 2010), or homographic transformation (de Franchis *et al.*, 2014). After completing the approximate epipolar rectification, the vertical disparities can be removed. In addition to the epipolar constraints, spatial constraints or the global smooth prior to a large context are commonly used in most DIM algorithms. These constraints or priors are used to remove the ambiguities of image matching in a small local window, which may be caused by repeated textures or textureless areas (Hu, Chen *et al.*, 2016; Scharstein and Szeliski, 2002). For the local methods, Delaunay triangulation is used to assist the propagation and searching of nearby matches (Furukawa and Ponce,

2010; Wu *et al.*, 2012). Expanding the randomly sampled best match to nearby areas also requires a larger overlapped region (Barnes *et al.*, 2009). Furthermore, the global methods generally formulate the image matching as a discrete optimization problem, which can be formulated as a Markov random field (Kolmogorov and Zabih, 2001) or dynamic programming (Scharstein and Szeliski, 2002) and can explicitly impose smooth constraints on the optimization in a larger context.

Due to the above-mentioned irregular overlapping of NACs, the performance of the DIM methods generally degrades or even fails when three or four stereo models are generated for one NAC stereo pair (Tran *et al.*, 2010). This problem is especially severe for SGM (Hu, Chen *et al.*, 2016), which aggregates the matching cues in the overall scan lines and adopts the contour information. However, the problem can be effectively handled by the coupled epipolar rectification method as proposed in this chapter, which not only reduces the matching problem to only one stereo model, but also removes the problem of the irregular overlap region and reduces the possible gaps in the final matches.

3 BORESIGHT CALIBRATION OF THE NACS

3.1 *Imaging geometry of the NACs*

The LROC NACs are designed to provide high-resolution monochrome imagery to locate safe landing sites for future robotic and human missions (Robinson *et al.*, 2010, 2012). As illustrated in Figure 10.2(a), the two NACs are separated and closely mounted on the optical bench. Each telescope has a diameter of 27 cm and the perspective centers of the two NACs are separated by about 33 cm. Each NAC uses a 5064-pixel charge-coupled device (CCD) line array providing a 2.85° FOV. The NAC-L is pointing 2.765° away (cross-track) from NAC-R so that the footprints of the two images overlap by about 135 pixels in the cross-track direction (Y-axis). This configuration allows the two NACs to provide a FOV of 5.7°, twice that of a single camera (Robinson *et al.*, 2010). The NAC-R is also mounted 0.106° forward of the NAC-L, with about 185 pixels separation in the along-track direction (X-axis) between the two images acquired by the NACs. The NAC-L and NAC-R are mounted such that pixel 0 for the NAC-L is at the −Y end of its CCD and pixel 0 for the NAC-R is at the +Y end of its

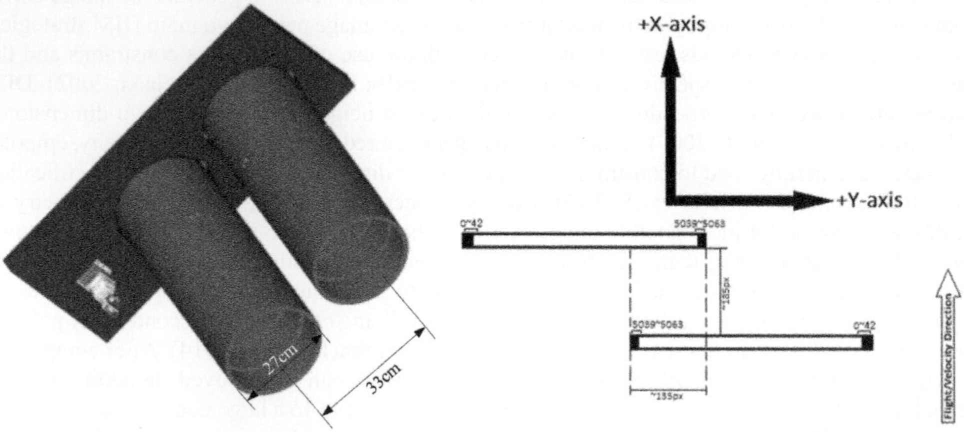

(a) (b)

Figure 10.2. Conceptual illustration of (a) the configuration of the LROC NACs and (b) the configurations of the NAC CCDs as mounted on the spacecraft (numbers show zero-based pixel addresses of the masked and non-imaging pixels) (Robinson *et al.*, 2010).

CCD (Figure 10.2(b)). This orientation requires that one of the NAC frames from a NAC-L and NAC-R paired observation must be transformed such that both images have the same ground orientation (Robinson *et al.*, 2010).

The coordinate frames in the LROC NAC sensor model include (1) the lunar body-fixed frame, (2) the LRO spacecraft frame, and (3) the NAC camera frame (Fig. 10.3). The lunar body-fixed frame is defined by the Mean Earth/Polar Axis (ME) reference system with the origin located at the center of mass of the Moon, the Z-axis along the mean rotational pole, and the prime meridian in the mean Earth direction. Geographic locations of surface features can be expressed in planetocentric coordinates. In the present research, the exterior orientation parameters of NAC images are defined in the ME reference frame and the coordinates of lunar surface objects are expressed in planetocentric coordinates. The position and pointing information of LROC NAC are archived in a series of binary and text based Spacecraft, Planet, Instrument, C-Matrix and Events (SPICE) kernels. NASA's Navigational and Ancillary Information Facility (NAIF) maintains the SPICE ancillary information system (Acton, 1996). A series of spacecraft position kernels (SPKs), C-matrix kernels (CKs), and a single frames kernel (FK) store the orientation parameters for the spacecraft and associated instruments (Speyerer *et al.*, 2016).

The image orientation includes interior orientation (IO), relative orientation between NAC-L and NAC-R, and exterior orientation (EO) of the images. IO parameters describe the intrinsic physical geometric properties of the sensor. They are provided in the SPICE instrument kernels as follows (Speyerer *et al.*, 2016): (1) focal length: 699.62 mm for NAC-L and 701.57 mm for NAC-R; (2) pixel size (single CCD cell size) for both NACs: 7 μm; and (3) principal point (pixel): 2548 for NAC-L, and 2496 for NAC-R.

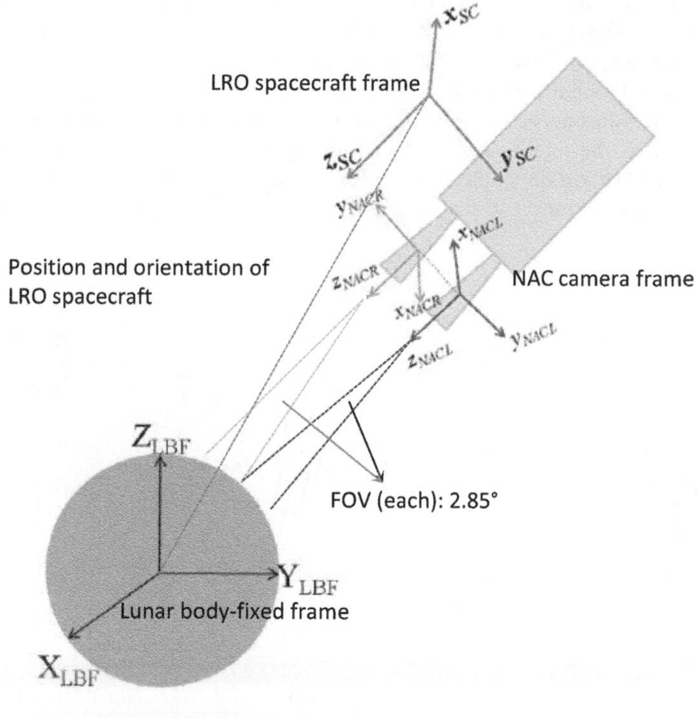

Figure 10.3. Coordinate frames including the lunar body-fixed frame with a subscript of "LBF", the LRO spacecraft frame with a subscript of "SC", and the NAC camera frame with subscripts of "NACL" or "NACR".

Using the EO and IO parameters, a 3D point in the lunar body-fixed frame can be projected to the image space using the standard collinear equation as the following:

$$x = f\Pi\big(OR(X - C)\big) + x_0 \tag{1}$$

where $\boldsymbol{x}$ is the image measurement, f is the focal length, Π: $\Re^3 \rightarrow \Re^2$ is the projection function from 3D to 2D, $\boldsymbol{R}$ is the rotation matrix defined by the three rotation angles in the EO parameters, $\boldsymbol{X}$ is the object point position, $\boldsymbol{C}$ is the camera perspective center defined by the three translations in the EO parameters, and $\boldsymbol{x}_0$ is the principal point. The auxiliary rotation matrix $\boldsymbol{O}$ is introduced to compensate the different image space geometry as described above. Different definitions of $\boldsymbol{O}$ for NAC-L and NAC-R are as follows:

$$O_L = R_z(\pi/2), O_R = R_x(\pi)R_z(\pi/2) \tag{2}$$

where $\boldsymbol{R}_x$ and $\boldsymbol{R}_z$ denote the rotation matrix around the x and z axis, respectively, for a specific angle. In addition, the rotation matrix $\boldsymbol{R}$ is only parametrized by three values $\boldsymbol{r} = (r_1, r_2, r_3)^T$, together with the three parameters for the translational components. The EO parameters are fitted using polynomials. In addition, a set of 78 rational polynomial coefficients (RPC) (Grodechi and Dial, 2003) can be fitted for fast projection and consistent description of the different camera models in the block adjustment, using both the intrinsic parameters of the NACs (Robinson *et al.*, 2010) and the extracted EO parameters for the images.

3.2 *Calibration of boresight offsets of the NACs*

The nine elements of the rotation matrix in Equation (1) are uniquely determined by three rotation components. Although there are many possible representations, such as the angle-axis, quaternion, Rodrigues rotations, the most intuitive method, by three Euler angles, is chosen in this research. Because the translational part of the boresight offsets is almost negligible, only the three Euler angles are refined during the boresight calibration. Figure 10.4 illustrates the three angular boresight offsets. ω_b is the rotation angle of the boresight of NAC-L with respect to that of NAC-R around the X-axis (flight direction) in the NAC camera frame. φ_b is the rotation angle of the boresight of NAC-R with respect to that of NAC-L around the Y-axis. κ_b is the rotation angle of the

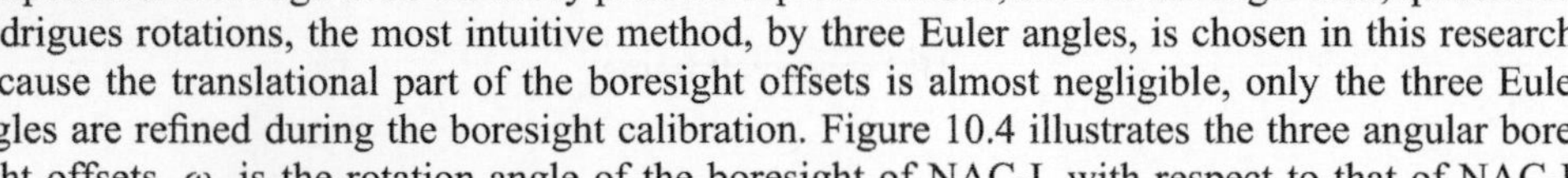

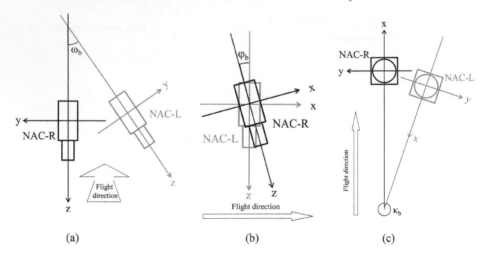

Figure 10.4. Conceptual illustration of the angular boresight offsets, (a) omega ω_b as viewing along the flight direction, (b) phi φ_b as viewing perpendicular to the flight direction, and (c) kappa κ_b as viewing pointing at the nadir.

boresight of NAC-L with respect to that of NAC-R around the Z-axis (nadir). Two sets of boresight angles are explored in this research, the fixed set as generated by the LRO Mission Operations Center (MOC) and the temperature-dependent SPICE kernels as described in Speyerer *et al.* (2016). This research retrieves the improved orientation parameters for NAC imagery from ISIS 3. The approach herein uses tie points obtained from the stereo images to calibrate the boresight angles through the least-squares adjustment.

Because the overlapping region among the four images (denoted as $O_1^L, O_1^R, O_2^L, O_2^R$, where $I = 1,2$ refer to different tracks and L, R refer to the NAC-L and NAC-R, respectively) of the stereo pairs is very narrow, it is rare to obtain quadruple-matching tie points from all the four images. Instead, triple-matching tie points are more common to appear on the images. Therefore, triple-matching tie points are mainly used in the approach. Quadruple-matching tie points are also used if they are available. Figure 10.5 illustrates the triple-matching tie points used in the approach. They can be obtained from four types of image combinations, $\left(O_1^L O_1^R O_2^L\right)$, $\left(O_1^L O_1^R O_2^R\right)$, $\left(O_1^L O_2^L O_2^R\right)$, $\left(O_1^R O_2^L O_2^R\right)$. We developed a self-adaptive triangle-constrained image matching (SATM) method for feature point matching from multiple images (Wu *et al.*, 2011, 2012; Zhu *et al.*, 2007, 2010), which has been used to automatically obtain the triple-matching tie points in this research. Normally, dozens of triple-matching tie points can be matched for a stereo pair of NAC images through the SATM method.

In the boresight calibration, for NAC-R, the EO parameters for O_1^R is fixed, O_2^R estimated; for NAC-L, the EO parameters of both O_1^L and O_2^L are determined by the calibrated boresight angles and the corresponding EO parameters for NAC-R. The observations of the boresight calibration include the tie points, the 3D ground coordinates of the tie points, the boresight parameters, and the image EO parameters. Specifically, the observations in the least-squares optimization consist of two types: (1) the re-projection errors of tie points, and (2) the regularized term determined by the initial values that constrains the optimization not to deviate too much from initial values. The boresight calibration is formulated as follows:

$$\min_{C_2^R, r_2^R, \Delta r} \frac{1}{|Z|} \sum_{i,j} \left(\boldsymbol{w}_x^T \left(f_i \Pi \left(\boldsymbol{O}_i \boldsymbol{R}_i \left(\boldsymbol{X}^j - \boldsymbol{C}_i \right) \right) + \boldsymbol{x}_{i0} \right) \right)^2 + \lambda \sum_i \left(\left(\boldsymbol{w}_C^T \left(\boldsymbol{C}_i - \hat{\boldsymbol{C}}_i \right) \right)^2 + \left(\boldsymbol{w}_r^T \left(\boldsymbol{r}_i - \hat{\boldsymbol{r}}_i \right) \right)^2 \right) \quad (3)$$

where $\boldsymbol{X} \in \mathbb{R}^3$ are the unknowns for the 3D object points; Z is the normalization factor, denoted as the number of feature points; i and j are the indices of the image and object point, respectively; $\boldsymbol{w}_x \in \mathbb{R}^2$, $\boldsymbol{w}_C \in \mathbb{R}^3$ and $\boldsymbol{w}_r \in \mathbb{R}^3$ are the weights for the observations of the tie points, the initial values of translation and rotation parameters, respectively; $\boldsymbol{x}$ is the coordinate of the tie points. Equation (3)

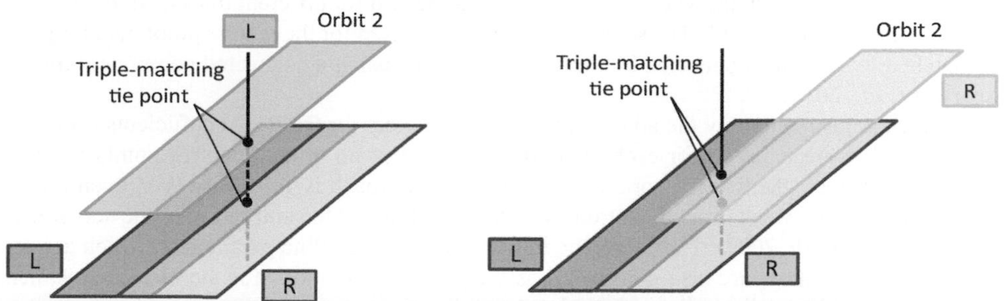

Figure 10.5. Illustration of triple-matching tie points, (a) triple-matching tie points for the triplet $\left(O_1^L O_1^R O_2^L\right)$, and (b) triple-matching tie points for triplet $\left(O_1^L O_1^R O_2^R\right)$.

is a regularized least-squares problem (Hastie *et al.*, 2009), and a standard solver is used for optimization (Agarwal and Mierle, 2010). The output consists of $\boldsymbol{C}_2^R, \boldsymbol{r}_2^R, \Delta\boldsymbol{r}$, which are the translation part, rotation part of the NAC-R in the second track, and the boresight angle between NAC-L and NAC-R. Using these sets of parameters, the EO parameters of all the four images can be obtained accordingly, which serve as initial values for the subsequent block adjustment as described below.

4 BLOCK ADJUSTMENT AND COUPLED EPIPOLAR RECTIFICATION OF NAC IMAGES

4.1 *Block adjustment of NAC stereo pairs*

After the boresight calibration as described above, the intra-track inconsistencies of NAC images will reduce. However, the inter-track inconsistencies may still exist, which can be explained in the following two reasons: (1) the boresight angles are fixed for both tracks, which may be affected by numerous reasons, such as temperature changes and mechanical rigidness; and (2) un-calibrated distortions or other factors may influence the geometrical accuracies. Therefore, in this research, a successive block adjustment of all the four images is also conducted. In order to compensate for the unknown distortions, an additional set of affine parameters is fitted for each image, which correct the fitted RPCs. The block adjustment of all the NAC images will remove the vertical disparity of the image pair between the NAC-L and NAC-R of the same orbiter and the images from different orbiters.

However, as shown in Figure 10.1, the overlapping region between NAC-L and NAC-R comprises a very small portion of the image. Thus, if the feature points are detected throughout the entire image, the search space for the correspondences will be quite narrow and will not be able to retrieve sufficient tie points for successful block adjustment. In addition, for the major part of the overlapping region, the tie points need to be distributed evenly on the entire image to attain effective block adjustment. Therefore, in this research, we separate the feature matches into two segregated parts, as shown in Figure 10.6. For the first set, feature points are detected on the entire images from both NAC-L and NAC-R, and only the two major overlap regions are searched for correspondences. For the second set, feature points are only detected in the overlapping region between NAC-L and NAC-R. In the same track, this region is fixed at about 150 pixels (Robinson *et al.*, 2010), although this region varies for different configurations, as shown in Figure 10.1. Therefore, the region is determined by the union of the geographic locations of the 150 pixels close to the border by back projection using the RPC parameters.

Figure 10.6(c) and (d) show the matching results between NAC-L and NAC-R for the same and different tracks, respectively. It can be seen that the region for the same track is almost a rectangular area with a fixed width, whereas the shape is a trapezoid for different tracks, which indicates the gap regions in Figure 10.1. The same SATM method is used for the feature point matching, and the matched features are connected into multiple tie points using a connected component method (Agarwal *et al.*, 2011).

For block adjustment, only the additional affine parameters for the RPC coefficients (Grodechi and Dial, 2003) are adjusted for each NAC image. Because no ground control points available on the lunar surface, the free network block adjustment approach is used. However, even for free network adjustment, weak ground controls are needed to remove the ambiguity of the solver (Fraser and Ravanbakhsh, 2009). Therefore, the initial values of the affine parameters, which are zero translation and rotation, are also considered as weighted observations for the block adjustment. Specifically, the block adjustment is formulated as follows:

$$\min_{\boldsymbol{A},\boldsymbol{X}} \frac{1}{|Z|} \sum_{i,j} \left(\boldsymbol{w}_x^T \left(\boldsymbol{A}_i f\left(\boldsymbol{X}^j\right) - \boldsymbol{x}_i^j \right) \right)^2 + \frac{\lambda}{|N|} \sum_i \left\| \boldsymbol{W}_A \circ \left(\boldsymbol{A}_i - \boldsymbol{A}_0 \right) \right\|_F^2 \tag{4}$$

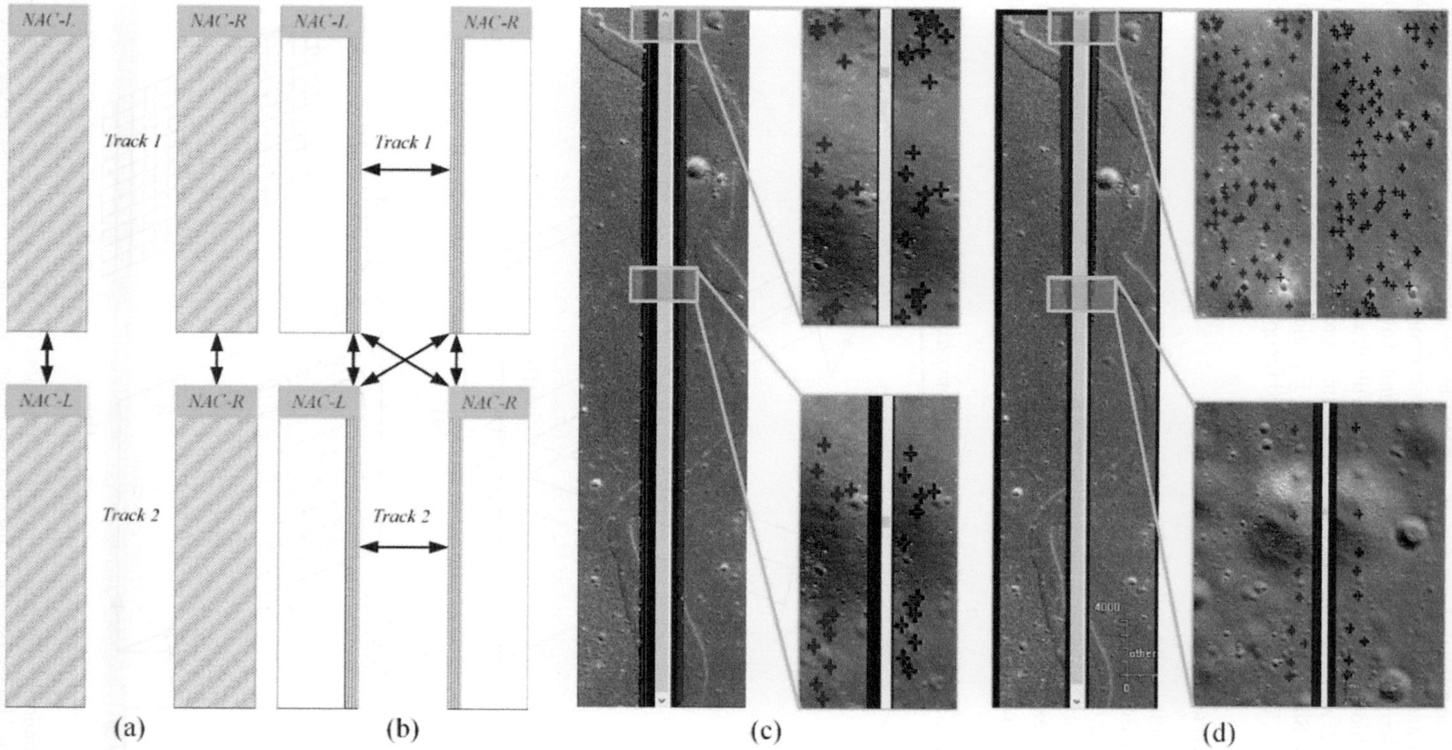

Figure 10.6 Separated feature matching strategy for block adjustment. Only the shaded areas in the image are used for matching for each set. (a) The first set of feature matches detected on the entire image, where the arrows indicate the pairs for searching the correspondences; (b) The second set of feature matches detected on the overlapping region of NAC-L and NAC-R; (c) Rectangular regions of the feature matches for NAC-L and NAC-R of the same track; (d) Trapezoid regions of the feature matches for NAC-L and NAC-R between different tracks.

where $A \in \mathbb{R}^{2\times 3}$ and $X \in \mathbb{R}^3$ are the unknowns for the affine parameters and 3D object points, respectively; Z and N are the normalization factors, denoted as the numbers of feature points and images, respectively; i and j are the indices of the image and object point, respectively; $w_x \in \mathbb{R}^2$ and $W_A \in \mathbb{R}^{2\times 3}$ are the weights for the observations of the tie points and initial affine parameters, respectively; $f(\bullet)$ is the RPC projection function from the 3D point to 2D image pixel; x is the coordinate of the tie points; A_0 is the initial value of the affine parameters; (^{o}) is the Hadamard product of matrices; and $\|\bullet\|_F$ is the Fresenius norm of matrix. Similar to the boresight calibration, Equation (4) is also a regularized least-squares problem (Hastie *et al.*, 2009), and a standard solver is used for optimization (Agarwal and Mierle, 2010).

4.2 *Coupled epipolar rectification*

As mentioned previously, the epipolar geometry for pushbroom satellite imagery is not a straight line. Because the sensor size for the NAC images is relatively small and the attitude variation during the collection in one orbit is negligible, we have found that the affine model (Wang *et al.*, 2011) is sufficient to describe the epipolar geometry. The basic idea is to determine a series of epipolar curves and to align each vertical scanline of the epipolar image parallel to the epipolar curve, as shown in the center of Figure 10.7(b). The generation of a single epipolar curve is an analog of

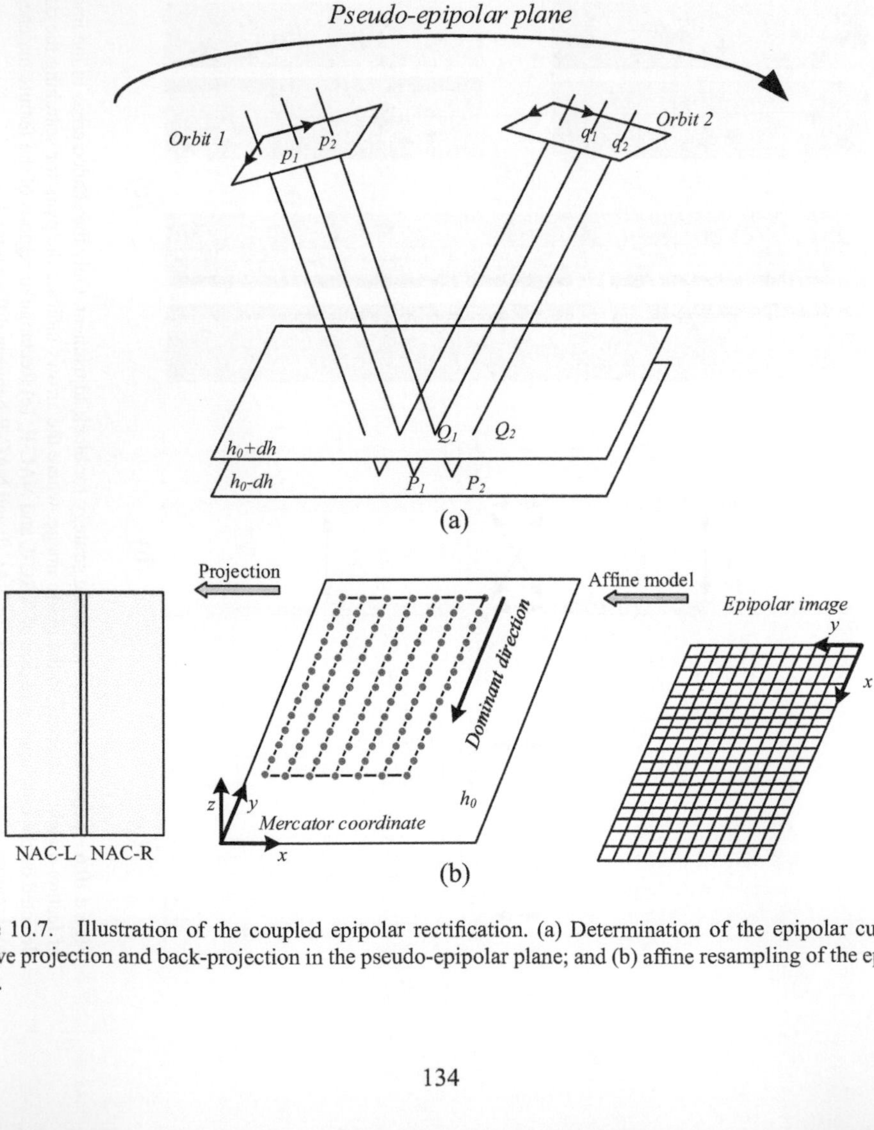

Figure 10.7. Illustration of the coupled epipolar rectification. (a) Determination of the epipolar curve by iterative projection and back-projection in the pseudo-epipolar plane; and (b) affine resampling of the epipolar image.

Table 10.1 Algorithm for tracing an epipolar curve

TraceEpipolarCurve
Input: The EO information of a pair of NACs after block adjustment and the center of the forward image as p_1.
Output: A epipolar curve on the forward images $\{p_i, I = 1,2,3...\}$
Procedure:
- *for* $I = 1,2,3,...$ *until* p_i falls out of the forward NAC
1. Back-project p_i to the minimum plane as P_i
2. Project P_i to the corresponding backward NAC as q_i
3. Back-project q_i to the maximum plane as Q_i
4. Project Q_i to the corresponding forward NAC as p_{i+1}
- Swap the minimum and maximum plane and repeat the above procedure, this will make the epipolar curve marching in the opposite direction

the intersection of the epipolar plane and the image for the frame camera, which is denoted as the pseudo-epipolar plane, as shown in Figure 10.7(a). We iteratively project and back-project inside the pseudo-epipolar plane to obtain a series of points, as described in Table 10.1. Although this series of points should be mathematically describable by a hyperbola (Kim, 2000), in practice, it is sufficient to fit the points by a straight line. The direction of this line will help to determine the dominant direction of the epipolar curve and, unlike the work of Wang *et al.* (2011), which only involves one curve, we offset the initial point p_1 orthogonal to the curve by a certain distance (512 pixels used in this research) and trace a series of curves for both NAC-L and NAC-R. The average direction is then used for the dominant direction.

After tracing each epipolar curve on the NACs, the image points are back-projected to the center plane h_0 of this area in a projected coordinate system, and the Mercator projection is used for simplicity. The dominant direction is estimated using all of the epipolar curves, and this direction determines the rotation part of the affine transformation between the epipolar image and the projected object space, by which the x axis of the epipolar image is aligned parallel to the dominant direction. Then, given a ground sample distance of the epipolar image, the scale part of the affine transformation can also be determined. In this way, a one-on-one map between the epipolar image and the object space is established, as shown in Figure 10.7(b).

To obtain the pixel value of the epipolar image, the object point is first projected onto the corresponding NAC-L, and if the image point falls outside of NAC-L, it is then projected onto NAC-R. A mask is also recorded, which indicates the source of each pixel in the epipolar space. In this way, we are able to use the correct camera model for space intersection.

5 DENSE IMAGE MATCHING

Following the coupled epipolar rectification, texture-aware SGM (Hu, Chen *et al.*, 2016) is used to generate a disparity image. The SGM algorithm is an extension of dynamic programming, which uses 16 aggregated directions rather than only one direction along the horizontal scanline. In each direction $\boldsymbol{r}$, SGM solves the following dynamic programming:

$$\begin{aligned} L_r(\boldsymbol{p},d) = C(\boldsymbol{p},d) + \min\big(& L_r(\boldsymbol{p}-\boldsymbol{r},d), \\ & L_r(\boldsymbol{p}-\boldsymbol{r}.d-1) + P_1, \\ & L_r(\boldsymbol{p}-\boldsymbol{r},d+1) + P_1, \\ & \min_i L_r(\boldsymbol{p}-\boldsymbol{r},i) + P_2 - \min_k L_r(\boldsymbol{p}-\boldsymbol{r},k)\big) \end{aligned} \tag{5}$$

where p is the current pixel; $C(p,d)$ measures the matching cost or dissimilarity at the disparity of d; and P_1 and P_2 control the smoothness of the discontinuities of one pixel and more than two pixels, respectively. These two parameters are tuned adaptively according to the image contours, which means that at the object boundaries, the texture-aware SGM allows larger fluctuations of the disparities (Hu, Chen *et al.*, 2016).

After applying the SGM, we obtain a disparity map $d(\boldsymbol{p})$ of the epipolar image. The disparity map indicates that for each pixel $\boldsymbol{p}(x,y)$ in the coupled epipolar image from the first orbit, the correspondence in the epipolar image from the second orbit is $\boldsymbol{p}'(x + d(\boldsymbol{p}),y)$. Furthermore, we can acquire the Boolean values $b(\boldsymbol{p})$ and $b'(\boldsymbol{p}')$ from the accompanying mask for each epipolar image, which describes whether to project onto NAC-L or NAC-R using the transformation indicated in Figure 10.7(b). The corresponding EO parameters are then used to triangulate an object point. After processing all of the valid disparities in the epipolar space, the 3D point clouds are interpolated using inverse distance weighting to the gridded DEM.

6 RESULTS AND DISCUSSION

Two NAC datasets are used for experimental evaluation. The first is a pair of images with a 0.5-m ground sample distance (GSD) on a long lunar rille area and the second is a pair of images with a 1.2-m GSD on Dorsa Whiston. To evaluate the quality of the epipolar rectification, anaglyphs with cyan-red stereo mode are used to visually validate the removal of the vertical disparity. Furthermore, the coupled epipolar rectification is validated by measuring the vertical and horizontal disparities of the subpixel tie points obtained using the commercial photogrammetric software Leica Photogrammetry Suite (LPS). To evaluate the quality of the mapping results, both the disparity map and DEM are generated and the NAC-DEM is compared with the SLDEM (Barker *et al.*, 2016), which is the most precise DEM publicly available for the lunar surface. SLDEM was generated by co-registering 4.5 x 10^9 LOLA heights measurements and 43,200 DEMs from the SELENE Terrain Camera (TC). After co-registration, the root mean square error of vertical residuals between TC DEMs and LOLA DEM is less than 5 m at 90% margin. The SLDEM has a typical vertical accuracy of about 3 to 4 meters and a horizontal resolution of 512 pixels per degree.

6.1 *Experiment on NAC images covering a lunar rille area*

The first stereo pair of NACs on the long lunar rille is used to evaluate the performance of the proposed method. The details of the stereo pair are listed in Table 10.2. It should be noted that the convergent angle of the two pairs of NACs is formulated by banking the satellite across the orbit direction and the average convergent angle of the two major overlapping regions is about 19°. In this configuration, the theoretical measurement accuracy at the corresponding altitude is about 1.5 m according to the baseline-height ratio estimation (Cook *et al.*, 1996) at a matching accuracy of 1 pixel for SGM (Scharstein *et al.*, 2017). All the images have the same dimension of 5064 × 52224 pixels. The overlapping region for NAC-L and NAC-R in the same orbit is a rectangle area with about 150 pixels wide. The overlapping regions for NAC-L and NAC-R from different

Table 10.2 Information on the experimental NAC stereo pair for the lunar rille area

Product ID	*Resolution (m)*	*Emission (°)*	*Incidence (°)*
M173246166L	0.52	16.32	49.04
M173246166R	0.52	19.14	49.07
M173252954L	0.51	11.62	48.75
M173252954R	0.50	8.80	48.78

orbits formulate two triangles with the widest part of about 400 pixels at the top (or bottom) of the images. Figure 10.8 overlays the NACs in the first track on top of the SLDEM. The center of the two NACs is around 44°N and 51.5°W. An enlarged view of the georeferenced images is also shown in the right part of Figure 10.8.

For the block adjustment, the feature matching and track connection produce 23,253 object points. In the bundle adjustment, we remove several outliers using the 68–95–99.7 rule of 3σ, and 23,043 object points contribute to the final bundle adjustment, with a subpixel level of σ_0 at 0.7 pixels. This indicates a successful bundle adjustment. The *a priori* standard deviation used for weight determination of the pixel observations is set to 1 pixel. And for the affine parameters, the six terms are split into rotation and translation parts, for which the standard deviation is set to 0.0001 and 50 pixels,

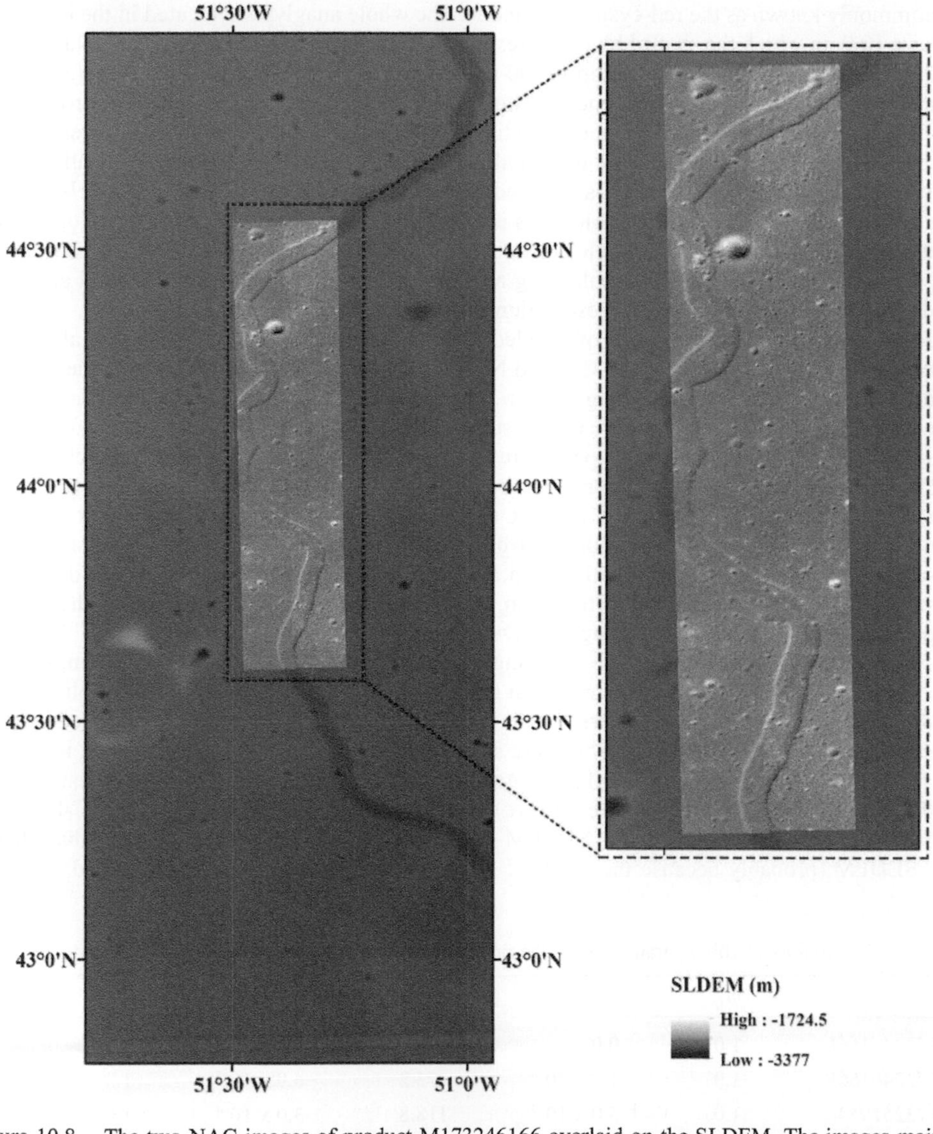

Figure 10.8. The two NAC images of product M173246166 overlaid on the SLDEM. The images mainly cover part of a long lunar rille. The legend indicates the elevation in meters of the SLDEM in the background.

respectively. The weights are the reciprocals of the *a priori* standard deviations. The balance parameter λ for the observations of feature points and affine parameters is fixed as 0.01. The affine parameters after block adjustment for all the images are listed in Table 10.3. It can be noticed that the inconsistency between NAC-L and NAC-R in the same orbit are all around −73 pixels for both orbits.

In theory, the epipolar rectification removes the vertical disparities in the images. However, it is possible to visually perceive the stereo effect formulated by two epipolar images, as long as each eye only observes a single epipolar image. This not only provides an effective method for direct 3D measurements, but also a quick approach for qualitative evaluation of the epipolar rectification. If the rectification is not satisfactory, large vertical disparities will be apparent in the two epipolar images together with other visually observable noises and the stereo effects will be degraded.

In this research, we create anaglyphs by overlaying the two grayscale images, with the red channel for the left epipolar image and the green and blue channels for the right epipolar image. This is commonly known as the red-cyan stereo mode. The whole anaglyph is located in the left part of Figure 10.9, in which the shaded blue squares indicate enlarged views of four representative areas along the rille. The squares from top to the bottom represent subfigures Figure 10.9(a) through (d), respectively. With the corresponding filters, the anaglyphs can be seen to show strong stereo effects, which indicates that the epipolar rectification is of good quality, with no apparent vertical disparities. The flat parts, which are outside the craters or rilles, present relatively small parallax; this is because the epipolar image is rectified in the horizontal plane of the average elevation in the covered area. In addition, the color bias in the middle part of the anaglyphs in Figure 10.9(c) and (d) is caused by the inherent color differences between the NAC-L and NAC-R, as shown in Figure 10.8. Although local color balancing may be required prior to the generation of anaglyphs and orthoimages for better visual presentations, this is out of the scope of this research.

Furthermore, a basic assumption of coupled epipolar rectification is that the horizontal disparity between the epipolar images of NAC-L and NAC-R should be small if not zero. This assumption can be visually validated in the anaglyphs in Figure 10.9 because, except for the intensity difference, there are no obvious geometric inconsistencies between NAC-L and NAC-R. In addition, the horizontal disparity and vertical disparity are validated by the subpixel tie points matching using LPS. The statistics of the matching results are recorded in Table 10.4. It should be noted that for the epipolar images of NAC-L and NAC-R in the same track, the horizontal disparity does not exceed 1 pixel and the root mean squared error (RMSE) is less than 0.3 pixels, which validates the assumption of coupled epipolar rectification. As expected, the RMSE for the vertical disparity is almost 0.1 pixels for the coupled rectification, which indicates that a bottleneck lies in the subpixel capabilities of least-squares matching (Hu, Ding *et al.*, 2016).

For the topographic mapping, the sampling distance for the NAC-DEM is set to 2 m, which is a commonly used DEM sampling strategy at three to five times the original image resolution. The matching results are shown in Figure 10.10, in which the disparity maps of the coupled epipolar images for the NACs of the first orbiter are shown in Figure 10.10(a) and the shaded relief map of the DEM is shown in Figure 10.10(b). In addition, two profile comparisons between SLDEM and NAC-DEM are shown in Figure 10.10(c) and (d). It should be noted that the NAC-DEM has more detailed information than the SLDEM and that the crater is obviously deeper than that of the SLDEM, probably because the SLDEM has few points sampled in the crater and the DEM

Table 10.3 Additional affine parameters for the first dataset after the combined block adjustment

Image	m_{11}	m_{12}	m_{13}	m_{21}	m_{22}	m_{23}
M173246166L	1.0	0.0	0.0	0.0	1.0	0.0
M173246166R	1.0	-1.9×10^{-6}	−73.2	-6.9×10^{-5}	1.0	27.4
M173252954L	1.0	3.0×10^{-4}	118.8	3.0×10^{-4}	1.0	−17.7
M173252954R	1.0	3.0×10^{-4}	46.5	-8.6×10^{-5}	1.0	18.8

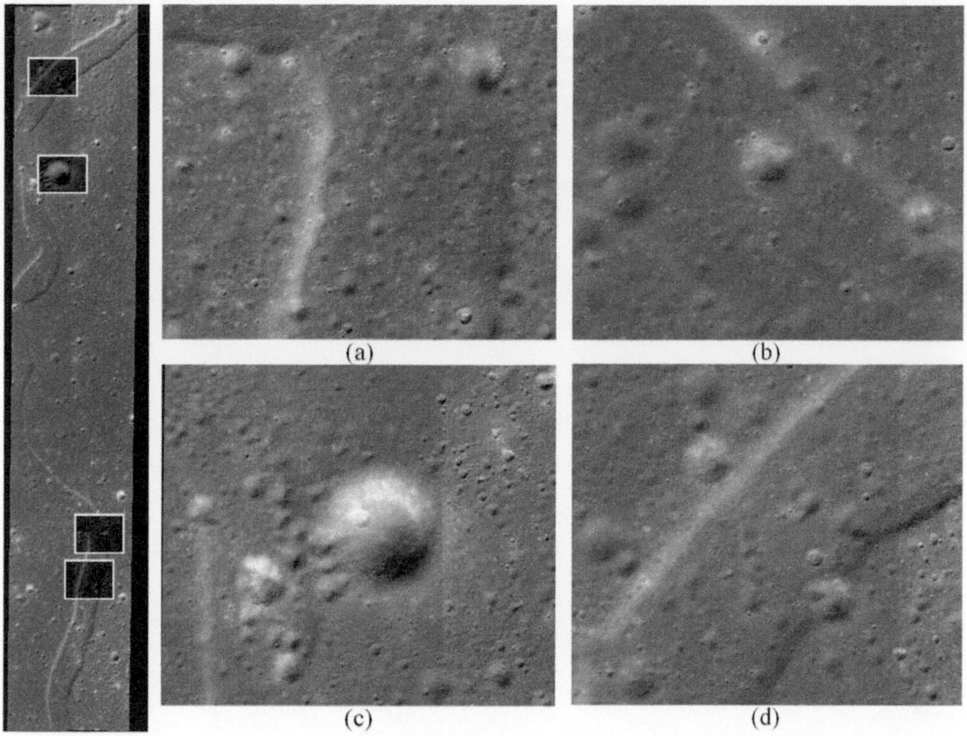

Figure 10.9. Anaglyphs of the whole epipolar image and selected regions. The enlarged views through (a) to (d) indicate the blue shaded boxes in the left thumbnails from bottom to top. The color bias in the middle of the anaglyphs is caused by the color inconsistency between NAC-L and NAC-R. The color modes of the anaglyphs are red/left to cyan/right. Please wear the appropriate filter glasses to observe the stereo effects.

Table 10.4 Statistics of the tie points on the epipolar image of the lunar rille area. The first two rows indicate the results for the epipolar images on the same track and the last row for the results of the coupled rectification.

Image pair	*Horizontal (pixels)*			*Vertical (pixels)*			*# points*
	Min	*Max*	*RMSE*	*Min*	*Max*	*RMSE*	
M173246166(L-R)	−0.68	0.73	0.27	−0.56	0.16	0.27	49
M173252954(L-R)	−0.69	0.77	0.26	−0.49	0.44	0.19	113
Coupled rectification	N/A			−0.21	0.31	0.09	453

is excessively interpolated, and thus the depth is shallower. Furthermore, we notice a slight shift, about 200 m in the horizontal direction, between the SLDEM and NAC-DEM, which is due to the systematic errors in the EO parameters of the NAC images. This phenomenon was also observed in a previous study (Tran *et al.*, 2010). The shift can be compensated using a translational model through several tie points interactively identified on the two DEMs.

6.2 *Experiment on NAC images covering Dorsa Whiston*

The second NAC dataset depicts a region of the Dorsa Whiston, which is a wrinkle ridge system in the Oceanus Procellarum on the Moon, as shown in the SLDEM and NAC images in the left

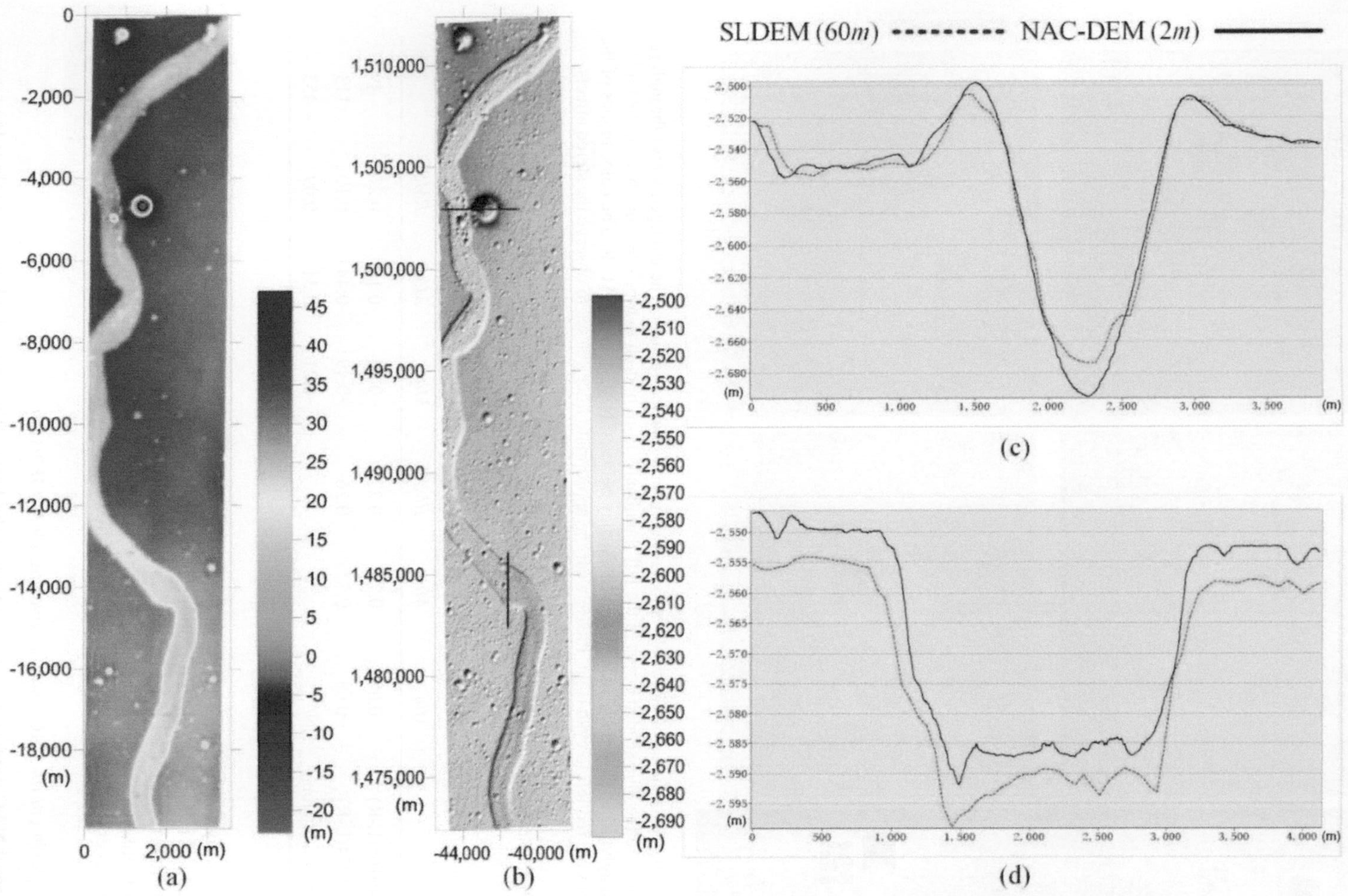

Figure 10.10. Disparity map and DEM of the stereo pair. (a) Disparity map of the coupled rectification images; (b) shaded relief map of the DEM, the elevation is exaggerated by five times to better reveal the terrain surface and relief shading effect; (c) and (d) the profile comparisons of the horizontal and vertical red lines in (b), respectively, between the SLDEM and NAC-DEM. The metrics in the legend indicate the resolution of the corresponding DEM.

part of Figure 10.11. Wrinkle ridges, which are a common feature on lunar maria, are assumed to have been created when the basaltic lava cooled and contracted. In addition to the wrinkle ridge system, the stereo image pair features a series of continuous craters, as shown in the subset of the NAC images in the right part of Figure 10.11. The information of the NAC images of the two tracks are summarized in Table 10.5. It should be noted that the resolutions of the two images are around 1.2 m and the convergent angle of the two orbits is about 12°. With this configuration, the theoretical measurement accuracy in the vertical direction is about 6 m. All the images have the same dimension of 5064 x 52,224 as the previous dataset. Although the convergent angle of this

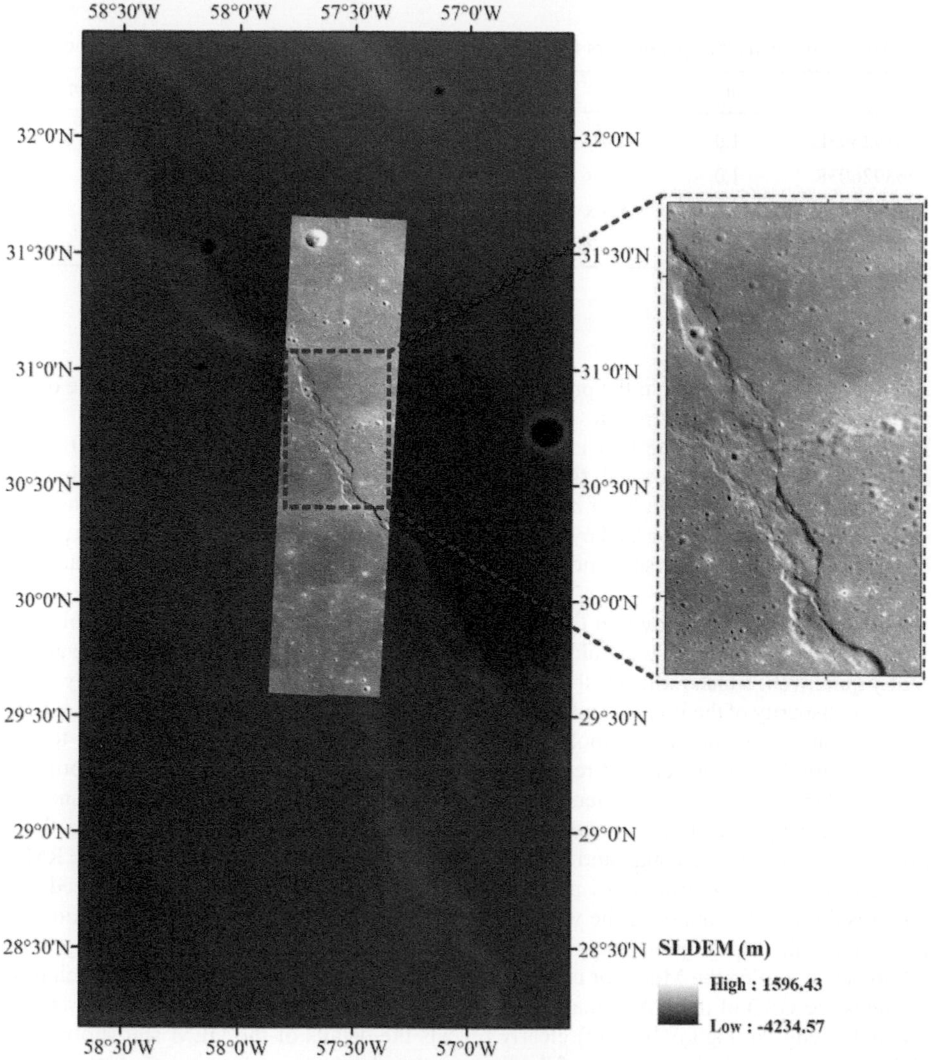

Figure 10.11. The two NAC images of product M1206392603 overlaid on the SLDEM. The images mainly cover part of the long wrinkle ridge, as shown in the background SLDEM, and a series of continuous craters, as shown on the highlighted right part. The legend indicates the elevation in meters of the SLDEM in the background.

Table 10.5 Information on the experimental NAC stereo pair for Dorsa Whiston

Product ID	*Resolution (m)*	*Emission (°)*	*Incidence (°)*
M1206392603L	1.24	18.13	45.91
M1206392603R	1.23	15.18	46.08
M1206406672L	1.20	9.03	47.32
M1206406672R	1.21	11.98	47.49

Table 10.6 Additional affine parameters for the second dataset after the combined block adjustment

Image	m_{11}	m_{12}	m_{13}	m_{21}	m_{22}	m_{23}
M1206392603L	1.0	0.0	0.0	0.0	1.0	0.0
M1206392603R	1.0	-3.4×10^{-6}	−95.4	1.4×10^{-4}	1.0	3.8
M1206406672L	1.0	1.3×10^{-5}	30.8	3.0×10^{-4}	1.0	-2.8×10^{-3}
M1206406672R	1.0	1.7×10^{-5}	−66.1	9.9×10^{-6}	1.0	−1.9

dataset is slightly different from the previous dataset, the overlapping region in the same orbit and between two orbits are quite similar.

The block adjustment of the four images begins with 30,222 object points. After 10 iterations of adjustments and outlier removal, the optimization results in 29,284 valid object points with a standard deviation of unit weight of $\sigma_0 = 0.8$ pixels, which is good enough, considering that the block adjustment is also regularized by the initial affine parameters. The weights and the parameter λ are the same as the first dataset. And the affine parameters after the combined block adjustment are listed in Table 10.6.

The cyan/red anaglyphs shown in Figure 10.12 are used to validate the coupled epipolar rectification for the Dorsa Whiston. It should be noted that the small craters in Figure 10.12(a) align horizontally quite well, which indicates that the combined block adjustment has successfully removed the vertical disparity of the images and that the coupled rectification model is capable of successful approximation of the complex epipolar geometry of the NAC images. Furthermore, the major crater in the top left of the selected region, as shown in Figure 10.12(c), has a more abrupt slope and greater depth than the other craters, as shown in Figure 10.12(a). To quantitatively analyze the quality of the coupled epipolar images, dozens of tie points are extracted separately in the LPS, and the statistics of the horizontal and vertical disparities are reported in Table 10.7. The RMSE of the horizontal disparity is around 0.1 pixels and even the maximum value does not exceed 1 pixel. Similar results are also found for the vertical disparity and the case of coupled rectification.

To evaluate the topographic mapping results, the GSDs of both the epipolar image and the DEM are set at 4 m on the Mercator coordinate system, as shown in Figure 10.13, which is about three times the GSD of the NAC images. The profile comparison of the two adjacent craters on the wrinkled edge in Figure 10.13(c) clearly reveals the details of the DEM and shows that the SLDEM presents the distinct effects of the linear interpolation caused by insufficient points. The NAC-DEM preserves much better terrain details. A noticeable vertical discrepancy of about 100 m is presented in Figure 10.13(c). This problem is caused by the initial EO parameters of the NAC images, because only free network block adjustment is used. Similar differences in the terrain details can also be observed in Figure 10.13(d), in which the terrain features in the SLDEM are smoothed and some small craters are even omitted.

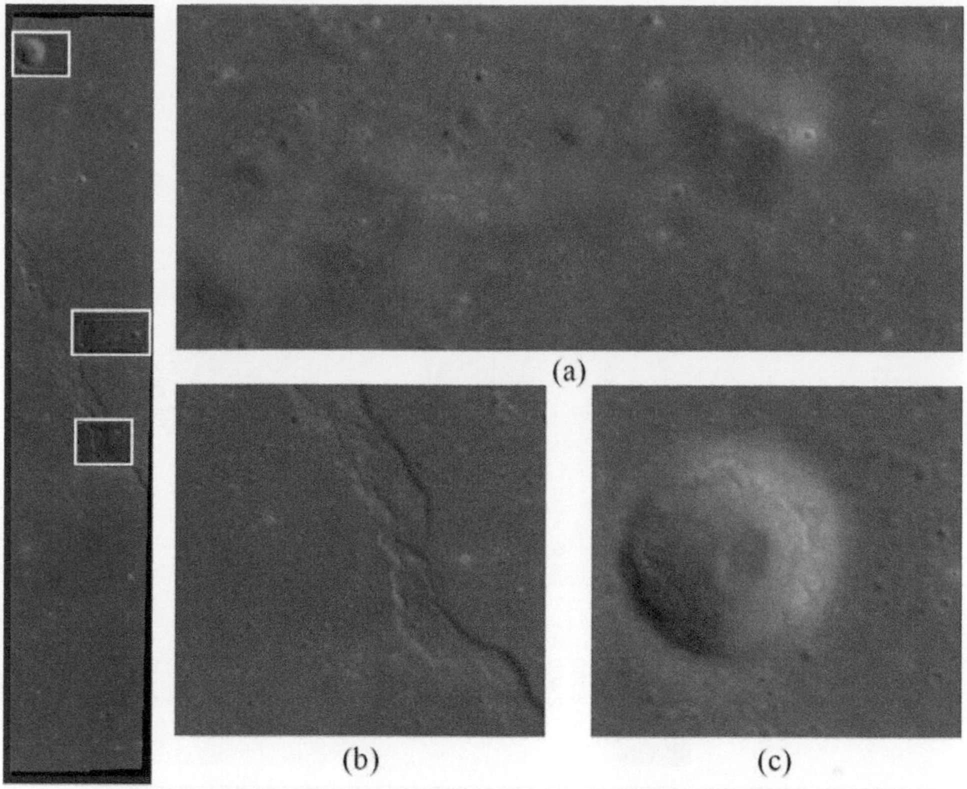

Figure 10.12. Anaglyphs of the whole epipolar image and selected regions. The enlarged views in (a) through (c) indicate the blue shaded boxes in the left thumbnails. The color modes of the anaglyphs are red/left to cyan/right. Please wear the appropriate filter glasses to observe the stereo effects.

Table 10.7 Statistics of the tie points on the epipolar image for Dorsa Whiston. The first two rows show the results for the epipolar image of the same track, and the last row shows the results of the coupled rectification.

Image pair	*Horizontal (pixels)*			*Vertical (pixels)*			*# points*
	Min	*Max*	*RMSE*	*Min*	*Max*	*RMSE*	
M1206392603(L-R)	−0.47	0.23	0.10	−0.15	0.22	0.06	70
M1206406672(L-R)	−0.43	0.15	0.09	−0.50	0.08	0.31	60
Coupled Rectification	N/A			−0.70	0.38	0.26	254

7 CONCLUSIONS

In this chapter, we present an effective photogrammetric processing pipeline for precision and high-resolution topographic mapping using LROC NAC images. The boresight calibration estimates the intra-track boresight offsets, which reduces the intra-track inconsistencies. Using the calibrated boresight parameters as initial values, the block adjustment effectively removes the internal inconsistencies among all the four NAC images. Furthermore, the proposed method resamples the NAC-L and NAC-R images from the same platform with two different lenses bundled

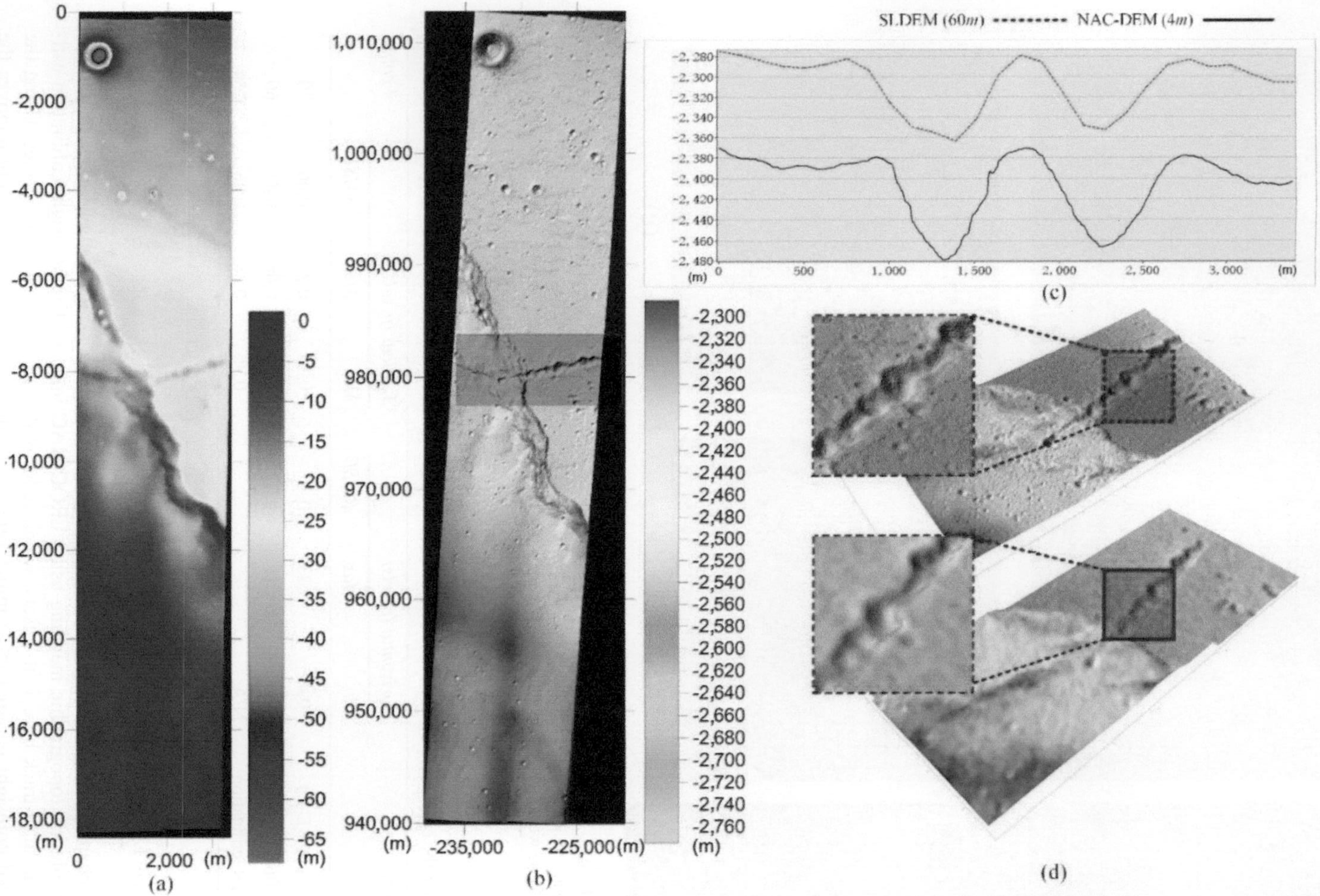

Figure 10.13. Disparity map and DEM of the stereo pair for Dorsa Whiston. (a) Disparity map of the coupled rectification images; (b) shaded relief map of the DEM (elevation is exaggerated by five times to better reveal the terrain surface and relief shading effect; (c) profile comparison of the red line in (b); and (d) enlarged views of the 3D surface for NAC-DEM (top) and SLDEM (bottom). The metrics in the legend indicate the resolution of the corresponding DEM.

into the same epipolar image. In this way, the number of stereo models required to reconstruct the entire overlapping area is reduced from three or four to only one. The basic assumption that only a negligible horizontal disparity exists between the epipolar images of NAC-L and NAC-R is confirmed by the tie points in the overlapping area. As evidenced by the anaglyphs and DEMs in the experimental analysis, no inconsistencies or noises are visually noticeable in the overlapping area between the NAC-L and NAC-R, which demonstrates that the block adjustment and coupled epipolar rectification model are tenable. Comparison of the SLDEM and NAC-DEM reveals that the details of the latter significantly exceed those of the former.

The approach presented in this chapter can facilitate high-resolution LROC NAC images to be used for high-quality 3D topographic mapping of the lunar surface. The approach and the generated results can also be used in a variety of applications, such as landing site mapping and selection, rover maneuvering, and geological analyses of the lunar surface that require high-quality topographic data.

ACKNOWLEDGEMENTS

This work was supported by a grant from the Research Grants Council of Hong Kong (Project No. PolyU 152086/15E) and grants from the Hong Kong Polytechnic University (1-ZEA4, 4-BCBW, and G-YBA5). The authors would like to thank the people who worked on the Planetary Data System archive for the LRO datasets to make the LROC NAC imagery publicly available.

REFERENCES

Acton Jr, C.H. (1996) Ancillary data services of NASA's navigation and ancillary information facility. *Planetary and Space Science*, 44(1), 65–70.

Agarwal, S. & Mierle, K. (2010) *Ceres solver*. Available from: http://ceres-solver.org/, [accessed: 30 August 2016].

Agarwal, S., Furukawa, Y., Snavely, N., Simon, I., Curless, B., Seitz, S.M. & Szeliski, R. (2011) Building rome in a day. *Communications of the ACM*, 54(10), 105–112.

Barker, M.K., Mazarico, E., Neumann, G.A., Zuber, M.T., Haruyama, J. & Smith, D.E. (2016) A new lunar digital elevation model from the Lunar Orbiter Laser Altimeter and SELENE Terrain Camera. *Icarus*, 273, 346–355.

Barnes, C., Shechtman, E., Finkelstein, A. & Goldman, D. (2009) PatchMatch: a randomized correspondence algorithm for structural image editing. *ACM Transactions on Graphics-TOG*, 28(3), 24.

Burns, K.N., Speyerer, E.J., Robinson, M.S., Tran, T., Rosiek, M.R., Archinal, B.A. & Howington-Kraus, E. (2012) Digital elevation models and derived products from LROC NAC stereo observations. *22nd Congress of the International Society for Photogrammetry and Remote Sensing, ISPRS 2012*, Melbourne, Australia.

Cook, A.C., Oberst, J., Roatsch, T., Jaumann, R. & Acton, C. (1996) Clementine imagery: selenographic coverage for cartographic and scientific use. *Planetary and Space Science*, 44(10), 1135–1148.

de Franchis, C., Meinhardt-Llopis, E., Michel, J., Morel, J. & Facciolo, G. (2014) On stereo-rectification of pushbroom images. *2014 IEEE International Conference on Image Processing (ICIP)*, Paris, France. pp. 5447–5451.

Di, K., Liu, Y., Liu, B., Peng, M. & Hu, W. (2014) A self-calibration bundle adjustment method for photogrammetric processing of Chang'E-2 Stereo Lunar Imagery. *IEEE Transactions on Geoscience and Remote Sensing*, 52(9), 5432–5442.

Fraser, C. & Ravanbakhsh, M. (2009) Georeferencing performance of GEOEYE-1. *Photogrammetric Engineering and Remote Sensing*, 75(6), 634–638.

Furukawa, Y. & Ponce, J. (2010) Accurate, dense, and robust multiview stereopsis. *IEEE Transactions on Pattern Analysis and Machine Intelligence*, 32(8), 1362–1376.

Fusiello, A., Trucco, E. & Verri, A. (2000) A compact algorithm for rectification of stereo pairs. *Machine Vision and Applications*, 12(1), 16–22.

Garvin, J.B., Sakimoto, S., Schnetzler, C. & Frawley, J.J. (1999) Global geometric properties of Martian impact craters: a preliminary assessment using Mars Orbiter Laser Altimeter (MOLA). *The Fifth International Conference on Mars*, Pasadena, CA, USA.

Grodechi, J. & Dial, G. (2003) Block adjustment of high-resolution satellite images described by rational polynomials. *Photogrammetric Engineering and Remote Sensing*, 69(1), 59–68.

Hastie, T., Tibshirani, R. & Friedman, J. (2009) *The Elements of Statistical Learning*, Springer, Berlin, Germany. 763 p.

Henriksen, M.R., Manheim, M.R., Burns, K.N., Seymour, P., Speyerer, E.J., Deran, A., Boyd, A.K., Howington-Kraus, E., Rosiek, M.R. & Archinal, B.A. (2017) Extracting accurate and precise topography from LROC narrow angle camera stereo observations. *Icarus*, 283, 122–137.

Hirschmuller, H. (2008) Stereo processing by semiglobal matching and mutual information. *IEEE Transactions on Pattern Analysis and Machine Intelligence*, 30(2), 328–341.

Hu, H., Chen, C., Wu, B., Yang, X., Zhu, Q. & Ding, Y. (2016) Texture-aware dense image matching using ternary census transform. *ISPRS Annals of the Photogrammetry, Remote Sensing and Spatial Information Sciences*, Prague, Czech Republic, Volume III-3. pp. 59–66.

Hu, H., Ding, Y., Zhu, Q., Wu, B., Xie, L. & Chen, M. (2016) Stable least-squares matching for oblique images using bound constrained optimization and a robust loss function. *ISPRS Journal of Photogrammetry and Remote Sensing*, 118, 53–67.

Kim, T. (2000) A study on the epipolarity of linear pushbroom images. *Photogrammetric Engineering and Remote Sensing*, 66(8), 961–966.

Kolmogorov, V. & Zabih, R. (2001) Computing visual correspondence with occlusions using graph cuts. *9th IEEE International Conference on Computer Vision (ICCV2001)*, 7–14 July, Vancouver, British Columbia, Canada. pp. 508–515.

Li, R., Hwangbo, J., Chen, Y. & Di, K. (2011) Rigorous photogrammetric processing of HiRISE stereo imagery for Mars topographic mapping. *IEEE Transactions on Geoscience and Remote Sensing*, 49(7), 2558–2572.

Loop, C. & Zhengyou, Z. (1999) Computing rectifying homographies for stereo vision. *IEEE Computer Society Conference on Computer Vision and Pattern Recognition,* Ft. Collins, CO, USA, Volume 1. p. 131.

Mattson, S., Ojha, L., Ortiz, A., McEwen, A.S. & Burns, K. (2012) Regional digital terrain model production with LROC-NAC. *Lunar and Planetary Science Conference*, 19–23 March, The Woodlands, TX, USA.

Oberst, J., Scholten, F., Matz, K., Roatsch, T., Wählisch, M., Haase, I., Gläser, P., Gwinner, K., Robinson, M.S. & Team, L. (2010) Apollo 17 landing site topography from LROC NAC stereo data – first analysis and results. *Lunar and Planetary Science Conference*, The Woodlands, TX, USA, p. 2051.

Oh, J., Lee, W.H., Toth, C.K., Grejner-Brzezinska, D.A. & Lee, C. (2010) A piecewise approach to epipolar resampling of pushbroom satellite images based on RPC. *Photogrammetric Engineering and Remote Sensing*, 76(12), 1353–1363.

Robinson, M.S., Speyerer, E.J., Boyd, A., Waller, D., Wagner, R.V. & Burns, K.N. (2012) Exploring the Moon with the lunar reconnaissance orbiter camera. *International Archives of the Photogrammetry, Remote Sensing & Spatial Information Sciences*, 39. p. B4.

Robinson, M.S., Brylow, S.M., Tschimmel, M., Humm, D., Lawrence, S.J., Thomas, P.C., Denevi, B.W., Bowman-Cisneros, E., Zerr, J. & Ravine, M.A. (2010) Lunar Reconnaissance Orbiter Camera (LROC) instrument overview. *Space Science Reviews*, 150(1), 81–124.

Scharstein, D. & Szeliski, R. (2002) A taxonomy and evaluation of dense two-frame stereo correspondence algorithms. *International Journal of Computer Vision*, 47(1), 7–42.

Scharstein, D., Szeliski, R. & Hirschmüller, H. (2017) *Middlebury Stereo Evaluation – Version 3*. Available from: http://vision.middlebury.edu/stereo/eval3/ [accessed: 19 December 2017].

Smith, D.E., Zuber, M.T., Frey, H.V., Garvin, J.B., Head, J.W., Muhleman, D.O., Pettengill, G.H., Phillips, R.J., Solomon, S.C. & Zwally, H.J. (2001) Mars Orbiter Laser Altimeter: experiment summary after the first year of global mapping of Mars. *Journal of Geophysical Research: Planets*, 106(E10), 23689–23722.

Smith, D. E., Zuber, M. T., Jackson, G. B., Cavanaugh, J. F., Neumann, G. A., Riris, H., Sun, X., Zellar, R. S., Coltharp, C., Connelly, J., Katz, R. B., Kleyner, I., Liiva, P., Matuszeski, A., Mazarico, E. M., McGarry, J. F., Novo-Gradac, A., Ott, M. N., Peters, C., Ramos-Izquierdo, L. A., Ramsey, L., Rowlands, D. D., Schmidt, S., Scott, V. S., Shaw, G. B., Smith, J. C., Swinski, J., Torrence, M. H., Unger, G., Yu, A. W. & Zagwodzki, T. W. (2010). The Lunar Orbiter Laser Altimeter Investigation on the Lunar Reconnaissance Orbiter Mission. Space Science Reviews, 150(1-4), 209–241.

Speyerer, E.J., Wagner, R.V., Robinson, M.S., Licht, A., Thomas, P.C., Becker, K., Anderson, J., Brylow, S.M., Humm, D.C. & Tschimmel, M. (2016) Pre-flight and on-orbit geometric calibration of the Lunar Reconnaissance Orbiter Camera. *Space Science Reviews*, 200(1–4), 357–392.

Tran, T., Rosiek, M.R., Howington-Kraus, E., Archinal, B.A., Anderson, E. & Team, L.S. (2010) *Generating Digital Terrain Models Using LROC NAC Images*. ASPRS/CaGIS 2010 Special Conference, Orlando, FL, USA, pp. 15–19.
Wang, M., Hu, F. & Li, J. (2011) Epipolar resampling of linear pushbroom satellite imagery by a new epipolarity model. *ISPRS Journal of Photogrammetry and Remote Sensing*, 66(3), 347–355.
Watters, W.A., Geiger, L.M., Fendrock, M. & Gibson, R. (2015) Morphometry of small recent impact craters on Mars: size and terrain dependence, short-term modification. *Journal of Geophysical Research: Planets*, 120(2), 226–254.
Wu, B., Hu, H. & Guo, J. (2014) Integration of Chang'E-2 imagery and LRO laser altimeter data with a combined block adjustment for precision lunar topographic modeling. *Earth and Planetary Science Letters*, 391, 1–15.
Wu, B., Zhang, Y. & Zhu, Q. (2011) A triangulation-based hierarchical image matching method for wide-baseline images. *Photogrammetric Engineering & Remote Sensing*, 77(7), 695–708.
Wu, B., Zhang, Y. & Zhu, Q. (2012) Integrated point and edge matching on poor textural images constrained by self-adaptive triangulations. *ISPRS Journal of Photogrammetry and Remote Sensing*, 68, 40–55.
Wu, B., Guo, J., Hu, H., Li, Z. & Chen, Y. (2013) Co-registration of lunar topographic models derived from Chang'E-1, SELENE, and LRO laser altimeter data based on a novel surface matching method. *Earth and Planetary Science Letters*, 364, 68–84.
Wu, B., Li, F., Ye, L., Qiao, S., Huang, J., Wu, X. & Zhang, H. (2014) Topographic modeling and analysis of the landing site of Chang'E-3 on the Moon. *Earth and Planetary Science Letters*, 405, 257–273.
Zhu, Q., Wan, N. & Wu, B. (2007) A filtering strategy for interest point detecting to improve repeatability and information content. *Photogrammetric Engineering & Remote Sensing*, 73(5), 547–553.
Zhu, Q., Zhang, Y., Wu, B. & Zhang, Y. (2010) Multiple close-range image matching based on a self-adaptive triangle constraint. *The Photogrammetric Record*, 25(132), 437–453.

Planetary Remote Sensing and Mapping – Wu et al.

Chapter 11

Mercury stereo topographic mapping using data from the MESSENGER orbital mission

F. Preusker, J. Oberst, A. Stark and S. Burmeister

ABSTRACT: We produce high-resolution (222 m grid element-1) Digital Terrain Models (DTMs) for Mercury using stereo images from the MESSENGER orbital mission. We have developed a scheme to process large numbers of images (typically more than 6000) by photogrammetric techniques, which include, multiple image matching, pyramid strategy, and bundle block adjustment. In this paper, we present models for map quadrangles of the northern hemisphere H3, H5, H6, and H7. The models are demonstrated to be in excellent agreement with data from MESSENGER's onboard laser altimeter.

1 INTRODUCTION

Shape and morphology constitute basic geodetic data for any planet. This is particularly true for Mercury, the smallest among the planets (Fig. 11.1). Mercury moves around the Sun in 88 Earth days and is locked in a 3:2 spin-orbit resonance. This implies that the planet rotates exactly three times, as it orbits the Sun twice. Owing to the Mercury's eccentric orbit, solar distance and thermal conditions on the surface (and on a spacecraft in the proximity) vary drastically during the planet's year.

NASA's Mariner 10 performed three flybys of Mercury in 1974 and 1975 and returned the first close-up images, covering about 45% of the planet. The MESSENGER (MErcury Surface, Space ENvironment, GEochemistry, and Ranging) spacecraft was launched in 2004 and performed three flybys (two in 2008 and one in 2009) before entering orbit around the planet in March 2011 to begin a comprehensive remote sensing mission (Solomon *et al.*, 2001, 2008), until the end of April 2015.

MESSENGER moved in a high-eccentricity polar orbit. During its first year, the spacecraft periapsis altitude was as low as 200 km at high northern latitudes, but as large as 15,300 km at high southern latitudes with a spacecraft orbit period of 12 h. Thus, consecutive orbit tracks are separated by 131 km (approximately 3.07°) near the equator. In April 2012 the orbit period was reduced to 8 h, which reduced the separation of consecutive orbit tracks to approximately 2.05° (~87 km near the equator). Furthermore with the new orbit, the eccentricity was reduced, with the result that the spacecraft periapsis altitude was approximately 278 km and spacecraft apoapsis altitude moved to approximately 10,300 km.

MESSENGER was equipped with the Mercury Dual Imaging System (MDIS), which consisted of two framing cameras, a wide-angle camera (WAC) and a narrow-angle camera (NAC), co-aligned on a pivot platform and equipped with identical 1024-x-1024-pixel charge-coupled device (CCD) sensors (Hawkins *et al.*, 2007). The WAC features 11 narrow-band filters from visible to near-infrared wavelengths and a broadband clear filter. We used images taken by the WAC filter 7 (WAC-G), which, at 750 nm (orange), is similar to performance and sensitivity of the panchromatic NAC. Both cameras consist of a compact off-axis optical system that has been geometrically calibrated using laboratory as well as in-flight data (Hawkins *et al.*, 2007). The harsh thermal environment of Mercury requires sophisticated models for calibrations of focal length and

Figure 11.1. Color image of Mercury obtained by MESSENGER on its flyby at Mercury in 2008 (From: © NASA/JHUAPL/Carnegie Institution of Washington)

distortion of the camera. In particular, the WAC camera and NAC camera were demonstrated to show a linear increase in focal length by up to 0.10% over the typical range of temperatures (−20 to +20 °C) during operation, which causes a maximum displacement of up to 1 pixel. Following methods described earlier (Oberst *et al.*, 2011) the focal length dependencies and geometric distortions for WAC and NAC were modeled using observations of star fields in different temperature regimes of the MESSENGER orbit (Denevi *et al.*, 2017). During the mission, MDIS acquired more than 200,000 images, most of them in the Mercury orbital phase. Owing to the eccentric orbit of the spacecraft with a pericenter at high northern latitudes an imaging strategy was chosen which combined the use of the WAC-G and NAC camera to cover both hemispheres at similar resolutions.

Several complementary techniques have been used to study Mercury's topography with MESSENGER data, including laser altimetry (Zuber *et al.*, 2012), measurements of radio occultation times (Perry *et al.*, 2015), and limb profiling (Elgner *et al.*, 2014). Owing to the character of the orbit (see above), the MLA (Mercury Laser Altimeter) could not cover the southern hemisphere of the planet. Hence, radio occultation and limb data were used to complement the limited MLA coverage to produce global planetary topography models. However, the models are not suited for morphological studies because of their limited resolution.

Topographic models on regional and local scales were derived with stereo photogrammetry using images from the early Mercury flybys and in MESSENGER's orbit phase (Oberst *et al.*, 2010; Preusker *et al.*, 2011, 2017). High-resolution models require the combination of larger numbers of images as well as a carefully designed data processing strategy to maximize data product quality within the constraints of data processing resources. In this paper, we demonstrate the production of gridded models, which have resolutions as high as 222 meters per grid element, and which involve more than 6000 input images. We also demonstrate the good agreement with the MLA data. The DTMs complement MLA data nicely, because of their high geometric rigidity over large areas.

2 PHOTOGRAMMETRIC PROCESSING

The construction of the DTMs follows procedures we have previously used for Mercury (Oberst *et al.*, 2010; Preusker *et al.*, 2011, 2017), which comprises five main tasks (see sections 2.2–2.6).

2.1 *Reference frame and map scheme*

All computations are carried out in the Mercury-fixed "MESSENGER" reference frame, which is defined through a parameter set through estimates of rotation rate from MESSENGER radio tracking (Mazarico *et al.*, 2014) as well as through measurements of spin axis orientation and librations from Earth-based radar observations (Margot *et al.*, 2012). While recently updated rotation parameters are available (Stark, Oberst, Hussmann *et al.*, 2015; Stark, Oberst, Preusker *et al.*, 2015), a decision was made to use the above mentioned frame to maintain consistency with other MESSENGER data products. The longitude grid for Mercury is defined through the crater Hun Kal, located at 340° E, compliant with the IAU definition (Archinal *et al.*, 2011, 2017). By combination of laser altimetry and radio occultation data, the mean radius of Mercury was derived as 2439.4 km (Perry *et al.*, 2015), which is used as a reference for our topographic models.

We use MESSENGER nominal orbit, pointing, and instrument alignment data, as provided by the mission project (http://naif.jpl.nasa.gov/pub/naif/pds/data/mess-e_v_h-spice-6-v1.0/messsp_1000/). All ancillary data are typically provided in the format of SPICE kernels (Acton, 1996).

2.2 *Image selection*

In order to manage the complexity and challenges of the global surface reconstruction, we chose to derive individual digital terrain models following a scheme of 15 quadrangles (Fig. 11.2)

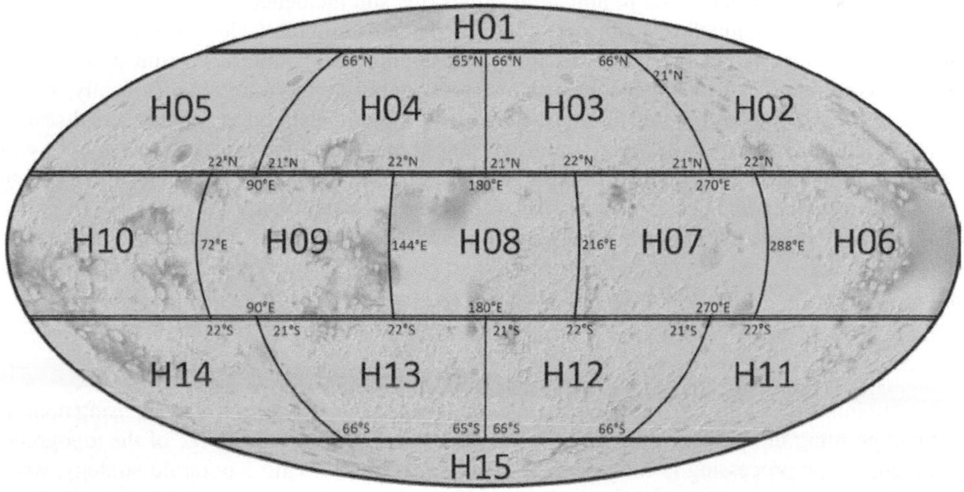

Figure 11.2. Mercury's 15 tiles quadrangle scheme

Table 11.1 Optimal, adequate, and minimal parameter ranges for key stereo observing attributes

Parameter	*Optimal*	*Adequate*	*Minimal*
Illumination variation	0°–10°	0°–10°	0°–10°
Stereo angle	15°, 15°–65°	5°, 15°–65°	5°, 12°–65°
Incidence angle	5°–55°	5°–80°	5°–90°
Emission angle	0°–55°	0°–65°	0°–70°
Sun phase angle	5°–135°	5°–145°	5°–160°

proposed for Mercury (Greeley and Batson, 1990). Using image footprint information, we identify all narrow-angle and wide-angle filter G images that had a resolution between 50 m to 350 m that fall into the area of a quadrangle. The stereo-photogrammetric analyses require a favorable image- and illumination geometry, which affect quantity and quality of matched tie points (see section 2.3) and resulting DTM points. As image coverage, image scale, and illumination conditions vary substantially during the MESSENGER mapping mission, the "quality" of stereo conditions vary accordingly. We use data from our previous MESSENGER image processing (Preusker *et al.*, 2011), to define optimal, adequate, and minimal criteria for surface reconstruction.

To fully exploit the image data by means of stereo-photogrammetric methods, specific requirements to illumination and viewing condition have to be fulfilled. Variable illumination within stereo combinations has a strong influence on accuracy and robustness of multi-image-matching (see section 2.3). Further small stereo angles reduce the quality of the finally reconstructed stereo image geometry. Here the two values for the minimum stereo angle describe the minimum angle between the master image to its pairs and the minimum angle between each pair to each other, respectively. Finally the incidence, emission and sun phase angles have influence on accuracy and robustness of multi-image-matching as well as on the quality of reconstructed image geometry. To optimize the coverage and quality of the final products, we begin in the stereo-photogrammetric processing with areas having images obtained under optimal stereo conditions, while remaining areas are filled with images taken under adequate and finally minimal conditions.

To identify stereo combinations we formed a latitude/longitude grid of 0.1° x 0.1°, in which we identified the images covering each grid element (typically: 3–300 images), for which we computed the relevant stereo and illumination angles (i.e, sun incidence, emission, and sun phase angles for each pair). Only those pairs were considered that had image scales differing by not more than a factor of three. All pairs were tested against the conditions of Table 1. We aim at the identification of "combinations" of images concatenated by favorable stereo conditions. Typically, we find combinations with five to eight members; combinations with less than three images were discarded from the subsequent analysis. Several groups of stereo image combinations extending over wide areas of the grid space are typically identified, which can be combined to "stereo networks". Large stereo networks containing several thousands of images were identified, i.e., networks of images tied through favorable conditions, but not sharing favorable conditions with images of the other network. Smaller stereo networks, having less than 50 images, are typically removed from the analysis as they contributed little new data to the area.

2.3 *Multi-image matching*

Images are initially rectified to a common map projection, i.e., to one common scale, using nominal orbit and pointing information, Mercury's mean radius and a priori knowledge of the topography as reference. The processing is carried out in several stages following a pyramid strategy, where map scale and a priori topography for each pyramid level are selected accordingly. A multi-image matching technique (Wewel, 1996) is applied in order to derive conjugate points in each of the stereo combinations. The algorithm makes use of area-based correlation to derive approximate values for the image coordinates, which are refined to sub-pixel accuracy by least-squares matching. The correlation is done for each master image as the reference image with all stereo partners, i.e. all overlapping images. After the matching, the derived image coordinates are transformed back to raw data geometry (Scholten *et al.*, 2005). For each stereo combination two kinds of tie-point datasets are generated. First, the images are matched in a sparse grid usually about every 20th pixel (used as input for the bundle adjustment), whereas in a second run a pixel-by-pixel grid is produced (used for DTM generation).

2.4 *Bundle block adjustment*

We carry out a least-squares inversion of image tie-point measurements to determine the unknown six camera orientation parameters (three metric parameters for the camera position and three angular parameters for the camera pointing) as well as three coordinates for each tie point in object space. The estimation of the interior orientation of the camera within bundle adjustment was not considered in this work, but is under investigation. The relation between tie-point coordinates and the corresponding surface point is mathematically defined through the so-called collinearity equations (Albertz and Wiggenhagen, 2009). Nominal navigation and reference frame data are used to begin the iterations. Here the nominal spacecraft positions and camera pointing were assumed to be correct within the random errors assumed to be +/−50 m and +/−1.0 mrad, respectively. We expect that any systematic offsets of the spacecraft trajectory from nominal will not affect the characteristics of surface reconstruction beyond overall positioning. Furthermore we estimate that the accuracy of the measured image coordinates was +/−0.2 pixel.

2.5 *Object point calculation and DTM interpolation*

Next, we compute the line of sight for each observation, defined by the image coordinates, the geometric calibration, and the orientation data. Lines of sight for tie points are combined to compute forward ray intersections using least-squares techniques. The redundancy given by multi-stereo capability allows us to accept only those object points that are defined by at least three stereo observations. Thus, we avoid occasional gross matching errors, typical for simple two-image matching. We obtain object points in Cartesian coordinates and their relative accuracies.

To form the DTM we combine the object points of all models. First, all points are transformed from Cartesian to spherical coordinates (latitude/longitude/radius), where height values are computed with respect to Mercury's adopted mean radius of 2439.4 km (see section 2.1.). The latitude/longitude coordinates are transformed to the standard map projection of our quadrangle. The gridded DTM is formed by interpolation, using a distance-weighted mean filtering technique (Gwinner *et al.*, 2009) and a gap-filling algorithm using data from preceding DTM pyramid levels of reduced resolution. Equatorial DTMs are produced in equirectangular projection centered at 0° degree latitude. They cover 7.9% (5.9 x 10^6 km^2) of Mercury's surface, each. Conversely the high-latitude northern quadrangles, such as H3 DTM, are produced in Lambert Conformal projection with the first standard parallel at 30°N and second standard parallel at 58°N. They cover 6.8% (5.08 x 10^6 km^2) of Mercury's surface, each and are thus smaller as equatorial quadrangles. All quadrangle DTMs are produced with a grid spacing of 192 pixels per degree (~222 m pixel^{-1}).

3 RESULTS

3.1 *H6 "Kuiper"*

The H6 Equatorial Quadrangle ("Kuiper", named after the Dutch astronomer) was chosen to produce a prototype model. Like the other equatorial quadrangles, H6 extends from 22.5°S to 22.5°N latitude. H6 ranges from 288°E to 360°E longitude and includes crater Hun Kal (0.5°S, 340°E), which defines Mercury's longitude system. This allows us a verification of correct alignment of our terrain model with the reference frame. Furthermore H6 as an equatorial quadrangle was chosen, as these quadrangles typically combine large numbers of WAC and NAC images, each requiring different (temperature-dependent) calibration schemes. Also, comparisons with MLA tracks are possible which extend to southern latitudes up to 16°S.

Within the H6 quadrangle we selected approximately 10,500 MDIS images (~9000 NAC and ~1500 WAC-G images) that have resolutions better than 350 m pixel^{-1}. We applied our stereo conditions (Table 1) to identify about ~21,000 independent stereo combinations (with at least three images each) which resulted in a connected image block including about 7300 MDIS images (6100 NAC and 1200 WAC-G images). We used a set of 250,000 tie-point observations from coarse image matching (see section 2.3) to adjust the nominal navigation data, i.e., pointing and position data (see section 2.4). From dense image matching (see section 2.3) we derived about 6.3 billion surface points with a mean accuracy of about 50 m. Finally, using interpolation, we generated a gridded DTM centered at 324°E degree longitude (see Fig. 11.3). The vertical accuracy of H6 DTM is about 35 m.

The quadrangle area is geologically rewarding, as it hosts several prominent (> 300 km) impact basins among other "Homer" (1.3°S/323.2°E) and "Sanai" (13.6°S/353.5°E), in addition to large numbers of large craters and tectonic scarps (see Fig. 11.3). The H6 area was already covered by images during the flybys by Mariner 10 in 1974 and MESSENGER in 2008, from which stereo topographic models had been obtained (Cook and Robinson, 2000; Preusker *et al.*, 2011). Furthermore the crater Hun Kal was chosen such that the 20°W meridian falls in its center and defined the longitude system of Mercury (Robinson *et al.*, 1999) and it was approved by Preusker *et al.* (2017).

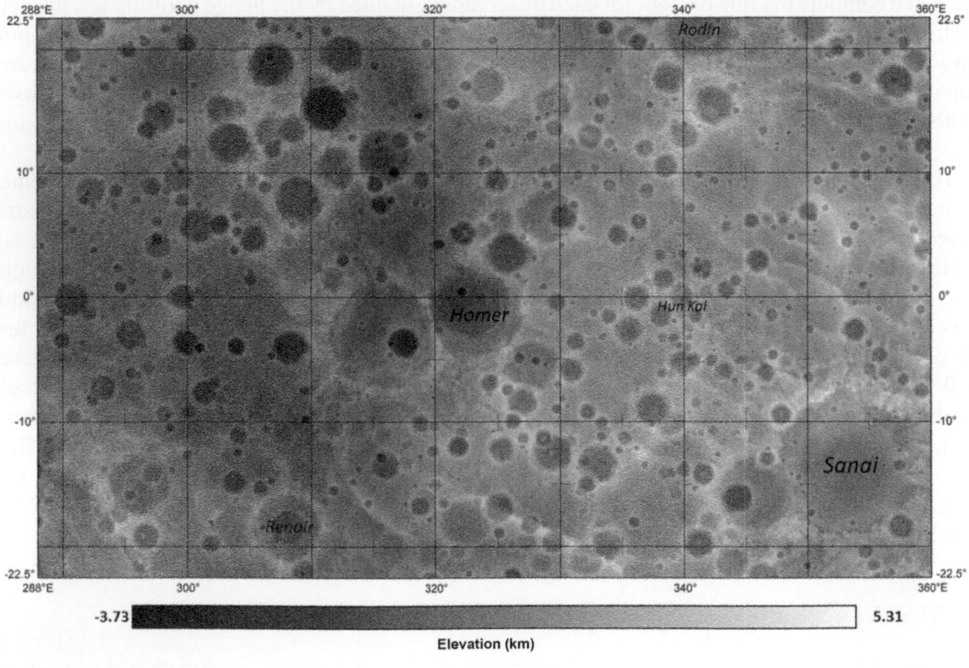

Figure 11.3. H6 "Kuiper" quadrangle DTM

3.2 *H7 "Beethoven"*

The H7 quadrangle is named after German composer Ludwig van Beethoven. As an equatorial quadrangle like H6 "Kuiper", H7 extends from 216°E to 288°E longitude. Thus, it is connected to H6 in the east direction (see Fig. 11.3). Similar to H6, H7 combines large numbers of WAC and NAC images. In total we have selected about 10,800 images (~9800 NAC images and 1000 WAC images) with resolutions ranging from 50 m pixel^{-1} to 350 m pixel^{-1}. We identified about ~28,200 independent stereo combinations which yield a connected block consisting of about 7700 MDIS images (7200 NAC and 500 WAC-G images). For the bundle block adjustment 150,000 tie-point observations were used, which improves the 3D point accuracy from ±920 m to ±50 m. From dense image matching we derived about 3.2 billion surface points with a mean accuracy of about ±50 m. The gridded DTM map is centered at 254°E degree longitude (see Figure 11.4). The vertical accuracy of the H7 DTM is about 35 m.

The H7 DTM comprises a total height range of 10.1 km. This model includes the large (see Fig. 11.4) Beethoven impact basin (~670 km diameter), the Vivaldi basin (~210 km diameter), and the Dürer basin (~190 km diameter).

3.3 *H3 "Shakespeare"*

The H3 quadrangle is named after English writer William Shakespeare. Like the other northern quadrangles H3 extends from 21°S to 66°N latitude. H3 extends from 180°E to 270°E longitude. Thus it comprises the eastern part of Caloris basin, the largest impact basin on Mercury. Furthermore it is connected with the H7 "Beethoven" quadrangle in the south (see Figs. 11.2, 11.4, and 11.5). Contrary to H6 and H7 the number of NAC and WAC images used in the processing is similar for this quadrangle. Here, we have selected about 3800 images (~1800

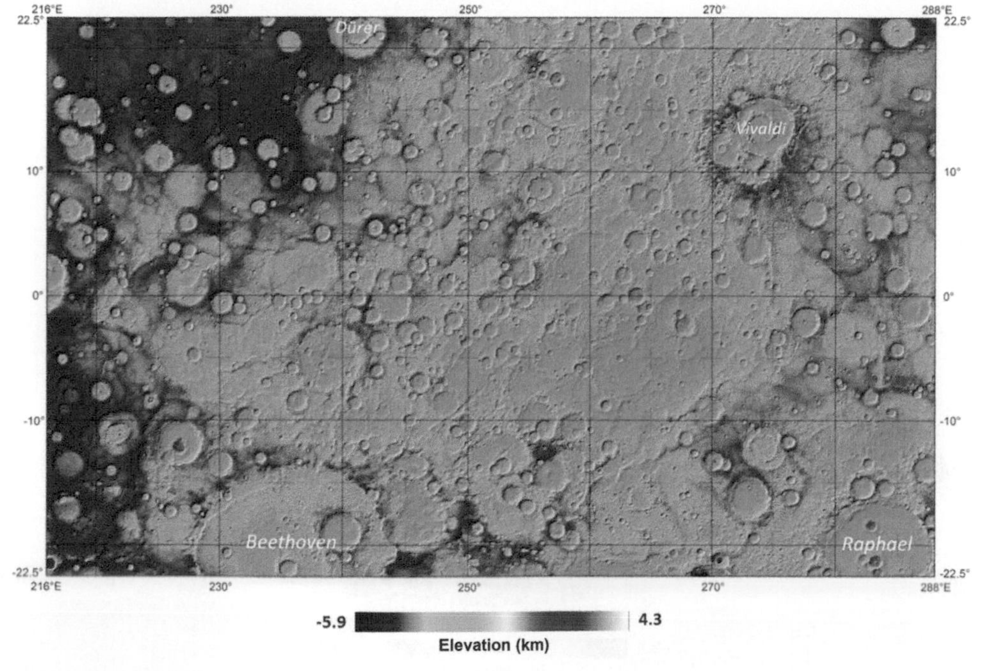

Figure 11.4. H7 "Beethoven" quadrangle DTM

NAC images and 2000 WAC images) that resolutions range from 50 m pixel^{-1} to 350 m pixel^{-1}. About 130,000 tie points were used for the bundle block adjustment, which improves the 3D point accuracy from ±800 m to ±40 m. From about 50,000 individual dense matching runs and forward ray intersection coordinates for about 5.6 billion object points were determined. The mean ray intersection errors of the object points were ±40 m. The final DTM vertical accuracy is about 30 m (see Fig. 11.5).

H3 DTM comprises a total height range of 7.0 km. The model features three large basins (see Fig. 11.5).), van Eyck (~250-km diameter), Shakespeare (~380-km diameter), and Sobkou (~780-km diameter).

3.4 *H5 "Hokusai"*

The H5 northern quadrangle, named after Japanese painter Katsushika Hokusai, ranges from 0°E to 90°E longitude. Thus, the H5 quadrangle is opposite to H3. Image counts are similar to the quadrangle H3 (section 3.3). Within H5 quadrangle we selected about 3000 images (~1000 NAC images and 2000 WAC images) with resolutions ranging from 50 m pixel^{-1} to 350 m pixel^{-1}. About 125,000 tie points were used for the block adjustment, which improves the 3D point accuracy from ±860 m to ±45 m. From about 45,000 individual dense matching runs and forward ray intersection about 4.7 billion object points were computed. The mean ray intersection errors of the object points were ±50 m. The final vertical DTM accuracy is about 35 m (see Fig. 11.6).

The H5 DTM comprises a total height range of 9.3 km. This model features the young Rachmaninoff basin (~310-km diameter) (Fig. 11.7), the floor of which is one of the lowest-elevation areas (about 5 km w.r.t. Mercury's standard sphere) on Mercury.

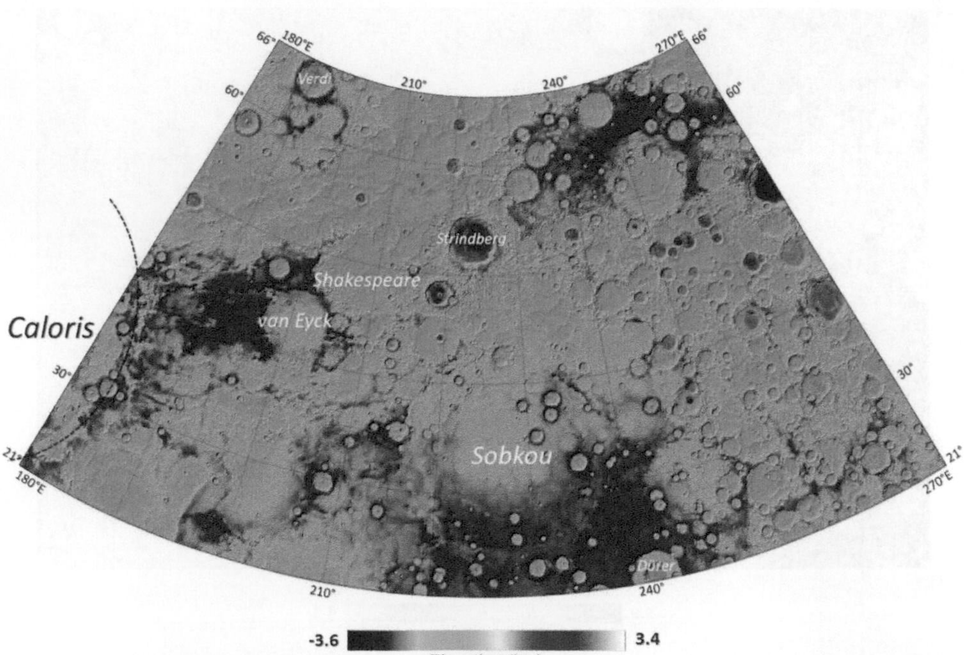

Figure 11.5. H3 "Shakespeare" quadrangle DTM

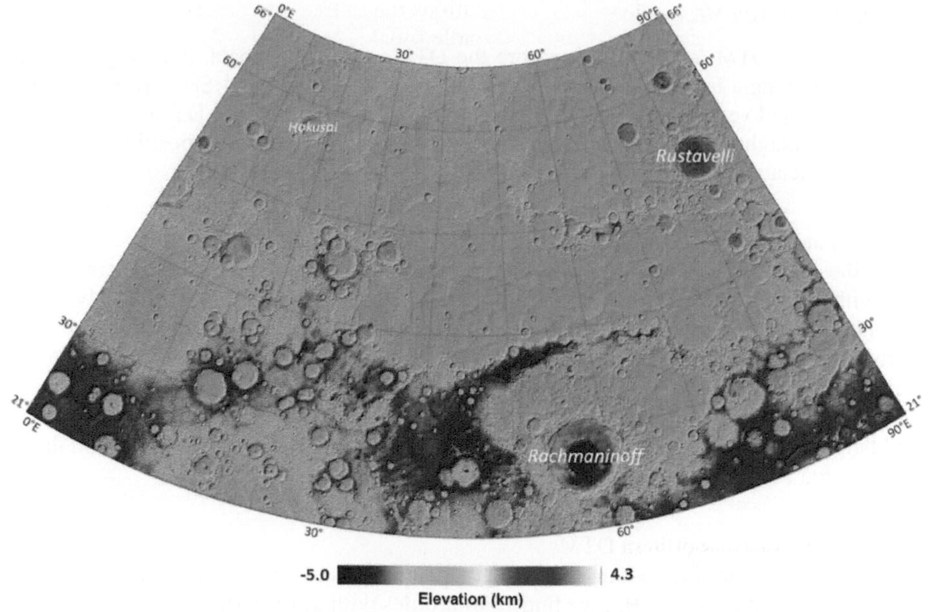

Figure 11.6. H5 "Hokusai" quadrangle DTM

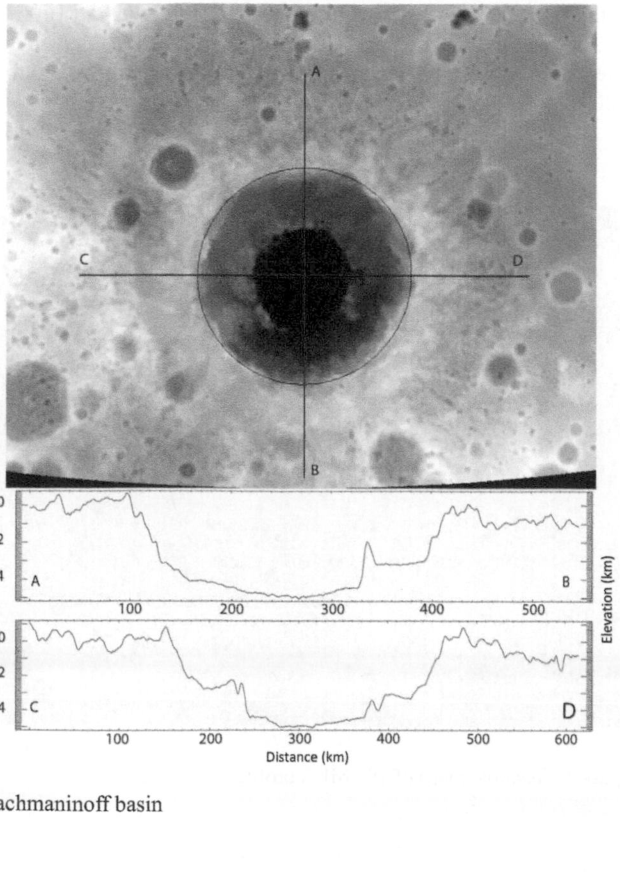

Figure 11.7. The Rachmaninoff basin

3.5 *Comparison with MLA*

In order to assess DTM quality, we compare the DTMs with MLA data. A co-registration was performed following a technique described by Gläser *et al.* (2013) and Stark (2015). Here, the optimum lateral and vertical positions of a laser profile with respect to the DTM are computed by minimizing the height differences between the two datasets. The laser spot position can be determined with an accuracy better than the size of one DTM grid element.

3.5.1 Comparison with equatorial DTMs

Owing to the eccentric spacecraft orbit and high southern apoapsis (cf. above), coverage of MLA profiles within the equatorial quadrangles H6 and H7 is sparse. For H7, we find only 198 laser tracks with at least 100 laser footprints covering the quadrangle area including a near-equatorial laser profile acquired during MESSENGER's second flyby (Fig. 11.8). For H6, there are only 130 profiles with at least 100 laser footprint spots covering the quadrangle area, which include a total of 156,371 spots. MLA and the DTMs show excellent agreement after co-registration. The root mean squared height difference for H6 and H7 are of 88 m and 86 m, respectively, well within the range of the estimated formal accuracy of the DTMs.

3.5.2 Comparison with northern DTMs

The coverage by laser profiles within the northern quadrangles is significantly denser compared to the equatorial quadrangles. For H3, we find 1590 profiles with at least 100 laser footprints, which cover the quadrangle area, involving a total of over four million laser spots. MLA data and the H3 DTM show also a good agreement after co-registration. The root mean squared height difference of 87 m is also within the range of the estimated formal accuracy of the H3 DTM.

The coverage of the H3 DTM by laser profiles is similar. Here, we find 1606 profiles with at least 100 laser footprints. In total about 4.8 million laser spots were selected. MLA and H5 DTM show also a good agreement after co-registration (see Fig. 11.9). The root mean squared height difference of 110 m is about 20% higher compared to H3 DTM.

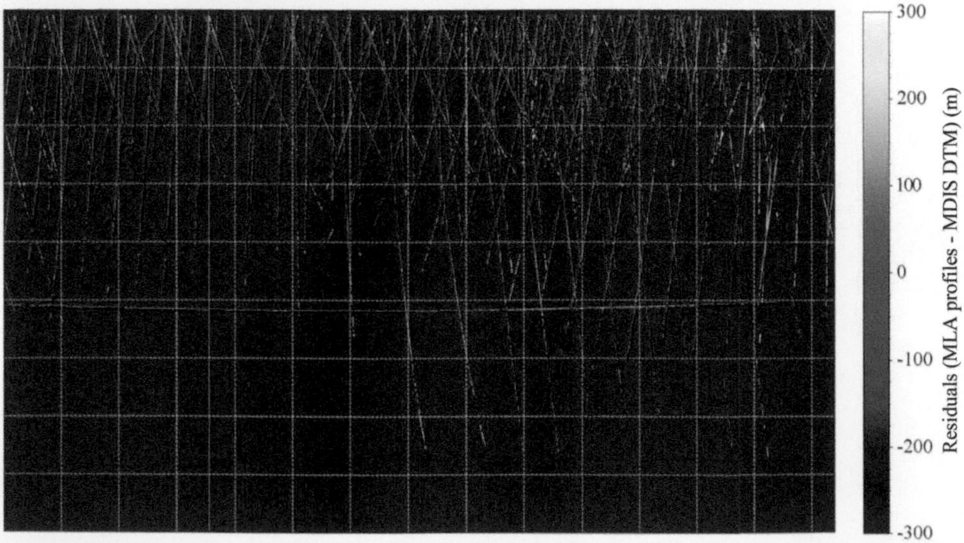

Figure 11.8. Comparison (residual map) of 198 MLA profile with H7 DTM

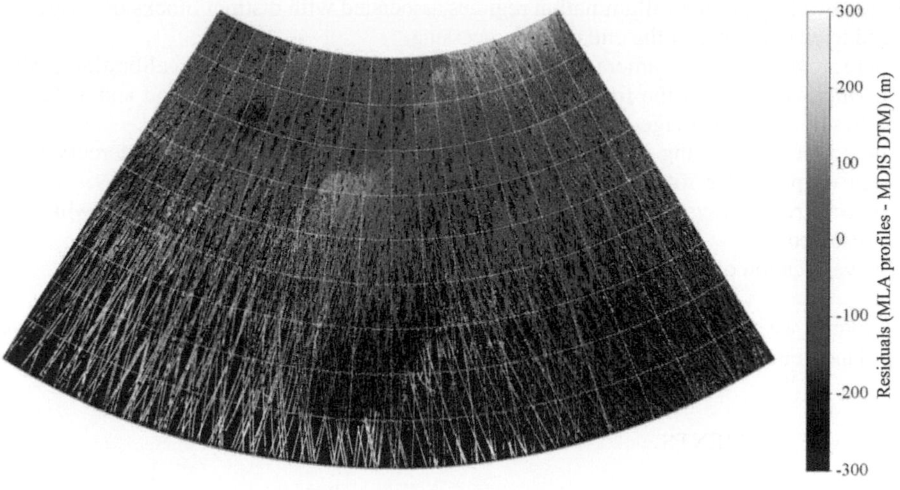

Figure 11.9. Comparison (residual map) of 1606 MLA profile with H5 DTM

4 OUTLOOK: BEPICOLOMBO MISSION

BepiColombo is Europe's first mission to Mercury, scheduled for launch in October 2018 and for arrival in late 2025. A nominal mission of one year is foreseen, with a possible one-year extension. Selected spacecraft instruments include the BepiColombo Laser Altimeter (BELA) as well as a sophisticated camera system SIMBIO-SYS (Spectrometer and Imagers for MPO BepiColombo Integrated Observatory System) (Flamini *et al.*, 2010). The STC channel of SIMBIO-SYS will provide global multispectral coverage of the surface in full stereo at 50 m pixel^{-1} resolution with the aim of characterizing the main geological and morphological units. The main objective of HRIC (High spatial Resolution Imaging Channel) is the characterization of selected surface targets with high resolution. Images will have ground pixel sizes of 5 m pixel^{-1} from altitude of 400 km, for more than 10% of the surface, in four different color bands (Flamini *et al.*, 2010).

The new data obtained by BepiColombo will greatly improve our understanding of planet Mercury as well as the formation and evolution of the terrestrial planets, including Earth. While all instruments have been mounted on the spacecraft the science teams are now anxiously awaiting BepiColombo's launch, this year (2018). The MESSENGER image and topographic data have set the stage for these future investigations.

5 SUMMARY AND DISCUSSION

We have produced Digital Terrain Models for Mercury from stereo images obtained by the MESSENGER mission. The work for every quadrangle represented a unique and very large data sorting and data processing task. In the process, we had to cope with various challenges:

- The high-resolution models require the processing of large numbers of images. Image matching runs require significant processing time. Besides, large image blocks are to be processed by bundle-block adjustments, requiring much computer memory and smart memory allocation schemes.
- The dataset is very heterogeneous in terms of stereo viewing, resolution, and illumination. Therefore, it is often not trivial to find the "optimum" image list.

- Often, we find different illumination regimes associated with distinct blocks of images, which need to be combined at the end of the processing.
- There are two different cameras (WAC/NAC), each with its own distinct calibration challenges.
- The thermal effect on the focal length for WAC is still poorly understood and added careful studies of available image data. Also, a camera modeling may be required.
- The correct referencing of the images still suffers from uncertainties of Mercury reference frame (in particular: rotation model).
- There are residual systematic errors in spacecraft orbit and instrument alignment which have to be resolved.
- The verification of the model by comparisons with MLA data is to be continued.

In the future, we will complete the processing for quadrangles still missing to finally cover the entire planet. Finally, models for all quadrangles will be merged.

ACKNOWLEDGEMENTS

The data DTM are available at: https://pdsimage2.wr.usgs.gov/archive/mess-h-mdis-5-dem-elevation-v1.0/MESSDEM_1001/DEM/QUAD/IMG/.

REFERENCES

Acton, C.H. (1996) Ancillary data services of NASA's navigation and ancillary information facility. *Planetary and Space Science*, 44(1), 65–70. doi:10.1016/0032-0633(95)00107-7.

Albertz, J. & Wiggenhagen, M. (2009) *Taschenbuch zur Photogrammetrie und Fernerkundung*. Wichmann, Heidelberg, Germany.

Archinal, B.A., *et al.* (2011) Report of the IAU Working Group on cartographic coordinates and rotational elements: 2009. *Celestial Mechanics and Dynamical Astronomy*, 109(2), 101–135. doi:10.1007/s10569-010-9320-4.

Archinal, B.A., *et al.* (2017) Report of the IAU Working Group on cartographic coordinates and rotational elements: 2015. *Celestial Mechanics and Dynamical Astronomy*, accepted for publication. Available from: ftp://ftpext.usgs.gov/pub/wr/az/flagstaff/barchinal/WGCCRE/WGCCRE2015preprint.pdf.

Cook, A.C. & Robinson, M.S. (2000) Mariner 10 stereo image coverage of Mercury. *Journal of Geophysical Research: Planets*, 105(E4), 9429–9443. doi:10.1029/1999je001135.

Denevi, B.W., *et al.* (2017) Calibration, projection, and final image products of MESSENGER's Mercury dual imaging system. *Space Science Reviews*, 214(1), 2. doi:10.1007/s11214-017-0440-y.

Elgner, S., Stark, A., Oberst, J., Perry, M.E., Zuber, M.T., Robinson, M.S. & Solomon, S.C. (2014) Mercury's global shape and topography from MESSENGER limb images. *Planetary and Space Science*, 103, 299–308. doi:10.1016/j.pss.2014.07.019.

Flamini, E., *et al.* (2010) SIMBIO-SYS: the spectrometer and imagers integrated observatory system for the BepiColombo planetary orbiter. *Planetary and Space Science*, 58(1–2), 125–143. doi:10.1016/j.pss.2009.06.017.

Gläser, P., Haase, I., Oberst, J. & Neumann, G.A. (2013) Co-registration of laser altimeter tracks with digital terrain models and applications in planetary science. *Planetary and Space Science*, 89, 111–117. doi:10.1016/j.pss.2013.09.012.

Greeley, R. & Batson, R.M. (1990) *Planetary Mapping*, Greeley, R. & Batson, R.M. (eds). Cambridge University Press, Cambridge, England.

Gwinner, K., Scholten, F., Spiegel, M., Schmidt, R., Giese, B., Oberst, J., Heipke, C., Jaumann, R. & Neukum, G. (2009) Derivation and validation of high-resolution digital terrain models from Mars Express HRSC data. *Photogrammetric Engineering & Remote Sensing*, 75(9), 1127–1142.

Hawkins, I., Eduard, S., *et al.* (2007) The Mercury dual imaging system on the MESSENGER spacecraft. *Space Science Reviews*, 131(1–4), 247–338. doi:10.1007/s11214-007-9266-3.

Margot, J.L., Peale, S.J., Solomon, S.C., Hauck, S.A., Ghigo, F.D., Jurgens, R.F., Yseboodt, M., Giorgini, J.D., Padovan, S. & Campbell, D.B. (2012) Mercury's moment of inertia from spin and gravity data. *Journal of Geophysical Research: Planets*, 117(E12), E00L09. doi:10.1029/2012je004161.

Mazarico, E., Genova, A., Goossens, S., Lemoine, F.G., Neumann, G.A., Zuber, M.T., Smith, D.E. & Solomon, S.C. (2014) The gravity field, orientation, and ephemeris of Mercury from MESSENGER observations after three years in orbit. *Journal of Geophysical Research: Planets*, 119(12), 2417–2436. doi:10.1002/2014JE004675.

Oberst, J., Preusker, F., Phillips, R.J., Watters, T.R., Head, J.W., Zuber, M.T. & Solomon, S.C. (2010) The morphology of Mercury's Caloris basin as seen in MESSENGER stereo topographic models. *Icarus*, 209(1), 230–238. doi:10.1016/j.icarus.2010.03.009.

Oberst, J., Elgner, S., Turner, F.S., Perry, M.E., Gaskell, R.W., Zuber, M.T., Robinson, M.S. & Solomon, S.C. (2011) Radius and limb topography of Mercury obtained from images acquired during the MESSENGER flybys. *Planetary and Space Science*, 59(15), 1918–1924. https://doi.org/10.1016/j.pss.2011.07.003.

Perry, M.E., *et al.* (2015) The low-degree shape of Mercury. *Geophysical Research Letters*, 42(17), 6951–6958. doi:10.1002/2015GL065101.

Preusker, F., Oberst, J., Head, J.W., Watters, T.R., Robinson, M.S., Zuber, M.T. & Solomon, S.C. (2011) Stereo topographic models of Mercury after three MESSENGER flybys. *Planetary and Space Science*, 59(15), 1910–1917. doi:10.1016/j.pss.2011.07.005.

Preusker, F., Stark, A., Oberst, J., Matz, K.-D., Gwinner, K., Roatsch, T. & Watters, T.R. (2017) Toward high-resolution global topography of Mercury from MESSENGER orbital stereo imaging: a prototype model for the H6 (Kuiper) quadrangle. *Planetary and Space Science*, 142, 26–37. doi:10.1016/j.pss.2017.04.012.

Robinson, M.S., Davies, M.E., Colvin, T.R. & Edwards, K. (1999) A revised control network for Mercury. *Journal of Geophysical Research: Planets*, 104(E12), 30847–30852. doi:10.1029/1999je001081.

Scholten, F., Gwinner, K., Roatsch, T., Matz, K.-D., Wählisch, M., Giese, B., Oberst, J., Jaumann, R., Neukum, G. & HRSC Co-I-Team (2005) Mars Express HRSC data processing – methods and operational aspects. *Photogrammetric Engineering and Remote Sensing*, 71(10), 1143–1152.

Solomon, S.C., *et al.* (2008) Return to Mercury: a global perspective on MESSENGER's first Mercury flyby. *Science*, 321(5885), 59–62. doi:10.1126/science.1159706.

Solomon, S.C., *et al.* (2001) The MESSENGER mission to Mercury: scientific objectives and implementation. *Planetary and Space Science*, 49(14–15), 1445–1465. doi:10.1016/S0032-0633(01)00085-X.

Stark, A. (2015) *Observations of Mercury's Rotational State From Combined MESSENGER Laser Altimeter and Imaging Data*. PhD thesis, Technische Universität Berlin, Berlin.

Stark, A., Oberst, J. & Hussmann, H. (2015) Mercury's resonant rotation from secular orbital elements. *Celestial Mechanics and Dynamical Astronomy*, 123(3), 263–277. doi:10.1007/s10569-015-9633-4.

Stark, A., Oberst, J., Preusker, F., Gwinner, K., Peale, S.J., Margot, J.-L., Phillips, R.J., Zuber, M.T. & Solomon, S.C. (2015) Mercury's rotational parameters from MESSENGER image and laser altimeter data: a feasibility study. *Planetary and Space Science*, 117, 64–72. doi:10.1016/j.pss.2015.05.006.

Wewel, F. (1996) Determination of conjugate points of stereoscopic three line scanner data of Mars 96 mission. *International Archives Photogrammetry and Remote Sensing*, 31, 3.

Zuber, M.T., *et al.* (2012) Topography of the northern hemisphere of Mercury from MESSENGER laser altimetry. *Science*, 336(6078), 217–220. doi:10.1126/science.1218805.

Section IV

Feature information extraction from planetary remote sensing data

Planetary Remote Sensing and Mapping – Wu et al.
© 2019 Taylor & Francis Group, London, ISBN 978-1-138-58415-0

Chapter 12

Automatic crater detection for mapping of planetary surface age

A. L. Salih, A. Lompart, P. Schulte, M. Mühlbauer, A. Grumpe, C. Wöhler and H. Hiesinger

ABSTRACT: Ages of planetary surfaces are typically obtained by manually determining the impact crater size-frequency distribution (CSFD) in spacecraft imagery, which is a very complicated and time-consuming procedure. In this chapter, a template-based crater detection algorithm (CDA) is used to analyze image data under known illumination conditions. For this purpose, artificially illuminated crater templates are used to detect and count craters and their diameters in the areas under investigation. Firstly, we investigate how well our CDA is suitable to determine the absolute model age (AMA) of different lunar regions. The sensitivity of the CDA is calibrated based on five different regions in Mare Cognitum on the Moon such that the age inferred from the manual CSFD measurements corresponds to the age inferred from the CDA results. The obtained detection threshold is used to apply our CDA to another five regions in Oceanus Procellarum. It is shown that the automatic age estimation yields AMA values that are consistent with values obtained by manual CSFD measurements. Then the image-based CDA is applied to the floor region of the lunar farside crater Tsiolkovsky. The obtained CSFD is then used to estimate the AMA of the surface. The detection threshold is calibrated based on a 100 km^2 test area for which the CSFD has been determined by manual CSFD measurements in a previous study. Furthermore, CSFDs and AMAs are computed for overlapping quadratic regions covering the complete floor of Tsiolkovsky. This results in a spatially resolved age map showing AMAs of typically 3.2–3.3 Ga, while for small regions lower and higher AMAs are found. It is well known that the CSFD may be affected by secondary craters. Hence, we present a method to refine our detection results by applying a secondary candidate detection (SCD) algorithm relying on Voronoi tessellation of the spatial crater distribution which searches for clusters of craters. The detected clusters are assumed to result from the presence of secondary craters which are then removed from the CSFD, where it was favorable to apply the SCD algorithm separately to each diameter bin of the CSFD histogram. In comparison to the original age map, the age map obtained after removal of secondary candidates has a more homogeneous appearance and does not exhibit regions of spuriously high age resulting from contamination by secondary craters.

This chapter has mostly been adopted and/or adapted from the works of Salih *et al.* (2016), Salih, Lompart *et al.* (2017), Salih, Schulte *et al.* (2017).

1 INTRODUCTION

The age of a planetary surface is of major importance for its subdivision into different geological units and the analysis of geological processes. There exists a well-established statistical approach for the estimation of the surface age that relies on the impact crater size-frequency distribution (CSFD) (e.g., Neukum *et al.*, 2001). Absolute ages of planetary surfaces can be determined by means of radiometric methods, which, however require the acquision and possibly also the return of surface samples. In contrast, the most practical methods for age estimation are those that do not involve sampling of the surface. They rely on the general observation that the number of craters increases with the time that has passed since the surface has been deposited. The extracted information regarding the crater rate per area may be directly used to derive the relative age of different planetary surface areas. The main advantage of the crater-based methods is that the required information may be extracted from images and or digital elevation models (DEM). The information is thus extracted only from remotely obtained data of the surface.

In order to derive absolute model ages (AMA) of the surface, the methods are combined, i.e., the CSFD of areas for which samples are available is determined and then calibrated with respect to the absolute age of the sample. This calibration allows for the estimation of absolute ages of individual areas on the Moon (e.g., Neukum *et al.*, 2001). The advantage of performing this procedure through applying an automatic crater detection algorithm (CDA) is to reduce the amount of time needed to calculate the AMAs for large areas on planetary surfaces (Salamunićcar and Lončarić, 2010).

Usually, the estimation of the CSFD is done by manual CSFD measurements and size determination in the available planetary images. Since the AMA highly depends on the CSFD, it is very important to detect all craters and to not include false detections into the CSFD. Although this manual process is highly demanding and time consuming for large areas, it is commonly still preferred over automatic crater detection algorithm, which often miss some craters and/or falsely detect craters. The automatic detection algorithms, however, operate comparably fast and may thus be easily applied to large high-resolution global image mosaics of various planetary surfaces. Plenty of automated methods for crater detection based on image data or topographic data have been developed recently (e.g., Salamunićcar and Lončarić, 2008, 2010; Stepinski *et al.*, 2012). Since the AMA is derived from the CSFD using a statistical approach, this work is focused on the analysis of the feasibility of AMA estimation based on an automatic CDA. We intend to analyze the behavior of CDA-based AMA estimates rather than to evaluate the CDA's per-crater accuracy or its true vs. false detections. The latter tends to be difficult to obtain anyway due to an inherent uncertainty of the required manually determined "ground truth", because significant variations between manual crater counts of the same area by different persons, even when performed by experts, may occur (Robbins *et al.*, 2014).

In this study an automatic crater detection algorithm (CDA) is applied to regions of the lunar surface (Mare Cognitum, parts of Oceanus Procellarum, and the floor of crater Tsiolkovsky) and configured in order to obtain crater counts of similar accuracy as manual crater counts. The AMA values obtained for Mare Cognitum and Oceanus Procellarum are compared with those derived from manual crater counts, and a spatially resolved AMA map is constructed for the floor of the crater Tsiolkovsky.

An important aspect in CSFD-based AMA estimation is the influence of secondary cratering on the CSFD. The process of secondary cratering usually occurs on planetary bodies having a gravitational acceleration which is high enough for the ejecta from an initial impact to fall back on the planet's surface. In the last decades, a number of researchers have focused on analyzing the role of secondary craters in the estimation of the AMA. They concluded that secondary craters "contaminate" the crater statistics, which results in inaccurate age estimation especially when craters smaller than about 1 km are considered (e.g., McEwen and Bierhaus, 2006; Robbins and Hynek, 2014; Werner *et al.*, 2009). To speed up the CSFD measurements process, a number of automatic detection and counting methods have already been developed. Some of them are fully automated but dependent on the choice of a detection threshold. According to the previous literature, none of them has the capability of automated detection of secondary crater candidates (e.g., Salamunićcar and Lončarić, 2010). We describe a statistically based automated approach to the discrimination between primary and secondary craters. The Tsiolkovsky crater area is used as a testing ground for this method because of its size and the presence of a homogeneous lava-flooded floor. As a final result, an AMA map of the floor of Tsiolkovsky corrected for secondary craters is presented.

2 CRATER DETECTION

Impact cratering is a very common geological process that occurs on all planets of the solar system, and craters are commonly analyzed to estimate the relative ages of surfaces that have not been sampled before (e.g., Michael and Neukum, 2010). Currently, manual counting still dominates the field of age estimation of planetary surfaces. However, a single standardized approach for the assessment of impact craters and measurement of their sizes has not been developed for use in the process of absolute age determination of planetary surfaces.

A number of tools have been developed to support the scientists in manual CSFD measurements. Examples are the CraterTools software developed by Kneissl *et al.* (2011) and also the CraterStats software package developed by Michael *et al.* (2012) that allow for estimating the AMA based on the CSFD. Most of these tools simplify the process of working with crater information, but CSFD measurements itself still has to be performed manually.

The availability of high-resolution images of the Moon provided by the Kaguya Terrain Camera (TC) (Haruyama *et al.*, 2012) or the Lunar Reconnaissance Orbiter Narrow Angle Camera (NAC) (Robinson *et al.*, 2010) provides an opportunity to analyze and extract more geological information than ever before. The resolution of those images may reach to up to < 1 m per pixel or better, which allows for the identification of craters down to ~10 m or less in diameter. That might provide unprecedented accuracy for age estimation of the lunar surface. However, the high resolution and huge amount of data also increases the time required to identify these geological features and suggests the development of an automated counting system.

Hence, it has become an important research topic in remote sensing to develop crater detection algorithms (CDAs), which automatically determine the locations and sizes of craters in images (e.g., Salamanićcar and Lončarić, 2008, 2010). Numerous CDAs have been developed as a consequence of the high importance of knowledge about the impact crater distribution for remote geologic studies of planetary surfaces (Stepinski *et al.*, 2012). However, many works about CDAs are limited to the demonstration of the accuracy of a particular algorithm on a small set of test images displaying quite simple and clearly pronounced craters, while meaningful studies in the field of planetary science require CDAs achieving a high performance also for less obvious, e.g., degraded, impact craters (Stepinski *et al.*, 2012).

In recent years, a number of computer vision techniques have been developed to address the task of automated CSFD measurements. They can be split in several groups, one group based on edges and Hough Transform as in Sawabe *et al.* (2006), who propose a multistage approach on edge detection and fuzzy Hough Transform. It uses multiple heuristics methods to detect impact craters. A crater detection rate of 80% compared to manual detection was obtained by Sawabe *et al.* (2006). Another group is the neural networks-based approach as by Cohen *et al.* (2016), who propose a CNN (Convolutional Neural Network) technique. In contrast to other approaches, CNNs do not have to use handcrafted features; instead they learn optimal filters and features from training examples (for an overview see, e.g., Simard *et al.*, 2003). A drawback of the CNN approach is the need for a large number of examples to be used for training, which needs to be extracted and labeled, as well as long training times. Recently, methods not based on images but on topographic maps have been developed to address CSFD measurements (e.g., Salamunićcar and Lončarić, 2010; Yin *et al.*, 2013). In contrast to image-based methods, they are not affected by illumination, visual surface properties, and atmospheric conditions.

In this study, the template matching based CDA of Grumpe and Wöhler (2013) with the ability of automatic detection of small craters (<10 image pixels in diameter) will be used for crater detection in lunar images. This template-based algorithm will first be applied to ten areas from lunar Mare Cognitum and Oceanus Procellarum for the purpose of calibration and evaluation. Secondly, it will be used for analyzing orbital image data of the mare-like floor region of lunar crater Tsiolkovsky. The main step in the estimation of the AMA is the statistical analysis of the CSFD constructed using the image-based CDA for determining the crater locations and diameters. The CDA utilized in this chapter relies on cross-correlation based template matching. It examines the similarity between the original image and a set of crater templates at each pixel position. It can be considered as a reliable yet simple method to automatically detect craters in orbital images of the lunar surface. Since this algorithm, like most other CDAs, has a parameter that defines its sensitivity, it is important to calibrate its detection threshold based on a small reference area for which manual crater counts are available (Salih *et al.*, 2016).

3 CRATER DETECTION VIA TEMPLATE MATCHING

Template matching is a technique that finds areas within an image that are similar to a template image, where usually the sliding window approach is employed. The window starts sliding from the initial position and is shifted by a given increment. For every step, a similarity measure is calculated. Common measures include cross-correlation and the sum-of-squared differences between template and image. See, e.g., Brunelli (2009) for an overview.

To detect craters within our testing area, the template-matching algorithm of Grumpe and Wöhler (2013) is applied. It uses six templates which represent six different 3D models of small craters. They were constructed using tracks of the Lunar Orbiter Laser Altimeter (LOLA) instrument (Smith *et al.*, 2010) that contain cross-sections of satellite craters of the lunar crater Plato. The craters are characterized by three basic shapes: the simple bowl shape, a transient shape with a flat floor, and a complex shape with a central peak. Each crater is split at its center point to extract two different profiles. Then these curvature models were rotated symmetrically around the crater center to generate two-dimensional surfaces in 3D space. From these surfaces, crater templates could be obtained by applying the Hapke model (Hapke, 1984, 2002) to render crater template images based on the known directions towards the sun and the viewer. The obtained templates are grey scale images as shown in Figure 12.1, which are then re-scaled to match a given crater diameter with respect to the image. The similarity between the template and the image is computed by normalized cross-correlation. All available templates were scaled to sizes between 5 and 200 image pixels in order to allow for the detection of craters across a broad range of diameters based on the normalized cross-correlation coefficient between the templates and the image parts being analyzed (Salih *et al.*, 2016). When the cross-correlation exceeds a given threshold, the crater is added to the list of candidates.

This procedure is repeated for a specified range of diameters. The method yields a list of observed craters and their positions and diameters in pixel units. The template matching stage

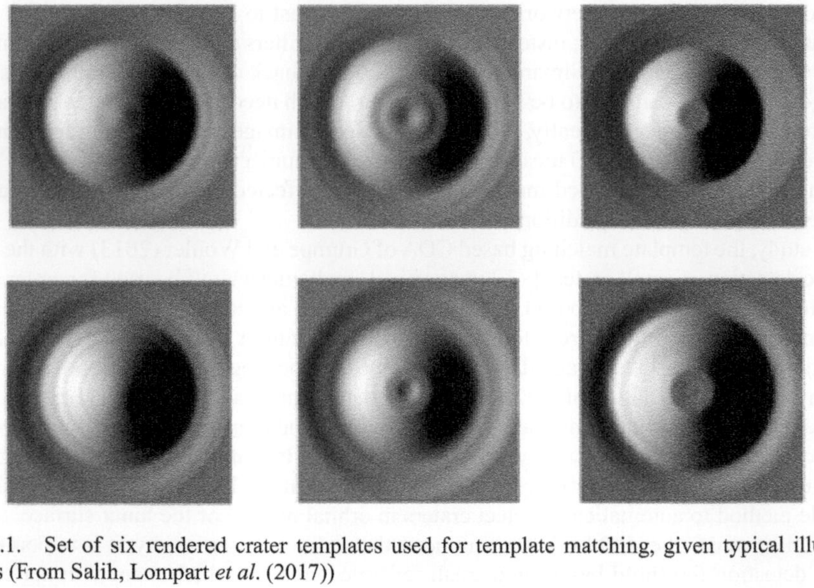

Figure 12.1. Set of six rendered crater templates used for template matching, given typical illumination conditions (From Salih, Lompart *et al.* (2017))

is followed by a fusion process that replaces multiple detections of the same crater at slightly different positions or with slightly different diameters by the average of these positions or diameters (Salih *et al.*, 2016).

Like virtually all other object detection algorithms, the applied template matching based CDA relies on a threshold value which controls the sensitivity of the detector. An appropriate detection threshold is determined such that for a given reference area with available manual crater counts it yields an AMA that comes as close as possible to the manually determined AMA.

4 SURFACE AGE ESTIMATION

Age dating by CSFD measurements is based on the assumption that a new surface forms without impact craters. Over time, it is bombarded by asteroids and comets, which results in an increase of the impact crater population. The age of a surface can be estimated by assessing the distribution of craters and fitting estimated crater size-frequency distribution (CSFD) to a so-called production function, which depends on the absolute age of the surface (Neukum, 1983). A function to describe the total number of craters with diameters that exceed a given diameter D per unit area has been introduced by Neukum (1983), which was formulated from CSFD data extracted from different lunar areas of various ages according to:

$$\log_{10}\left[N(D)\right] = \sum_{i=0}^{11} a_i \left[\log_{10}(D)\right]^i \quad (1)$$

In Equation (1), $N(D)$ is the cumulative crater frequency that denotes the number of craters per km^2 having diameters exceeding D, and a_i are the coefficients for the age estimation on the lunar surface. The logarithm of $N(D)$ is given in polynomial form (Neukum, 1983). In the case of the Moon, Neukum *et al.* (2001) provide the set of coefficients a_i (see Table 12.1) resulting in a good match with the known data.

Table 12.1 Coefficients of Equation (1) for age estimation of the lunar surface according to (Neukum *et al.* (2001))

	Neukum et al. (2001)
a_0	−3.0876
a_1	−3.557528
a_2	0.781027
a_3	1.021521
a_4	−0.156012
a_5	−0.444058
a_6	0.019977
a_7	0.086850
a_8	−0.005874
a_9	−0.006809
a_{10}	0.000825
a_{11}	0.0000554

5 MARE REGIONS IN MARE COGNITUM AND OCEANUS PROCELLARUM

According to previous studies, the volcanic activity on the Moon has been high between 1.2 to 4 billion years (Ga) ago (Hiesinger *et al.*, 2003). These activities and eruptions formed a large variety of lunar mare basalt units in different areas. Figure 12.2 shows the primary mare areas on the lunar nearside. For the study of lunar volcanic processes, it is very important to know the temporal sequence of individual volcanic activities and how they affect the extents of the corresponding basalt areas.

The determination of the CSFD is the main approach to estimate the AMAs for various individual geologic regions on the Moon (e.g., Neukum *et al.*, 2001). We derive the CSFD from a mosaic of Wide Angle Camera (WAC) images of the Lunar Reconnaissance Orbiter Camera (LROC) (Robinson *et al.*, 2010; Wagner *et al.*, 2015). The spatial resolution of the mosaic is about 100 m per pixel. We assume that the craters can be detected if their diameter equals or exceeds 300 m (3 pixels in the image).

The generated crater templates from the above explained template-based CDA are then scaled to a given diameter range of 0.3–20 km and the normalized cross-correlation with the image is computed, respectively, where local maxima of the normalized cross-correlation are set to be crater candidates. Based on a specified threshold value, the determined cross-correlation value indicates whether or not the detected candidate is a crater. Since the same crater may be found by multiple templates and diameters, a fusion procedure is applied to the detected candidates such that multiple detections are removed and a unique diameter value can be determined. In case of several

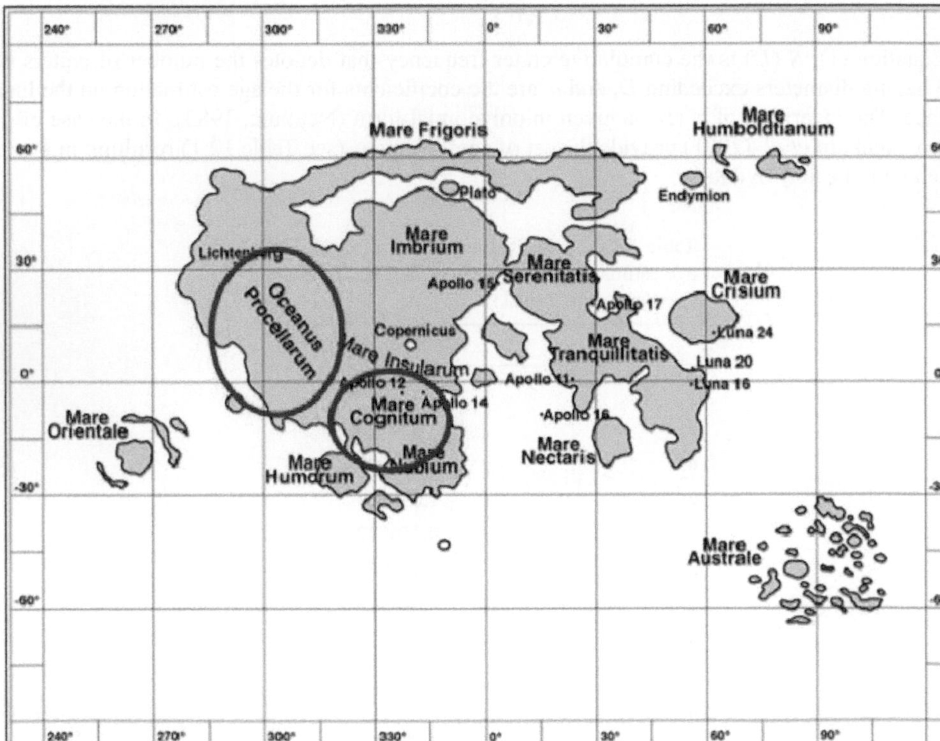

Figure 12.2. Lunar nearside maria (From Hiesinger *et al.* (2003))

templates corresponding to different crater diameters, the diameter values are averaged to obtain the final diameter.

The determined crater positions and diameters are then used to derive the CSFD for the study region. This study focuses on the five areas C1–C5 defined by Hiesinger *et al.* (2003) located in Mare Cognitum and further five areas P5, P41, P49, P50, and P51 defined by Hiesinger *et al.* (2003) located in Oceanus Procellarum. All 10 study areas are located in a latitude range between 15°S and 30°N and in a longitude range between 285°E and 345°E.

Figure 12.3 shows the obtained craters for the five study areas C1–C5 located in Mare Cognitum by applying the described template matching–based CDA. We adapted the sensitivity threshold of the template matching such that the squared difference between the AMA inferred for C1–C5 and the AMA given by Hiesinger *et al.* (2003) is minimized.

Figure 12.4 shows the detected craters for the other five study areas P5, P41, P49, P50, and P51 located in Oceanus Procellarum. Here the template matching–based CDA has been applied with the optimal threshold obtained for the Mare Cognitum test areas.

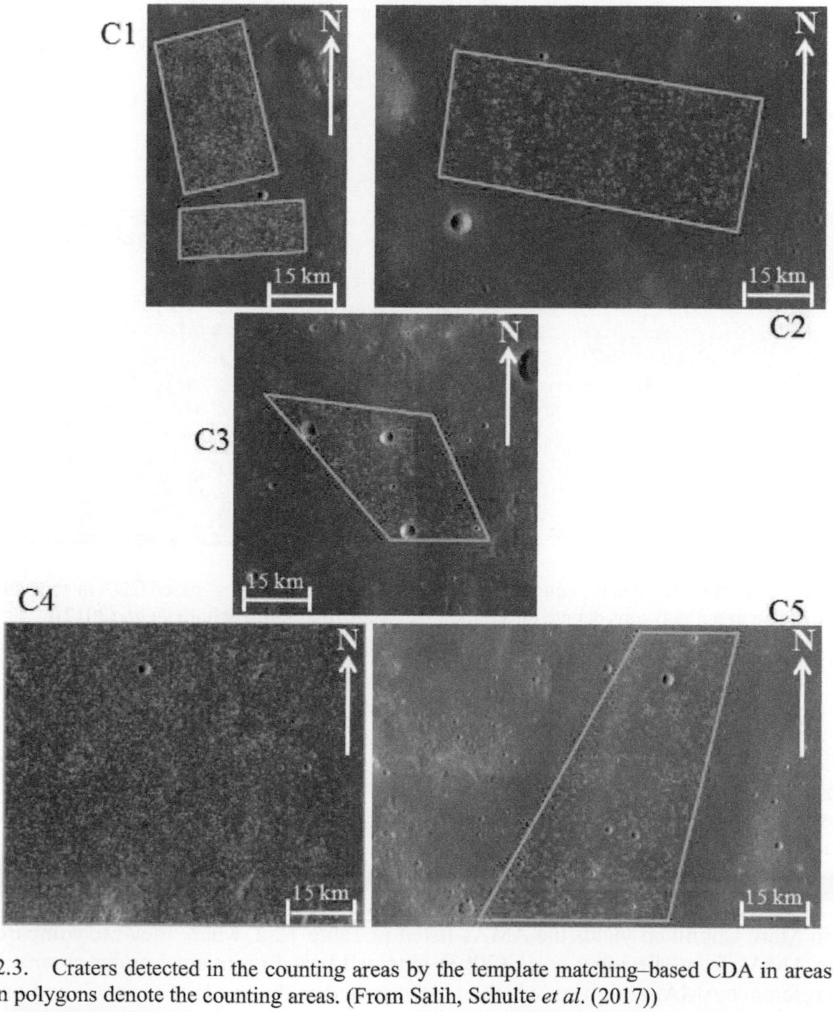

Figure 12.3. Craters detected in the counting areas by the template matching–based CDA in areas C1–C5. The green polygons denote the counting areas. (From Salih, Schulte *et al.* (2017))

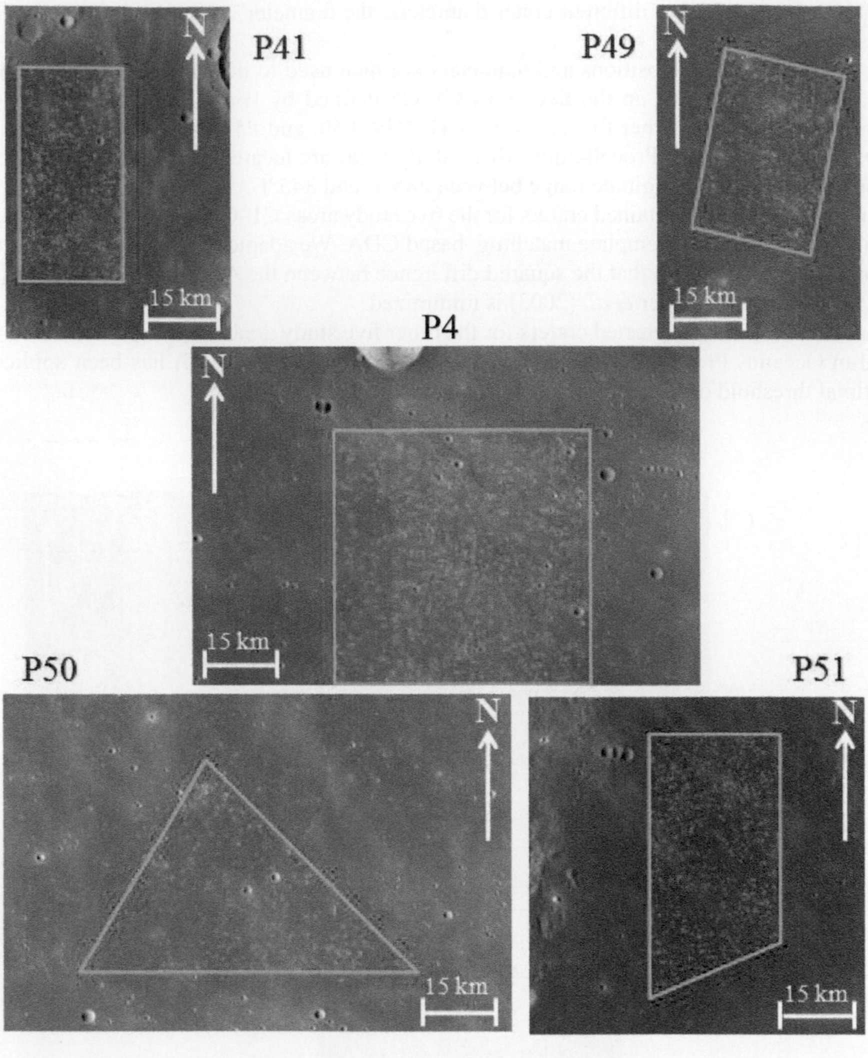

Figure 12.4. Craters detected in the counting areas by the template matching–based CDA in areas of Oceanus Procellarum. The green polygons denote the counting areas. (From Salih, Schulte *et al.* (2017))

Firstly, the optimal cross-correlation thresholds for the regions C1–C5 were calculated automatically by comparing the CDA-based AMAs with the reference AMAs from Hiesinger *et al.* (2003). The obtained threshold values are shown in Table 12.1.

The arithmetic mean of the five calibrated detection thresholds amounts to 0.6592. Area C1 consists of two polygons for the CSFD measurements. The other investigated areas have always been defined by a single polygonal shape. By applying this "optimal" threshold value to the areas C1–C5 in Mare Cognitum yields the AMAs listed in Table 12.2, where they are compared to the reference AMAs from Hiesinger *et al.* (2003) obtained based on manual crater counts. For area C5, two reference AMAs are given, which are interpreted as the result of a resurfacing process by Hiesinger *et al.* (2003).

Table 12.1 Calibrated detection threshold values for the individual regions C1–C5 in Mare Cognitum

Area	*Calibrated threshold*
C1	0.6637
C2	0.6455
C3	0.6541
C4	0.6677
C5	0.6648
Average value	0.6592

Table 12.2 CDA-based and reference AMAs for the study areas in Mare Cognitum after applying the optimal detection threshold

Area	*CDA-based AMA (Ga)*			*Reference AMA (Ga) (Hiesinger et al., 2003)*		
	AMA	*− error*	*+ error*	*AMA*	*− error*	*+ error*
C1	3.52	−0.008	+0.007	3.49	−0.10	+0.08
C2	3.46	−0.004	+0.004	3.45	−0.06	+0.09
C3	3.63	−0.024	+0.018	3.41	−0.08	+0.08
C4	3.43	−0.002	+0.002	3.36	−0.11	+0.10
C5	3.52	−0.006	+0.005	3.32 3.65	−0.14 −0.08	+0.10 +0.08

Table 12.3 CDA-based and reference AMAs for the study areas in Oceanus Procellarum, obtained using the optimal detection threshold inferred from areas C1–C5

Area	*CDA-based AMA [Ga]*			*Reference AMA [Ga] (Hiesinger et al., 2003)*		
	Age	*− error*	*+ error*	*Age*	*− error*	*+ error*
P49	1.71	−0.140	+0.110	2.01	−0.43	+0.37
P5	3.45	−0.019	+0.015	3.48	−0.06	+0.08
P41	3.33	−0.009	+0.009	2.13	−0.85	+0.75
P50	1.90	−0.166	+0.128	1.87	−0.25	+0.56
P51	2.01	−0.205	+0.166	1.85	−0.34	+0.37

In the next step, the study areas in Oceanus Procellarum are again delineated by polygons. The optimal threshold value from Table 12.1 was applied to the areas P5, P41, P49, P50, and P51 in Oceanus Procellarum as defined by Hiesinger *et al.* (2003), resulting in the AMAs listed in Table 12.3. It can be seen that all except one (P41) of the CDA-derived AMAs are very close to the manually based reference AMAs from Hiesinger *et al.* (2003).

The largest deviations between CDA-based and reference AMAs occurs for region P41 with a reference AMAs of 2.13 Ga. Due to the nearly horizontal chronology function for AMAs between about 2 and 3 Ga (Neukum *et al.*, 2001), ages in this range are particularly sensitive to changes in the *N(1)*, and a small number of falsely detected or missed craters can make a large difference in AMA. Hence, this area P41 has relatively large errors of the reference AMA value in Table 12.3. Despite the similarly large AMA estimation errors, the inferred AMAs of the regions P50 and P51 are in good agreement with the AMAs given by Hiesinger *et al.* (2003). This implies that the AMA estimation is less sensitive to small errors for younger surfaces. For all study regions except area P41, the deviations of the CDA-based AMA and the manual count based AMA are within the statistical error intervals.

6 TSIOLKOVSKY CRATER

The CDA-based AMA estimation approach has also been applied to the floor of the crater Tsiolkovsky which is located at 20°S and 129°E on the farside of the Moon. It has a diameter of about 180 km. Tsiolkovsky is partially filled by lava and thus has a dark and smooth floor as shown in Figure 12.5. According to previous studies, ages of the crater floor of Tsiolkovsky of approximately 3.8 Ga (Walker and El-Baz, 1982) and 3.5 Ga (Tyrie, 1988) have been found. A recent age estimation of basalts in Tsiolkovsky crater has been performed by Pasckert *et al.* (2015), who obtained $3.19^{+0.08}_{-0.12}$ Ga. In this work we use images acquired by the Terrain Camera (TC) of the lunar spacecraft Kaguya (Haruyama *et al.*, 2012). The resolution of the image is 7.4 m per pixel.

Choosing the optimal detection threshold of the CDA is again of crucial importance. It is obtained by comparison to the AMA inferred from manual crater counts of a region of about 100 km² size performed by Pasckert *et al.* (2015). The area was split into three equally sized parts, for each of

Figure 12.5. Image of the crater Tsiolkovsky, extracted from the LROC WAC global mosaic (Speyerer *et al.*, 2011)

which the absolute difference between CDA-based and manually determined AMA was minimized by a quasi-Newton method. A threshold for each area is estimated so as to match the corresponding manual count as closely as possible. The result consists of three slightly different thresholds, whose mean value of 0.6568 was chosen as the "optimal" threshold value for this area. Applying the CDA to the total 100 km² reference area with this detection threshold resulted in an AMA of 3.21±0.13 Ga, being very similar to the value of $3.19^{+0.08}_{-0.12}$ Ga obtained by Pasckert *et al.* (2015) based on manual counting (Salih *et al.*, 2016).

We applied the template matching technique to the 7.4-m-per-pixel resolution Kaguya image with a 900-x-900-pixel (6.6 x 6.6 km²) wide sliding window moving over the whole image of crater Tsiolkovsky with a step width of 10 pixels (74 m) (Salih *et al.*, 2016). Based on the optimal threshold value of 0.6568, CSFDs were computed for these overlapping quadratic sub-regions in order to estimate an AMA value for each sub-region, leading to a spatially resolved AMA map (Fig. 12.6). The AMA values were obtained using our Matlab implementation of the CSFD-based method for AMA estimation (Michael, 2013; Michael and Neukum, 2010; Michael *et al.*, 2012; Neukum, 1983), where the production function and chronology function of Neukum *et al.* (2001) were applied. Most AMAs of the map lie between 2.9 and 3.6 Ga, with the majority of AMAs centered around 3.3 Ga. In the AMA map, the central peak, the crater walls, and parts of the crater floor not covered by mare basalt were excluded from the analysis. According to stratigraphic considerations, the age of these steeply sloped regions exceeds that of the dark mare floor (Salih *et al.*, 2016).

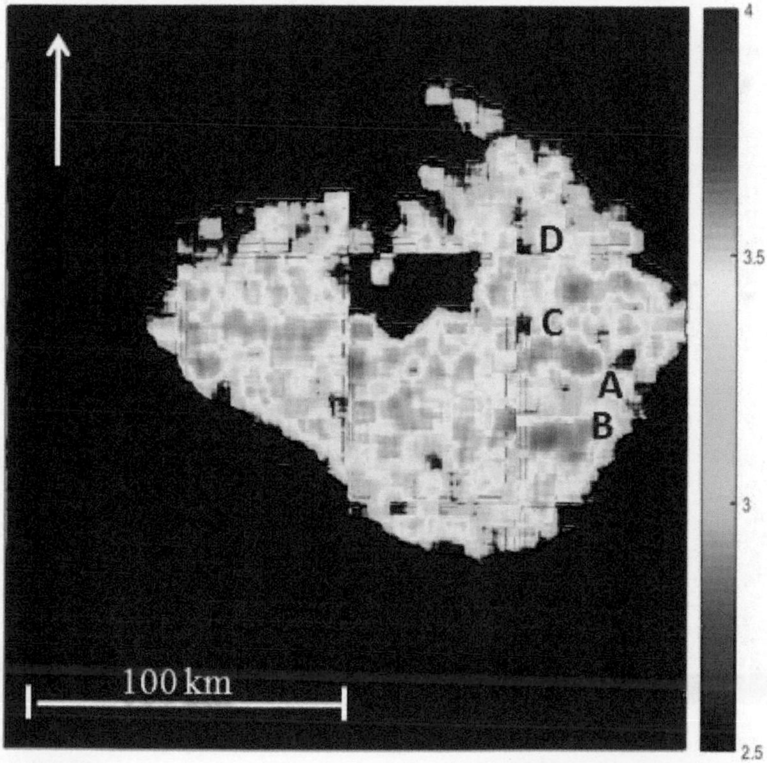

Figure 12.6. AMA map of the floor of the crater Tsiolkovsky. The mare-like floor region has been extracted manually. Black: no data. (From Salih, Lompart *et al.* (2017))

The color ratio map of crater Tsiolkovsky (Fig. 12.7) was derived from Clementine UV/VIS multispectral data (red channel: 750 nm/415 nm reflectance; green channel: 750 nm/950 nm reflectance; blue channel: 415 nm/750 nm reflectance, as proposed by Pieters *et al.* (1994) and www.mapaplanet.org/explorer/help/data_set.html#moon_clementine_ratio). It can be used to identify compositionally distinct geological units on the mare-like crater floor. Although frame-specific calibration artifacts make it hard to unambiguously divide the color ratio image into geological units, some similarities between structures visible in Figures 12.6 and 12.7 are clearly apparent through visual comparison. Specifically, four localized AMA anomalies can be found in the eastern part of the crater floor. They are marked by letters A, B, C, and D in Figures 12.6 and 12.7, where each letter also denotes a small spectrally distinct region in the color ratio image. Regions A, C, and D have lower AMAs than the surrounding surface area, respectively (about 2.9–3.2 Ga), while for region B a higher AMA of about 3.7 Ga has been estimated. According to Figure 12.7, these small regions can all be distinguished spectrally from the surrounding surface, indicating compositional variations of the mare basalts (Salih *et al.*, 2016).

A more detailed view of the eastern and southeastern part of crater Tsiolkovsky is shown in Figure 12.8. It reveals that the two small low-AMA anomalies marked in Figures 12.6 and 12.7 by letter D are located at the same locations as two bright details in Figure 12.5, corresponding to small mountains of the original crater floor that penetrate through the dark mare basalt layer. Hence, the lower estimated AMA is probably due to small craters being destroyed more easily on the steep slopes on the flanks of these mountains. The low-AMA anomalies marked by letters A and C cannot be linked to any conspicuous surface details but correspond

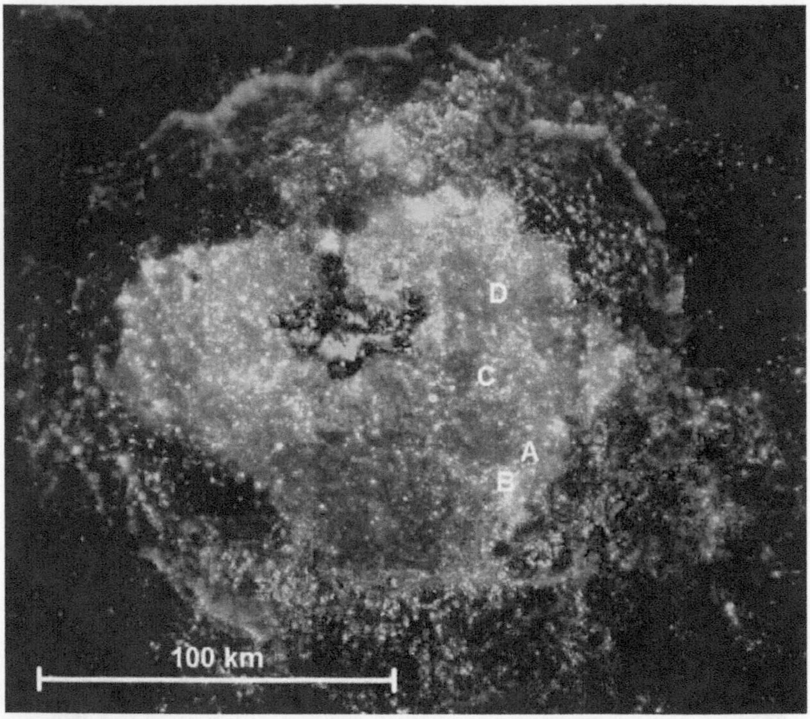

Figure 12.7. Clementine UV/VIS color ratio image of crater Tsiolkovsky (downloaded from www.mapaplanet.org/explorer/moon.html, definition of color channels according to Pieters *et al.*, 1994) (From Salih *et al.* (2016))

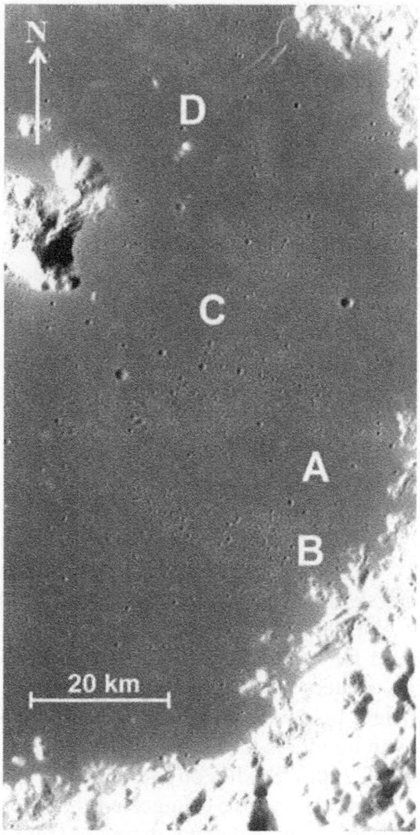

Figure 12.8. Detailed view of the eastern and southeastern part of Tsiolkovsky. The regions A–D marked in Figures 12.5 and 12.6 are indicated. Contrast-enhanced excerpt from the global LROC WAC mosaic (Speyerer *et al.*, 2011). (From Salih *et al.* (2016))

to dark surface parts with a visually apparent low abundance of small impact craters. In contrast, the high-AMA anomaly marked by letter B corresponds to an area of slightly increased albedo characterized by a visually apparent exceptionally high abundance of small craters. The distribution of these craters as a cluster, however, suggests that they are secondary craters (McEwen and Bierhaus, 2006). Thus, the seemingly increased AMA of region B is probably does not reflect a real surface age but is due to the presence of these secondary craters (a more detailed analysis of the effect of secondary craters on the estimated surface age is provided later in this chapter).

Hence, the AMA anomalies B and D are not due to lava flows having an age differing from the mean age of the crater floor. The negative AMA anomalies A and C might be interpreted as lava flows of a composition and age different from the rest of the crater floor, which would imply the occurrence of small eruptions on the floor of Tsiolkovsky several hundred (Ma)s after the emplacement of the majority of the mare basalts. An alternative (and probably more realistic) interpretation for regions A and C is that the distribution of small craters on the floor of Tsiolkovsky may be affected in a spatially variable manner by secondary craters, which would then cause the observed AMA variations across the crater floor (Salih *et al.*, 2016).

7 SECONDARY CRATERS

Many planetary studies have considered the occurrence of craters on solid planetary surfaces to be the result of direct (primary) impacts, and based the estimation of AMAs on this assumption.

By definition, secondary craters were formed by the impact of material which has been ejected by a primary impact (e.g., McEwen and Bierhaus, 2006; Shoemaker, 1965). Early observations of the lunar surface have shown that the density of small craters is much higher on the rays surrounding large and fresh primary craters than outside these structures (McEwen and Bierhaus, 2006). These craters have been considered as secondary craters by Shoemaker (1965), who provided an estimate that many craters with diameters smaller than ~200 m are secondary craters located far away from the location of the primary impact.

Many secondary craters are formed in proximity to the primary crater. They do not have a circular outline and are commonly less deep than primary craters of the same size. They often exhibit spatial patterns such as linear chains, clusters, and herringbone patterns, whereas remote secondary craters formed by material ejected at high speed are less elongated and more evenly distributed and thus appear more similar to primary craters (e.g., McEwen and Bierhaus, 2006). Large clusters of likely secondary craters are apparent on the floor of crater Tsiolkovsky (Fig. 12.8).

Commonly, it is assumed that small craters on the terrestrial planetary bodies, except those in close proximity to primary craters, are mostly primaries, so that they were used as a basis for surface dating (e.g., Basilevsky *et al.*, 2005). However, studies on Europa by Bierhaus *et al.* (2005), Mars by McEwen *et al.* (2005), and the Moon by Dundas and McEwen (2007) argued that this assumption is incorrect for smaller craters and that it is necessary to take into account the effect of remote secondary craters for accurate age estimation, especially when relying on CSFDs of craters with diameters below several kilometers, because one primary impact may be able to form millions or even hundreds of millions of secondary craters at distances of up to 1000 km (McEwen and Bierhaus, 2006). Studies on Mercury, Europa, the Earth, Moon, and Mars (Bierhaus *et al.*, 2001, 2005; Shoemaker, 1965) suggested that the upper limit of diameters for secondary craters is less than 5% of the parent crater diameter (McEwen and Bierhaus, 2006). As a general conclusion, McEwen and Bierhaus (2006) stated that the existence of secondary craters is not an impediment against using crater statistics for estimating relative surface ages and those small craters are appropriate for estimating the AMA of young surface areas. They also suggested a technique for reducing the contamination from secondary craters by mapping the possible secondary craters of all large recent craters in a study area. The absence of craters larger than a limiting diameter of about 1 km would allow for calculating an upper limit of the surface age. Further refinement of the surface age is then possible based on the so-called crossover diameter, i.e., the diameter value at which the CSFDs of the primary and secondary craters in the study area intersect. Hence, for diameters exceeding the crossover diameter primary craters dominate the total CSFD while at smaller diameters secondary craters dominate the CSFD, where the crossover diameter decreases with decreasing surface age. Hence, combining the upper-limit age together with the mapped secondary craters would allow for an estimation of the crossover diameter and thus a refined surface age (McEwen and Bierhaus, 2006).

In contrast, more recent work by Speyerer *et al.* (2016) shows based on impact craters formed during the LRO mission that the one-year isochron inferred from the production function of Neukum *et al.* (2001) is consistent with the directly observed crater production without the need for invoking secondary craters. However, the slope of the CSFD in the 10–20 m diameter range is steeper than predicted by Neukum *et al.* (2001), indicating that the steep CSFD slope at small crater diameters described, e.g., by Bierhaus *et al.* (2005) is not necessarily due to secondary craters. Similarly, the study by Malin *et al.* (2006) about recent cratering on Mars suggested that the present-day production of Martian craters cannot be described well by the assumption of most small craters being secondary craters.

8 STATISTICAL METHOD USED FOR SECONDARY CANDIDATE DETECTION

In this study, we will at least partially account for the effect of secondary craters on the CSFD by following a direct statistical approach and proposing a technique to remove secondary craters from the CSFD based on the assumption that they do not exhibit a uniform spatial distribution but tend to appear in chains and clusters. Hence, secondary craters are distinguished from the surrounding primary population of similar diameter based on their distinctly non-uniform spatial distribution. Of course, this approach will not recognize a possible uniformly distributed population of secondary craters. Furthermore, clustered craters may also be of primary origin when old and thus strongly cratered surface areas are surrounded by younger surfaces. A well-known example is the kipuka Darney χ, a strongly cratered and relatively bright area in Mare Cognitum surrounded by presumably younger mare lava exhibiting a lower crater density and a significantly lower albedo (Nichols *et al.*, 1974). Hence, an automatic detection of clusters of putative secondary craters always has to be accompanied by a geological analysis of the region under study in terms of, e.g., surface texture and spectral behavior. For this reason, we will apply the method described in this section to mare-flooded floor of the lunar farside crater Tsiolkovsky, which shows a high degree of homogeneity with respect to its surface texture and spectral appearance.

In most previous studies, secondary craters were removed from the CSFD manually. An attempt to develop a method for estimation of the secondary crater population was made by Bierhaus *et al.* (2005), who utilized the single-linkage clustering algorithm together with Monte Carlo simulations in order to distinguish clusters of secondary craters from the spatially uniform distribution of primary craters. A more recent approach by Michael *et al.* (2012) is based on the distribution of distances between the craters, in particular the "mean 2nd-closest neighbor distance". Michael *et al.* (2012) concluded that the crater density obtained from an ideal random distribution is related to this statistical distance value.

The main idea of our Secondary Candidate Detection (SCD) algorithm is to detect secondary craters based on deviations of their spatial distribution from the uniform distribution of the surrounding primary craters (which is similar to Michael *et al.*, 2012, but using a different criterion for detecting crater clusters). Hence, the SCD algorithm determines whether the crater population is uniformly distributed or clustered, which allows for removing the secondary crater candidates from a crater population that is used to estimate the age of a surface part.

The first step is to obtain crater locations using the template matching results of the given region. Primary craters are created from a random distribution of bodies hitting the surface. Their distribution should appear uniform and homogeneous. In contrast, secondary craters usually appear as high-density regions. To separate the secondary crater population from the background population, we developed an algorithm that removes secondary crater candidates from any spatial distribution. This method combines a Voronoi tessellation, Monte Carlo simulations of a uniform distribution, and a one-tailed test of clustering, which divides the detected craters into two groups based on the probability that they exhibit a non-uniform spatial distribution. The technique of Voronoi tessellation has also been used in the work of Honda *et al.* (2014) to detect non-uniformly distributed craters.

The SCD algorithm recognizes distal secondary craters from the surrounding primary population based on their clustering with respect to an ideal random distribution. Similar to Bierhaus *et al.* (2005), our algorithm generates a new uniformly distributed population, which has the same average density as the detected craters. Then the Voronoi tessellation is applied to each population, and distribution parameters are calculated for each iteration of the simulation process.

The Voronoi diagram (Voronoi tessellation) is a method of subdivision of a plane into regions that comprise the part of the plane around each point within a distance that is shorter than the distances to the neighboring points (Aurenhammer, 1991). Clustering can be inferred based on deviations of the local spatial density from the mean spatial density indicated by variations of the Voronoi cell area. Unfortunately, the statistical distribution of the Voronoi cell area for a uniform distribution cannot be derived from the observed spatial crater distribution due to its possible contamination by clustered secondary craters, such that the Voronoi cell area distribution should be inferred from simulations (Chiu, 2003). The fact that the distribution of points is usually computed for a bounded observation area introduces edge effects to the polygons close to the boundary of the observation area. To avoid these undesirable effects, polygons at the boundary have to be ignored.

In each iteration of the algorithm, craters are uniformly redistributed on the surface. A new Voronoi tessellation is computed along with the areas of Voronoi polygons. After simulation of *n* iterations, a distribution model of Voronoi cell areas is obtained and the median and standard derivation are computed. The clustering values, i.e., areas of Voronoi polygons, of the original crater distribution are compared to the threshold value obtained from Monte Carlo simulations of random impacts. Our algorithm detects a crater as a secondary crater candidate if its Voronoi cell area is below a threshold value, which resembles a one-tailed test of clustering.

We have also applied the SCD algorithm such that the statistical analysis is applied to several diameter intervals separately. This is expected to provide a better detection performance as the statistical spatial distribution of craters strongly depends on their size. In this work, we have used eight intervals of equal width with limiting diameters of 80, 170, 260, 350, 440, 530, 620, 710, and 800 m.

9 RESULTS AND CONCLUSIONS FOR THE REFERENCE AREA

The SCD algorithm has been applied to different regions of crater Tsiolkovsky in order to analyze the impact of the secondary craters on the crater density, as well as the difference of the estimated ages obtained for the CSFDs with and without secondary craters. As a first step, the template-based crater detection algorithm has been applied to a small area in the middle of crater Tsiolkovsky for which manual crater counts are available as reference data (Pasckert *et al.*, 2015). This reference area has a size of 99.867 km^2 with a total number of 1967 craters, ranging from 23 m to 905 m in diameter. Figure 12.9 presents the resulting CSFD of this test area, based on a range of 128–1000 m, thus excluding the rollover apparent at small diameters. This results in an age of 3.2 ± 0.132 Ga, this is very similar to the value obtained by Pasckert *et al.* (2015).

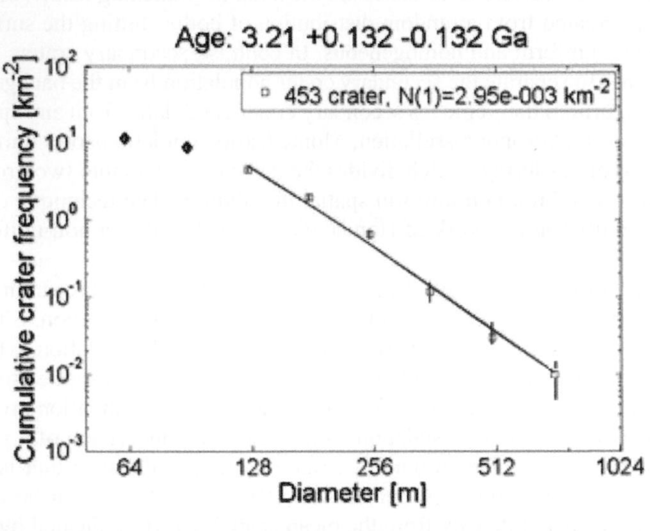

Figure 12.9. AMA of the reference region obtained based on craters in the diameter range 128–1000 m (From Salih, Lompart *et al.* (2017))

Crater detection has been performed using the previously described template-matching algorithm to the given region. The diameter range of 128–1000 m is used for the template-matching algorithm, as craters with smaller diameters would result in a large number of false positive detections. A detection threshold value of the template-matching algorithm of 0.6525 has been computed by minimizing the difference between the age obtained from the reference data and the age resulting from template matching combined with secondary crater removal.

To obtain the optimal clustering threshold, we redistributed the detected craters 1000 times by Monte Carlo modeling. The Voronoi tessellation and Voronoi areas for every crater were computed for each iteration. The results of the application of our algorithm to the test area are shown in Figure 12.10.

To illustrate the effect of the clustering threshold value, we applied our algorithm with different threshold values to the detected craters. Lower threshold values can detect only a very small number of secondary crater candidates, and the impact of the SCD algorithm on the age estimate becomes insignificant.

Both SCD algorithms (with and without application in bins) correspond well on the number of detected craters with diameters less than 150 m, while in the range of 150–180 m the number of automatically detected craters is significantly lower without application of bins. The binned SCD algorithm detects more craters in the larger diameter range, which would probably be considered primary craters by a human expert (Fig. 12.11). Separating craters into diameter bins shows an increase in the detection of secondary craters although most craters with diameters exceeding 500 m are falsely denoted as secondary craters due to their more irregular distribution resulting from their relatively small number. Hence, the diameter intervals exceeding 500 m were excluded from the AMA estimation when applying the binned SCD algorithm. The final AMA results are displayed in Figure 12.12. The binned SCD algorithm shows an age estimate, which is closer to the reference value than the SCD algorithm applied to all craters.

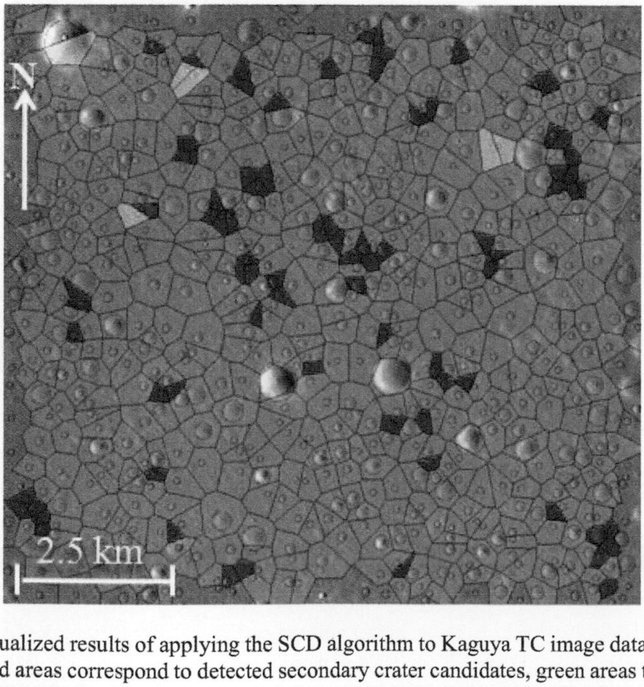

Figure 12.10. Visualized results of applying the SCD algorithm to Kaguya TC image data of the floor of crater Tsiolkovsky. Red areas correspond to detected secondary crater candidates, green areas to detected primary crater candidates. (From Salih, Lompart *et al.* (2017))

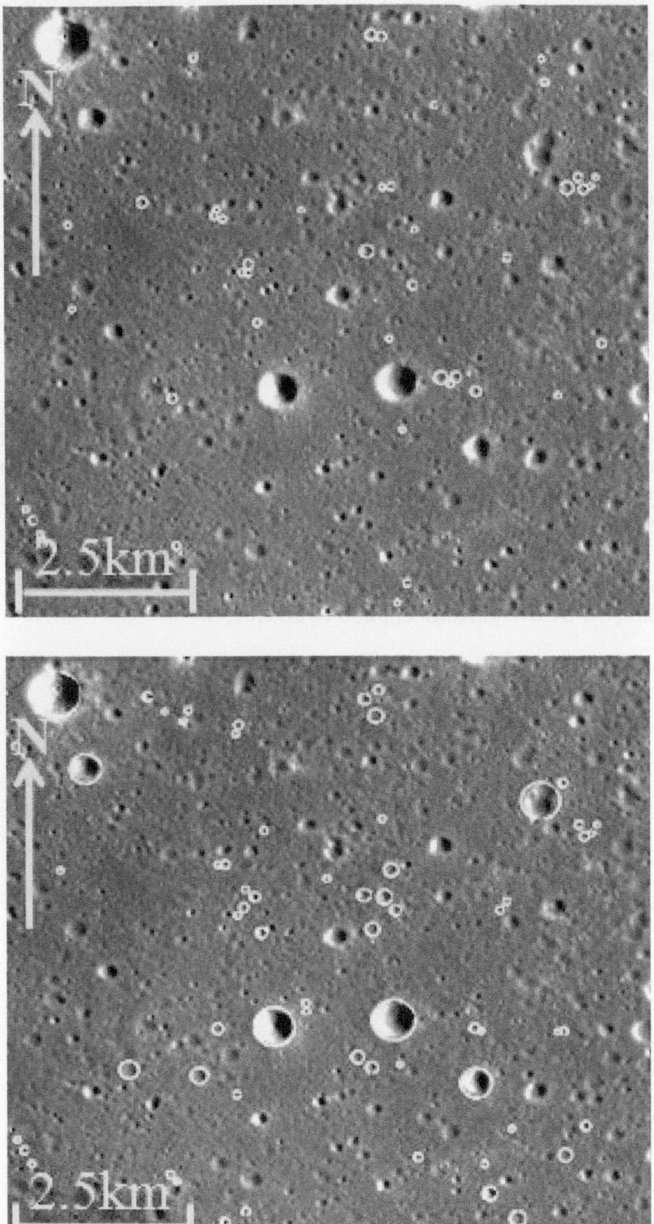

Figure 12.11. Top: Craters detected in Kaguya TC image data by the SCD algorithm (top) and the binned SCD algorithm (bottom) (From Salih, Lompart *et al.* (2017))

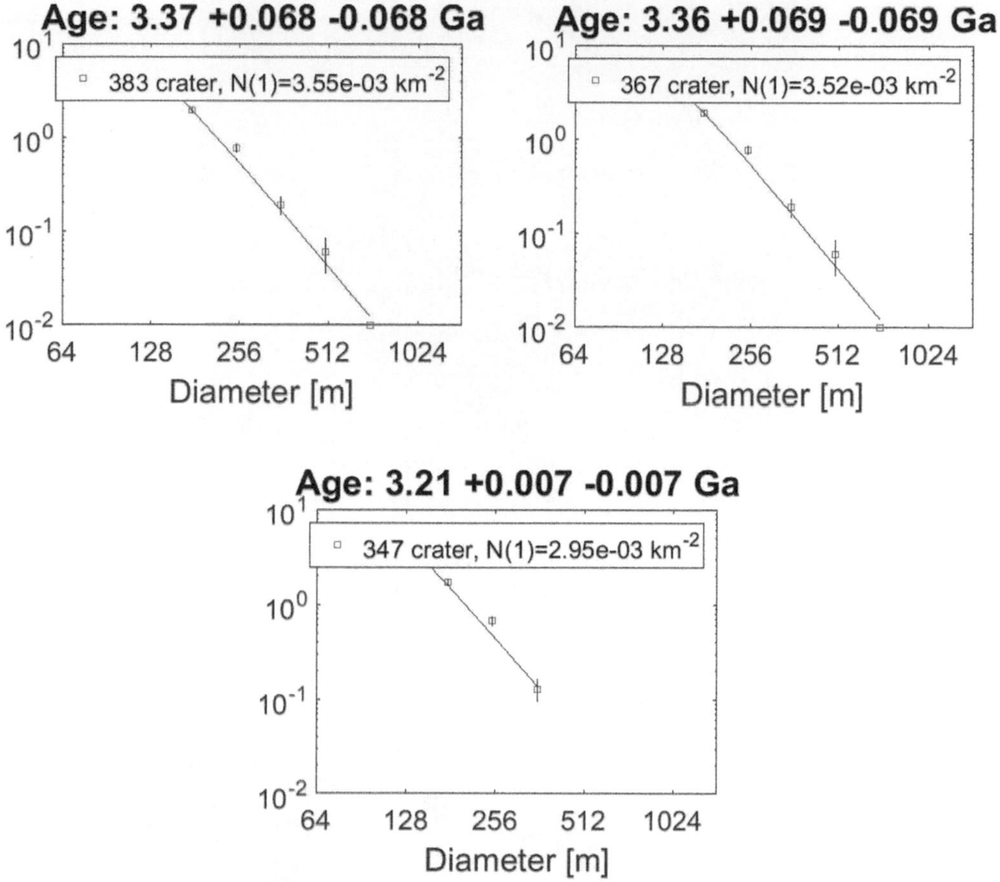

Figure 12.12. AMAs for the reference area obtained using the template-matching algorithm without SCD (upper left), with unbinned SCD algorithm (upper right), and with binned SCD algorithm (bottom) (From Salih, Lompart *et al.* (2017))

10 RESULTS FOR NON-REFERENCE AREAS

For further evaluation of our method, three larger areas on the crater floor of Tsiolkovsky, denoted as E, F, and G in Figure 12.13, have been analyzed with our algorithm. Each selected region has an area of about 2700 km^2 (52 km x 52 km). Unfortunately there are no reference data for these regions except the small part manually counted by Pasckert *et al.* (2015). This count area belongs to region E in Figure 12.13 and, consequently, age estimates have to be similar. Because these regions represent parts of the mare-flooded crater floor of Tsiolkovsky, the previously determined optimal threshold value was applied. We used a 600-x-600-pixel window for constructing crater density and age maps with a step width of 10 pixels. Central peak material was excluded from the counting area.

Three maps were plotted in Figure 12.14, representing the densities of detected craters before and after removal of secondary candidates. There are visible fluctuations of the crater density in the maps obtained without SCD algorithm. Applying the SCD algorithm reduces the crater density fluctuations especially for the binned version.

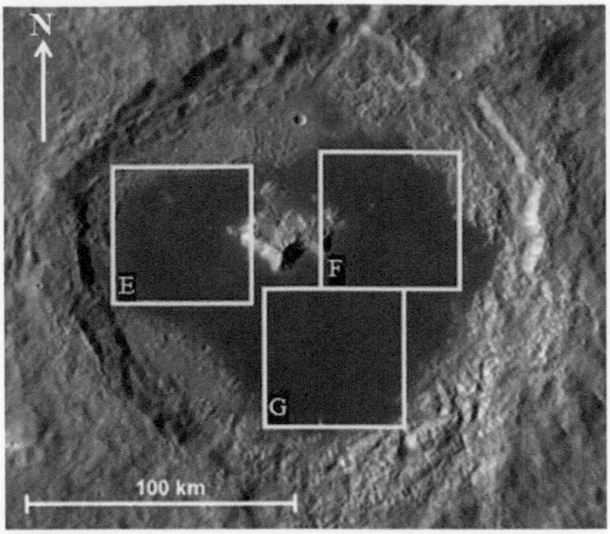

Figure 12.13. Test regions E, F, and G in crater Tsiolkovsky superposed on the LROC WAC mosaic image (Speyerer *et al.*, 2011) (Adapted from Salih, Lompart *et al.* (2017))

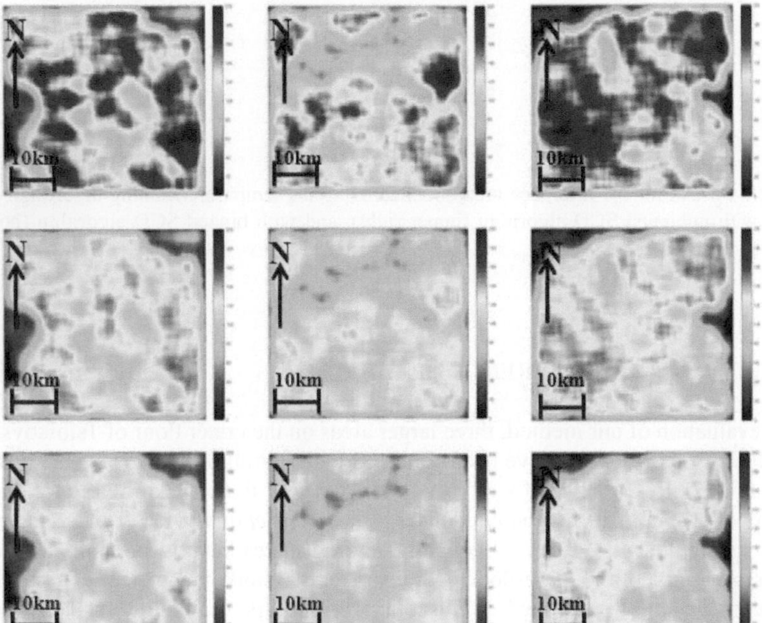

Figure 12.14. Effect of removal of secondary candidates on the crater densities of test regions E, F, and G. First row: Craters detected by template matching. Second row: Crater density after applying the SCD algorithm. Third row: Crater density after applying the binned SCD algorithm. (From Salih, Lompart *et al.* (2017))

Table 12.4 AMAs in Ga for test regions E, F, and G

Method	*E*	*F*	*G*
Template matching	3.297±0.002	3.446±0.003	3.371±0.001
With SCD	3.271±0.002	3.435±0.0004	3.353±0.001
With binned SCD	3.149±0.004	3.37±0.0002	3.249±0.001

Table 12.5 Number of detected craters in the test regions E, F, and G

	E	*F*	*G*
Craters detected without SCD	17,571	16,292	17,938
With SCD	15,349 (87%)	14,341 (88%)	15,633 (87%)
With binned SCD	14,441 (82%)	13,430 (82%)	14,651 (81%)

The estimated AMAs of the regions before removing secondary candidates range from 3.29 to 3.44 Ga without SCD and from 3.14 to 3.37 Ga with binned SCD (Table 12.4). Although the SCD algorithm has a relatively strong effect on the crater density, the unbinned SCD algorithm only has a weak effect on the AMA (around 0.2 Ga), while removing more than 10% of the craters.

The SCD and binned SCD remove 12% and 18% percent of all craters, respectively (Table 12.5), but the binned SCD has a much stronger effect on the AMA (Table 12.4). The obtained age estimates are closer to the 3.21 Ga value than those obtained using plain template matching.

11 REFINED AGE MAP OF THE FLOOR OF TSIOLKOVSKY

Using the template-matching algorithm with a detection threshold of 0.6568 and without SCD algorithm, the age map of the floor of Tsiolkovsky shown in Figure 12.6 was obtained. The area surrounding crater Tsiolkovsky consists of rough highland surfaces that cannot be taken into account because the template matching threshold has been optimized for the flat basaltic lava surface of Tsiolkovsky's floor. Although the Tsiolkovsky floor region looks homogeneous, AMA anomalies are clearly visible in Figure 12.6, having estimated AMAs of up to 3.7 Ga. This value is significantly higher than the AMA estimated based on manually counted craters by Pasckert *et al.* (2015). We propose that these age fluctuations occur as a result of clusters of secondary craters.

We applied the template matching algorithm together with the SCD algorithm using a local detection threshold derived for each test region and using a regional threshold corresponding to the mean of the thresholds of regions E, F, and G. The regional threshold produced more consistent results. The age map obtained with the unbinned SCD algorithm shows some reduction in age within high-age regions but the overall effect was nearly invisible in comparison to Figure 12.6. To construct the age map of Figure 12.15, the binned SCD algorithm has been applied based on the same detection threshold. In comparison to Figure 12.6, this map shows visible changes in the estimated AMA. Especially, the ages of the localized regions of high age in Figure 12.6 are reduced to the value of the surrounding surface, and the application of the binned SCD algorithm results in an age map that strongly reduces the fluctuations in age of the surface of the floor of crater Tsiolkovsky. Indeed, the apparent morphological homogeneity of the floor of Tsiolkovsky points at a low probability of significant surface age variations on large spatial scales (small-scale age variations may occur, e.g., due to the burial of old small craters by the ejecta of fresh craters). The large positive AMA anomalies on the eastern floor of Tsiolkovsky obtained without applying the SCD algorithm (Fig. 12.6) can thus be considered as being due to clusters of secondary craters,

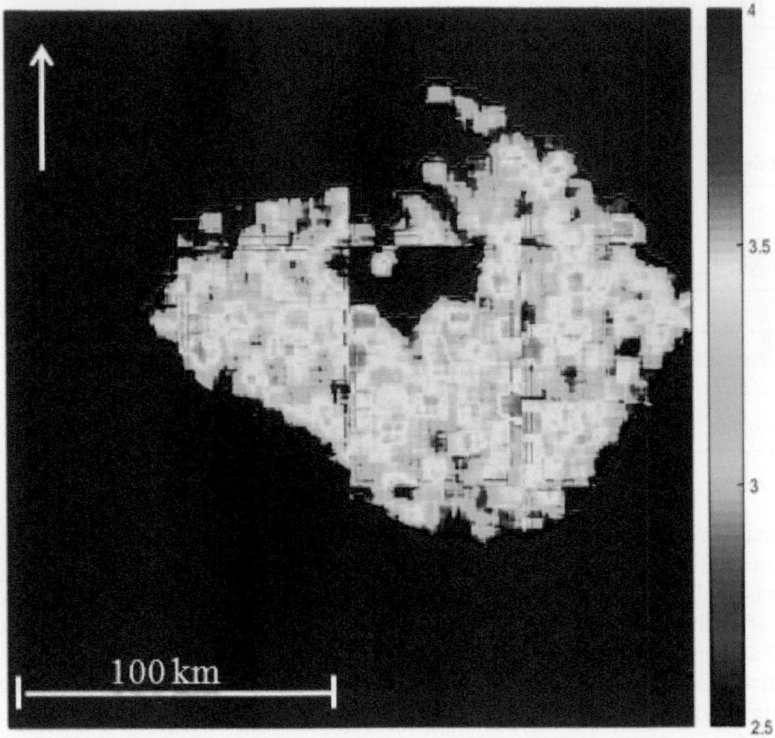

Figure 12.15. Age map of the floor of crater Tsiolkovsky using the CSFD obtained by the template matching method with an optimal threshold value in combination with the binned SCD algorithm. (From Salih, Lompart *et al.* (2017))

and the successful removal of these anomalies by the binned SCD algorithm (but not the unbinned SCD algorithm) indicates that the binned version of the algorithm yields plausible results. A more quantitative evaluation of the SCD algorithm is difficult because for this purpose a classification of all craters according to their primary or secondary origin would be required.

12 SUMMARY, CONCLUSION, AND FUTURE WORK

An automatic template matching–based CDA has been applied to study areas in Mare Cognitum on the Moon. The AMA has been computed for each of the 10 study areas based on the obtained CDA-based CSFD. The calibrated CDA has then been applied to another five study areas in Oceanus Procellarum with well-known AMAs. All in all, the applied template matching method yields similar results between CDA-based AMAs and AMAs manually derived by Hiesinger *et al.* (2003) with minor deviations. In most areas, the difference was within the one standard deviation error interval. The CDA-based AMAs are especially reliable for surfaces older than 3.0 Ga or younger than 2.0 Ga, encouraging the application of CDAs as a valuable tool for planetary AMA estimation. In the range of 2–3 Ga, the error bars of the AMAs are large due to an almost flat chronology function in logarithmic-linear space. Thus we caution researchers to be careful when using automatic CDAs on lunar surface areas with ages in this range. A disadvantage of the applied template matching method is that the illumination direction of the examined area needs to be known for each analyzed position in the image. If the regions do not show a uniform illumination, crater templates need to be generated for all occurring illumination conditions. This can be counteracted by (manually or automatically) detecting such variations and applying the CDA with the appropriate setting to each sub-region of homogeneous illumination.

The next step has been to apply the CDA to Kaguya TC image data of the mare-like floor of the lunar farside crater Tsiolkovsky. The AMA of the surface has been estimated based on the CSFD inferred from the detection results of the CDA. The detection threshold has been calibrated based on manual crater counts performed for a small sub-region of the crater floor. The average CDA-AMA of the crater floor is consistent with previous surface age estimates. Furthermore, a spatially resolved AMA map has been constructed by applying the CDA to overlapping sub-regions of the crater floor. We could also establish correlations between structures in the AMA map and local variations of spectral properties in Clementine UV/VIS color ratio data. Two small local negative anomalies in the AMA map could be identified as steep-sloped highland-like remnants of the original crater floor penetrating the mare fill, while positive anomalies are probably due to the presence of clusters of secondary craters. Independent of such interpretations, the proposed surface ages and mapping technique allow for a distinction between different geologic units in terms of their AMAs. It can be applied in a straightforward manner to different surface regions as long as manually determined AMAs are available for small sub-regions.

In order to account for secondary craters that might affect the CSFD, the SCD algorithm for removing secondary crater candidates from the CSFD has been presented and applied to the floor region of crater Tsiolkovsky. This algorithm is based on the statistical analysis of the Voronoi diagram of the detected crater centers. In its binned version, the SCD algorithm result in an increased homogeneity of the constructed age map and eliminates local areas of significantly increased apparent ages, which are characterized by clusters of presumable secondary craters. The SCD method does not guarantee the detection of all secondary craters because secondary craters may also be distributed in a more or less uniform way (Bierhaus *et al.*, 2001; McEwen and Bierhaus, 2006). This means that any algorithm that depends on the detection of unusual spatial crater distributions as a criterion for secondary craters will not be able to detect them completely. Because of the ambiguous nature of secondary craters, there is no definitive way to validate the actual origin of those craters. Due to the lack of reference data for the whole crater floor region, our algorithm could not be tested in a strictly quantitative and rigorous way. Nevertheless, the secondary crater fraction estimated with the binned SCD algorithm varies between 12% and 18%, and thus, is consistent with the estimated range of 5–25% of Werner *et al.* (2009) for surfaces of similar age and craters of similar diameter on Mars (notably, such a comparison may be distorted by the difference in gravity and the presence of an atmosphere on Mars).

To further improve the process, a combination of different automatic crater detectors could be used. This might provide a way to eliminate misdetections and achieve better detection results. Thus, a larger number of craters could be detected and a smaller amount of false detections would affect the AMAs. Nevertheless, our relatively simple template matching–based method has shown that it is applicable to crater detection on the Moon and thus in the future might also achieve good results on other planets, such as Mars or Mercury.

REFERENCES

Aurenhammer, F. (1991) Voronoi diagrams – a survey of a fundamental geometric data structure. *ACM Computing Surveys*, 23(3), 345–404.

Basilevsky, A.T., Neukum, G., Ivanov, B.A., Werner, S.K., van Gasselt, S., Head, J.W., Denk, T., Jaumann, R., Hoffmann, H., Hauber, E. & McCord, T. (2005) Morphology and geological structure of the western part of the Olympus Mons volcano on Mars from the analysis of the Mars Express HRSC imagery. *Solar System Research*, 39(2), 85–101.

Bierhaus, E.B., Chapman, C.R. & Merline, W.J. (2005) Secondary craters on Europa and implications for cratered surfaces. *Nature*, 437(7062), 1125–1127.

Bierhaus, E.B., Chapman, C.R., Merline, W.J., Brooks, S.M. & Asphaug, E. (2001) Pwyll secondaries and other small craters on Europa. *Icarus*, 153(2), 264–276.

Brunelli, R. (2009) *Template Matching Techniques in Computer Vision: Theory and Practice*. Wiley Publishers, Torquay, UK.

Chiu, S.N. (2003) Spatial point pattern analysis by using Voronoi diagrams and Delaunay tessellations – a comparative study. *Biometrical Journal*, 45(3), 367–376.

Cohen, J.P., Lo, H.Z., Lu, T. & Ding, W. (2016) Crater detection via convolutional neural networks. *arXiv preprint arXiv:1601.00978*.

Dundas, C.M. & McEwen, A.S. (2007) Rays and secondary craters of Tycho. *Icarus*, 186(1), 31–40.

Grumpe, A. & Wöhler, C. (2013) Generative template-based approach to the automated detection of small craters. *European Planetary Science Congress*, London, UK, Volume 8, abstract #EPSC2013-685-1.

Hapke, B. (1984) Bidirectional reflectance spectroscopy: 3. Correction for macroscopic roughness. *Icarus*, 59(1), 41–59.

Hapke, B. (2002) Bidirectional reflectance spectroscopy: 5. The coherent backscatter opposition effect and anisotropic scattering. *Icarus*, 157(2), 523–534.

Haruyama, J., Hara, S., Hioki, K., Iwasaki, A., Morota, T., Ohtake, M., Matsunaga, T., Araki, H., Matsumoto, K., Ishihara, Y. & Noda, H. (2012) Lunar global digital terrain model dataset produced from SELENE (Kaguya) terrain camera stereo observations. *Lunar and Planetary Science Conference*, The Woodlands, TX, USA, Volume XXXXIII, abstract #1200.

Hiesinger, H., Head, J.W., Wolf, U., Jaumann, R. & Neukum, G. (2003) Ages and stratigraphy of mare basalts in Oceanus Procellarum, Mare Nubium, Mare Cognitum, and Mare Insularum. *Journal of Geophysical Research: Planets*, 108(E7). doi:10.1029/2002JE001985.

Honda, C., Kinoshita, T., Hirata, N. & Morota, T. (2014) Detection abilities of secondary craters based on the clustering analysis and Voronoi diagram. *European Planetary Science Congress 2014*, Cascais, Portugal, Volume 9, abstract #EPSC2014-119.

Kneissl, T., van Gasselt, S. & Neukum, G. (2011) Map-projection-independent crater size-frequency determination in GIS environments – new software tool for ArcGIS. *Planetary and Space Science*, 59(11), 1243–1254.

Malin, M.C., Edgett, K.S., Posiolova, L.V., McColley, S.M. & Noe Dobrea, E.Z. (2006) Present-day impact cratering rate and contemporary gully activity on Mars. *Science*, 314, 1573–1577.

McEwen, A.S. & Bierhaus, E.B. (2006) The importance of secondary cratering to age constraints on planetary surfaces. *Annual Review of Earth and Planetary Science*, 34, 535–567.

McEwen, A.S., Preblich, B.S., Turtle, E.P., Artemieva, N.A., Golombek, M.P., Hurst, M., Kirk, R.L., Burr, D.M. & Christensen, P.R. (2005) The rayed crater Zunil and interpretations of small impact craters on Mars. *Icarus*, 176(2), 351–381.

Michael, G.G. (2013) Planetary surface dating from crater size – frequency distribution measurements: multiple resurfacing episodes and differential isochron fitting. *Icarus*, 226(1), 885–890.

Michael, G.G. & Neukum, G. (2010) Planetary surface dating from crater size-frequency distribution measurements: partial resurfacing events and statistical age uncertainty. *Earth and Planetary Science Letters*, 294(3), 223–229.

Michael, G.G., Platz, T., Kneissl, T. & Schmedemann, N. (2012) Planetary surface dating from crater size-frequency distribution measurements: spatial randomness and clustering. *Icarus*, 218(1), 169–177.

Neukum, G. (1983) *Meteoritenbombardement und Datierung planetarer Oberflächen*. Habilitation Dissertation for Faculty Membership, Ludwig-Maximilians-University, Munich, Germany.

Neukum, G., Ivanov, B.A. & Hartmann, W.K. (2001) Cratering records in the inner solar system in relation to the lunar reference system. *Space Science Reviews*, 96(1–4), 55–86.

Nichols, D.J., Young, R.A. & Brennan, W.J. (1974) Lunar kipukas as evidence for an extended tectonic and volcanic history of the maria. *Lunar and Planetary Science Conference*, Houston, TX, USA, Volume V. pp. 550–552.

Pasckert, J.H., Hiesinger, H. & van der Bogert, C.H. (2015) Small-scale lunar farside volcanism. *Icarus*, 257, 336–354.

Pieters, C.M., Staid, M.I., Fischer, E.M., Tompkins, S. & He, G. (1994) A sharper view of impact craters from clementine data. *Science*, 266, 1844–1848.

Robbins, S.J. & Hynek, B.M. (2014) The secondary crater population of Mars. *Earth and Planetary Science Letters*, 400, 66–76.

Robbins, S.J., Antonenko, I., Kirchoff, M.R., Chapman, C.R., Fassett, C.I., Herrick, R.R., Singer, K., Zanetti, M., Lehan, C., Huang, D. & Gay, P.L. (2014) The variability of crater identification among expert and community crater analysts. *Icarus*, 234, 109–131.

Robinson, M.S., Brylow, S.M., Tschimmel, M., Humm, D., Lawrence, S.J., Thomas, P.C., Denevi, B.W., Bowman-Cisneros, E., Zerr, J., Ravine, M.A. & Caplinger, M.A. (2010) Lunar Reconnaissance Orbiter Camera (LROC) instrument overview. *Space Science Reviews*, 150(1–4), 81–124.

Salamunićcar, G. & Loncaric, S. (2008) GT-57633 catalogue of Martian impact craters developed for evaluation of crater detection algorithms. *Planetary and Space Science*, 56(15), 1992–2008.

Salamuniccar, G. & Loncaric, S. (2010) Method for crater detection from Martian digital topography data using gradient value/orientation, morphometry, vote analysis, slip tuning, and calibration. *IEEE Transactions on Geoscience and Remote Sensing*, 48(5), 2317–2329.

Salih, A.L., Lompart, A., Grumpe, A., Wöhler, C. & Hiesinger, H. (2017) Automatic detection of secondary craters and mapping of planetary surface age based on Lunar Orbital Images. *Proceedings of the International Symposium on Planetary Remote Sensing and Mapping, International Archives of the Photogrammetry, Remote Sensing & Spatial Information Sciences*, XLII-3/W1. pp. 125–132.

Salih, A.L., Schulte, P., Grumpe, A., Wöhler, C. & Hiesinger, H. (2017) Automatic crater detection and age estimation for mare regions on the lunar surface. *Proceedings of the IEEE European Signal Processing Conference*, Kos Island, Greece. pp. 518–522.

Salih, A.L., Mühlbauer, M., Grumpe, A., Pasckert, J.H., Wöhler, C. & Hiesinger, H. (2016) Mapping of planetary surface age based on crater statistics obtained by an automatic detection algorithm. *Proceedings of the XXIII ISPRS Congress, ISPRS International Archives of the Photogrammetry, Remote Sensing and Spatial Information Sciences*, XLI-B4. pp. 479–486.

Sawabe, Y., Matsunaga, T. & Rokugawa, S. (2006) Automated detection and classification of lunar craters using multiple approaches. *Advances in Space Research*, 37(1), 21–27.

Shoemaker, E.M. (1965) Chapter 2: Preliminary analysis of the fine structure of the lunar surface in Mare Cognitum. In: *International Astronomical Union Colloquium*, Volume 5. Cambridge University Press. Cambridge, UK. pp. 23–77.

Simard, P.Y., Steinkraus, D. & Platt, J. (2003) Best practices for convolutional neural networks applied to visual document analysis. *Proceedings of tje 12th International Conference on Document Analysis and Recognition*.

Smith, D.E., *et al.* (2010) Initial observations from the Lunar Orbiter Laser Altimeter (LOLA). *Geophysical Research Letters*, 37(18). doi:10.1029/2010GL043751.

Speyerer, E.J., Robinson, M.S. & Denevi, B.W. (2011) Lunar Reconnaissance Orbiter Camera global morphological map of the Moon. *Lunar and Planetary Science Conference*, Volume XXXXII, abstract #2387. Available from: http://wms.lroc.asu.edu/lroc/view_rdr/WAC_GLOBAL.

Speyerer, E.J., Poviliatis, R.Z., Robinson, M.S., Thomas, P.C. & Wagner, R.V. (2016) Quantifying crater production and regolith overturn on the Moon with temporal imaging. *Nature*, 538, 215–218.

Stepinski, T.F., Ding, W. & Vilalta, R. (2012) Detecting impact craters in planetary images using machine learning. In: *Intelligent Data Analysis for Real-Life Applications: Theory and Practice*. IGI Global, Hershey, PA, USA. pp. 146–159.

Tyrie, A. (1988) Age dating of mare in the lunar crater Tsiolkovsky by crater-counting method. *Earth, Moon, and Planets*, 42, 245–264.

Wagner, R.V., Speyerer, E.J., Robinson, M.S. & LROC Team (2015) New mosaicked data products from the LROC team. *Lunar and Planetary Science Conference*, The Woodlands, TX, USA, Volume XXXXVI, abstract #1473.

Walker, A.S. & El-Baz, F. (1982) Analysis of crater distributions in mare units on the Lunar far side. *The Moon and the Planets*, 27, 91–106.

Werner, S.C., Ivanov, B.A. & Neukum, G. (2009) Theoretical analysis of secondary cratering on Mars and an image-based study on the Cerberus Plains. *Icarus*, 200(2), 406–417.

Yin, J., Xu, Y., Li, H. & Liu, Y. (2013) A novel method of crater detection on digital elevation models. *Proceedings of IEEE International Geoscience and Remote Sensing Symposium*, Melbourne, Australia. pp. 2509–2512.

Planetary Remote Sensing and Mapping – Wu et al.
© 2019 Taylor & Francis Group, London, ISBN 978-1-138-58415-0

Chapter 13

Small craters on the Moon: Focus on Tycho's crater rays

R. Bugiolacchi

ABSTRACT: The size-frequency distribution (SFD) of small lunar craters ≤ 30 metres in diameter is not random but follows predictable trends described by power law functions. Three prominent ejecta rays belonging to the Tycho impact crater show analogous cumulative SFDs within a radius of 1500 km. Likewise, neighbouring and distal mature mare regions display communalities in craters distribution as described by similar power law equations and representing the crater saturation equilibrium for lunar maria. A comparison of the relative bin-size representation within the range of 9–100 metres also suggests correlations, but this time linked to locality and the physical properties of the target. Derived Absolute Model Ages (AMAs) for all units are around 30 ± 10 Ma, with secondaries contributing up to 20 Ma to the AMA of a mature mare target, thus restricting the usefulness of this dating technique when using small craters to surfaces not older than around 20 Ma.

1 INTRODUCTION

The estimation of the time when a planetary surface was last resurfaced, named Absolute Model Age (AMA), is calculated from the size-frequency distribution (SFD) of impact craters compared to those from dated lunar locations (e.g., Neukum, 1977), after adjustments due to different planetary characteristics (e.g., Hartmann and Neukum, 2001). In the early days of planetary studies high-resolution imagery of the Moon was only available for a small fraction of the lunar surface (typically the landing sites), thus most crater size-frequency distribution surveys had focused on diameters in the kilometre range and above. Consequently, N(1), which represents the crater cumulative density of all craters ≥1 km per square kilometre, was adopted as a comparative standard (i.e., Hartmann, 1977), although several studies had also considered the distribution of sub-kilometre craters as temporal markers, notably Basilevsky (1976) and Schultz *et al.* (1977).

However, image coverage of the lunar surface (and beyond) has increased dramatically in the last decade, much at sub-metre spatial resolutions. This has redirected the focus to relatively small geological features such as Ina pit craters (Qiao *et al.*, 2017), ring-moat dome structure (RMDS) (Zhang *et al.*, 2017) and impact melt deposits (Krüger *et al.*, 2016), which all require a temporal frame for a correct geological interpretation. Consequently, the application of AMA techniques has progressively expanded to include the sub-kilometre or even sub-hectometre crater range.

Nevertheless, the scientific value of small craters data (<1 km diameter) on the Moon and other planetary bodies divides the planetary research community: some interpret the sub-kilometre population mainly as the product of secondary impacts, thus not suitable for cratering chronology based on size-frequency distribution (McEwen and Bierhaus, 2006; Xiao and Strom, 2012); others claim it represents mostly primary events, although with caveats (Neukum and Ivanov, 1994; Williams *et al.*, 2014). Most detractors indicate that small crater populations are theoretically in equilibrium across ancient lunar surfaces below a critical size (Gault, 1970; Hartmann, 1984; Xiao and Werner, 2015) and that the actual density of the production population is no longer preserved in the countable crater population. Furthermore, the morphologic evolution of small crater populations is strongly dependent on the physical properties of the impact site, including regolith thickness, slope and soil compaction (e.g., Basilevsky, 1976; van der Bogert *et al.*, 2017), although Richardson (2009) concluded that even on heavily saturated terrains on the Moon, such as highlands, the crater population continues to reflect the distribution shape of the Main Belt impactor population (i.e., the primary production function). However, the equilibrium model is not well constrained at the sub-hectometre crater sizes and, although targeted surveys of small craters have been carried out in selected areas (i.e., Xiao and Werner, 2015), only a few have stretched to geologically heterogeneous regions as comparison.

Consequently, the diagnostic value of the small crater fraction as chronological markers, especially for the <100 m sizes remains controversial and this paper will present new evidence regarding the usefulness of small crater data for resurfacing age derivation (Fig. 13.1). The SFD of craters within the rays originating from the Tycho impact (~110 Ma; Stöffler and Ryder, 2001) compared

Figure 13.1. Areas under investigation. NAC images (Robinson *et al.*, 2005) M1180249808RE, M111722186LE, M183288763RE, and M1111570864RE, res. ~0.9, 0.5, 0.8 and 1.6 px m^{-2}; illumination inc. angle (deg.) ~31, 39, 44 and 74 respectively. Background WAC image (Robinson *et al.*, 2005) and edited Clementine FC mosaic (Nozette *et al.*, 1994). 'Km' in figures relate to the distance between NAC centre and Tycho's central peak.

to their neighbouring areas will also help to constrain key geochronological variables such as cumulative slope indexes and the steady-state equilibrium diameter.

1.1 *Background – craters*

The craters' frequency and size distribution in a region are dependent on production rates (impactor flux), a reasonably well constrained estimation (i.e., Bottke *et al.*, 2005; Ivanov *et al.*, 2002; O'Brien and Greenberg, 2005; Richardson, 2009), but also by 'contamination' of secondary craters, which most of the time is not easily quantifiable (e.g. McEwen and Bierhaus, 2006) and the primary source identifiable. Further, the location and marking of the craters are subject to:

1 Marking uncertainties – The survey of small craters is not a straightforward process, depending on several factors such as the subjective observational skills and experience of the marker (Bugiolacchi *et al.*, 2016; Robbins *et al.*, 2014): until a reliable approach to automated crater detection is developed, probably based on high-resolution topographic data, the validity of the data will remain dependent on these factors.
2 Observational conditions – Detectability of small simple craters is also connected to varying observational conditions, such as angle of illumination, albedo (bright highlands against dark maria), shadowing and, obviously, image spatial resolution (van der Bogert *et al.*, 2017). These and many other uncertainties on the temporal evolution of craters highlight some of the fundamental problems in detection of the smaller sizes, even bordering on philosophy: does a crater's existence depend on the variable observational parameters (mainly spatial resolution of detector and illumination conditions)? Varying any of these two parameters slightly will make a(n) (unknowable) number of craters either appear or disappear (Wilcox *et al.*, 2005).

Their inevitable entropic journey to oblivion will be governed by several factors:

3 Space weathering – The unprotected lunar surface materials are fully exposed to space weathering (in the shape of micrometeorite, solar and cosmic wind bombardment) that contributes (along with the effects of from thermal expansion and contraction) to the disaggregation and modification of the exposed rocks. On a geological timescale these slow erosion rates are very small, estimated between 0.2 to 0.4 mm Ma^{-1} (Craddock and Howard, 2000; Fassett and Thomson, 2014, respectively), which means that the rim of a 100-m crater (~1.2 m high) could in theory survive to the present time since lunar formation, if left undisturbed. However, it is highly unlikely that any crater of that size would have withstood thousands of millions of years of larger impacts, including those forming the largest basins. Indeed, for instance it has been calculated that the Imbrium impact event would have covered the whole lunar surface by an ejecta blanket between 280 to 850 m (Haskin, 1997), thus resetting the crater population (to the rim height) less than around 6 km in diameter.
4 Target's physical properties – There is the strong dependency of small crater erosion on the target's physical attributes, such as the thickness of the regolith and its cohesion and compactness. On sloped terrains mass wasting is a leading degrading factor (Xiao *et al.*, 2013). Moonquakes or even relatively minor ground tremors (from a nearby large impact) would also accelerate the process of disaggregation, leading to the inevitable flattening of the crater edifice.
5 Steady state – The 'saturation equilibrium' (Gault, 1970; Hartmann, 1984; Shoemaker *et al.*, 1968), reached when new craters form at the same rate of destruction, sets a temporal limit on the accumulation (and observation) of craters with diameter below an upwardly shifting threshold size (D_{eq}), thus rendering their survey and comparison of limited value for age estimation of planetary surfaces. However, the onset of this equilibrium stage is not well constrained (both temporarily and in crater size) and it is likely that not a single lunar-wide estimate, let alone planetary, can be defined in terms of onset equilibrium diameter (Xiao and Werner, 2015) and size-frequency distribution.

6 Crater-size dependence – Basilevsky (1976) proposed a relationship between a crater's apparent diameter (D) and its lifetime in millions of years as $T_{Ma} = D_m/0.4$. Consequently, we would expect craters smaller than 40 m to become undetectable within 100 million years or so. Fassett and Thomson (2014) estimated that 20 m craters survive around 70 Ma whereas 50 m craters survive 400 Ma, but they rightly stress that these survival rates refer to the nearly complete obliteration of the excavation (99% of the original depth). Clearly, for small craters the detectable lifetime must be much shorter than the theoretical one, since a 40-m crater after 70 Ma would have a maximum depth on the order of a few centimetres. This aspect will hold important implications when comparing the craters SDF within Tycho's surge regions (estimated at ~109 Ma) and unaffected neighbouring ones, which in theory should have reached back to background levels at least for crater sizes <~70 m.

1.2 *Background – pixel size and illumination conditions*

Given the different spatial resolution of the various images, ranging from 0.5 to 1.6 m, the detectability and reliability of the data of the smallest craters would be expected to vary considerably. For instance, a 5-m diameter crater would be represented by 10-x-10-pixel grid on the highest-resolution image (186) but only 3 x 3 on the lowest one (864), below the common (although not scientifically tested yet) 5-x-5-px minimum. To address this question each image was investigated on the smallest distinguishable crater that could be marked. As we can see in Figure 13.2 a combination of very high spatial resolution and low illumination angle allows craters down to around 3 m diameter ('A') to be confidently drawn (top left panel), with 'full' pixel size given a diameter of 2.4 m or using ArcGIS projection (as used in this study) of 3.1 m.

At the other end of the scale (864) craters 'A' (bottom panels) represent my limit of confidence and it translates to crater sizes between 5 and 6 metres. Therefore, for this work I have decided to set a lower threshold of 7–8 metres (depending on the analytical approach) and discard the smaller sizes. Clearly, the full set will be made freely accessible for further investigations.

In terms of size marking errors, these are very tricky to compute and translate into helpful information. The boundary of a 7-metre crater can be estimated using the high-resolution image measuring around 15 pixels, thus with an error margin of maybe one or two (± 0.5–1 m), but the same crater in a 1.5 m px^{-1} image would by around 5 pixels, with a potential error margin of ± 1.5–3 m. Computing an acceptable error margin for every crater dataset derived from different images is therefore impractical and, in the opinion of the author, of potentially limited scientific value (in interpreting CSFD that is). At this point the choice of bin size in the analysis of the data is fundamental, given that the 7-metre crater example above would fall in either the 6 to 8 metres bin in the first case (when using 1-metre binning), or the 5 to 9 metres in the latter. This potentially would skew a fair comparison. However, crater sizes using a 'high' threshold compound the larger potential errors to three NAC image sets but not on the crucial 'mare vs ray' areas (808R, 186R and 763R) with pixel resolution better than 0.8 m.

It is worth stressing the importance of using a 'zooming technique' when 'hunting' for the smaller crater size population. As we can see from all examples, using a high magnification is surprisingly detrimental to the correct marking of a crater, since their boundaries become blurred and pixelated. This is one of the main difficulties in automated computer-based crater recognition approaches based on pixel values relationships alone, since they lack the human ability of reconstructing complex geometrical features from few 'pixel' clues (see magnifications losing any resemblance of a circular excavation when zoomed in close as per Fig. 13.2). However, often the mind reconstruction can backfire and get its elaboration wrong, like 'seeing' human faces in stains or random patterns. In short, when the eye is trained to look for craters, every cluster of a few brighter pixels (opposite the illumination direction) against low pixels is interpreted as depression and circularity attributed. Machine learning (AI) techniques could soon incorporate the best of human 'flexibility' with machine objectivity to match or even improve on the best of human efforts.

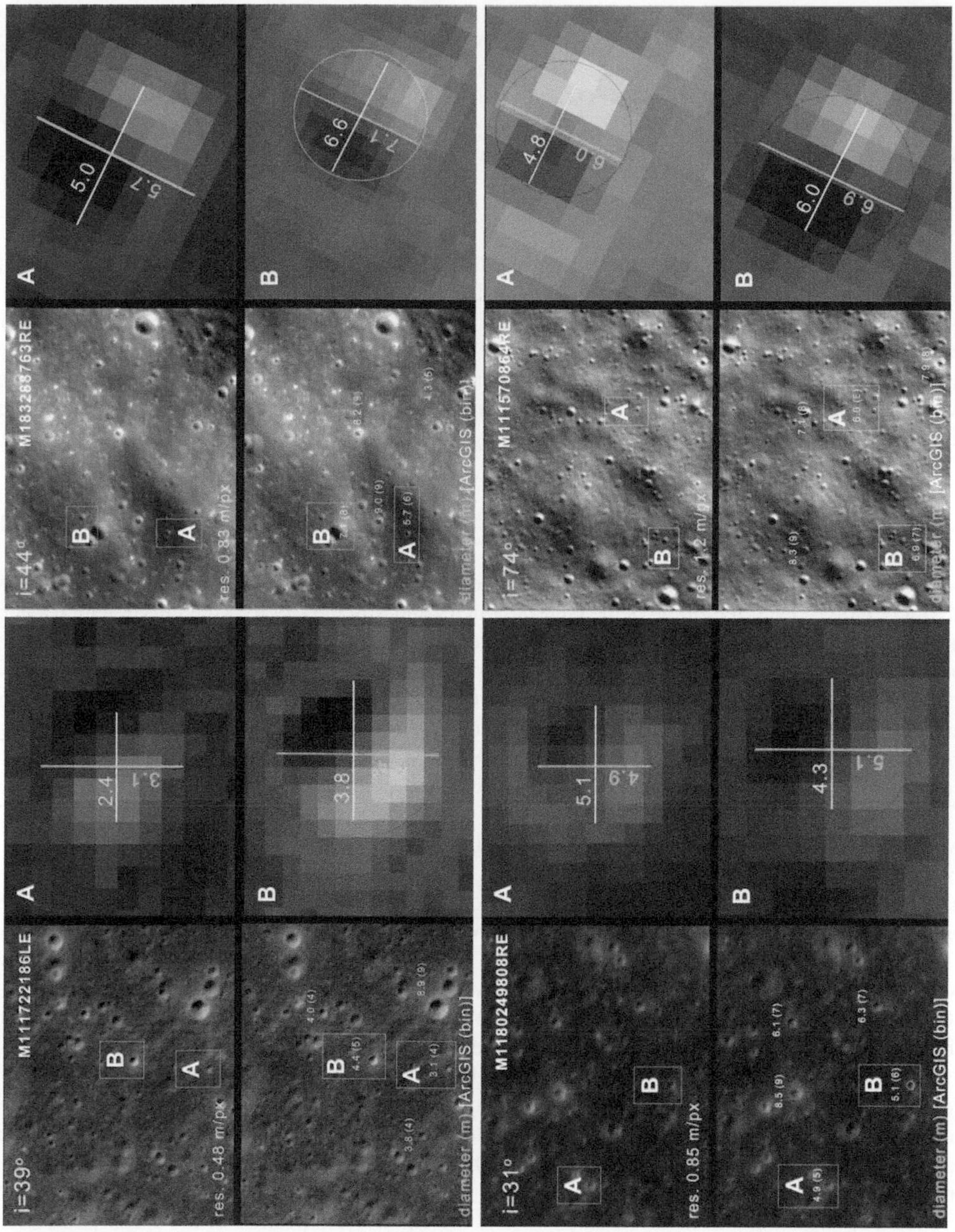

Figure 13.2. Examples of crater notations using different images with varying illumination angles and spatial resolutions. On the left of each inset, two of the smallest craters that can be marked with confidence are then zoomed in at pixel level to compare pixel size against ArcMap 'free' measurement (against lunar projection). Thus, we see on the top left frame that crater A's diameter can be described by 5 pixels (thus a diameter of 2.4 m), the hand-drawn circle gives us an effective diameter of 3.1 m. This crater when binned in the narrowest size (1 m) will be marked as a 4 m (3–4 m bin).

The range of illumination conditions (incidences from 30 to 74 degrees) does not appear to affect or bias the quantitative yield of craters, as can be deduced by the sample frames in Figure 13.2. Again, the choice of setting a 'safe' threshold level would minimise missed craters (by definition, the smallest size) in images with high illumination angles (and lower spatial resolution).

However, if systematic errors and bias arise by differences in resolution and illumination, they would surface during data analysis and considered further.

1.3 *Background – crater populations – power law comparison*

Crater populations are conventionally displayed in a log-log plot format (Crater Analysis Techniques Working Group *et al.*, 1979) as a reverse differential density of the number of craters within a size interval (bin) or as its integral, a Cumulative Size-Frequency Distribution (i.e., CSFD, the crater density per square kilometre of all craters above a certain diameter size). Since there is an inverse proportionality between number of craters and their relative sizes, data usually form lines where steepness reflects the relative density between the larger and smaller bins. Both the production and distribution of craters sizes on planetary surfaces can be expressed by an equation (e.g., Trask, 1967) of the form:

$$\mathrm{M} = C\mathrm{D}^{\pm} \qquad (1)$$

where M is the N_{cum} km^{-2} (CSFD), C is a parameter sensitive to the crater production/destruction characteristic of each region, D is the crater diameter and α is the slope of the curve, which steepness reflects the relative density of smaller craters against larger ones. Most investigated crater populations of ages <~3.8 Ga are approximated by a cumulative power law index α of -3 ± 1.

Geochronologic studies based on CSFD use the index as a classifier to compare populations and constrain their range to a specific production function, such as young/old last resurfacing, enhanced (i.e., spallation) and/or punctuated size-dependent destruction (i.e., ground shaking) or contamination from primary craters products (secondary cratering and debris surges).

Crater populations eventually reach a point of equilibrium (Gault, 1970 and references within), where the number of new craters statistically balances the number of those destroyed (reaching a 'steady-state'). Schultz *et al.* (1977) for instance found that even craters <20 m in diameter follow a –2 (α) slope. Indeed, they employ the expression $M = 0.0794\ D^{-2}$ (e.g., Shoemaker *et al.*, 1968) as an approximation of the steady-state cumulative crater densities on lunar maria.

The 'geometric saturation' represents an ideal case of crater equilibrium (theoretical maximum number of circles per square unit that do not overlap) with parameter C equal to 1.54 and α –2 (Gault, 1970); in literature we often find equilibrium expressed as a percentage of this value, in the range of 2–4% for young mare surfaces and up to ~7% for old ones (Schultz *et al.*, 1977). However, Xiao and Werner (2015) revisited the field of SFD of crater populations in equilibrium (e.g., Gault, 1970; Shoemaker *et al.*, 1968) and supported by new data concluded that the –2 index not to be universally applicable. They state that the only method to reliably evaluate equilibrium is not to refer to any empirical equilibrium density because each location holds a unique history: first it is necessary to determine the production SFD, evaluate the geological history to exclude possible local resurfacing, and then estimate which is the equilibrium portion of the crater population. According to Richardson (2009) secondaries populations would show a steep slope of–3.5 to–4.0.

2 DATA AND METHOD

2.1 *Base images and boundaries*

Aiming to maintain the highest level of uniformity in the marking environment, each NAC image was layered on one ArcGIS project page (ArcMap 10.2) using the same coordinate system, map projection (Equirectangular Moon) and geographic projection (GCS_MOON). About 100,000 craters ≤ 30 m in diameter were surveyed by the author and marked using 'CraterTools' (Kneissl *et al.*, 2011). The lowest size-confidence threshold was set to 6 pixels in diameter, translating into a 'ground' resolution between 4 and 9 metres, according to each NAC's pixel resolution. Each area was selected to minimise cross-over terrains (Fig. 13.3) and named according to the last three digits in the NAC image adding either 'R' (Ray) or 'M' (Mare) suffixes. For this work abbreviated names for the NAC images are used, using only the last three digits to aid clarity (Table 13.1). The rays

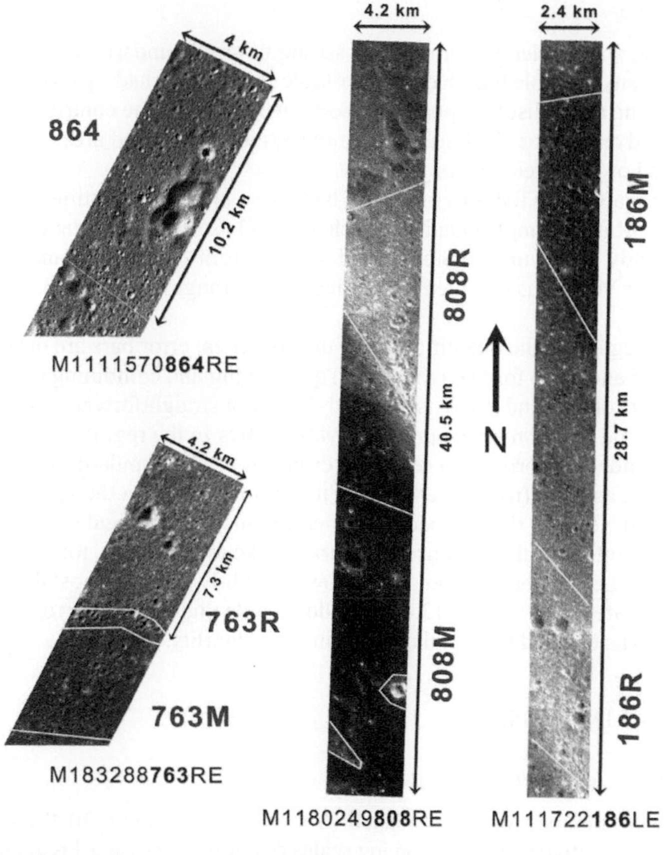

Figure 13.3. Selected areas of interest and nomenclature.

Table 13.1 NAC images details, annotated number of craters for each and areas of the surveyed regions

NAC image	*M1180249 808RE*	*M111722 186LE*	*M183288 763RE*	*M1111570 864RE*	*Total.*
Centre latitude	–26.05	–27.56	–17.87	–16.49	
Centre longitude	338.82	340.05	35.68	36.12	
Resolution (px/m)	0.86	0.50	0.82	1.56	
Inc. angle (deg.)	31	39	44	74	
# craters	35,539	25,708	12,588	8,694	**82,529**
Area MARE (km²)	63.24	16.98	21.64	37.42	
Area RAY	38.61	16.93	31.61		
Tot. Area (km²)	101.85	33.92	53.24	37.42	**226.43**

in images 808 and 186 (shown in detail in Fig. 13.3) are clearly demarked and used to define each region of interest. A buffer zone was employed to minimise 'gradation'.

Further north-east, image 763 offers the opportunity of sampling in detail the crater populations across both the crater cluster portion of the ray (763R) and south of it (763M). Centred some tens of km north, NAC 864 was selected to represent the 'tail' end of the surge and explore eventual differences in populations.

2.2 *Crater marking*

The crater marking was carried out prior to the setting of any boundaries and at varying zooming levels to minimise any possible bias due to larger-scale features or shading (NAC images were uniformly calibrated and normalised, as per Bugiolacchi *et al.*, 2016). The contribution of secondaries, outside of the rayed areas, was minimised by removing clearly affected areas (as in NAC 864) and discarding chained or clustered craters.

The sub-100 m craters, the focus of this work, have been labelled sometime Small Lunar Craters (SLCs) and given arbitrary range boundaries such as 35<D<250 m, for instance, by Mahanti *et al.*, (2018). In an already acronym-saturated field, I abstain from coining yet another nomenclature, although VSLC (for Very) was admittedly tempting. Crater ranges will be specified in each section and defined in plots.

A point of note regarding the quantification of uncertainties: error bars are omitted on some histograms (Figs. 13.5 and 13.6) for two reasons: (1) quantifying and combining the subjective bias in (a) marking a crater (or not) and (b) to its 'correct' size is not straightforward process. Indeed, work carried out comparing expert markers produced variabilities in the region of 20% (Robbins *et al.*, 2014); (2) the 'standard' approach to represent errors in crater populations statistics is 1-sigma error on N(1), taken as 1/sqrt(n), where 'n' is the number of craters in the selected range (Michael and Neukum, 2010). Clearly, this method quantifies automated statistical uncertainties only based on the number of points in each bin. Since this work, unlike many others, focuses on large numbers (n) of small craters, the error bars become so narrow to slightly smear at most the data points on the graphs (see small craters plots in Fig. 13.4 which do include undetectable error bars). Therefore, in some instances it was decided to omit them to gain in readability.

3 RESULTS AND DISCUSSION

3.1 *Chronology fits (absolute model ages)*

The crater data were plotted using CraterstatsII (Michael and Neukum, 2010), an IDL-based program, as reverse cumulative histograms on log scales (Cumulative Crater Frequency, CCF), where the size-frequency distribution is normalised to 1 km^2 area N(1).

Crater fits were produced for two ranges, one covering 9–30 m and the other 100–300 m diameters (Fig. 13.4). The lowest bin size was set to minimise potential comparative bias rising from different NAC spatial resolutions given that flattening of the curve for NACs 808, 186 and 763 occur below the 5-m diameter, against the ~8 m for the lower resolution NAC image (864). Likewise, the upper size bin was set at 300 m since most high-resolution images (i.e., 808 and 186) only contain a handful of craters above this diameter.

CCFs results for all units in the range below 30 m show four broadly common trends. The most consistent across the range is represented by the (clear) rays (186R, 763R and 808R). As we can see the modelled ages are similar both within the smaller and the larger ranges (~33.5 ± 5.5 Ma and ~145 ± 40 Ma, respectively). 864, a terrain just north of surge/visible ray shares the same distribution pattern on the smaller sizes but with a trend close to the equilibrium function (Hartmann, 1984) for the larger craters. The Cumulative Crater Frequency (CCF) for the mare, or less visibly affected bordering areas, show a broader range of values, with both units in Mare Nubium below ~30 Ma and 763M with the highest density in this survey. It is worth noting that except for 763M (probably affected by Tycho's secondary as in the higher albedo 'R' areas), all AMAs fall within the range of ~20 Ma for the 'M' regions and ~35 Ma for the 'R'. Craters >100 m are denser in the darker maria translating into ages >300 Ma (up to 1.1 Ga for 864). Clearly the survey of small craters helps in identifying the level of 'contamination' from secondary impacts even when they do not show clearly in images (Fig. 13.3). Thus, the results from the CCF plots indicate that the secondary surge from Tycho reset or modified the craters populations 100–300 m in diameter to a degree according to the proximity to the main surge. Incidentally, judging by the regional setting, 808M could represent a relatively unaffected (by secondaries) region of mature lunar maria, and the age of ~20 Ma an AMA typical of a crater population in equilibrium. At the other end of the spectrum, an age of ~40 Ma could be interpreted as the maximum AMA on the lunar maria, given the very high influx of secondaries from a relative young major primary impact. Also, the distance between the ray portions investigated, around 1000 km (Fig. 13.1) suggests either or both a uniformity in the production population or, again, an upper limit to the visible number (density) of craters.

3.2 *One-meter binning – cumulative*

Craters' cumulative size-frequency distributions are commonly plotted as histograms with diameter boundaries set within ranges derived from only a few approaches (mostly, pseudo-log and root-2). Binning always represents a compromise between a comprehensive and fair representation against ease of interpretation, given the potential diameter range spanning at least five orders of magnitude. Further, binning helps in averaging distributions within investigated ranges and aid comparison. Since this study focuses on a narrow diameter range and based on high-resolution images, the smallest binning size of 1 metre could be applied.

The highest spatial resolution of NAC images 186 (0.5 m px^{-1}) and 808 (0.86 m px^{-1}) allows for the analysis of smaller crater sizes than those used throughout this work (≥ 9 m). When all craters are plotted (except from these two NACs), we notice a general 'flattening' of the cumulative curves starting at around 10 m diameter which was originally interpreted as the result of imagery resolution limitation (i.e., Gault, 1970; Schultz *et al.*, 1977) but in later work often described as resulting from the relative higher destruction rate of small craters due to several physical factors (from shaking to target's properties). NAC 186 allows mapping of craters down to around 3 m in size; Figure 13.5-a shows a 'linear' cumulative trend (in log-antilog space) for the 'mare' area which can be described as in (1), a least square fit:

$$N_{cum}=0.007*D^{-2.14}, (R^2>0.99) \quad (2)$$

which holds true at least down to craters of 4 m in diameter.

The differences in image resolution make affects the number/density of surveyed craters: M*186, with the best spatial resolution of 0.5 m px^{-1} against ≥0.8 m px^{-1} for the others, hints at a potentially 'straight' power law cumulative trend for all other regions too (Fig. 13.5(a)). Further, the SFD similarities in the M*186 area suggest that the surface brightness is not correlated to the density of small 'new' craters but it is most likely due to the surge-led surface upturning, which exposed brighter, i.e. less weathered regolith.

NAC*808 instead (Fig. 13.5(b) shows a clearer difference in density between the upper (R) and lower (M) areas, although they appear to converge around the 30-m diameter. The rollover at ~8 m is consistent with the lower image resolution. The three areas in Mare Nectaris, although representing very different surface morphologies (Fig. 13.2), share very similar SFDs (Fig. 13.5-c), in particular, 864 and 763R.

Cumulative fits have indexes between −2.6 and−2.4 except for 808M and 186M (−2.0 and −2.1 respectively, see Table 13.2), confirming that secondary craters 'steepen the slope', i.e., preferentially boosting the population of smaller craters sizes. The data point to a threshold at around 20 m in diameter.

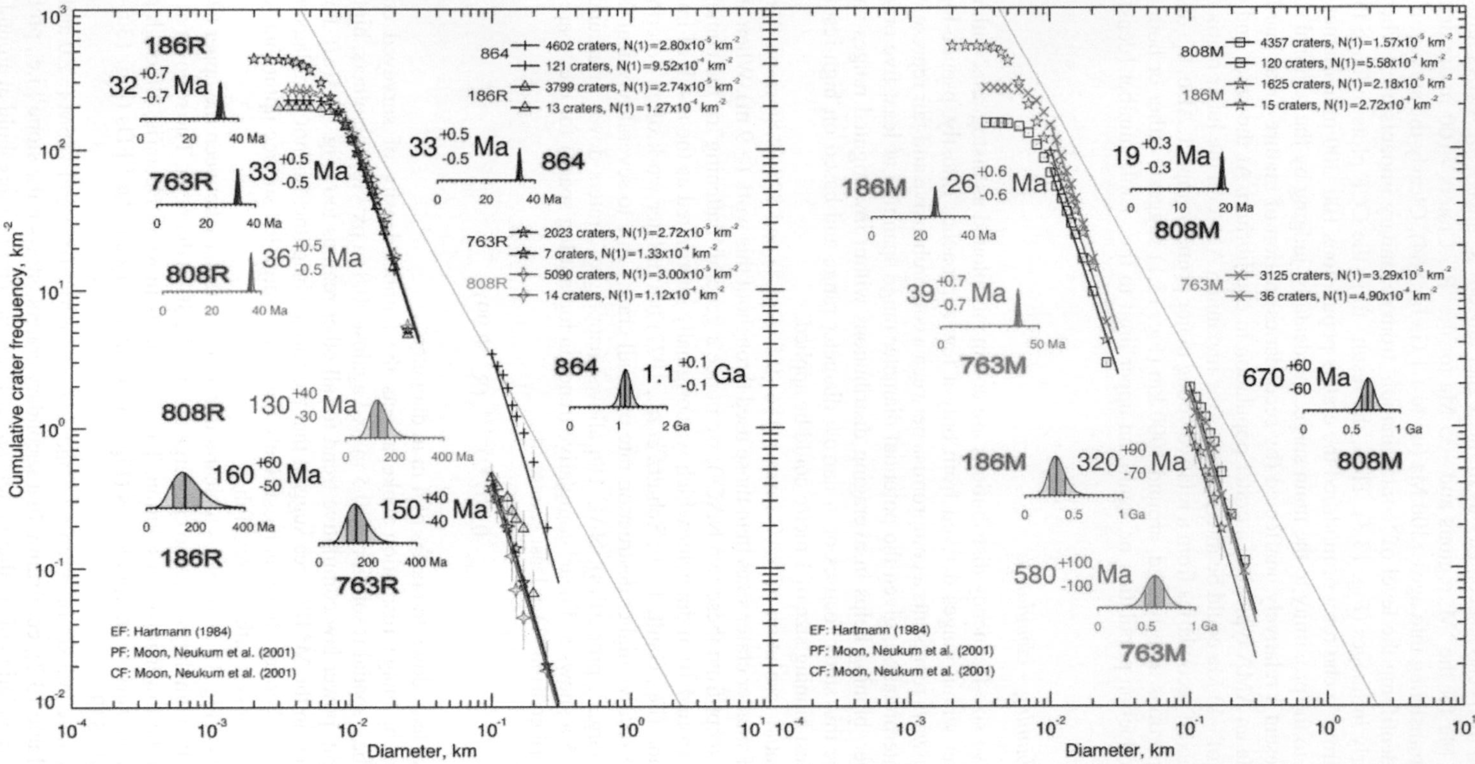

Figure 13.4. Cumulative size-frequency plots showing the number of craters greater than a given size per km^2 in a log-antilog histogram. Absolute Model Ages (AMA) are calculated by matching crater slopes with isochrones slopes calibrated against returned lunar samples. Poisson timing analysis (Michael *et al.*, 2016) and cumulative resurfacing correction are employed (Michael and Neukum, 2010). Pseudo-log binning. AMAS are produced for two bin sizes, the smallest 9–30 m, and in the range of 100–300 m, depending on individual dataset. Modelled ages for the smaller sizes are all within 29±10 Ma (see Table 13.2).

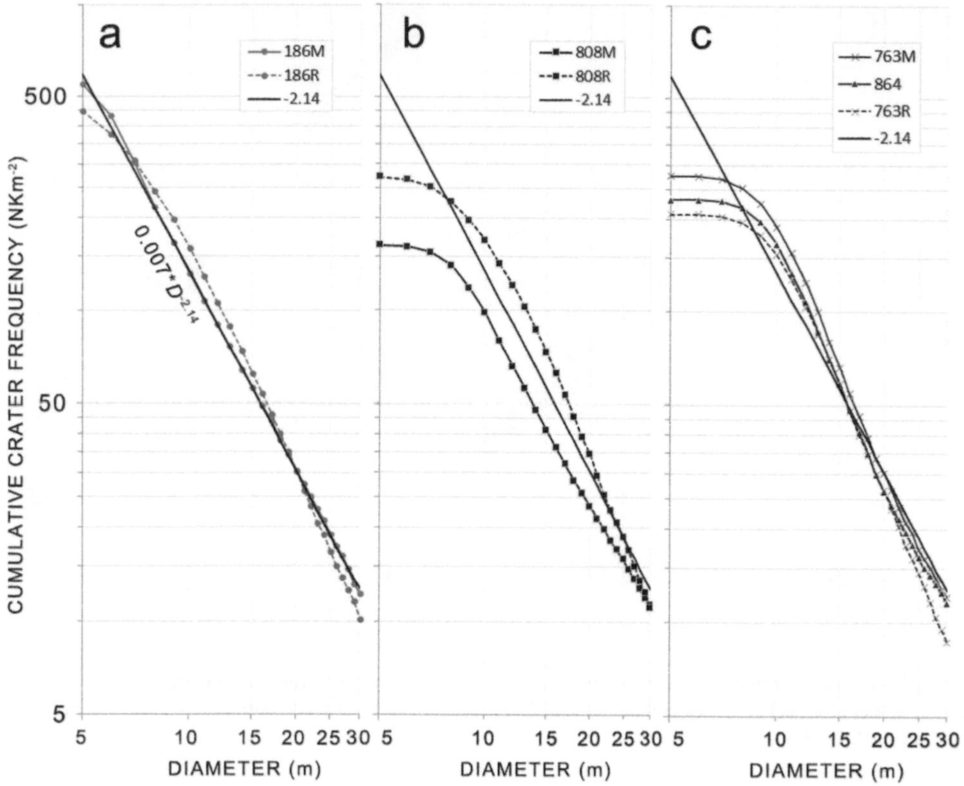

Figure 13.5. Cumulative craters SFD classified according to NAC.

Table 13.2 Fit lines represent: $M_{cum} = C*D^{\alpha}$, *where M is cumulative frequency* (Nkm^{-2}), *C* a parameter sensitive to the crater production/destruction of the area, D is the crater diameter (km), and α the slope of the curve. AMAs from plots in Figure 13.4.

Unit	*Cum. C*	*Cum. α*	*AMA (Ma)*	*AMA (Ma)*
763R	8.00E-04	–2.7	33 ± 0.5	150 ± 40
763M	1.30E-03	–2.6	39 ± 0.7	580 ± 100
808R	1.50E-03	–2.6	36 ± 0.5	130 ± 35
864	1.80E-03	–2.5	33 ± 0.5	1100 ± 200
186R	2.50E-03	–2.4	32 ± 0.7	160 ± 55
186M	7.00E-03	–2.1	26 ± 0.6	320 ± 80
808M	1.20E-02	–2.0	19 ± 0.3	670 ± 60

3.3 *Crater populations – relative comparison*

The relative representation of craters bin sizes can be used to explore further several aspects of size-frequency distributions that cumulative methods have highlighted. The extended range size 9–100 m was selected and the number of craters (i.e., not cumulative) in each 1-m bin normalised to 100, i.e. percentage. Thus, each plot reveals the relative 'weight' in terms of number craters in each bin for each unit.

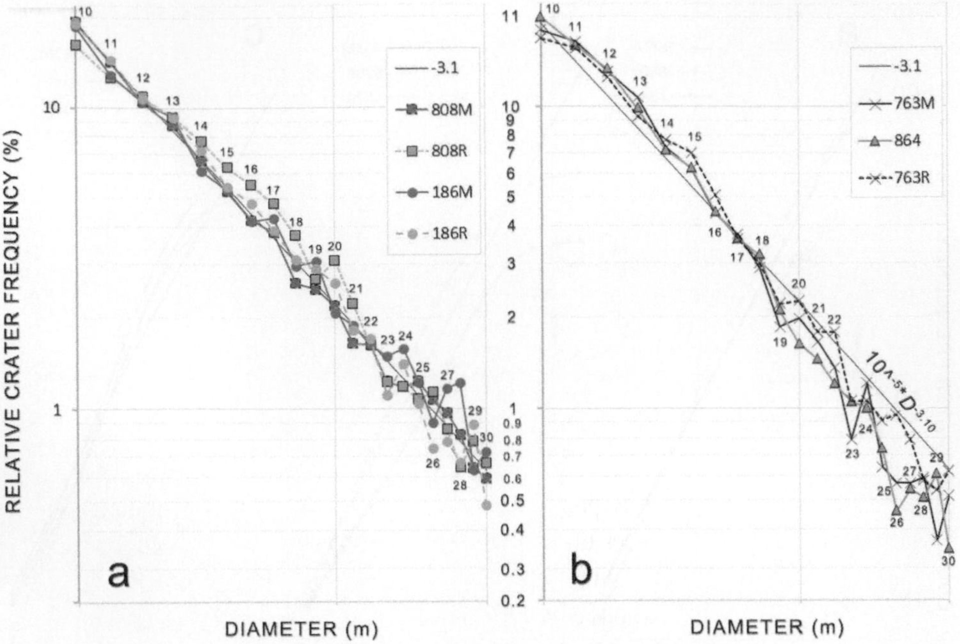

Figure 13.6. (a–b). Relative crater distribution expressed as percentage: for instance, in 'a' the 11–12-m craters represent around 12% of all craters in the 9–100 m range. For clarity and comparison, the range displayed here is cut off at 30 m.

The relative representation reveals first-order crater-size frequency distributions communalities not related to secondaries input or relative age of the impact area, rather to geographical locations, i.e., Mare Nubium and Mare Nectaris (Figs. 13.6(a) and 13.6(b), respectively).

The relation between bin sizes is nearly identical for NAC 186, irrespective to the degree of observable secondaries (see for instance 16–20 m). These three units share a power relation with a –3.1 slope (least-square fit). The 'signature' trend is even more evident within Mare Nectaris (Fig. 13.6(b)) with the three units showing similar relative distributions. Also, note that the plotting is in a log-log system so the differences between the larger craters are much exaggerated (fractions of 1%) in comparison to the smaller ones.

The geographical connection with crater distribution ranges could be explained by the two factors mentioned previously: a quantised projectile size distribution and/or preferential erosional decay/destruction according to crater diameter governed by the stratigraphy of the target region.

CONCLUSIONS

This study compared the size-frequency distribution of craters across lunar maria in the Nubium and Nectaris basins in areas clearly affected by secondary cratering related to the Tycho impact with those nearby which appear relatively unaffected. Several analytical approaches were applied to seek common patterns among the size-frequency distribution of the smaller crater fractions (including sub-30 m in diameter), which are generally considered either void of scientific relevance (due to secondaries and production equilibrium factors) or, contrariwise, meaningful in deriving Absolute Modelled Ages (AMAs).

These are the key findings of this work (Table 13.2 offers a quantitative summary):

1 *Size-frequency distribution data of sub-hectometer craters should only be employed with extreme caution to derive modelled ages.* AMAs for the units (fits to those of Neukum *et al.*, 2001) are within a very narrow range of 30±10 Ma (Section 3.1). The most cratered areas (by secondaries) are 'the oldest', and the least affected maria the 'youngest'. Clearly this cannot be the case, although the cumulative SFD slopes are good fits to the Neukum ones. Indeed, a growing number of planetary surfaces studies (including Mars and Mercury) are producing surface ages for limited geographical areas of around 20–40 Ma (e.g., Krüger *et al.*, 2016; Zhang *et al.*, 2017). This study proposes similar crater size-frequency distributions estimations for all lunar mare surfaces of ages > ~20 Ma, when not 'contaminated' by significant secondary surges. This might be due to the smaller craters reaching a steady-state production level within a few tens of millions of years of resurfacing, which then produces young AMAs. Thus, evidence for the proposed 'young ages' for INAs, ejecta pools, flow slopes, etc. should be sought elsewhere instead of geochronological methods based on small crater statistics. Also, we need to stress that AMAs for craters <1 km in size are calculated based on extrapolation of the production function expressed as a polynomial (e.g., Neukum *et al.*, 2001), calibrated for diameters ≥ 1 km, which may lead to gross errors and unrealistic derivations. Accordingly, one can predict that even the most heavily 'contaminated' (by small secondaries) regions of the Moon would probably produce similar CSFD to 808R or 186R, showing a good AMA fit of up to 40 Ma, while mature maria would have instead AMAs close to 20 Ma.

Larger sizes (here sampled up to ~300 m in diameter) again do not approach the real age estimate for the surface maria (all >3 Ga). However, the ray populations show a clear communality in size-frequency distributions, suggesting a uniform size-dependent production function of ejecta debris from the Tycho impact.

2 *Small craters display a non-stochastic distribution.* The analysis of small craters indicates that their SFD is not random but showing patterns reflecting locality and secondaries' density (in this case mostly from Tycho); an analysis of the cumulative distribution in terms of density per square kilometre suggest that the slope reflects the influx of secondaries which steepen the slope (Section 3.2). This relationship has been known for a long time, but here we offer an alternative quantitative estimation expressed as power laws (Table 13.2). It is likely that, in the size range under investigation, the crater production equilibrium for mature mare surfaces approaches a cumulative slope of −2 and around −2.5 for the saturated rays.

There is evidence that lunar craters might not display a cumulative 'production rollover', as found in many studies, and it could be due solely to image resolution or illumination factors.

3 *Usefulness of a Relative Fraction analysis of CSFD.* The use of a relative fraction analysis (i.e. factorised crater density in a given range to 100, expressed as percentage) offers a different prospective to CSFD studies and use actual Nkm^{-2} values, not cumulative (Section 3.3). While the other classification methods used in this work tend to group similar units in terms of craters SFD (i.e., rayed vs 'background' mare), the factorised distribution instead finds similarities based on location (Mare Nectaris, NACs 763 and 864, and Mare Nubium, 808 and 186). Overall the plots (Fig. 13.6) tell us something unexpected, i.e., that the relative crater-size distribution of small craters on most mare surfaces can be described by a standard power law. For instance, based on my findings, one could predict that, for instance, in the crater-size range of 9–100 m, 7–8% of all the craters will be 14 m in diameter, including the heavily 'contaminated' rayed regions (see Fig. 13.6).

Further work is planned to extend this investigation to the Apollo 17 landing region and the secondary crater clusters in the Taurus-Littrow valley allegedly related to the Tycho impact (e.g., Wolfe *et al.*, 1975).

This study was supported by the Macau Science and Technology Development Fund. 119/2017/A3

REFERENCES

Basilevsky, A.T. (1976) On the evolution rate of small lunar craters. In: Merrill, R.B. (ed) *Lunar and Planetary Science Conference Proceedings*. pp. 1005–1020.

Bottke, W.F., Durda, D.D., Nesvorný, D., Jedicke, R., Morbidelli, A., Vokrouhlický, D. & Levison, H. (2005) The fossilized size distribution of the main asteroid belt. *Icarus*, 175, 111–140. doi:10.1016/j.icarus.2004.10.026.

Bugiolacchi, R., Bamford, S., Tar, P., Thacker, N., Crawford, I.A., Joy, K.H., Grindrod, P.M. & Lintott, C. (2016) The Moon Zoo citizen science project: preliminary results for the Apollo 17 landing site. *Icarus*, 271, 30–48. doi:10.1016/j.icarus.2016.01.021.

Craddock, R.A. & Howard, A.D. (2000) Simulated degradation of lunar impact craters and a new method for age dating farside mare deposits. *Journal of Geophysical Research*, 105(E8), 20387–20401. doi:10.1029/1999JE001099.

Crater Analysis Techniques Working Group, Arvidson, R.E., Boyce, J., Chapman, C., Cintala, M., Fulchignoni, M., Moore, H., Neukum, G., Schultz, P., Soderblom, L., Strom, R., Woronow, A. & Young, R. (1979) Standard techniques for presentation and analysis of crater size-frequency data. *Icarus*, 37, 467–474. doi:10.1016/0019-1035(79)90009-5.

Fassett, C.I. & Thomson, B.J. (2014) Crater degradation on the lunar maria: topographic diffusion and the rate of erosion on the Moon. *Journal of Geophysical Research: Planets*, 119, 2255–2271. doi:10.1002/2014JE004698.

Gault, D.E. (1970) Saturation and equilibrium conditions for impact cratering on the lunar surface: criteria and implications. *Radio Sci*ence, 5, 273–291. doi:10.1029/RS005i002p00273.

Hartmann, W.K. (1977) Relative crater production rates on planets. *Icarus*, 31, 260–276. doi:10.1016/0019-1035(77)90037-9.

Hartmann, W.K. (1984) Does crater "saturation equilibrium" occur in the solar system? *Icarus*, 60, 56–74. doi:10.1016/0019-1035(84)90138-6.

Hartmann, W.K. & Neukum, G. (2001) Cratering chronology and the evolution of Mars. *Space Science Reviews*, 96, 165. doi:10.1023/A:1011945222010.

Haskin, L.A. (1997) The distribution of Th on the moon's surface. *Lunar and Planetary Science Conference Proceedings*, 28, 519.

Ivanov, B.A., Neukum, G., Bottke, W.F. & Hartmann, W.K. (2002) The comparison of size-frequency distributions of impact craters and asteroids and the planetary cratering rate. In: *Asteroids III*, University of Arizona Press, Tucson, AZ, USA. pp. 89–101.

Kneissl, T., van Gasselt, S. & Neukum, G. (2011) Map-projection-independent crater size-frequency determination in GIS environments: new software tool for ArcGIS. *Planetary and Space Science*, 59(11–12), 1243–1254. doi:10.1016/j.pss.2010.03.015.

Krüger, T., van der Bogert, C.H. & Hiesinger, H. (2016) Geomorphologic mapping of the lunar crater Tycho and its impact melt deposits. *Icarus*, 273, 164–181. doi:10.1016/j.icarus.2016.02.018.

Mahanti, P., Robinson, M.S., Thompson, T.J. & Henriksen, M.R. (2018) Small lunar craters at the Apollo 16 and 17 landing sites – morphology and degradation. *Icarus*, 299, 475–501. doi:10.1016/j.icarus.2017.08.018.

McEwen, A.S. & Bierhaus, E.B. (2006) The importance of secondary cratering to age constraints on planetary surfaces. *Annual Review of Earth and Planetary Science*, 34, 535–567. doi:10.1146/annurev.earth.34.031405.125018.

Michael, G.G. & Neukum, G. (2010) Planetary surface dating from crater size-frequency distribution measurements: partial resurfacing events and statistical age uncertainty. *Earth and Planetary Science Letters*, 294, 223–229. doi:10.1016/j.epsl.2009.12.041.

Michael, G.G., Kneissl, T. & Neesemann, A. (2016) Planetary surface dating from crater size-frequency distribution measurements: Poisson timing analysis. *Icarus*, 277, 279–285. doi:10.1016/j.icarus.2016.05.019.

Neukum, G. (1977) Lunar cratering. *Philosophical transactions of the Royal Society of London. Series A, Mathematical and Physical Sciences*, 285(1327), 267–272.

Neukum, G. & Ivanov, B.A. (1994) Crater size distributions and impact probabilities on earth from Lunar, terrestrial-planet, and asteroid cratering data. In: Gehrels, T., Matthews, M.S. & Schumann, A.M. (eds) *Hazards Due to Comets and Asteroids*, University of Arizona Press, Tucson, AZ, USA. p. 359.

Neukum, G., Ivanov, B.A. & Hartmann, W.K. (2001) Cratering records in the inner Solar System in relation to the lunar reference system. *Space Science Reviews*, 96(1), 55–86. doi:10.1023/A:1011989004263.

Nozette, S., *et al.* (1994) The clementine mission to the Moon: scientific overview. *Science*, 266, 1835–1839. doi:10.1126/science.266.5192.1835.

O'Brien, D.P. & Greenberg, R. (2005) The collisional and dynamical evolution of the main-belt and NEA size distributions. *Icarus*, 178, 179–212. doi:10.1016/j.icarus.2005.04.001.

Qiao, L., Head, J., Wilson, L., Xiao, L., Kreslavsky, M. & Dufek, J. (2017) Ina pit crater on the Moon: extrusion of waning-stage lava lake magmatic foam results in extremely young crater retention ages. *Geology*, 45(5), 455–458. doi:10.1130/G38594.1.

Richardson, J.E. (2009) Cratering saturation and equilibrium: a new model looks at an old problem. *Icarus*, 204, 697–715. doi:10.1016/j.icarus.2009.07.029.

Robbins, S.J., Antonenko, I., Kirchoff, M.R., Chapman, C.R., Fassett, C.I., Herrick, R.R., Singer, K., Zanetti, M., Lehan, C., Huang, D. & Gay, P.L. (2014) The variability of crater identification among expert and community crater analysts. *Icarus*, 234, 109–131. doi:10.1016/j.icarus.2014.02.022.

Robinson, M.S., Eliason, E.M., Hiesinger, H., Jolliff, B.L., McEwen, A.S., Malin, M.C., Ravine, M.A., Roberts, D., Thomas, P.C. & Turtle, E.P. (2005) LROC – Lunar Reconnaissance Orbiter Camera. In: Mackwell, S. & Stansbery, E. (eds) *Lunar and Planetary Science Conference*, League City, TX, USA.

Schultz, P.H., Greeley, R. & Gault, D. (1977) Interpreting statistics of small lunar craters. *Lunar and Planetary Science Conference*, Volume 3. pp. 3539–3564.

Shoemaker, E.M., Batson, R.M., Holt, H.E., Morris, E.C., Rennilson, J.J. & Whitaker, E.A. (1968) Television observations from Surveyor 3. *Journal of Geophysical Research*, 73(12), 3989–4043. doi:10.1029/JB073i012p03989.

Stöffler, D. & Ryder, G. (2001) Stratigraphy and isotope ages of lunar geologic units: chronological standard for the inner solar system. *Space Science Reviews*, 96(1/4), 9–54.

Trask, N.J. (1967) Distribution of Lunar craters according to morphology from Ranger VIII and IX photographs. *Icarus*, 6(1–3), 270–276. doi:10.1016/0019-1035(67)90023-1.

Van der Bogert, H., Hiesinger, H., Dundas, C.M., Krüger, T, Ewen, A.S., Zanetti, M. & Robinson, M.S. (2017) Origin of discrepancies between crater size-frequency distributions of coeval lunar geologic units via target property constraints. *Icarus*, 298, 49–63. doi:10.1016/j.icarus.2016.11.040.

Wilcox, B.B., Robinson, M.S., Thomas, P.C. & Hawke, B.R. (2005) Constrains on the depth and variability of the lunar regolith. *Meteoritics & Planetary Science*, 40(5), 695–710.

Williams, J.P., Pathare, A.V. & Aharonson, O. (2014) The production of small primary craters on Mars and the Moon. *Icarus*, 235, 23–36. doi:10.1016/j.icarus.2014.03.011.

Wolfe, E.W., Lucchitta, B.K., Reed, G.E., Ulrich, G.E. & Sanchez, A.G. (1975) Geology of the Taurus-Littrow valley floor. *Lunar and Planetary Science Conference*, Houston, TX, USA, Volume 3. pp. 2463–2482.

Xiao, Z. & Strom, R.G. (2012) Problems determining relative and absolute ages using the small crater population. *Icarus*, 220, 254–267. doi:10.1016/j.icarus.2012.05.012.

Xiao, Z. & Werner, S.C. (2015) Size-frequency distribution of crater populations in equilibrium on the Moon. *Journal of Geophysical Reserch: Planets*, 120, 2277–2292. doi:10.1002/2015JE004860.

Xiao, Z., Zeng, Z., Ding, N. & Molaro, J. (2013) Mass wasting features on the Moon – how active is the lunar surface? *Earth and Planetary Science Letters*, 376, 1–11. doi.org/10.1016/j.epsl.2013.06.015.

Zhang, F., Head, J.W., Basilevsky, A.T., Bugiolacchi, R., Komatsu, G., Wilson, L. & Zhu, M.-H. (2017) Newly discovered ring-moat dome structures in the lunar maria: possible origins and implications. *Geophysical Research Letters*, 44, 9216–9224. doi:10.1002/2017GL074416.

Planetary Remote Sensing and Mapping – Wu et al.
© 2019 Taylor & Francis Group, London, ISBN 978-1-138-58415-0

Chapter 14

Elemental and topographic mapping of lava flow structures in Mare Serenitatis on the Moon

A. Grumpe, C. Wöhler, D. Rommel, M. Bhatt and U. Mall

ABSTRACT: In this study we present elemental abundance maps, a petrological map and a digital terrain model (DTM) of a lava flow structure in northern Mare Serenitatis at (18.0°E, 32.4°N) and two possible volcanic vents at (11.2°E, 24.6°N) and (13.5°E, 37.5°N), respectively. The abundance maps of the refractory elements Ca, Mg, Fe, Al and the petrological map are obtained based on hyperspectral image data from the Moon Mineralogy Mapper (M3) instrument. The DTM is constructed using the Lunar Reconnaissance Orbiter (LRO) LOLA and Kaguya Terrain Camera (TC) merged SLDEM512 data in combination with a shape-from-shading based method using M3 and LRO Narrow Angle Camera (NAC) image data. The obtained NAC-based DTM has a very high effective resolution of about 1–2 m, which comes close to the resolution of the utilized NAC images without requiring intricate processing of NAC stereo image pairs. As revealed by our elemental maps and DTM, the examined lava flow structure occurs on a boundary between basalts consisting of low-Ca/high-Mg pyroxene and high-Ca/low-Mg pyroxene lithologies, respectively. The total thickness of the lava flow is about 100 m, which is composed of two or more layers based on our DTM.

1 INTRODUCTION

The detection of lunar lava flows based on local morphology highly depends on the available high-resolution images (Hiesinger *et al.*, 2002). The thickness of lava flows, however, has been studied by many researchers using different available data sources, e.g., multispectral Clementine images and Telescopic images. Lunar lava flows are shown to be as thick as 200 m (e.g., Gifford and El-Baz, 1981; Hiesinger *et al.*, 2002; Neukum and Horn, 1976), e.g., Imbrian-aged flow units in Mare Imbrium. Lunar lava flows are supposed to be concentrated on the northwestern lunar nearside (Hiesinger *et al.*, 2002 citing Gifford and El-Baz, 1981). Hiesinger *et al.* (2002) contributed this finding to the tendency of mascon mare to undergo subsidence and lava units are thus likely to form ponds. Furthermore, Hiesinger *et al.* (2002) suggested that later mare flows are emplaced on surfaces, which have been smoothed by previous events. These later mare flows are thus more likely to be spread laterally rather than forming ponds. Neither Hiesinger *et al.* (2002) nor Gifford and El-Baz (1981) found lunar lava flows in Mare Serenitatis.

The detection of lava flows is not a simple task since they are thin structures that do not show a strong shading signature unless the incidence angle of the solar rays is quite large. Hiesinger *et al.* (2002) presented a method to detect lava flows and to estimate the lava flow thickness using crater-size frequency distributions.

In this study we apply a digital terrain model (DTM) refinement method to very high-resolution Lunar Reconnaissance Orbiter (LRO) Narrow Angle Camera (NAC) (Chin *et al.*, 2007) images and the combined LRO Lunar Orbiter Laser Altimeter (LOLA) and Kaguya Terrain Camera (TC) DTM Merger termed SLDEM512 (Barker *et al.*, 2016). The resulting DTM inherits the lateral resolution of about 1 m per pixel from the NAC images while the lower resolution lateral resolution of the SLDEM512 is about 70 m per pixel at the equator. Using this method, w compute high-resolution

DTMs of a lava flow in Mare Serenitatis (18.0°E, 32.7°N). The same technique is applied to two possible volcanic vents from the same basin. Additionally, elemental abundance maps and a petrological map of the lava flow structure in northern Mare Serenitatis are computed based on hyperspectral image data of the Moon Mineralogy Mapper (M^3) instrument (Pieters *et al.*, 2009). This chapter is a significant extension of the conference proceedings paper of Wöhler *et al.* (2017).

2 DATASETS

2.1 *LRO Narrow Angle Camera (NAC)*

The Narrow-Angle Camera (NAC) (Robinson *et al.*, 2010) is a subsystem of LROC. It provides photometric measurements of the lunar surface at a resolution of up to 0.5 m per pixel, which is the highest resolution available. There are two different but parallel monochrome pushbroom cameras, the left NAC and the right NAC, which cover a surface width of about 5 km per swath in total. The distributed calibrated data records contain the so-called "intensity over flux" (IoF), which is the measured and calibrated radiance divided by the solar flux, i.e., the estimated reflectance of the surface. The NAC images are thus suitable for photometric surface refinement methods such as photoclinometry or SfS. The IoF data are accompanied by the selenographic coordinates of the image center and the image corners. Additionally, the position of the sun is specified in sub-solar longitude, sub-solar latitude and solar distance.

2.2 *Merged LRO Lunar Orbiter laser altimeter and Kaguya Terrain Camera DTM (SLDEM512)*

The merged LRO Lunar Orbiter Laser Altimeter (LOLA) and Kaguya Terrain Camera (TC) DTM (SLDEM512) has been created by the LOLA and Kaguya teams (Barker *et al.*, 2016). It covers latitudes within ±60°. The horizontal resolution of the DTM is 512 pixels per degree, i.e., about 59 m at the equator. The SLDEM512 was derived by co-registering 43,200 stereo analyses of TC image based DTMs to about 4.5×10^9 topographic height measurements of LOLA. Additionally, the co-registered DTMs were used to correct orbital and pointing geolocation errors from LOLA measurements. The root-mean-square residuals are less than 5 m for 90% of the TC DTMs after the co-registration procedure. Consequently, the typical vertical accuracy is given as 3–4 m (Barker *et al.*, 2016).

2.3 *Moon mineralogy mapper (M^3)*

The Moon Mineralogy Mapper (M^3) instrument carried by the Indian lunar spacecraft Chandrayaan-1 provided spectral radiance data in the range between 450 and 3000 nm. The M^3 sensor is a pushbroom camera which scans lines with a lateral extent of about 140 m per pixel (Pieters *et al.*, 2009). The spectral radiance is measured in 85 channels. The filter width of each channel is about 10 nm in the so-called "global mode". The "target mode" provides a higher spectral resolution. However, it is restricted to a few regions of interest (Pieters *et al.*, 2009) and thus not used -in this study. The M^3 data are published as images, i.e., a concatenation of scan lines. In addition to the radiance data, the spectral solar irradiance, the pixel-wise coordinates in selenographic longitude and latitude, the solar distance, the solar zenith and azimuth angles with respect to a spherical body, as well as the direction towards the spacecraft are given for each image of the M^3 sensor. These allow for the computation of sun and satellite positions with respect to local topography.

3 METHODS

In order to construct elemental and topographic maps of lava flow structures in the Mare Serenitatis region of the Moon, we use several different techniques. Section 3.1 reviews the Shape from

Shading (SfS) based method for the refinement of low-resolution DTMs. The construction of elemental abundance maps is presented in section 3.2.

3.1 *Refinement of digital terrain models*

A general illumination geometry is shown in Figure 14.1. The vectors $\vec{s}$, $\vec{v}$, and $\vec{n}$ denote the direction towards the light source, the viewer and the surface normal, respectively. Consequently, the incidence angle, the emission angle and the phase angle are denoted by ϑ_i, ϑ_e and α, respectively. The surface normal $\vec{n}$ is obtained based on the surface gradient field, i.e.:

$$\vec{n} = \frac{1}{\sqrt{\left(\frac{\partial z}{\partial x}\right)^2 + \left(\frac{\partial z}{\partial y}\right)^2 + 1}} \begin{bmatrix} -\frac{\partial z}{\partial x} \\ -\frac{\partial z}{\partial y} \\ 1 \end{bmatrix} \tag{1}$$

where $\frac{\partial z}{\partial x}$ and $\frac{\partial z}{\partial y}$ are the partial derivatives of the surface z with respect to the lateral coordinates x and y, respectively. The observed reflectance depends on the normal vector of the surface, the direction to the light source and the viewing direction. It is thus possible to determine the surface gradient field if the direction to the light source, the viewing direction and the reflectance behavior of the surface are known. Due to sensor noise and model inaccuracies the recovered gradient field, however, may be erroneous and non-integrable. It has become common practice in photometric surface recovery methods to introduce the independent photometric estimates p and q of the partial derivatives, i.e.:

$$\vec{n} = \frac{1}{\sqrt{p^2 + q^2 + 1}} \begin{bmatrix} -p \\ -q \\ 1 \end{bmatrix} \tag{2}$$

and to determine the surface z from the estimated gradient field in a subsequent step (e.g., Horn, 1990).

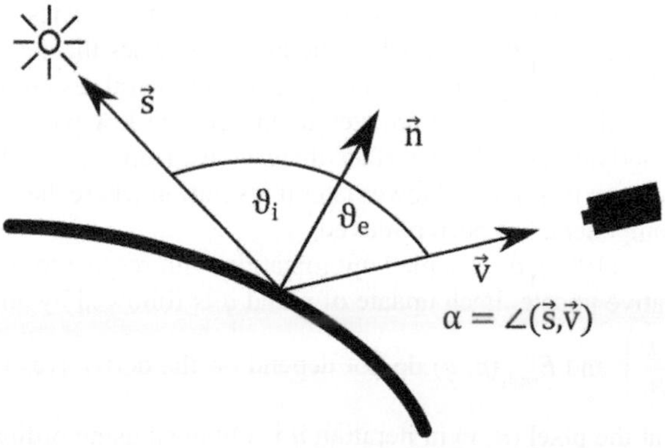

Figure 14.1. Illustration of the illumination and observation geometry (From Grumpe *et al.* (2014))

The minimization of the error:

$$E_{\mathrm{SfS}} = E_I(p,q) + \gamma E_{\mathrm{int}}\left(p,q,\frac{\partial z}{\partial x},\frac{\partial z}{\partial y}\right) + \delta E_{\mathrm{grad}}(p,q) + \tau E_{\mathrm{DTM}}(z) \tag{3}$$

with respect to p, q and z is a combination of the methods proposed by Grumpe *et al.* (2014) and Grumpe and Wöhler (2014). The intensity error

$$E_I(p,q) = \int\limits_{\vec{x}}\int\limits_{\vec{y}} \frac{1}{2}(R(p,q) - I)^2 dx\, dy \tag{4}$$

describes the squared deviation of the reflectance model $R(p, q)$ from the reflectance image I within a region defined by $\vec{x}$ and $\vec{y}$. The factors γ, δ and τ are arbitrary weights. This error term is responsible for the recovery of the surface gradient field from the photometric information contained in I. The term:

$$E_{\mathrm{int}}\left(p,q,\frac{\partial z}{\partial x},\frac{\partial z}{\partial y}\right) = \int\limits_{\vec{x}}\int\limits_{\vec{y}} \frac{1}{2}\left(\frac{\partial z}{\partial x} - p\right)^2 + \frac{1}{2}\left(\frac{\partial z}{\partial y} - q\right)^2 dx\, dy \tag{5}$$

describes the deviation of the estimated surface's gradient field from the photometric gradient estimates. $E_{\mathrm{int}}\left(p,q,\frac{\partial z}{\partial x},\frac{\partial z}{\partial y}\right)$ thus aims at finding the surface z that corresponds to the estimated gradient field. An a-priori known DTM of lower lateral resolution is included using soft constraints, i.e., a violation of the constraints is allowed but penalized by the error terms:

$$E_{\mathrm{grad}}(p,q) = \int\limits_{\vec{x}}\int\limits_{\vec{y}} \frac{1}{2}\left(f_{\mathrm{grad}}\left(\frac{\partial z_{\mathrm{DTM}}}{\partial x}\right) - f_{\mathrm{grad}}(p)\right)^2 + \frac{1}{2}\left(f_{\mathrm{grad}}\left(\frac{\partial z_{\mathrm{DTM}}}{\partial y}\right) - f_{\mathrm{grad}}(q)\right)^2 dx\, dy \tag{6}$$

and:

$$E_{\mathrm{DTM}}(z) = \int\limits_{\vec{x}}\int\limits_{\vec{y}} \frac{1}{2}\left(f_{\mathrm{DTM}}(z) - f_{\mathrm{DTM}}(z_{\mathrm{DTM}})\right)^2 dx\, dy \tag{7}$$

The former describes the deviation of the photometric gradient estimates from the gradient field of the a-priori known surface model z_{DTM}, while the latter describes the deviation of the estimated surface from z_{DTM} itself. Since z_{DTM} is assumed to be of lower lateral resolution than the image and thus of lower lateral resolution than the recovered surface, the low-pass filters f_{grad} and f_{DTM} are applied to the partial derivatives and the surface themselves, respectively. This operation may be interpreted as a comparison in a space of lower lateral resolution where the small-scale details, i.e., the high-frequency component, has been removed.

As proposed by Horn (1990), the coupled optimization with respect to p, q and z is carried out by an alternating iterative update. Each update of p and q is followed by an update of z. Since E_1 (p, q), $E_{\mathrm{int}}\left(p,q,\frac{\partial z}{\partial x},\frac{\partial z}{\partial y}\right)$ and $E_{\mathrm{grad}}(p, q)$ do not depend on the derivatives of p and q, the update equation for p and q at the pixel $(u; v)$ in iteration n is obtained using ordinary calculus (Grumpe *et al.*, 2014):

$$p_{u,v}^{(n+1)} = \left.\frac{\partial z}{\partial x}\right|_{u,v}^{(n)} - \frac{1}{\gamma}\left(R\left(\left.\frac{\partial z}{\partial x}\right|_{u,v}^{(n)}, \left.\frac{\partial z}{\partial y}\right|_{u,v}^{(n)}\right) - I_{u,v}\right)\left.\frac{\partial R}{\partial p}\right|_{\left.\frac{\partial z}{\partial x}\right|_{u,v}^{(n)}, \left.\frac{\partial z}{\partial y}\right|_{u,v}^{(n)}}^{(n)} + \frac{\delta}{\gamma}\left(\mathrm{a}_{\sigma_{\mathrm{grad}}} \circ \left(\mathrm{a}_{\sigma_{\mathrm{grad}}} * \left(\frac{\partial z}{\partial x} - \frac{\partial z_{\mathrm{DTM}}}{\partial x}\right)\right)\right), \quad (8)$$

$$q_{u,v}^{(n+1)} = \left.\frac{\partial z}{\partial y}\right|_{u,v}^{(n)} - \frac{1}{\gamma}\left(R\left(\left.\frac{\partial z}{\partial x}\right|_{u,v}^{(n)}, \left.\frac{\partial z}{\partial y}\right|_{u,v}^{(n)}\right) - I_{u,v}\right)\left.\frac{\partial R}{\partial q}\right|_{\left.\frac{\partial z}{\partial x}\right|_{u,v}^{(n)}, \left.\frac{\partial z}{\partial y}\right|_{u,v}^{(n)}}^{(n)} + \frac{\delta}{\gamma}\left(\mathrm{a}_{\sigma_{\mathrm{grad}}} \circ \left(\mathrm{a}_{\sigma_{\mathrm{grad}}} * \left(\frac{\partial z}{\partial y} - \frac{\partial z_{\mathrm{DTM}}}{\partial y}\right)\right)\right). \quad (9)$$

The low-pass function f_{grad} is set to be a Gaussian $\mathrm{a}_{\sigma_{\mathrm{grad}}}$ of width σ_{grad} and applied using the correlation operator $\circ$ and the convolution operator $*$. Following Horn (1990), the reflectance model is evaluated using the gradient field of the current surface estimate rather than the photometric estimates p and q. The error terms containing z, i.e., $E_{\mathrm{int}}\left(p, q, \frac{\partial z}{\partial x}, \frac{\partial z}{\partial y}\right)$ and $E_{\mathrm{DTM}}(z)$, also contain its partial derivatives and thus the solution is obtained by the calculus of variations in form of a partial differential equation:

$$\Delta z - \frac{\tau}{\gamma}\left(f_{\mathrm{DTM}}(z) - f_{\mathrm{DTM}}(z_{\mathrm{DTM}})\right)\frac{\partial}{\partial z} f_{\mathrm{DTM}}(z) = \frac{\partial p}{\partial x} + \frac{\partial q}{\partial y} \quad (10)$$

Where Δz denotes the Laplacian of z. Due to the lack of a computationally efficient solver, Grumpe *et al.* (2014) did not include $E_{\mathrm{DTM}}(z)$ into the SfS scheme. Grumpe and Wöhler (2014) proposed an iterative relaxation approach to solve this equation. The update of z is obtained from the quadratic equation:

$$\mathrm{c}_2 z^2 + c_1 z + c_0 = 0 \quad (11)$$

Where f_{DTM} is a discrete convolution with the Filter matrix $\mathbf{F}$ and:

$$\mathrm{c}_2 = -\hat{\tau} F_{\mathrm{c}} \frac{Z_x\left(\frac{F_{\mathrm{r}}}{2h_{\mathrm{l}}} - \frac{F_{\mathrm{l}}}{2h_{\mathrm{r}}}\right) + Z_y\left(\frac{F_{\mathrm{u}}}{2h_d} - \frac{F_{\mathrm{d}}}{2h_{\mathrm{u}}}\right)}{z_x^2 + z_y^2}, \quad (12)$$

$$\begin{aligned}\mathrm{c}_1 = {} & -\hat{\tau}\left(f_{-0}(z) - f_{\mathrm{DTM}}(z_{\mathrm{DTM}})\right)\frac{z_x\left(\frac{F_{\mathrm{r}}}{2h_{\mathrm{l}}} - \frac{F_{\mathrm{l}}}{2h_{\mathrm{r}}}\right) + z_y\left(\frac{F_{\mathrm{u}}}{2h_d} - \frac{F_{\mathrm{d}}}{2h_{\mathrm{u}}}\right)}{z_x^2 + z_y^2} - \\ & \hat{\tau} F_{\mathrm{c}} \frac{Z_x\left(\frac{F_{\mathrm{l}}}{2h_{\mathrm{r}}} z_{2\mathrm{r}} - \frac{F_{\mathrm{r}}}{2h_{\mathrm{l}}} z_{2\mathrm{l}} + f_{-\mathrm{h}}(z_x)\right) + Z_y\left(\frac{F_{\mathrm{d}}}{2h_{\mathrm{u}}} z_{2\mathrm{u}} - \frac{F_{\mathrm{u}}}{2h_{\mathrm{d}}} z_{2\mathrm{d}} + f_{-\mathrm{v}}(z_y)\right)}{z_x^2 + z_y^2} - \frac{2\left(h_{\mathrm{cx}}^2 + h_{\mathrm{cy}}^2\right)}{h_{\mathrm{cx}}^2 h_{\mathrm{cy}}^2},\end{aligned} \quad (13)$$

and

$$\begin{aligned}\mathrm{c}_0 = {} & -\hat{\tau}\left(f_{-0}(z) - f_{\mathrm{DTM}}(z_{\mathrm{DTM}})\right)\cdot \\ & \frac{Z_x\left(\frac{F_{\mathrm{l}}}{2h_{\mathrm{r}}} z_{2\mathrm{r}} - \frac{F_{\mathrm{r}}}{2h_{\mathrm{l}}} z_{2\mathrm{l}} + f_{-\mathrm{h}}(z_x)\right) + Z_y\left(\frac{F_{\mathrm{d}}}{2h_{\mathrm{u}}} z_{2\mathrm{u}} - \frac{F_{\mathrm{u}}}{2h_{\mathrm{d}}} z_{2\mathrm{d}} + f_{-\mathrm{v}}(z_y)\right)}{z_x^2 + z_y^2} \frac{h_{\mathrm{cx}}^2 z_{\mathrm{h}} + h_{\mathrm{cy}}^2 z_{\mathrm{v}}}{h_{\mathrm{cx}}^2 h_{\mathrm{cy}}^2} - \left(p_x + q_y\right).\end{aligned} \quad (14)$$

For brevity, the partial derivatives $\frac{\partial z}{\partial x}$, $\frac{\partial z}{\partial y}$, $\frac{\partial p}{\partial x}$ and $\frac{\partial q}{\partial y}$ are denoted by z_x, z_y, p_x and q_y, respectively, and $\hat{\tau} = \frac{\tau}{\gamma}$. The introduced symbols are summarized in Figure 14.2. The symbols $z_{2\mathrm{u}}$, z_{u}, z_{2d},

z_d, z_{2l}, z_l, z_{2r}, and z_r denote the adjacent pixels to $z_c = z|_{u,v}$ as shown in Figure 14.2a. $z_h = z_l + z_r$ and $z_v = z_u + z_d$ denote the sum of the horizontally and vertically edge-adjacent pixels, respectively. The horizontal extents h_l and h_r are edge-adjacent to $h_{cx} = h_x|_{u,v}$ (cf. Fig. 14.2b) and the vertical extents h_u and h_d are edge-adjacent to $h_{cy} = h_y|_{u,v}$ (cf. Fig. 14.2c). The filter coefficients F_u, F_d, F_l and F_r of the filter matrix **F** are edge-adjacent to $F_c = F_y|_{0,0}$ (cf. Fig. 14.2d). The filter functions f_{-0} $F_0 = 0$, f_{-h} ($F_l = 0$ and $F_r = 0$) and f_{-v} ($F_u = 0$ and $F_d = 0$) are obtained by setting the corresponding filter elements to zero. We set f_{DTM} to be a Gaussian g_{DTM} of width σ_{DTM}.

The update equations for $p_{u,v}^{(n+1)}$ and $q_{u,v}^{(n+1)}$ and the update equation for $z_{u,v}^{(n+1)}$ are computed in an alternating iteration, i.e., at first the values of $p_{u,v}^{(n+1)}$ and $q_{u,v}^{(n+1)}$ are computed from Equations (8) and (9) and the updated value $z_{u,v}^{(n+1)}$ is then computed from Equation (11) using the updated values for $p_{u,v}^{(n+1)}$ and $q_{u,v}^{(n+1)}$. Since the quadratic equation has two solutions and may cause an oscillating behavior, we adopt the update rule and the convergence criteria from Grumpe and Wöhler (2014). The solution to Equation (11) which is closer to the current value of $z_{u,v}^{(n)}$ is selected. Furthermore, we allow for an increase in the overall error. The number of steps since the last successful reduction in the overall error is termed n_{step} Consequently, each update of $p_{u,v}^{(n+1)}$, $q_{u,v}^{(n+1)}$, and $z_{u,v}^{(n+1)}$ is termed as a step. The algorithm stops if $N_{step} = 250$ steps are exceeded. In contrast, the iteration counter n_{iter} is increased after each step that successfully decreases the overall error. Additionally, the step counter n_{step} is reset to zero after each successful iteration and the current optimal surface estimate $z_{u,v}^{(*)}$ is saved. After termination of the algorithm, $z_{u,v}^{(*)}$ is returned. If the maximum number of iterations

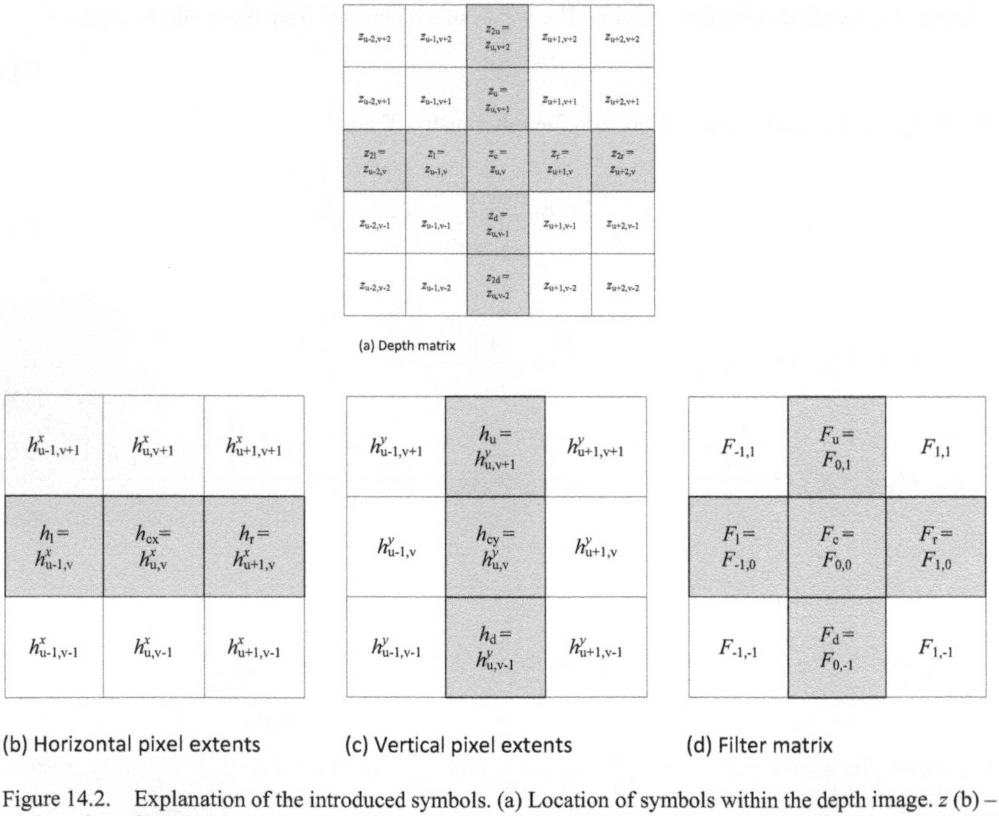

Figure 14.2. Explanation of the introduced symbols. (a) Location of symbols within the depth image. z (b) – (c) Location of symbols within the matrices of the horizontal and vertical extents h^x and h^y, respectively. (d) Location of symbols within the discrete Filter matrix **F**. (From Grumpe and Wöhler (2014))

N_{iter} = 1000 is reached the algorithm stops. Furthermore, we set a threshold E_{div} that causes the algorithm to stop if the overall error equals or exceeds E_{div}. In our experiments, the algorithm did not diverge and the threshold was set to the numerical representation of infinity.

In order to initialize the surface refinement methods, the SLDEM512 (cf. section 3.2) is selected. Its nominal resolution, however, is only 70 m per pixel at the equator and the effective lateral resolution can be assumed to be lower. To bridge the gap in the spatial resolution of NAC images (cf. section 3.1) and the SLDEM512, a pyramidal approach is applied by downscaling the image by a factor of 2 until the resolution is less than the resolution of the SLDEM, i.e., the NAC image is downscaled $n_{pyr} = 6$ times by a factor of 2 which yields an image at a resolution of about 64 m per pixel. The resulting image is then input into the SfS method with $\delta = 10^{-3}$, $\gamma = 10^{-4}$ and $\tau = 10^{-2}$. Finally, the resulting DTM is resized by a factor of two using a bicubic interpolation (Key, 1981) and input to the next surface refinement step at a higher pyramid level $n_{pyr} \leftarrow n_{pyr} - 1$, i.e., refined by an image of increased spatial resolution. The sequence is repeated until the resolution of 2 m per pixel is reached, i.e., $n_{pyr} = 1$. Notably, the SLDEM512 is used only in the pyramid level of the lowest lateral resolution. Afterwards, the refined DTM of the previous pyramid level is used as the initial DTM and as a soft constraint, which allows one to keep the low-pass filters σ_{grad} = 15 pixels and σ_{DTM} = 15 pixels constant while maintaining a pull towards the inherited low-frequency components. The Hapke model (Hapke, 2002) including the correction for macroscopic roughness (Hapke, 1984) is applied to model the reflectance of the surface (cf. section 2.2). Following Grumpe *et al.* (2014), only the single-scattering albedo is estimated by a pixel-wise minimization of the squared difference between the low-pass filtered image and the reflectance model evaluated for low-pass filtered versions of the vectors $\vec{s}$, $\vec{v}$, and $\vec{n}$ (see section 2.2 for details). The width of the Gaussian low-pass filter of the albedo computation algorithm was set to σ_{refl} = 15 pixels.

3.2 *Construction of elemental abundance maps*

Since the M^3 images correspond to a concatenation of pushbroom sensor lines (cf. section 3.3), we map all images to the same cylindrical projection, i.e., the data are resampled to the selenographic coordinate system using a fixed grid with 300 pixels per degree latitude and longitude, respectively. The resulting resolution is about 100 m per pixel and avoids subsampling effects. To avoid an extrapolation, the natural neighbor interpolation method is applied to obtain the values at each grid point. The resulting images were then manually co-registered to achieve subpixel accuracy.

For the derivation of elemental abundance maps, the spectral radiance data need to be converted to bi-directional reflectance data by dividing the measured spectral radiance data by the solar spectrum. Since the bi-directional reflectance values were acquired at varying illumination and viewing geometries, all data are converted to bi-directional reflectance at the standard illumination and viewing geometry of 30° incidence angle and 0° emission angle as proposed by Pieters (1999). The normalization is performed using the method by Grumpe *et al.* (2015). To model the bi-directional reflectance, we use the Hapke model (Hapke, 2002):

$$R_{\text{Hapke}}(\vec{s},\vec{v},\vec{n}) = \frac{w}{4\pi}\frac{\mu_0(\vec{s},\vec{n})}{\mu_0(\vec{s},\vec{n})+\mu(\vec{v},\vec{n})}\left[p(\alpha(\vec{s},\vec{v}))B_{\text{SH}}(\alpha(\vec{s},\vec{v}))+M(\mu_0(\vec{s},\vec{n}),\mu(\vec{v},\vec{n}))\right]B_{\text{CB}}(\alpha(\vec{s},\vec{v})) \quad (15)$$

Where w is the single-scattering albedo, μ_0 is the cosine of the incidence angle, μ is the cosine of the emission angle and α is the phase angle. The phase function $p(\alpha)$ models reflectance of a spherical particle and the term ($M(\mu_0, \mu)$ models the light which is multiply scattered within the layer of regolith. The terms $B_{SH}(\alpha)$ and $B_{CB}(\alpha)$ model the shadow hiding opposition effect and the coherent backscatter opposition effect. Additionally, the correction for macroscopic roughness as proposed by Hapke (1984) is included, i.e., μ_0 and μ are transformed and a correction factor modeling the fraction of shadowed area within the observed pixel is applied. Since the number of M^3 observations for each area are limited, we restrict the normalization to a pixel-wise adaption of the single-scattering albedo w by minimizing:

$$E_{\text{refl}} = \int_{\vec{x}} \int_{\vec{y}} \frac{1}{2}\left(g_{\text{refl}} * I - R_{\text{Hapke}}(\vec{s}_g, \vec{v}_g, \vec{n}_g)\right)^2 dx\, dy \tag{16}$$

Where g_{refl} is a Gaussian low-pass filter of width σ_{refl}. The vectors $\vec{s}_g$, $\vec{v}_g$ and $\vec{n}_g$ are low-pass filtered versions of $\vec{s}$, $\vec{v}$ and $\vec{n}$, respectively. The same low-pass filter g_{refl} is applied to each element of the vectors, respectively. E_{refl} is minimized with respect to w. The remaining parameters are adopted from Warell (2004). The parameters of the double Henyey-Greenstein phase function:

$$p_{\text{DHG}}(\alpha)=\frac{1+c_{\text{DHG}}}{2}\frac{1-b_{\text{DHG}}^2}{\left(1+2b_{\text{DHG}}\cos(\alpha)+b_{\text{DHG}}^2\right)^{\frac{3}{2}}}+\frac{1-c_{\text{DHG}}}{2}\frac{1-b_{\text{DHG}}^2}{\left(1-2b_{\text{DHG}}\cos(\alpha)+b_{\text{DHG}}^2\right)^{\frac{3}{2}}} \tag{17}$$

are set to $c_{\text{DHG}} = 0.7$ and $b_{\text{DHG}} = 0.21$ (Warell, 2004). Following Warell (2004), the terms describing the coherent backscatter opposition effect and the shadow hiding opposition effect are combined, i.e., $B_{\text{CB}}(\alpha) = 1$ and:

$$B_{\text{SH}}(\alpha)=1+\frac{B_{\text{opp}}}{1+\left(\frac{1}{h_{\text{opp}}}\right)\tan\left(\frac{\alpha}{2}\right)} \tag{18}$$

where $B_{\text{opp}} = 3.1$ and $h_{\text{opp}} = 0.11$. The macroscopic roughness parameter is set to $\bar{\theta} = 11°$. The normalization procedure requires a DTM of high resolution. To obtain this DTM, the extended photoclinometry of Grumpe *et al.* (2014) is applied and the surface is recovered from the derived gradient field using the method of Grumpe and Wöhler (2014).

The normalized reflectance is then obtained by inserting the inferred single-scattering albedo and a standard illumination geometry into the Hapke model. To reduce the effect of channel noise, the normalized reflectance data are smoothed along the wavelength axis using a smoothing spline (Marsland, 2009). The smoothing spline simultaneously minimizes the mean squared deviation from the normalized reflectance values and the curvature which is measured by the squared second derivative. The method of Akima (1970) is then applied to interpolate the reflectance data to a wavelength grid with 1-nm spacing and the convex hull of the smoothed spectrum is computed. The division of the smoothed interpolated spectrum by the convex hull yields the continuum-removed spectrum (Fu *et al.*, 2011).

Two absorption bands centered around 1000 nm and 2000 nm typically characterize the continuum-removed lunar reflectance spectra. The absorption bands are termed "band I" and "band II" within this study. Iron oxide (FeO) contained in lunar minerals such as pyroxene and olivine (Burns *et al.*, 1972; Matsunaga *et al.*, 2008; Smrekar and Pieters, 1985) or quenched glass (Tompkins and Pieters, 2010) causes absorption band I. Similarly, electron transitions in the minerals pyroxene (Burns *et al.*, 1972), spinel (Cloutis *et al.*, 2004) or in quenched glass (Tompkins and Pieters, 2010) result in absorption band II. Furthermore, a weak absorption band around 1300 nm and an absorption trough around 2800 nm may be observed in some lunar regions. These bands are caused by the mineral plagioclase (Adams and Goullaud, 1978; Cheek *et al.*, 2009) and water molecules/hydroxyl ions (e.g., Clark *et al.*, 2010), respectively. These two absorption bands, however, are not considered in this study. To measure the absorption troughs we use the spectral parameters as defined by Wöhler *et al.* (2014). The slopes of the continuum at band I (CSL1) and band II (CSL) are obtained from the convex hull (Fig. 14.3a), respectively. The wavelength of the absorption minimum of band I (LMIN1), the relative band depth of band I (BD1) and band II (BD2), and the full width at half maximum (FWHM1) of band I are inferred from the continuum-removed spectrum (Fig. 14.3b). Since the noise level of the M^3 spectra around 2000 nm leads to strong variations of the wavelength of the absorption minimum and the boundaries of the absorption trough, the absorption minimum and the full width at half maximum of band II are not computed. The relative depth of band II, however, is more robust with respect to variations and thus included in our analysis.

The elemental abundance maps are then computed following Wöhler *et al.* (2014). Based on the GLD100 DTM (Scholten *et al.*, 2012), a low-resolution topographically corrected reflectance map ranging from 60°S to 60°N latitude was produced and maps of the band I and II spectral parameters were inferred at a resolution of 0.2 pixels per degree latitude and longitude, respectively. Similarly, Lunar Prospector Gamma Ray Spectrometer (LP GRS) (Lawrence *et al.*, 1998) derived abundance maps of Ca, Al, Fe and Mg were mapped to the same 0.2 pixels per degree grid. Afterwards, a linear mapping from the spectral parameters to the LP GRS elemental abundance maps is computed. The inferred mapping parameters are then applied to spectral parameter maps at the full resolution of the M^3 images. This regression-based approach captures correlations between absorption bands caused by specific electron transitions in the crystal structure of the lunar minerals and LP GRS derived elemental abundances (Wöhler *et al.*, 2014).

In addition to the elemental abundance maps, we compute the petrological map according to Berezhnoy *et al.* (2005). Based on an analysis of LP GRS elemental abundances, Berezhnoy *et al.* (2005) found that the elements Al, Fe and Mg can be described at high accuracy by a plane in the three-dimensional (Al, Fe, Mg) abundance space. Wöhler *et al.* (2014) constructed near-global elemental abundance maps of 20 pixels per degree using M^3-inferred spectral parameters and the LP GRS–derived linear mapping. Based on these near-global abundance maps, the finding of Berezhnoy *et al.* (2005) was confirmed. Consequently, a lunar soil is sufficiently described by the abundances of Fe and Mg in terms of the petrological model suggested by Berezhnoy *et al.* (2005), which is defined by the three endmembers mare basalt (or pyroxene), Mg-rich rock (including, e.g., norite) and feldspathic rock (ferroan anorthosite, FAN). The relative fractions of the three

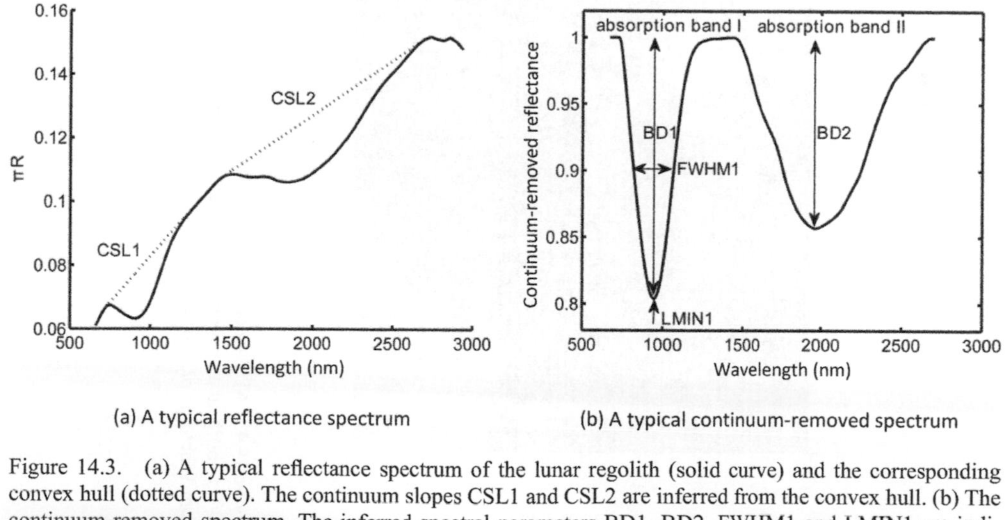

Figure 14.3. (a) A typical reflectance spectrum of the lunar regolith (solid curve) and the corresponding convex hull (dotted curve). The continuum slopes CSL1 and CSL2 are inferred from the convex hull. (b) The continuum-removed spectrum. The inferred spectral parameters BD1, BD2, FWHM1 and LMIN1 are indicated. (From Wöhler *et al.* (2014))

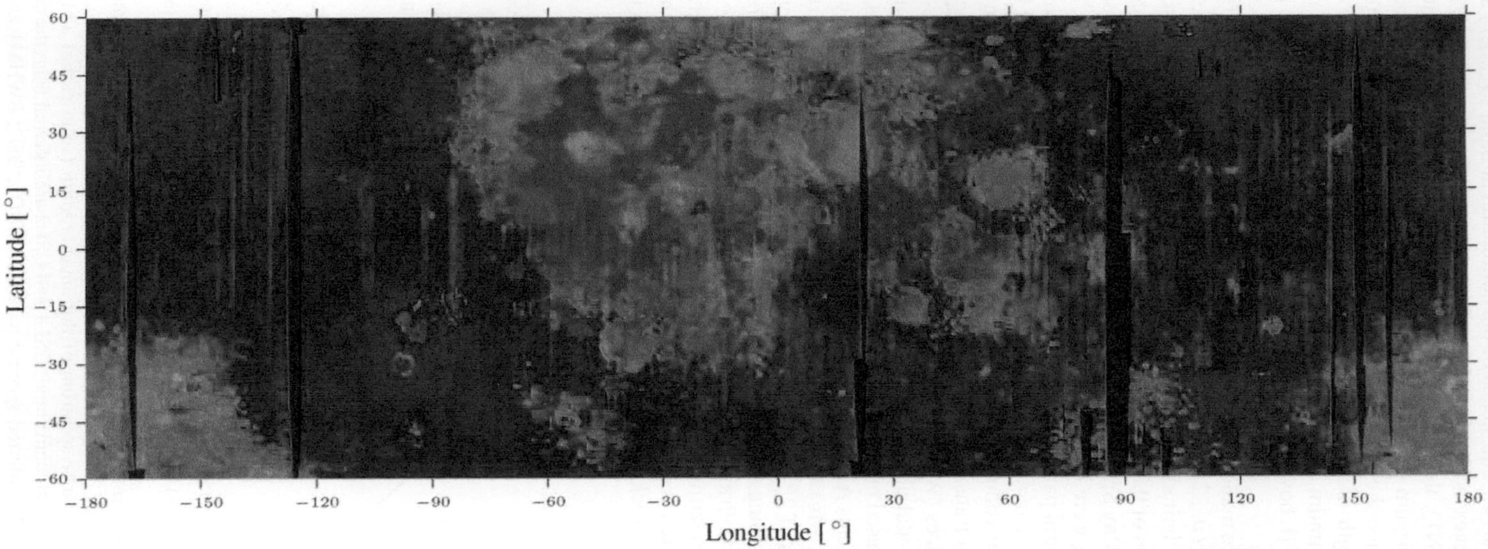

Figure 14.4. M³ data derived near-global petrological map of the Moon according to the model of Berezhnoy *et al.* (2005). The map covers the latitude range from 60°S to 60°N at a resolution of 2 pixels per degree. Red channel: mare basalt/pyroxene. Green channel: Mg-rich rock. Blue channel: feldspathic material. (Adapted from Wöhler *et al.* (2014))

endmembers are read from a ternary diagram in Fe-Mg space and visualized in the form of a RGB image. The mare basalt fraction is assigned to the red channel, the Mg-rich rock fraction to the green channel and the FAN fraction to the blue channel of RGB mosaic. A near-global petrological map ranging from 60°S to 60°N at a resolution of 2 pixels per degree is shown in Figure 14.4. The highlands mainly consist of feldspathic rock and thus appear in deep blue color. The mare regions appear in red to orange depending on their Mg-rich rock content. Small localized areas of high Mg-rich rock contents appear in green color.

4 RESULTS AND DISCUSSION

Figure 14.5 shows the petrological map of Mare Serenitatis. We located three interesting structures in Mare Serenitatis that are also covered in LROC NAC images as seen in Figure 14.5. They are captured by LROC NAC images M181052395LE/RE, M104447576LE/RE and M1108188835LE/RE, respectively. Area 1, Area 2 and Area 3 are centered at (18.0° E, 32.4° N), (11.2° E, 24.6° N),

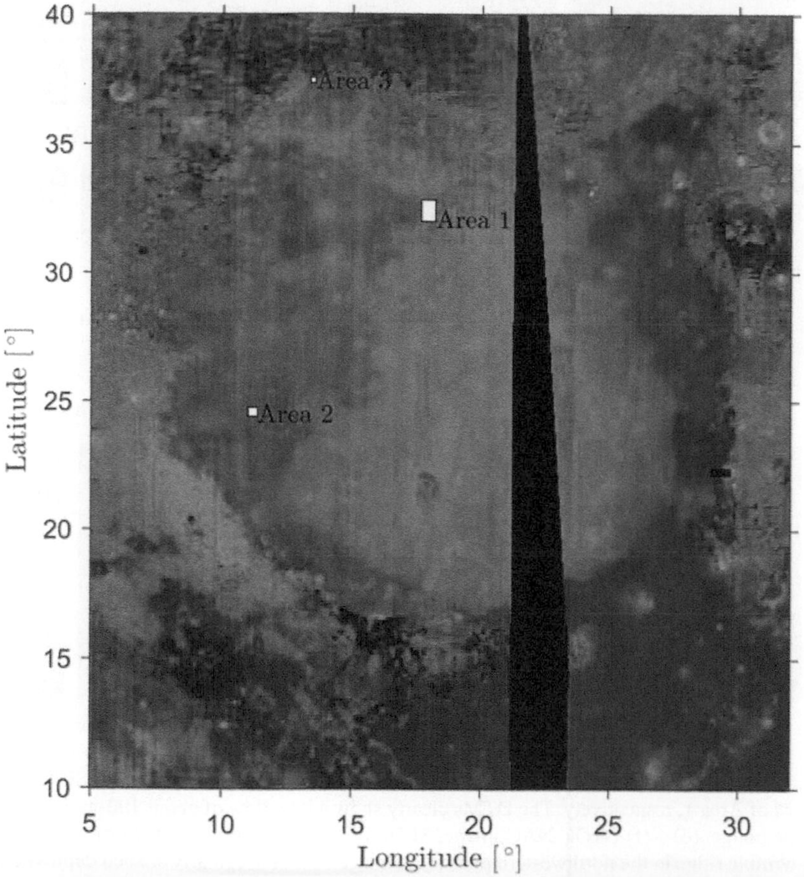

Figure 14.5. Petrographic map of Mare Serenitatis. The petrological map is based on the Fe and Mg content of the surface following three-endmember model of Berezhnoy *et al.* (2005). The three endmembers are used to create a color composite (Red: basaltic material, Green: Mg-rich rock, Blue: feldspathic material). The three regions considered in this study are shown as white rectangles. Black denotes missing data. (From Wöhler *et al.* (2017))

and (13.5° E, 37.5° N), respectively. The LROC NAC images, the SLDEM512 data and our refined DTMs are shown in Figure 14.6. The SLDEM512 clearly shows noise patterns which are typical for the stereo analysis of images. The refined DTMs, in contrast, show a surface that shows the same surface features as those shown in the LROC NAC images. Notably, the elemental abundance maps show stripe artifacts which correspond to scan lines of the M^3 pushbroom camera. Although it is possible to remove these artifacts by applying a low-pass filter along each scan line of the

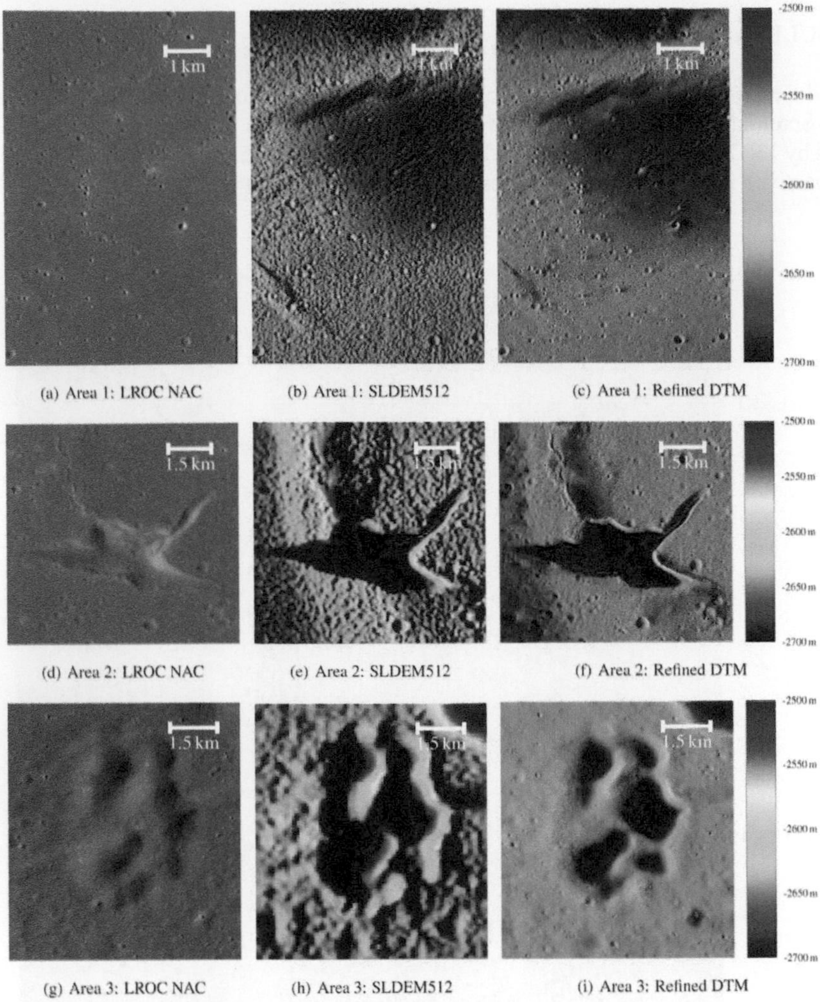

Figure 14.6. Refined DTMs of the lava flow structures. (a) – (c) LROC NAC image, SLDEM512 and the refined DTM of Area 1, respectively. The DTMs clearly show a lava flow of about 100-m height which is not visible in the image. (d) – (f) LROC NAC image, SLDEM512 and the refined DTM of Area 2, respectively. There is a wrinkle ridge in the northwestern part of the image and an irregularly shaped depression which is a possible volcanic vent. (g) – (i) LROC NAC image, SLDEM512 and the refined DTM of Area 3, respectively. The smooth depression is assumed to be another possibly volcanic vent. In all cases, the refined DTM captures the image details while the SLDEM512 constrains the solution and successfully reduces systematic errors originating from the integration of a noisy shading information. The visualization of the DTM was computed by color-coding the depth information and weighting the intensity channel of the image by a Lambertian shading of the DTMs, respectively.

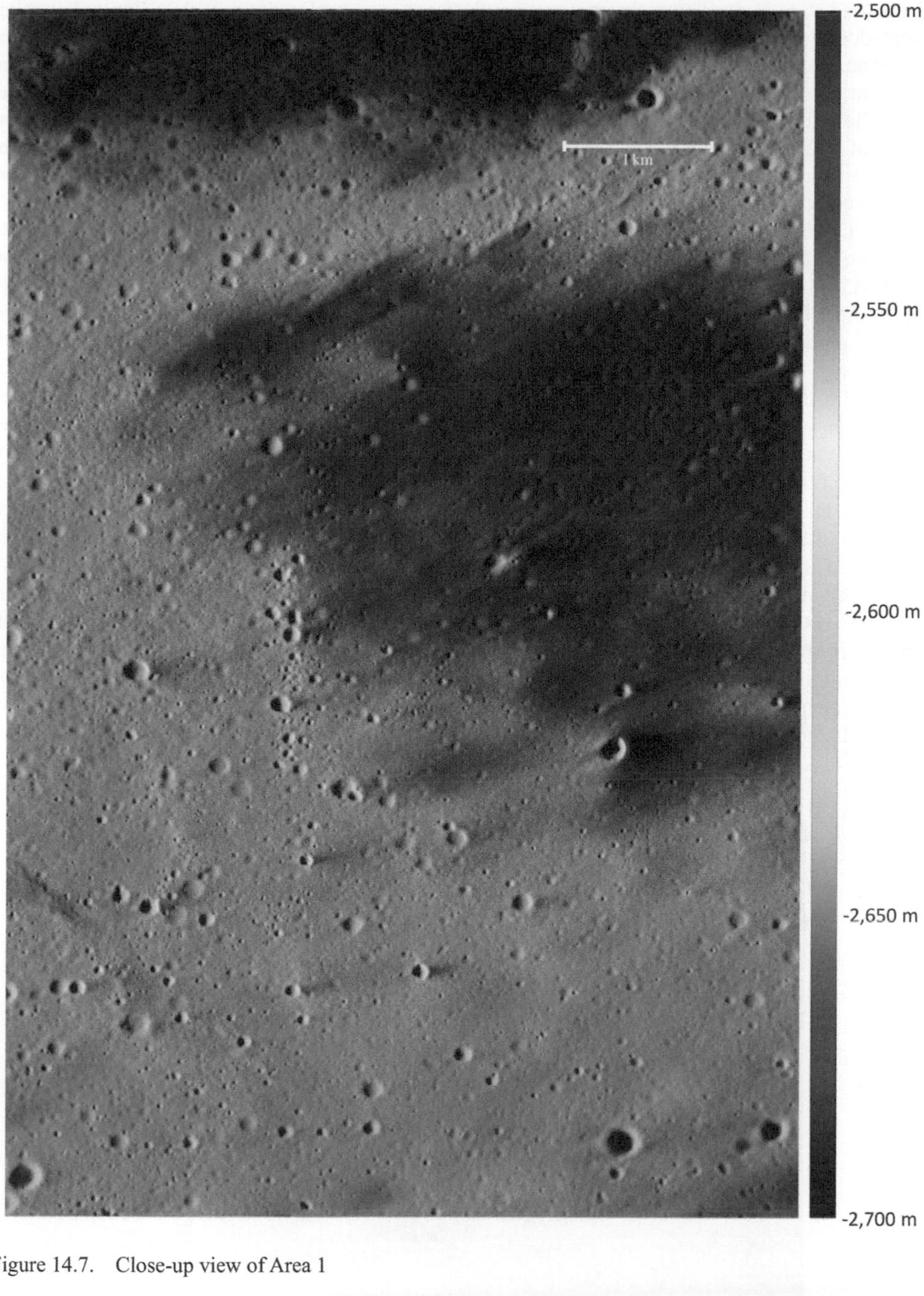

Figure 14.7. Close-up view of Area 1

sensor and thus normalizing the measured reflectance, we prefer the raw data derived version which does not produce an alteration of the measurements. Since the observed lava flow structures are larger than one pixel, the effect is not important for the results of this study. In general, this effect may also be observed for the refined DTMs, which are derived from pushbroom cameras as well. The effect produces height differences between scan lines which are smaller than the lateral extent of an image element and thus of the order of a few centimeters. The lava flow structures of interest show height differences of a few hundred meters and this effect is thus less visible and considered negligible.

Area 1 clearly reveals a lava flow while the Area 2 and Area 3 exhibit possible volcanic vents as shown in Figure 14.6. In all cases, the refined DTMs show the same large-scale trends as the SLDEM512. However, the spatial resolution is strongly improved. Even small craters are visible within the refined DTMs and all structures occurring in the image may be detected in the refined DTM. Figures. 14.7–14.9 show enlarged versions of Figure 14.6c, Figure 14.6f and Figure 14.6i, respectively. These enlarged versions illustrate that many of the details may not be displayed in Figure 14.6. The lava flow is found to have a thickness of about 100 m, which is relatively large in comparison with the average lava flow thickness of 30–60 m (Hiesinger *et al.*, 2002). The refined DTM of Area 1 suggests that the lava flow consists of at least two layers (Fig. 14.7). The irregularly shaped depression in Area 2 has a depth of 150–200 m, while the possible volcanic vents in Area 3 have an average depth of 100 m. While the volcanic vents are smaller than 3 km in extent, the lava flow extends by more than 5 km. Consequently, the volcanic vents are barely visible in the M^3 data.

The NAC and the M3 image of Area 1 are shown in Figure 14.10, whereas Figure 14.11 shows the derived elemental abundance maps of Ca, Mg, Fe and Al of the lava flow in Area 1. Unfortunately, the resolution of the M^3 sensor is about 140 m per pixel (Pieters *et al.*, 2009) and thus much lower than the resolution of the refined DTMs. Since the construction of elemental abundance maps is limited to the resolution of the spectral reflectance data, all elemental abundance maps are of a much lower resolution than the refined DTMs. The lava flow is clearly associated with an increased amount of Ca while the Mg content is decreased, corresponding to a composition richer

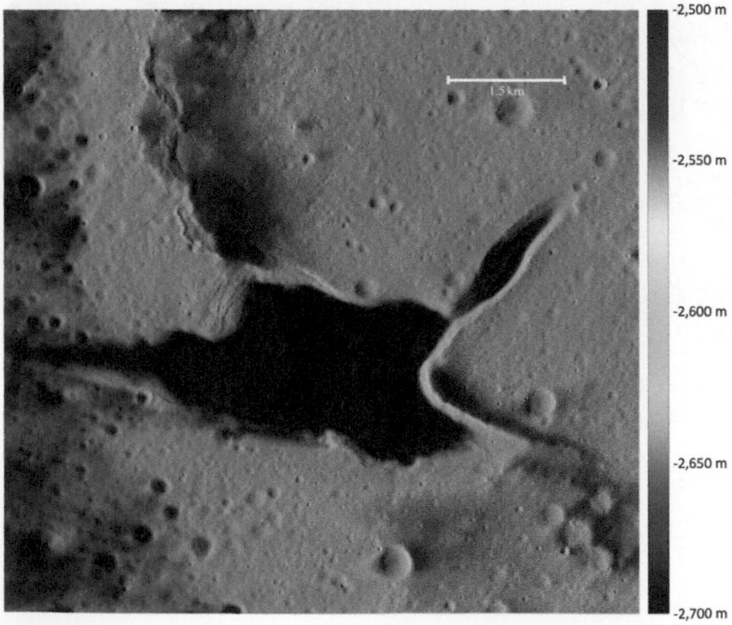

Figure 14.8. Close-up view of Area 2

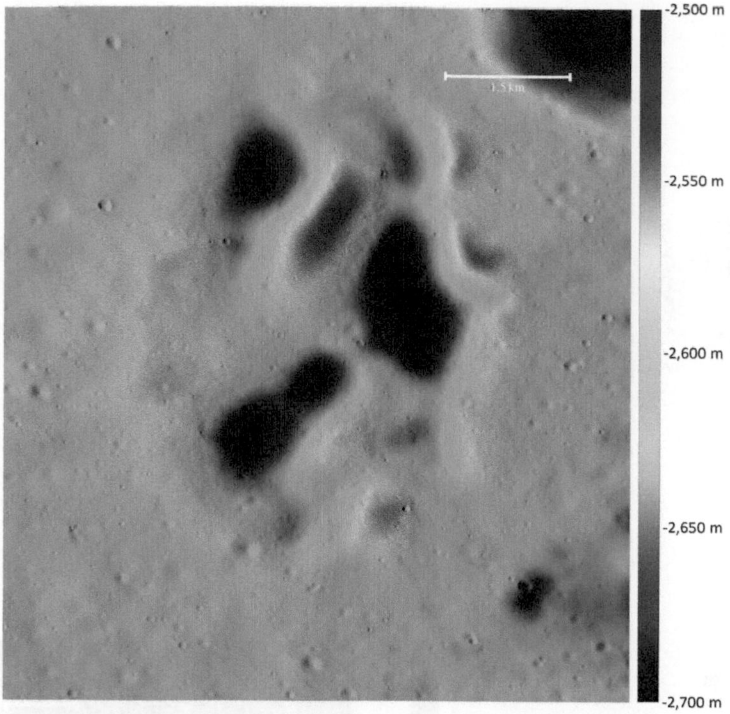

Figure 14.9. Close-up view of Area 3

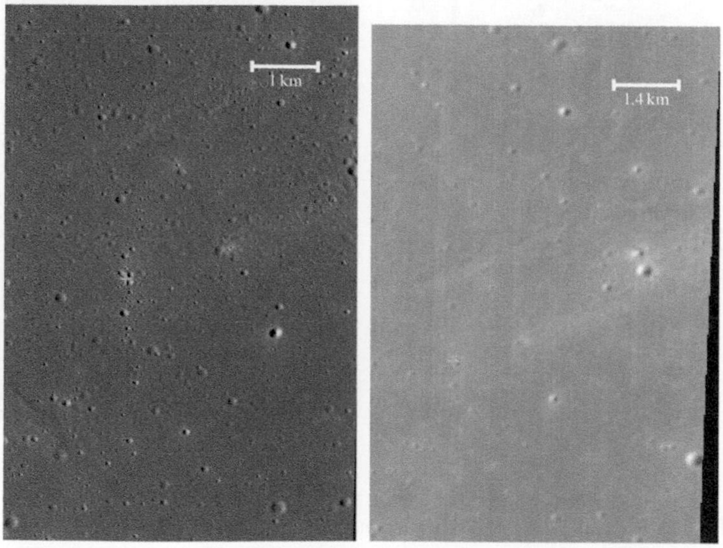

(a) Area 1: NAC image

(b) Area 1: M^3 radiance at 1579 nm

Figure 14.10. Images of Area 1. Notably, the NAC image (a) is not perfectly aligned to the M^3 radiance image (b). Due to problems with the star sensors, the M^3 data was imperfectly geo referenced and has been improved afterwards (Boardman *et al.*, 2011). However, it is evident that the alignment of the M^3 data still shows significant offsets of the order of one kilometer. For comparison, the same surface feature consisting of three craters are marked by the "plus" mark in both images.

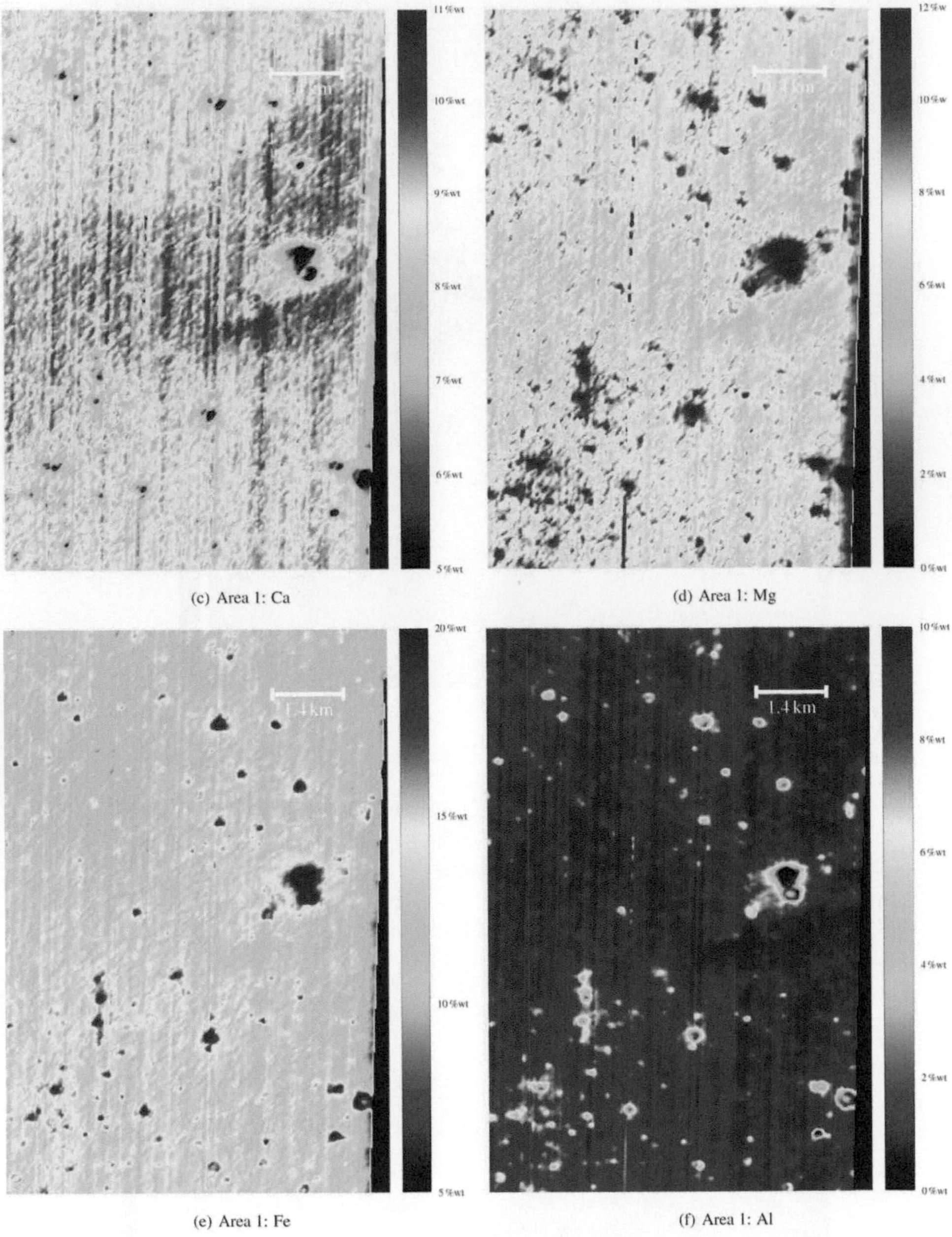

Figure 14.11. Elemental abundance maps of Area 1. The lava flow shows (a) an enrichment in Ca and (b) a depletion in Mg. The elements Fe (c) and Al (d) do not show distinctive features. Black denotes missing data.

in clinopyroxene than the basalt to the north and south. The elemental maps of Fe and Al do not exhibit distinctive features. The strong Ca depletion and Mg increase occurring at small fresh craters is an artifact of the method resulting from its sensitivity to low optical maturity due to the lack of sufficiently large low-maturity regions in the LP GRS data. Similarly, Fe appears to be systematically increased for immature craters while Al is systematically depleted.

5 SUMMARY AND CONCLUSION

In this study, we showed a method to refine DTMs of low lateral resolution, e.g., the SLDEM512, using high-resolution images, e.g., LROC NAC images. The method produces DTMs of a lateral resolution which comes close to the image resolution. All image details can also be detected in the DTM while the low-resolution DTM effectively regularizes the trend and thus prevents systematic errors. The described method is applied to the Mare Serenitatis and we detected new lava flow structures utilizing high-resolution DTMs not previously reported due to restricted spatial resolution. The elemental and petrologic maps of these newly reported features are presented in order to find the compositional variations. The possible volcanic vents do not show exceptional elemental compositions while the lava flow exhibits an increase in the Ca and a decrease in the Mg content, corresponding to a composition richer in clinopyroxene than the basalts north and south of the lava flow.

The methodology described in this chapter is an important tool to investigate lava flow structures. The DTM refinement method allows for the detection of small-scale surface features, e.g., small craters, wrinkle ridges, lava flows and volcanic domes/vents. Furthermore, it allows for an estimation of surface feature heights, e.g., the thickness of a lava flow. The mapping of elemental abundance maps allows for the detection of similar mineralogical compositions of the surface and thus may be used to detect surface features of anomalous composition.

REFERENCES

Adams, J.B. & Goullaud, L.H. (1978) Plagioclase feldspars: Visible and near infrared diffuse reflectance spectra as applied to remote sensing. *Lunar and Planetary Science Conference*, Houston, TX, USA, Volume IX. pp. 1–2.

Akima, H. (1970) A new method of interpolation and smooth curve fitting based on local procedures. *Journal of the Association for Computing Machinery*, 17(4), 589–602.

Barker, M.K., Mazarico, E., Neumann, G.A., Zuber, M.T., Haruyama, J. & Smith, D.E. (2016) A new lunar digital elevation model from the Lunar Orbiter Laser Altimeter and SELENE Terrain Camera. *Icarus*, 273, 346–355.

Berezhnoy, A.A., Hasebe, N., Kobayashi, M., Michael, G.G., Okudaira, O. & Yamashita, N. (2005) A three end-member model for petrologic analysis of lunar prospector gamma-ray spectrometer data. *Planetary and Space Science*, 53, 1097–1108.

Boardman, J.W., Pieters, C.M., Green, R.O., Lundeen, S.R., Varanasi, P., Nettles, J., Petro, N., Isaacson, P., Besse, S. & Taylor, L.A. (2011) Measuring moonlight: an overview of the spatial properties, lunar coverage, selenolocation, and related Level 1B products of the Moon Mineralogy Mapper. *Journal of Geophysical Research: Planets*, 116(E6), E00G14.

Burns, R.G., Abu-Eid, R.M. & Huggins, F.E. (1972) Crystal field spectra of lunar pyroxenes. *Lunar Science Conference*, Houston, TX, USA, Volume 2. pp. 533–543.

Cheek, L.C., Pieters, C.M., Dyar, M.D. & Milam, K.A. (2009) Revisiting plagioclase optical properties for lunar exploration. *Lunar and Planetary Science Conference*, The Woodlands, TX, USA, Volume XXXX, abstract #1928.

Chin, G., Brylow, S., Foote, M., Garvin, J., Kasper, J., Keller, J., Litvak, M., Mitrofanov, I., Paige, D., Raney, K., Robinson, M., Sanin, A., Smith, D., Spence, H., Spudis, P., Stern, S.A. & Zuber, M. (2007) Lunar Reconnaissance Orbiter overview: the instrument suite and mission. *Space Science Reviews*, 129(4), 391–419.

Clark, R., Pieters, C.M., Green, R.O., Boardman, J., Buratti, B.J., Head, J.W., Isaacson, P.J., Livo, K.E., McCord, T.B., Nettles, J.W., Petro, N.E., Sunshine, J.M. & Taylor, L.A. (2010) Water and hydroxyl on the moon as seen by the Moon Mineralogy Mapper (m^3). *Lunar and Planetary Science Conference*, The Woodlands, TX, USA, Volume XXXXI, abstract #1533.

Cloutis, E.A., Sunshine, J.M. & Morris, R.V. (2004) Spectral reflectance compositional properties of spinels and chromites: implications for planetary remote sensing and geothermometry. *Meteoritics and Planetary Science*, 39(4), 545–565.

Fu, Z., Robles-Kelly, A., Caelli, T. & Tan, R.T. (2011) On automatic absorption detection for imaging spectroscopy: a comparative study. *IEEE Transactions on Geoscience and Remote Sensing*, 45(11), 3827–3844.

Gifford, A.W. & El-Baz, F. (1981) Thicknesses of lunar mare flow fronts. *The Moon and the Planets*, 24(4), 391–398.

Grumpe, A. & Wöhler, C. (2014) Recovery of elevation from estimated gradient fields constrained by digital elevation maps of lower lateral resolution. *ISPRS Journal of Photogrammetry and Remote Sensing*, 94, 37–54.

Grumpe, A., Belkhir, F. & Wöhler, C. (2014) Construction of lunar DEMs based on reflectance modelling. *Advances in Space Research*, 53(12), 1735–1767.

Grumpe, A., Zirin, V. & Wöhler, C. (2015) A normalisation framework for (hyper-)spectral imagery. *Planetary and Space Science*, 111, 1–33.

Hapke, B. (1984) Bidirectional reflectance spectroscopy: 3. Correction for macroscopic roughness. *Icarus*, 59(1), 41–59.

Hapke, B. (2002) Bidirectional reflectance spectroscopy: 5. The coherent backscatter opposition effect and anisotropic scattering. *Icarus*, 157(2), 523–534.

Hiesinger, H., Head III, J.W., Wolf, U., Jaumann, R. & Neukum, G. (2002) Lunar mare basalt flow units: Thicknesses determined from crater size-frequency distributions. *Geophysical Research Letters*, 29(8), 89-1–89-4.

Horn, B.K.P. (1990) Height and gradient from shading. *International Journal of Computer Vision*, 5(1), 37–75.

Keys, R. (1981) Cubic convolution interpolation for digital image processing. *IEEE Transactions on Acoustics, Speech, and Signal Processing*, 29(6), 1153–1160.

Lawrence, D.J., Feldman, W.C., Barraclough, B.L., Binder, A.B., Elphic, R.C., Maurice, S. & Thomsen, D.R. (1998) Global elemental maps of the moon: the lunar prospector gamma-ray spectrometer. *Science*, 281(5382), 1484–1489.

Marsland, S. (2009) *Machine Learning: An Algorithmic Introduction*. CRC Press, Mahwah, NJ, USA.

Matsunaga, T., Ohtake, M., *et al.* (2008) Discoveries on the lithology of lunar crater central peaks by SELENE spectral profiler. *Geophysical Research Letters*, 35, L23201.

Neukum, G. & Horn, P. (1976) Effects of lava flows on lunar crater populations. *The Moon*, 15(3), 205–222.

Pieters, C.M. (1999) The Moon as a spectral calibration standard enabled by lunar samples: the Clementine example. *Proceedings of the Workshop on New Views of the Moon II*, Flagstaff, AZ, USA, abstract #8025.

Pieters, C.M., Boardman, J., Buratti, B., Chatterjee, A., Clark, R., Glavich, T., Green, R., Head III, J., Isaacson, P., Malaret, E., McCord, T., Mustard, J., Petro, N., Runyon, C., Staid, M., Sunshine, J., Taylor, L., Tompkins, S., Varanasi, P. & White, M. (2009) The Moon Mineralogy Mapper (M^3) on Chandrayaan-1. *Current Science*, 96(4), 500–505.

Robinson, M.S., Brylow, S.M., Tschimmel, M., Humm, D., Lawrence, S.J., Thomas, P.C., Denevi, B.W., Bowman-Cisneros, E., Zerr, J., Ravine, M.A., Caplinger, M.A., Ghaemi, F.T., Schaner, J.A., Malin, M.C., Mahanti, P., Bartels, A., Anderson, J., Tran, T.N., Eliason, E.M., McEwen, A.S., Turtle, E., Jolliff, B.L. & Hiesinger, H. (2010) Lunar Reconnaissance Orbiter Camera (LROC) instrument overview. *Space Science Reviews*, 150(1–4), 81–124.

Scholten, F., Oberst, J., Matz, K.-D., Roatsch, T., Wählisch, M., Speyerer, E.J. & Robinson, M.S. (2012) GLD100: the near-global lunar 100 m raster DTM from LROC WAC stereo image data. *Journal of Geophysical Research*, 117, E00H17.

Smrekar, S. & Pieters, C.M. (1985) Near-infrared spectroscopy of probable impact melt from three large lunar highland craters. *Icarus*, 63, 442–452.

Tompkins, S. & Pieters, C.M. (2010) Spectral characteristics of lunar impact melts and inferred mineralogy. *Meteoritics and Planetary Science*, 45(7), 1152–1169.

Warell, J. (2004) Properties of the Hermean regolith: IV. Photometric parameters of Mercury and the Moon contrasted with Hapke modelling. *Icarus*, 167(2), 271–286.

Wöhler, C., Grumpe, A., Berezhnoy, A., Bhatt, M.U. & Mall, U. (2014) Integrated topographic, photometric and spectral analysis of the lunar surface: application to impact melt flows and ponds. *Icarus*, 235, 86–122.

Wöhler, C., Grumpe, A., Rommel, D., Bhatt, M.U. & Mall, U. (2017) Elemental and topographic mapping of lava flow structures in mare serenitatis on the moon. *Proceedings of the ISPRS International Symposium on Planetary Remote Sensing and Mapping, Hong Kong, 2017, The International Archives of the Photogrammetry, Remote Sensing and Spatial Information Sciences*, Volume XLII-3/W1. pp. 163–170.

Planetary Remote Sensing and Mapping – Wu et al.
© 2019 Taylor & Francis Group, London, ISBN 978-1-138-58415-0

Chapter 15

Mineral abundance and particle size distribution derived from in situ spectra measurements of the Chang'E-3 Yutu rover

H. Lin, X. Zhang, X. Wu, Y. Yang and D. Guo

ABSTRACT: An important issue in planetary exploration and sciences is the determination of mineral abundances and particle size distributions from visible – near-infrared spectra. Such analyses can help elucidate which geological processes have been active on the surfaces of the Moon and planets. The imaging spectrometer on board the Yutu rover of the Chang'E-3 mission measured the reflectance spectra of lunar soil at a height of approximately 1 m, providing new insights into the lunar surface. A new method combining a Hapke radiative transfer model and the sparse unmixing algorithm was proposed to retrieve the mineral abundance and particle size distribution and applied to the in-situ measurements from the Yutu rover. The imaginary part of the refractive index of each endmember was first calculated by solving the Hapke model. The single-scattering albedos of each endmember with different particle sizes were then obtained based on the Hapke slab model, allowing the construction of an endmember library. The single-scattering albedos of mineral mixtures, which were computed using the Hapke bidirectional equation, were then unmixed using the sparse unmixing algorithm with the aid of the endmember library. Laboratory measurements obtained from the Reflectance Experiment Laboratory were used to validate the proposed methodology. The results revealed that the methodology exhibits a good performance in retrieving the mineral abundances and particle sizes from mixtures. Finally, the methodology was applied to the Yutu rover measurements. At one of the locations, Node E, the agglutinate abundance was 71% and the abundances of both clinopyroxene and olivine were approximately 10%. The particle size distributions of each mineral exhibited almost normal distributions with different mean particle sizes and variance, possibly indicating the distinct responses of each of the components at this site to space weathering.

1 INTRODUCTION

Knowledge of the types, abundances and particle size distributions (PSDs) of minerals is fundamental for planetary sciences, allowing us to elucidate which geological processes have been active on the planetary surface (Clark *et al.*, 1990). The minerals on the lunar surface mainly include clinopyroxene, orthopyroxene, olivine and plagioclase, etc. Lucey (2004) first used data from the Clementine probe to identify lunar minerals and obtain their distribution map. Shuai *et al.* (2013) quantitatively mapped the minerals on the lunar surface using the data collected by the imaging interferometer (IIM) on board the Chang'E-1. However, the spatial resolution of the IIM was limited, hindering the precise determination of the mineral components and their abundances and particle sizes. The imaging spectrometer on board the Yutu rover of the Chinese Chang'E-3 (CE-3) mission was used to collect reflectance spectra at four different sites, providing new insights into the lunar surface (Zhang *et al.*, 2015).

The reflectance of mixtures of minerals at visible to shortwave-infrared wavelengths is nonlinear. However, the abundance and effective particle size of the minerals can be derived by radiative transfer modelling (Hapke, 1981; Shkuratov *et al.*, 1999). For example, Li and Li (2011) determined the mineral abundance and effective particle size in lunar samples by solving the Hapke model using Newton's iteration and the least-squares method. The PSD of minerals in each pixel can better reveal the lunar geological processes and environment. However, attempts to simultaneously obtain the abundance and PSD from hyperspectral data have been limited. The geological environment can be specified using the PSD. Sparse unmixing (Iordache *et al.*, 2011) is an algorithm that selects, from a large spectral library, the spectra to best model each mixed pixel and estimates the abundance of the selected spectra based on an efficient linear sparse regression technique. Hapke radiative transfer models provide a means to calculate the single-scattering albedo, which can be added linearly. These models can be used to calculate not only the single-scattering albedo of minerals with different particle sizes but also that of mineral mixtures.

Here we propose a method to derive the abundance and PSD of minerals by combining a radiative transfer model and the sparse unmixing algorithm. A Hapke radiative transfer model was used to calculate the single-scattering albedo of pure minerals with various particle sizes, thereby affording a spectral library. Sparse unmixing was then applied to select the optimal subset from the library to model each of the mixed pixels. The proposed method was first validated using spectra of minerals measured in the laboratory and then applied to the measurements from the CE-3 mission.

2 DATA

The data used in this study include lunar surface spectra measured by the Yutu rover and mineral spectra obtained from the Reflectance Experiment Laboratory (RELAB) database.

2.1 *Spectra measured by the Yutu rover*

Chang'E-3 landed in northern Mare Imbrium at 44.1205°N, 19.5102°W (Wang *et al.*, 2017). Reflectance spectra were measured at four locations using the imaging spectrometer on board the Yutu rover (Fig. 15.1a and Fig. 15.1b). The data contain visible – near-infrared (VNIR) spectra and shortwave-infrared (SWIR) spectra with spectral resolutions of 2–7 nm and 3–12 nm, respectively. The noise was first removed using a wavelet transform (Fig. 15.1c).

2.2 *Mineral spectra obtained from the RELAB database*

Two sets of laboratory-measured mineral spectra were used in this study: (1) endmembers including clinopyroxene, orthopyroxene, olivine, plagioclase and agglutinate for unmixing of the CE-3 data (Fig. 15.2a; Table 15.1), and (2) montmorillonite with various particle sizes and mixtures of olivine and bronzite (Fig. 15.2b). All of the spectra were measured using an incidence angle of 30° and an emergence angle of 0°.

3 METHODOLOGY

The mineral abundance and PSD were derived using a three-step process. First, the k value of each endmember was calculated by solving the Hapke radiative transfer model. Second, the single-scattering albedos for various particle sizes were derived based on the Hapke slab model to afford an endmember library. The CE-3 reflectance measurements were also converted to single-scattering albedos. Finally, the abundance and PSD of each endmember were derived using the sparse unmixing algorithm.

3.1 *Determination of single-scattering albedo*

Minerals are generally considered to be intimate mixtures, and the optical characteristics of these mixtures depend on numerous parameters, such as the absorption and scattering characteristics, grain size of each component and the average optical distance between reflections, which renders

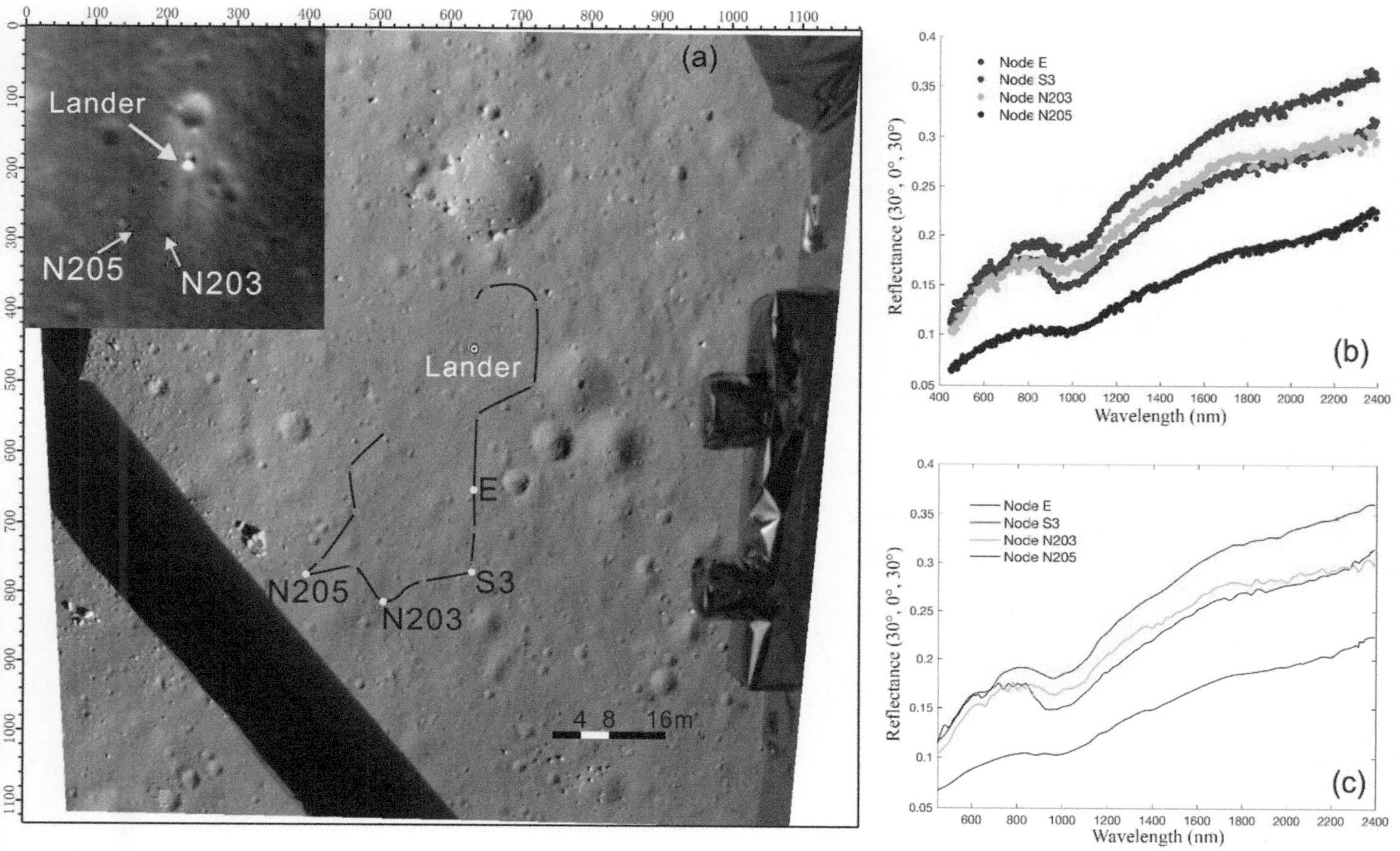

Figure 15.1. (a) Landing site of CE-3 and locations of the spectra measurements. (b) Four reflectance spectra measured by the Yutu rover after photometric correction (Jin *et al.*, 2015; Zhang *et al.*, 2015). (c) Reflectance spectra after noise removal.

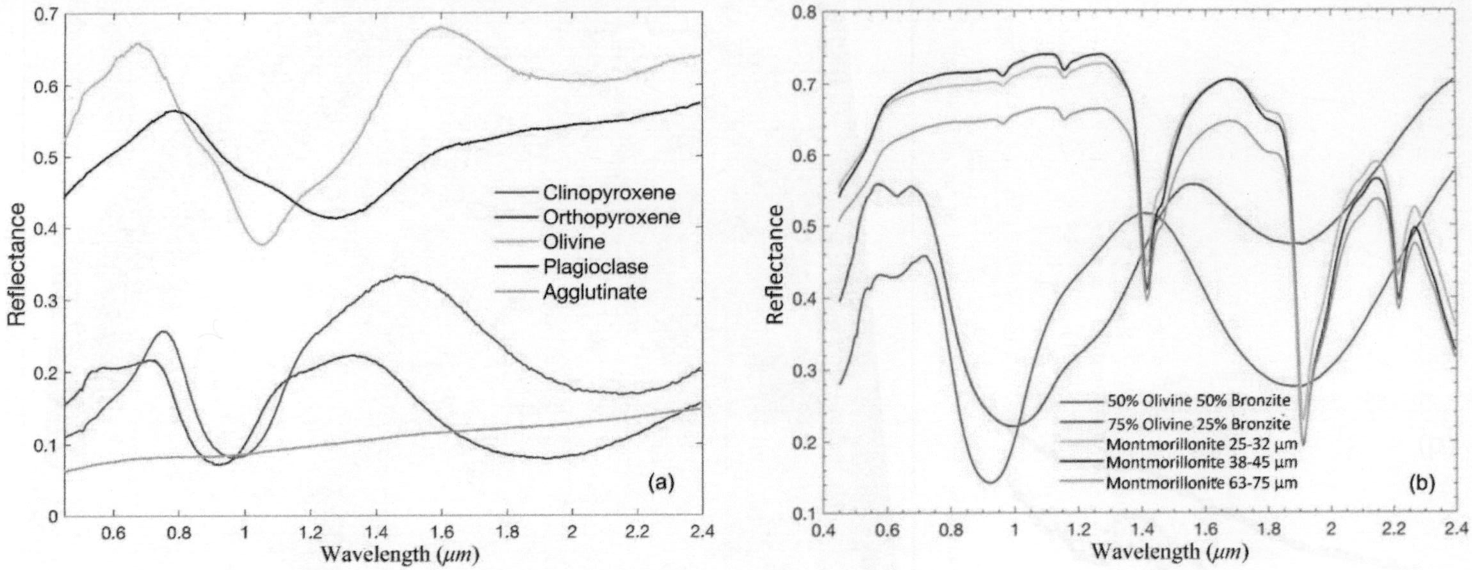

Figure 15.2. (a) Endmember spectra used in this study (Li and Li, 2011; Shuai *et al.*, 2013). (a) Data used to test the methodology.

Table 15.1 Mineral endmembers used in this study

Mineral name	*Sample ID*	*Particle size (μm)*	*Real part of refraction, n*
Clinopyroxene	LS-CMP-009	0–250	1.73
Orthopyroxene	LS-CMP-012	0–250	1.77
Olivine	LR-CMP-014	0–45	1.83
Plagioclase	LS-CMP-086	0–20	1.56
Agglutinate	LU-CMP-007–1	100–1000	1.49

the linear model inapplicable to reflectance data (Heylen *et al.*, 2014). According to the Hapke model, the single-scattering albedo (the ratio of scattered light to total extincted light) of a mineral mixture is a linear combination of the endmember single-scattering albedos in proportion to their relative geometric cross-sections (Hapke, 1981). This allows us to utilise linear unmixing techniques in single-scattering albedo space instead of reflectance space (Heylen and Gader, 2014). Hapke proposed a function to describe the relationship between the single-scattering albedo ω and reflectance r, assuming that the particles are larger than the wavelength of the light (Hapke, 1981):

$$r = \frac{\omega}{4(\mu_0 + \mu)}[(1 + B(g))P(g) + H(\mu_0)H(\mu)\text{-}1] \tag{1}$$

where r is the reflectance of the mineral, μ_0 and μ are the cosines of the angles of incidence and emergence, respectively, g is the phase angle, and $B(g)$ is the backscattering:

$$B(g) = 1/[1 + (1/h)\tan(g/2)],\ h = \frac{3}{8}\ln(1\text{-}\varphi) \tag{2}$$

where φ is a filling factor, which has a value of 0.41 for lunar regolith (Mustard and Pieters, 1989).

$P(g)$ is the phase function, which is expressed as:

$$P(g) = 1 + bP_1(g) + cP_2(g),\ P_1(g) = \cos(g), P_2(g) = \frac{3}{2}\cos^2(g) - \frac{1}{2}, \tag{3}$$

where b and c are set as −0.4 and 0.25, respectively.

H is a multiple-scattering function:

$$H(x) = \{1 - (1-\gamma)x[r_0 + (1 - r_0/2 - r_0 x)\ln((1+x)/x)]\}^{-1},\ \gamma = \sqrt{1-\omega},\ r_0 = (1-\gamma)/(1+\gamma) \tag{4}$$

where x is $\cos(\mu_0)$ or $\cos(\mu)$.

3.2 *Conversion of imaginary part of optical constants*

The optical constants and grain sizes of each mineral are the key variables for calculating the single-scattering albedo. The optical constants of the minerals can be obtained using the Hapke or Shkuratov radiative transfer models (Hapke, 1981; Shkuratov *et al.*, 1999). Here, we used a Hapke radiative transfer model to retrieve the imaginary indices k of the optical constants for the self-consistency of the model. The real part n of the optical constants was considered constant, because they do not vary by more than 0.1 in the VNIR wavelength region (Roush, 2003). The single-scattering albedos of the endmembers were calculated using the Hapke slab model for a given grain size:

$$\omega = S_e + (1 - S_e)\frac{(1 - S_i)}{1 - S_i\Theta}\Theta \tag{5}$$

where S_e is the reflectivity for external incident light:

$$S_e \approx \frac{(n-1)^2 + k^2}{(n+1)^2 + k^2} + 0.05 \tag{6}$$

S_i is the reflectivity for internal incident light:

$$S_i \approx 1 - \frac{4}{n(n+1)^2} \tag{7}$$

Θ is the particle internal transmission coefficient:

$$\Theta = \frac{r_i + \exp\left(\sqrt{\alpha(\alpha+s)}\langle D\rangle\right)}{1 + r_i \exp\left(\sqrt{\alpha(\alpha+s)}\langle D\rangle\right)} \tag{8}$$

where r_i is the internal diffuse reflectance inside the particle diameter:

$$r_i = \frac{1 - \sqrt{\alpha/(\alpha+s)}}{1 + \sqrt{\alpha/(\alpha+s)}} \tag{9}$$

and s is the internal volume scattering coefficient. Assuming that the internal volume scattering coefficient is 0, then:

$$\Theta = \exp(-\alpha\langle D\rangle) \tag{10}$$

$$\langle D\rangle = \frac{2}{3}n^2 - \frac{1}{n}(n^2 - 1)^{3/2} D \tag{11}$$

The absorption coefficient α can be expressed as:

$$\alpha = \frac{4\pi k}{\lambda} \tag{12}$$

According to the above equations, the single-scattering albedo is a function of n, k, λ and particle diameter D. Thus, the imaginary indices can be calculated if n, D and λ are known.

The single-scattering albedos of mineral endmembers with various particle sizes were calculated using the obtained k values. An endmember library was then constructed for spectral unmixing.

3.3 SPARSE UNMIXING ALGORITHM

The average single-scattering albedo of mixture is a linear combination of the K component:

$$\omega_{mix} = \sum_{i=1}^{K} F_i \omega_i \tag{13}$$

where ω_i and F_i are the single-scattering albedo and relative geometric cross-section, respectively of the *i*th component in the mixture. Assuming that the particles have approximately the same shape, F_i can be considered as the volumetric abundance.

Sparse unmixing is designed to determine the optimal subset of signatures from a large spectral library that best models the mixed spectrum. It is a linear model that can be expressed as follows (Iordache *et al.*, 2011):

$$\min_{X} \frac{1}{2}\|AX - Y\|_2^2 + \lambda\|X\|_1 \text{ subject to } X \geq 0 \tag{14}$$

where X is the abundance vector corresponding to the spectral library A, Y is the measured spectrum of the pixel, and λ is a regularisation parameter that controls the relative weight of the sparsity of the solution. The number of endmembers present in a mixed pixel is usually substantially smaller than the availability of the spectral library, which makes the sparsity constraint useful. The sparse unmixing algorithm was used to determine the optimal abundance and particle size from the endmember library.

4 RESULTS

4.1 *Laboratory test*

The proposed methodology was validated using laboratory-measured spectra in two steps: (1) samples of montmorillonite with known PSD (Robertson *et al.*, 2016) were used to test the validity of PSD retrieval, and (2) mixtures of bronzite and olivine with known fractions (Mustard and Pieters, 1989) were used to test the validity of abundance retrieval.

Three montmorillonite samples with different particle sizes were used to validate the method. Figure 15.3 shows that the actual and retrieved PSDs were very similar. For the 25–32-μm sample, the modelled and actual mean particle sizes were 31.20 μm and 31 μm, respectively. For the 38–45-μm sample, the modelled and actual mean particle sizes were 43.84 μm and 38 μm, respectively. For the 63–75-μm sample, the modelled and actual mean particle sizes were 60.40 μm and 62 μm, respectively.

Mixtures of bronzite and olivine (RELAB IDs: XT-CMP-31and XT-CMP-32) were used to test the accuracy of abundance retrieval. The particle sizes of the mixtures were in the range of 45–75 μm. Figure 15.4 shows the modelled abundances and PSDs. The average abundance retrieval error was 5.16%, demonstrating the good performance of our methodology. Although the PSD results cannot be validated owing to the lack of true PSD information, the retrieved particle sizes were mostly located in the range of 45–75 μm.

4.2 *Analysis of Yutu rover measurements*

The *k* values (Fig. 15.5a) of all of the endmembers were first obtained as described in section 3.2. Then, the single-scattering albedos of all of the minerals were calculated for different particle sizes. Thus, an endmember library including various minerals with different grain sizes was constructed. Finally, the single-scattering albedos of the four reflectance spectra measured on the CE-3 mission (Fig. 15.5b) were calculated and unmixed using the sparse unmixing algorithm to determine the abundances and PSDs of each mineral.

Figure 15.6 shows the retrieved mineral PSDs and abundances from the spectrum obtained at Node E. A good fit between the measured and modelled single-scattering albedo spectra was observed at wavelengths of less than 1.6 μm. The model showed a worse fit after 1.6 μm owing to thermal effects. The abundances of agglutinate, clinopyroxene and olivine were 71%, 13.08% and 10.58%, respectively. Plagioclase was found to be the least abundant mineral with a value of 4.82%. The PSDs of all of the endmembers exhibited almost normal distributions, and the observed differences in the mean particle sizes indicate differences in the space weathering rates of these endmembers. Although agglutinate may be finer, particle sizes smaller than 5 μm were not considered owing to the assumption of the Hapke model that particles are larger than the wavelength of the light.

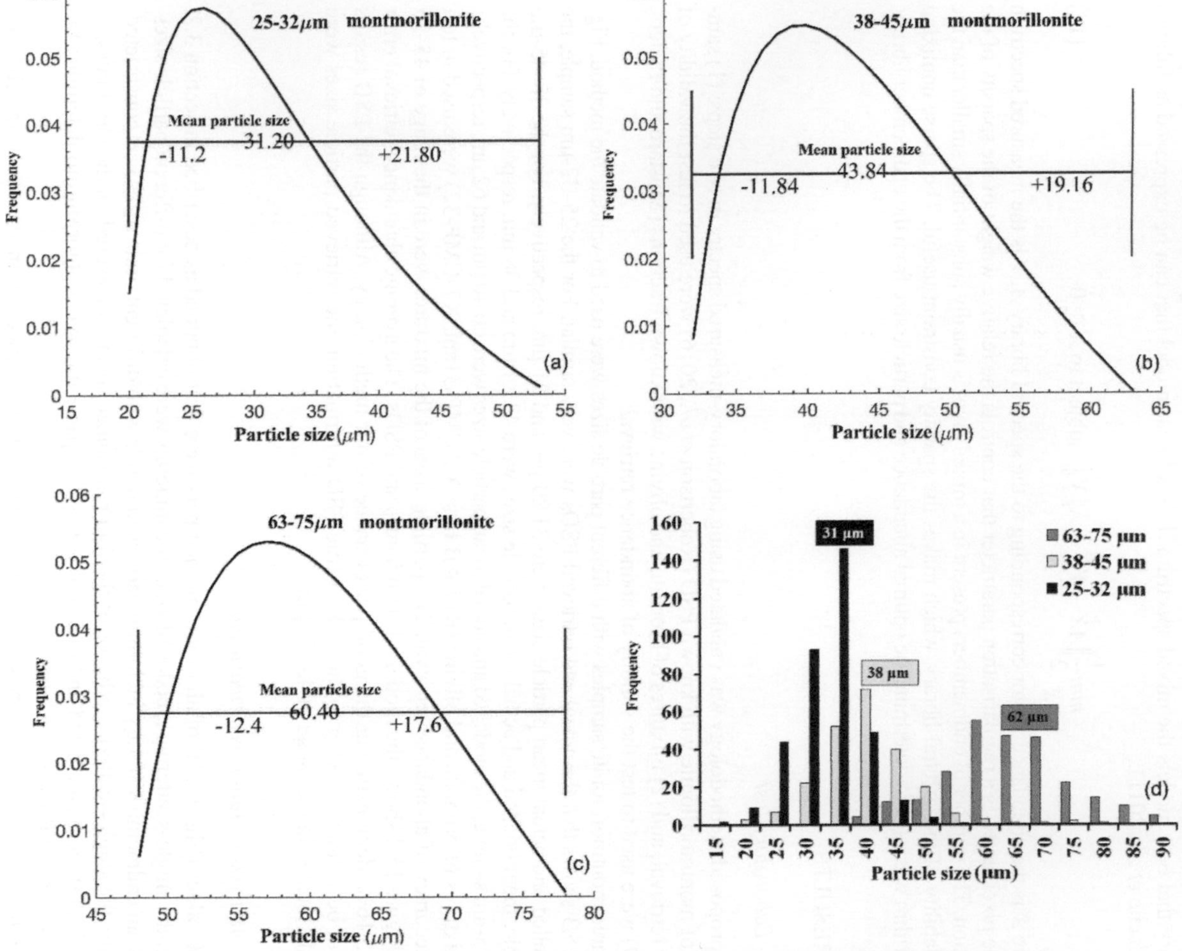

Figure 15.3. (a) Modelled PSD of 25–32-μm montmorillonite sample. (b) Modelled PSD of 38–45-μm montmorillonite sample. (c) Modelled PSD of 63–75-μm montmorillonite sample. (d) True PSDs of the three samples.

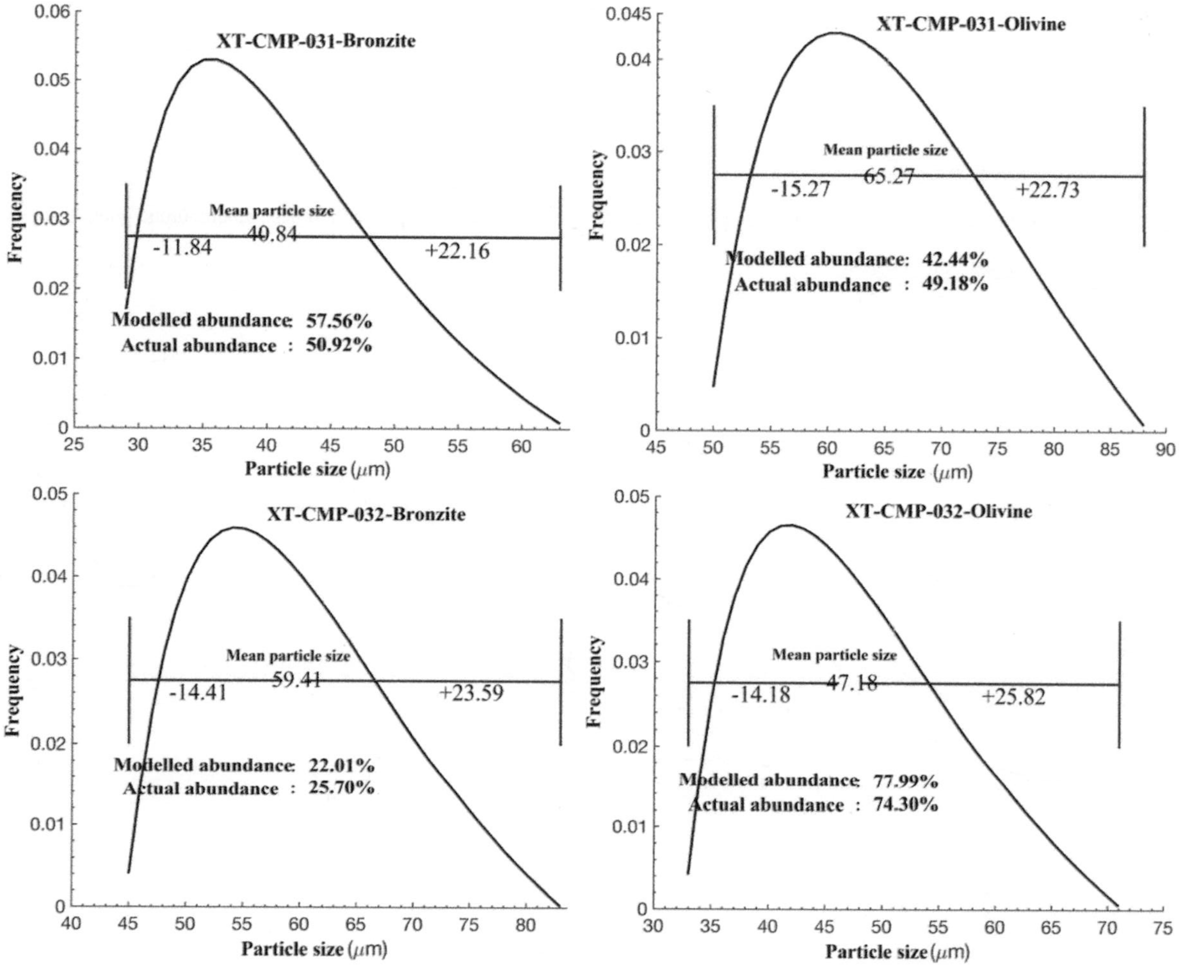

Figure 15.4. Abundances and PSDs of bronzite and olivine in mixtures

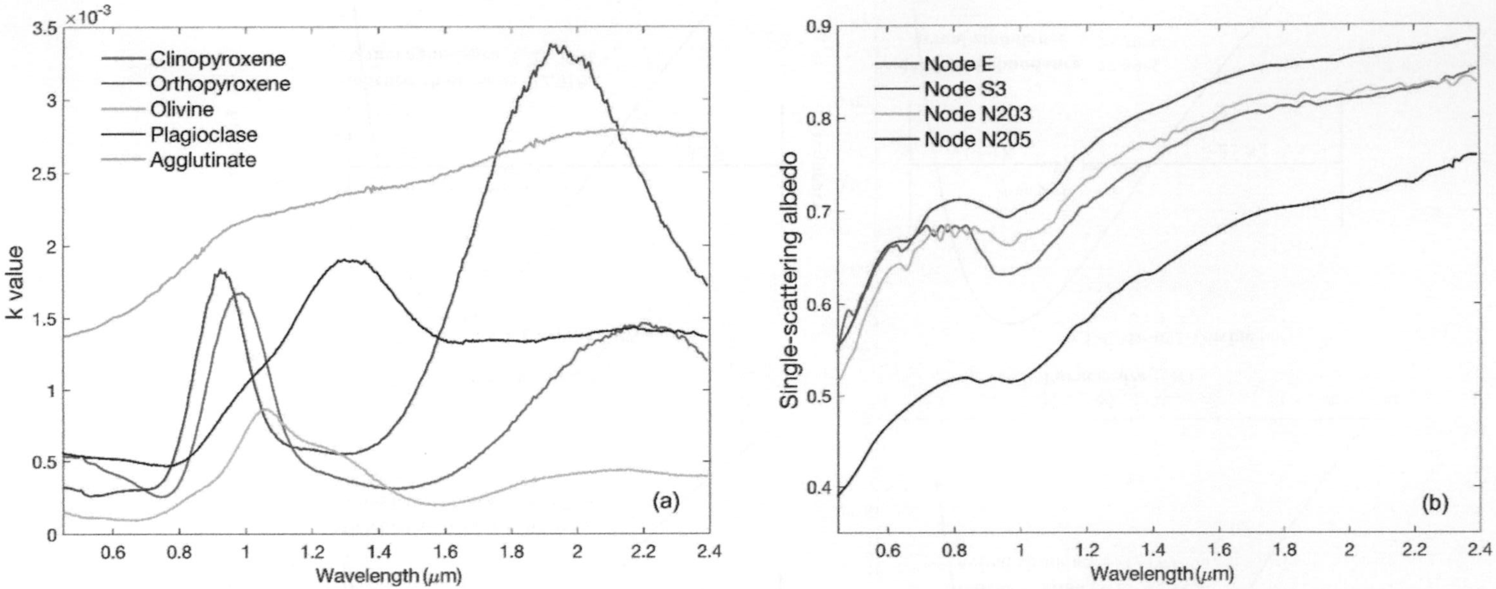

Figure 15.5. (a) k values for mineral endmembers used in this study. (b) Single-scattering albedos of the four reflectance spectra measured at different locations on the CE-3 mission.

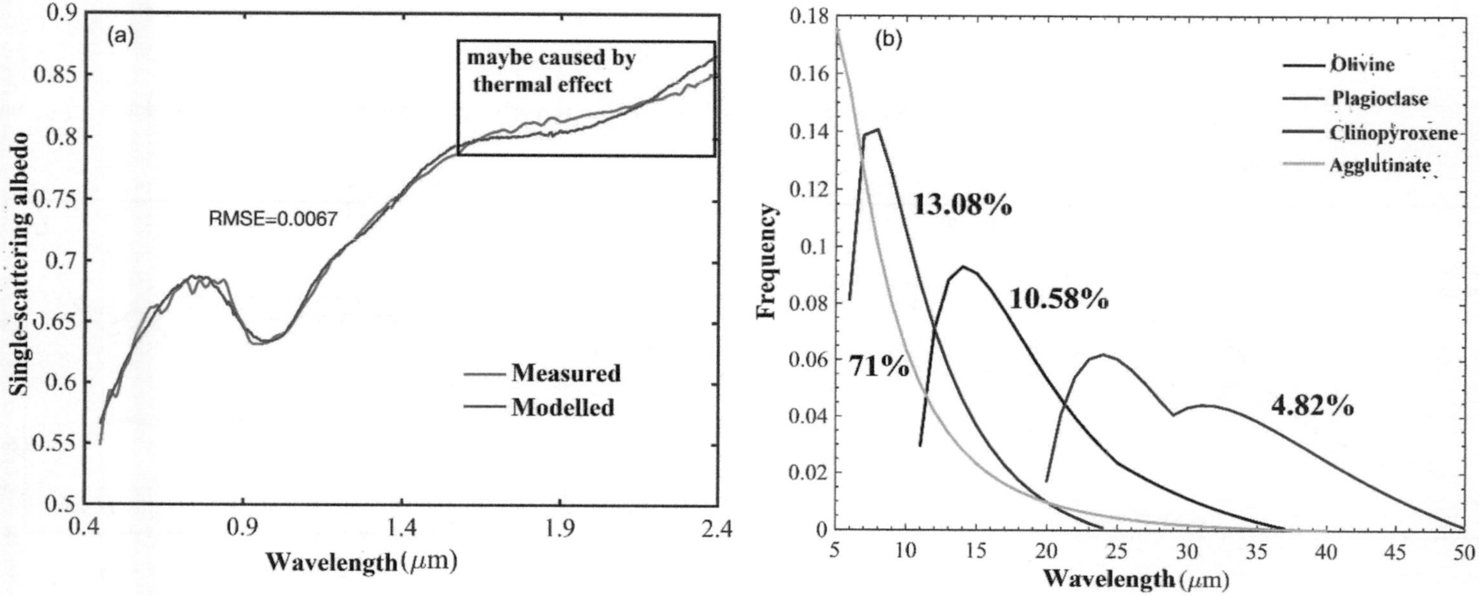

Figure 15.6. (a) Comparison between measured and modelled single-scattering albedos of Node E. (b) PSDs of different minerals.

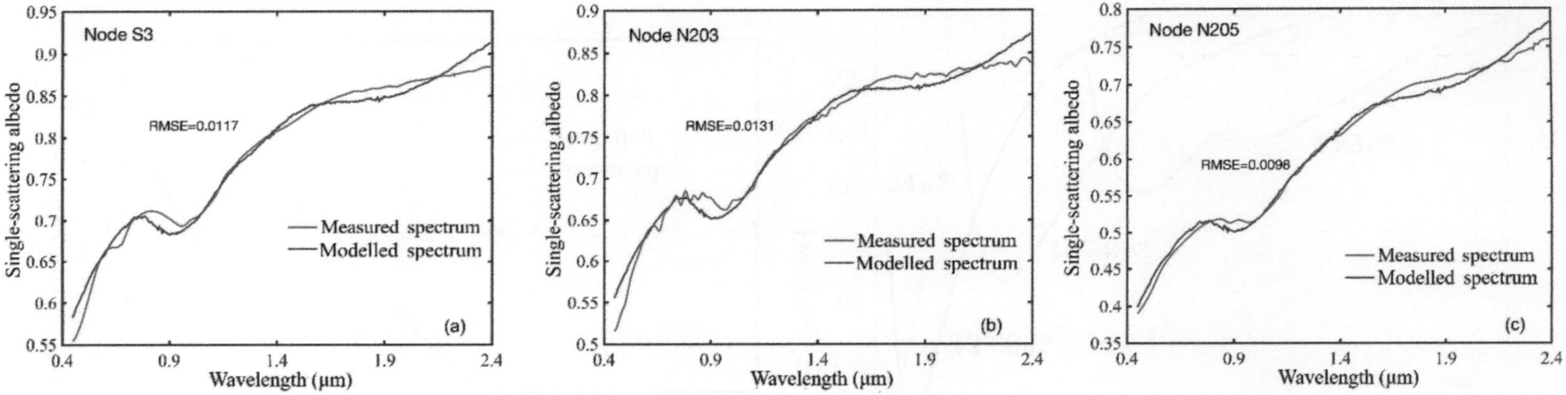

Figure 15.7. Comparisons of modelled and measured single-scattering albedo spectra obtained at Node S3, Node N203 and Node N205.

5 DISCUSSION

Of the four experiments, the model afforded the best results for the spectrum measured at Node E. For the other three measurement locations, the modelled and measured spectra also did not match very well at around 0.9 μm, in addition to the aforementioned deviation at wavelengths above 1.6 μm (Fig. 15.7). This can probably be ascribed to two main reasons: (1) The endmembers used in this study were inappropriate. The spectra may vary significantly for the same mineral reflecting surfaces owing to differences in chemistry or crystallinity. (2) The thermal effect. The temperature at the lunar surface can reach up to 100°C during the daytime (Combe *et al.*, 2011). In the long-wavelength region (beyond 1.6 μm), the energy emitted from the surface becomes more intense and cannot be neglected. This thermal effect from the reflectance spectra needs to be corrected. Thus, our future research will involve the consideration of additional endmembers, including olivine samples with different Mg contents, pyroxene with different calcium, iron and magnesium contents, and plagioclase with different iron contents. The thermal effects will also be corrected by using the data from other missions (e.g. LRO Diviner) (Li and Milliken, 2016).

6 SUMMARY

Quantitative determination of mineral abundance and particle size using VNIR spectroscopy is a fundamental goal of planetary exploration and science. A method combing a Hapke radiative transfer model and a sparse unmixing algorithm was proposed to derive the mineral abundances and PSDs from spectra measured by the imaging spectrometer on board the Yutu rover of the CE-3 mission. The Hapke model provides a means to convert reflectance into single-scattering albedo and construct an endmember library composed of mineral spectra with different particle sizes. The sparse unmixing algorithm then allows the pixels to be decomposed using the library.

The proposed method was tested using mineral mixtures obtained from the RELAB spectral database. The retrieved and actual PSDs were very similar, and the average absolute error of the retrieved abundances was 5.16%. For the data measured at Node E, the results revealed that the abundance of agglutinate reached up to 70% and the abundances of clinopyroxene and olivine were approximately 10%. The PSDs for each mineral exhibited almost normal distributions with different mean particle sizes and variance, possibly indicating the different susceptibilities of each component to space weathering at this site. For the data measured at Node S3, Node 203 and Node 205, the fit between the measured and modelled spectra was inferior to that for the data measured at Node E, which may be attributable to a combination of inappropriate endmember selection and thermal effects. However, the performance of laboratory mixtures and the measurements at Node E demonstrate that the proposed methodology has the potential to be applied to orbital hyperspectral remote sensing data.

REFERENCES

Clark, R.N., King, T.V., Klejwa, M., Swayze, G.A. & Vergo, N. (1990) High spectral resolution reflectance spectroscopy of minerals. *Journal of Geophysical Research: Solid Earth and Planets*, 95(B8), 12653–12680.

Combe, J.-P., McCord, T., Hayne, P. & Paige, D. (2011) Mapping of lunar volatiles with Moon Mineralogy Mapper spectra: a challenge due to thermal emission. *EPSC-DPS Joint Meeting*, Nantes, France,, 1644.

Hapke, B. (1981) Bidirectional reflectance spectroscopy theory. *Journal of Geophysical Research*, 86(Nb4), 3039–3054.

Heylen, R. & Gader, P. (2014) Nonlinear spectral unmixing with a linear mixture of intimate mixtures model. *IEEE Geoscience and Remote Sensing Letters*, 11, 1195–1199.

Heylen, R., Parente, M. & Gader, P. (2014) A review of nonlinear hyperspectral unmixing methods. *IEEE Journal of Selected Topics in Applied Earth Observations and Remote Sensing*, 7, 1844–1868.

Iordache, M.D., Bioucas-Dias, J.M. & Plaza, A. (2011) Sparse unmixing of hyperspectral data. *IEEE Transactions on Geoscience and Remote Sensing*, 49(6), 2014–2039.

Jin, W.D., Zhang, H., Yuan, Y., Yang, Y.Z., Shkuratov, Y.G., Lucey, P.G., Kaydash, V.G., Zhu, M.H., Xue, B., Di, K.C., Xu, B., Wan, W.H., Xiao, L. & Wang, Z.W. (2015) In situ optical measurements of Chang'E-3 landing site in Mare Imbrium: photometric properties of the regolith. *Geophysical Research Letters*, 42(20), 8312–8319.

Lucey, P.G. (2004) Mineral maps of the moon. *Geophysical Research Letters*, 31(8), 289–291.

Li, S.A. & Li, L. (2011) Radiative transfer modeling for quantifying lunar surface minerals, particle size, and submicroscopic metallic Fe. *Journal of Geophysical Research-Planets*, 116(E9).

Li, S.A. & Milliken, R.E. (2016) An empirical thermal correction model for Moon Mineralogy Mapper data constrained by laboratory spectra and Diviner temperatures. *Journal of Geophysical Research-Planets*, 121(10), 2081–2107.

Mustard, J.F. & Pieters, C.M. (1989) Photometric phase functions of common geologic minerals and applications to quantitative analysis of mineral mixture reflectance spectra. *Journal of Geophysical Research-Solid Earth*, 94(B10), 13619–13634.

Robertson, K., Milliken, R. & Li, S. (2016) Estimating mineral abundances of clay and gypsum mixtures using radiative transfer models applied to visible-near infrared reflectance spectra. *Icarus*, 277, 171–186.

Roush, T.L. (2003) Estimated optical constants of the Tagish Lake meteorite. *Meteoritics & Planetary Science*, 38(3), 419–426.

Shkuratov, Y., Starukhina, L., Hoffmann, H. & Arnold, G. (1999) A model of spectral albedo of particulate surfaces: implications for optical properties of the moon. *Icarus*, 137(2), 235–246.

Shuai, T., Zhang, X., Zhang, L.F. & Wang, J.N. (2013) Mapping global lunar abundance of plagioclase, clinopyroxene and olivine with Interference Imaging Spectrometer hyperspectral data considering space weathering effect. *Icarus*, 222(1), 401–410.

Wang, Z.C., Wu, Y.Z., Blewett, D.T., Cloutis, E.A., Zheng, Y.C. & Chen, J. (2017) Submicroscopic metallic iron in lunar soils estimated from the in situ spectra of the Chang'E-3 mission. *Geophysical Research Letters*, 44(8), 3485–3492.

Zhang, H., Yang, Y., Yuan, Y., Jin, W., Lucey, P.G., Zhu, M.H., Kaydash, V.G., Shkuratov, Y.G., Di, K. & Wan, W. (2015) In situ optical measurements of Chang'E-3 landing site in Mare Imbrium: mineral abundances inferred from spectral reflectance. *Geophysical Research Letters*, 42(17), 6945–6950.

Section V

Planetary remote sensing data fusion

Planetary Remote Sensing and Mapping – Wu et al.
 ISBN 978-1-138-58415-0

Chapter 16

Fusion of multi-scale DEMs from Chang'E-3 descent and Navcam images using compressed sensing method

M. Peng, W. Wen, Z. Liu and K. Di

ABSTRACT: The multi-source digital elevation models (DEMs) generated using images acquired during Chang'e-3's descent and landing phases and after landing contain supplementary information that allows a higher-quality DEM to be produced by fusing multi-scale DEMs. The proposed fusion method consists of three steps. First, source DEMs are split into small DEM patches, which are classified into a few groups by local density peak clustering. Next, the grouped DEM patches are used for sub-dictionary learning by stochastic coordinate coding. The trained sub-dictionaries are combined to form a dictionary for sparse representation. Finally, the simultaneous orthogonal matching pursuit algorithm is used to achieve sparse representation. We use real DEMs generated from Chang'e-3 descent images and navigation camera stereo images to validate the proposed method. Through our experiments, we reconstruct a seamless DEM with the highest resolution and the broadest spatial coverage of all of the input data. The experimental results demonstrate the feasibility of the proposed method.

1 INTRODUCTION

Chang'e-5 will be China's first lunar sample return mission. High-precision topographic mapping of the Chang'e-5 landing site yields detailed terrain information that will help to ensure the safety of the lander and support the tele-operated sampling of lunar soils and rocks. The lander will acquire descent images during its descent and landing phases and stereo images of the sampling area after landing. Various digital elevation model (DEM) products will be generated using image data acquired by different sensors in different phases, which often differ in spatial coverage, resolution and accuracy. Multi-source DEMs contain supplementary information, allowing a higher-quality DEM to be produced by fusing multi-scale DEMs. During previous Mars and lunar landing missions, descent images have been used to localize the lander and/or map the landing area. Li *et al.* (2002) conducted experiments with simulated descent images and Field Integrated Design and Operations Rover data. Ma *et al.* (2001) developed an integrated bundle adjustment system incorporating both descent and rover-based images to localize the rover along the traverse. Approaches to visual localization using overhead and ground images have been reviewed and the particular capabilities under development at the Jet Propulsion Laboratory discussed (Matthies *et al.*, 1997). During the Mars Exploration Rover mission, sequential Descent Image Motion Estimation System images were taken and used for lander localization (Li *et al.*, 2005). During the Mars Science Laboratory mission, descent images acquired by the Mars Descent Imager in its entry, descent and landing phases were compiled into image mosaics to provide colour coverage of the landing site and scientific target regions; these mosaics were incorporated into the landing base map (Parker *et al.*, 2013). During the Chang'e-3 (CE-3) mission, descent images were used to generate high-precision topographic products of the landing site with different resolutions. The images were also used for lander localization (Liu *et al.*, 2015).

In planetary exploration as well as Earth observation applications, DEMs with a high spatial resolution have limited spatial coverage, due to the high cost of data acquisition. They may also experience data quality problems such as data voids and noise. Relatively low-resolution DEMs provide less spatial information but usually cover larger areas. In the last few years, many studies have been conducted on ways of improving the quality of DEMs, and fusion ideas have thereby been introduced to DEM reconstruction. For example, a Shuttle Radar Topography Mission DEM with a resolution of 90 m was fused with data from the Advanced Spaceborne Thermal Emission and Reflection Radiometer Global Digital Elevation Model with a spatial resolution of 30 m in the frequency domain, filling data voids and thereby improving the overall accuracy of the fused data (Karkee *et al.*, 2008). Multi-scale modelling has been adopted to fill the voids in high-resolution DEM data, and a multi-scale Kalman smoother has been used to remove blocky artefacts in DEM fusion (Jhee *et al.*, 2013). However, these methods cannot be used to fuse DEMs that differ in resolution, coverage or vertical accuracy.

To address the inhomogenity of available DEM products, several methods of fusing DEMs have been developed to obtain complete DEM coverage with improved quality. Recently, sparse representation based methods, as a subset of transform-domain fusion methods, have been applied to DEM fusion. Ersoy (2009) created a dictionary that relates high-resolution image patches from a panchromatic image to the corresponding filtered low-resolution versions, and proposed two algorithms which directly use the dictionary and its low-resolution version to construct a fused image. Papasaika *et al.* (2011) presented a generic algorithmic approach to fusing two arbitrary DEMs, based on a framework of sparse representation, and conducted experiments with real DEMs from different earth observation satellites. Boufounos *et al.* (2011) introduced a new sparsity model for fusion frames, which provides a promising new set of mathematical tools and signal models with a variety of applications. Probabilistic analysis shows that under very mild conditions, the probability of recovery failure decays exponentially as the dimensions of the subspace increase. Tao and Qin (2011) proposed a new image fusion algorithm in the compressive domain, using an improved sampling pattern. Yue *et al.* (2015) proposed a regularised framework for the production of high-resolution DEM data with extended coverage. However, these studies focused on DEMs based on Earth observations; to date, compressed sensing methods of DEM fusion have not been investigated in planetary mapping contexts.

In this paper, we present a compressed sensing method of fusing multi-scale DEMs to produce a high-resolution DEM with extended coverage. In an experiment using Chang'e-3 DEMs, we reconstruct a seamless DEM with the highest resolution and the largest spatial coverage of all of the input data.

The rest of this paper is structured as follows. Section 2 briefly describes compressed sensing; section 3 presents and specifies the proposed method; and section 4 presents the experimental results. Finally, conclusions are drawn and suggestions for future work are offered in section 5.

2 COMPRESSED SENSING

Compressed sensing is a signal processing technique used to efficiently acquire and reconstruct signals by finding solutions to underdetermined linear systems. Through optimisation, the sparsity of a signal can be exploited to recover it from far fewer samples than required by the Shannon-Nyquist sampling theorem (Donoho, 2006).

Suppose that the signal f can be recovered from a set of M measurements. This compressive measurement vector can be formulated as follows:

$$y = \Phi f \tag{1}$$

Where $y \in R^M (M << N)$, N denotes the dimensions of the original signal and $\Phi \in R^{M \times N}$ is a measurement matrix. As M << N, the recovery of the signal vector f from the measurement vector

y is a highly underdetermined problem. However, recovery is possible under two conditions. The first is sparsity: the signal must be sparse in some domain. The signal f can be represented sparsely on an orthogonal basis. The second condition is incoherence, which is applied through the isometric property sufficient for sparse signals. The orthogonal basis Ψ and the compressive measurement matrix are incoherent.

3 METHOD

The proposed fusion method consists of three steps. First, source DEMs are split into small DEM patches, which are then classified into a few groups. Next, the grouped DEM patches are used for dictionary learning with the K-SVD algorithm. Finally, the simultaneous orthogonal matching pursuit (OMP) algorithm is used to achieve sparse representation. The resulting sparse coefficients are obtained using the max L1-norm rule. Using the learned dictionary, thc fused coefficients can be inversely transformed into a high-resolution DEM with extended coverage.

3.1 *Problem formulation*

Suppose that y_k represents DEMs with different resolutions and areas of coverage. The model used to generate a multi-DEM can be defined as follows:

$$y_k = F_k x + \varepsilon_k \tag{2}$$

where ε_k represents the noise vectors and F_k represents the multiplication of the transformation and downsampling matrices. The measurements y_k (k = 1, 2, . . ., n) must be fused to recover DEM x.

Assume that the fused result x can be represented as a sparse linear combination of elements from a dictionary D. The elements of D are called atoms. The result x is sparsely represented over D if $x = Da$, where a denotes a sparse coefficient vector with the most zeros. Therefore, the generative model can be represented as follows:

$$y_k = F_k D_k \alpha + \varepsilon_k \tag{3}$$

where D_k defines dictionaries of different resolutions. If D_k and y_k are available, the sparse coefficients a can be recovered and the fused DEM x can be determined by computing $D_k\alpha$. This function can be expressed as follows:

$$\min_{a \in R^N} \left[\sum_k \left\| F_k D_k \alpha - y_k \right\|_2^2 + \tau \left\| \alpha \right\|_1 \right] \tag{4}$$

The first term corresponds to the reconstruction error with respect to the observed DEMs, y_k. The second term is associated with the L1 norm of the candidate solution vector α. The parameter τ controls the trade-off between data fitting and sparsity. However, as points in different DEMs have different levels of accuracy, weight parameters should be included in the problem formulation. Therefore, the optimisation function is modified as follows:

$$\min_{a \in R^N} \left[\sum_k w_k \left\| F_k D_k \alpha - y_k \right\|_2^2 + \tau \left\| \alpha \right\|_1 \right] \tag{5}$$

where w_k denotes the weight for the kth DEM.

To improve computational efficiency, local DEM patches are used in Equation (5), as represented by y_k. To deal with blocking artefacts along the patch borders, consistency is imposed between neighbouring patches. Assuming that operator P extracts the overlap region between patches and y_p is a vector containing DEM values in the overlap region, adding to (5) a regularisation term that minimises the discrepancy between overlapping patches gives the following final formulation:

$$\min_{\alpha \in R^N} \left[\sum_k w_k \left\| F_k D_k \alpha - y_k \right\|_2^2 + \beta \left\| P F_k D_k \alpha - y_p \right\|_2^2 + \tau \left\| \alpha \right\|_1 \right] \tag{6}$$

where parameter β controls the influence of the patch overlap factor.

3.2 *Dictionary construction*

The raw DEM patches are randomly sampled, similar to the approach used by Yang *et al.* (2008), and the samples are classified into several new groups based on feature vectors. As K-SVD is one of the most popular dictionary learning algorithms, it is used to train the sub-dictionaries. As the DEM patches in each cluster have similar structures, a sub-dictionary learning scheme can provide a more accurate structural description of the input DEM patches. The processes of K-SVD-based sub-dictionary learning and combination are shown in detail below.

Step 1: A few sub-dictionaries, $S_1, S_2, \ldots, S_n$, are learned for each DEM patch group.

Step 2: The sub-dictionaries in each cluster are trained using the K-SVD algorithm. Next, the sub-dictionaries are combined to form a dictionary for fusion, as follows:

$$\Phi = [S_1, S_2, \ldots, S_n] \tag{7}$$

where Φ is the combined dictionary and $S_1, S_2, \ldots, S_n$ are the trained sub-dictionaries.

3.3 *Adaptive weight determination for DEM fusion*

The fusion is completed using weight maps that reflect the estimated relative accuracy of the source DEMs at each grid point. First, the datasets undergo geometric registration. After registration, the transformation matrix M_k can be obtained and the cropping region can be derived according to the coordinates.

A data-driven strategy is used to find the weights based on geomorphological characteristics, similar to the approach used by Boufounos *et al.* (2011). Based on the slope and entropy, the accuracy maps for both input DEMs at each overlapping point can be determined; finally, the reciprocal values are used as weights in the fusion (Papasaika *et al.*, 2011).

3.4 *Fusion of DEMs*

As optimisation problems of this form constitute the main computational kernel of compressed sensing applications, a wide range of algorithms have been developed to solve them. Due to its simplicity and computational efficiency, the OMP algorithm (Mallat, 1998) is used in our experiment. The overlap parameter β is set between 0.5 and 1.0. The number of non-zero atoms is set between 5 and 10. The minimum patch size is 5 x 5 and the maximum patch size is 9 x 9.

4 RESULTS AND DISCUSSION

4.1 *Landing site mapping using CE-3 descent images*

CE-3 began to descend from lunar orbit at an altitude of around 15 km, and at 2 km above the lunar surface the descent camera started to take images. During the phases of descent, hovering

Table 16.1 Technical parameters of CE-3 descent camera

Image size	*Actual imaging distance*	*Focal length*	*Pixel size*
1024 x 1024 pixels	4 m ~ 2000 m	8.3 mm	6.7 μm

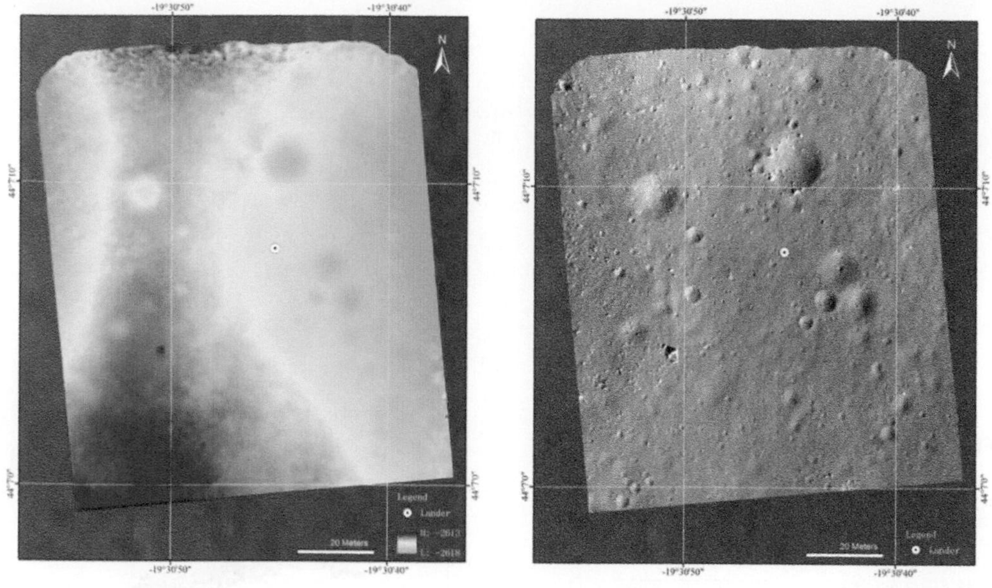

Figure 16.1. DEM (left) and DOM (right) generated from descent images

and obstacle avoidance, and landing, the descent camera acquired 4672 images with resolutions higher than 1 m in an area of 1 x 1 km and as high as 0.1 m within 50 m of the landing point. The main technical parameters of the CE-3 descent camera are listed in Table 16.1. One hundred and eighty images with equal time intervals are selected and incorporated into a self-calibration free network bundle adjustment, and the initial trajectory of the camera is recovered (including position and attitude measurements). Next, 26 ground control points (GCPs) are selected from the rectified Chang'e-2 DEM and a digital orthophoto map (DOM) for absolute orientation. The root mean square errors (RMSEs) of these GCPs are 0.724 m, 0.717 m and 0.602 m in three directions. The RMSEs of 18 of the check points are smaller than 1 m. Figure 16.1 shows the DEM and DOM generated from 80 descent images with a resolution of 0.05 m. The dots in the maps represent the lander position. The maps cover an area of 97 m x 115 m. Using more descent images at higher altitudes generates additional topographic products with broader coverage.

4.2 *Mapping with Navcam images*

The geometric parameters of the Yutu rover's Navcam are listed in Table 16.2. Figure 16.2 shows the DEM at waypoint D. It has a resolution of 0.02 m and is automatically generated from the Navcam images. To fuse the DEMs with the descent images, the ground DEM is transformed into a lunar body fixed coordinate system.

4.3 *Integration of mapping products*

The topographic products of the descent and ground images show discrepancies. The two DEMs are co-registered using the iterative closet point algorithm. This reduces the mean differences

Table 16.2 Parameters of Yutu Navcam camera

Stereo base	*Focal length*	*Image size*	*Field of view*
27 cm	1189 pixels	1024 x 1024 pixels	46.4° x 46.4°

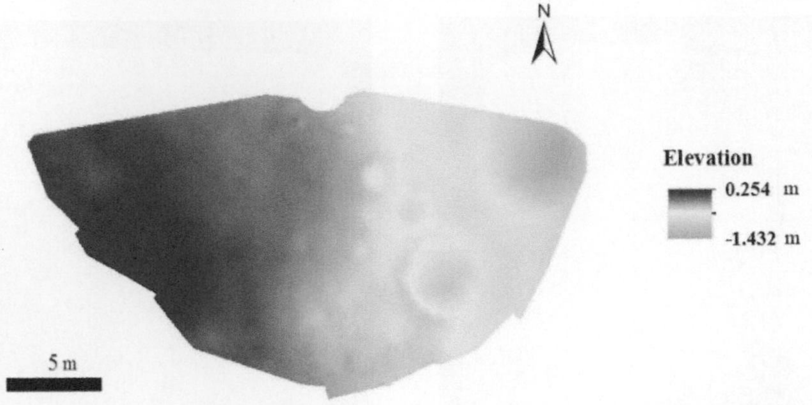

Figure 16.2. DEM of Navcam images at site D

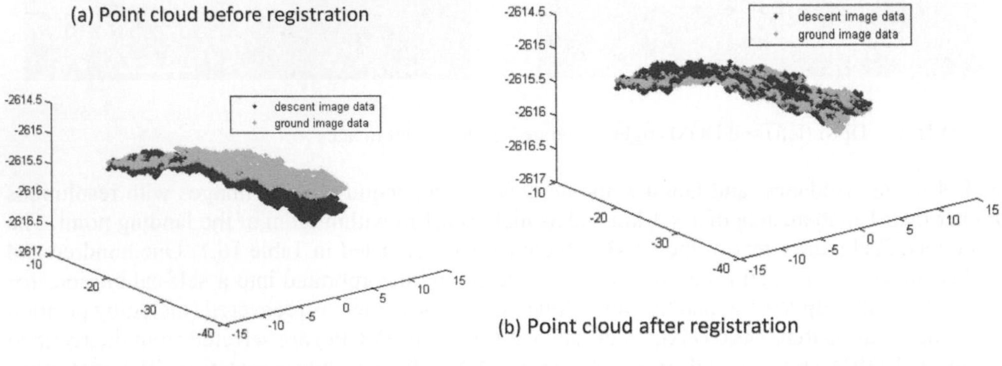

Figure 16.3. Co-registration of two datasets

between the two datasets from 0.142 m, 0.096 m and 0.765 m to 0.003 m, 0.004 m and 0.221 m in three directions, respectively, as shown in Figure 16.3.

We conduct two sets of experiments to test and evaluate the feasibility of the proposed method. A DEM with a resolution of 0.05 m from lander images with a small selected area, as shown in Figure 16.4, supplies the original data and the ground truth for the quantitative evaluation. The other test DEMs are derived after downsampling and cropping, as shown in Figure 16.5.

Our purpose is to reconstruct seamless DEM data with a 0.05 m resolution and the same coverage as the 0.4 m DEM by fusing the supplementary information between the two DEMs. The fusion result obtained using the proposed method is compared with the results of bilinear and kriging interpolation, which are commonly used for DEM densification. In addition, the quantitative indexes of the mean square error (MSE) and peak noise signal ratio (PNSR) are used to evaluate the vertical accuracy of the results:

$$RMSE = \frac{1}{mn}\sum_{i=0}^{m-1}\sum_{j=0}^{n-1}[I(i,j) - K(i,j)]^2 \tag{8}$$

$$PSNR = 20\log_{10}\left(\frac{MAX_I}{\sqrt{MSE}}\right) \tag{9}$$

As shown in Equation (8), K represents the reconstructed measurement, I denotes the reference data and MAX_I is the maximum possible pixel value of the image. A smaller RMSE and a larger PNSR correspond to better performance. Figure 16.6 shows that our proposed method performs better than the other two methods, and that reconstruction using this method produces an enhanced elevation result. These findings are confirmed by the quantitative results in Table 16.3.

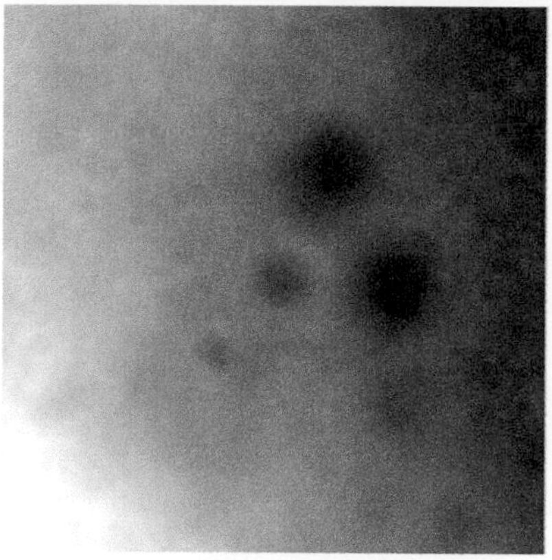

Figure 16.4. Ground truth

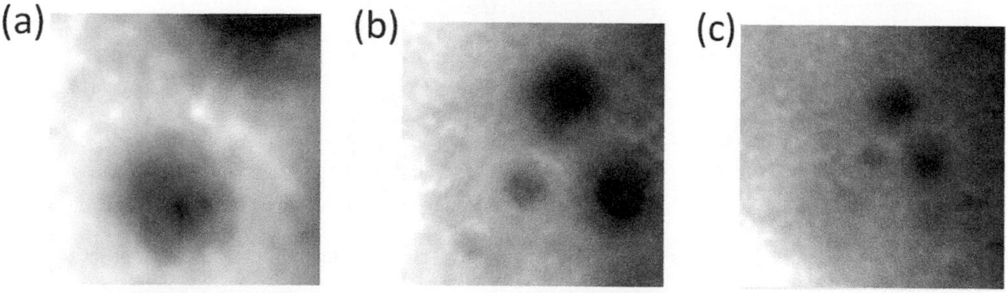

Figure 16.5. DEMs with resolutions of 0.05 m, 0.2 m and 0.4 m and different areas of coverage

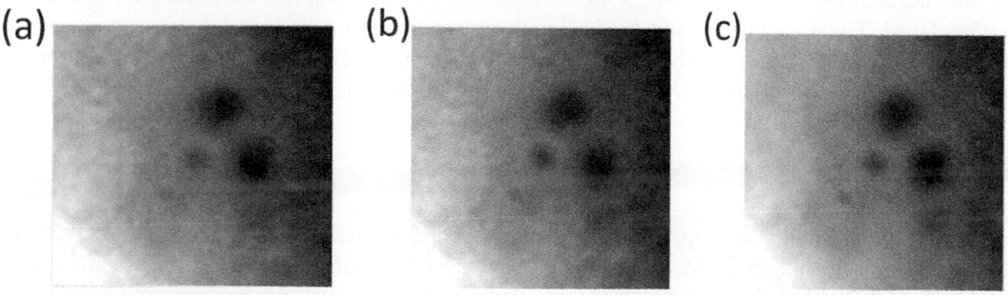

Figure 16.6. Reconstruction results for bilinear interpolation, kriging interpolation and the proposed method

Table 16.3 Quantitative results

	Bilinear	*Kriging*	*Proposed method*
RMSE	15.772 7.886		1.718
PNSR	24.17 30.193		43.43

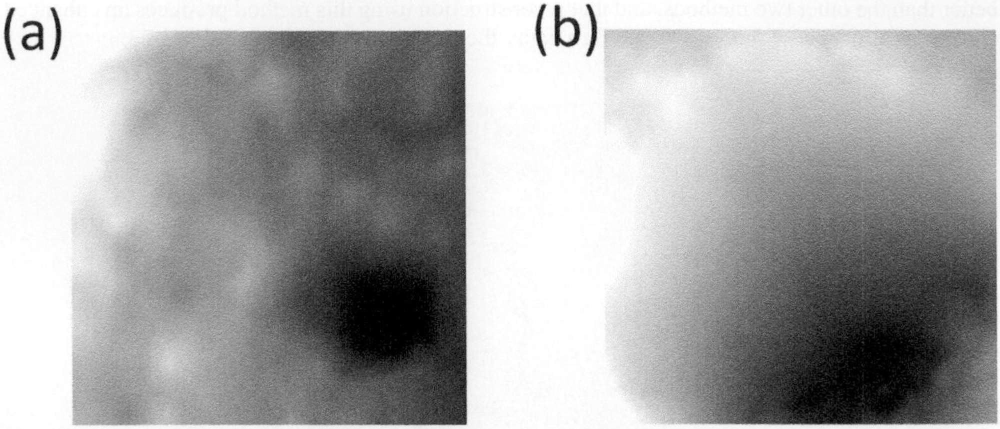

Figure 16.7. DEMs with resolutions of 0.05 m and 0.02 m and different areas of coverage

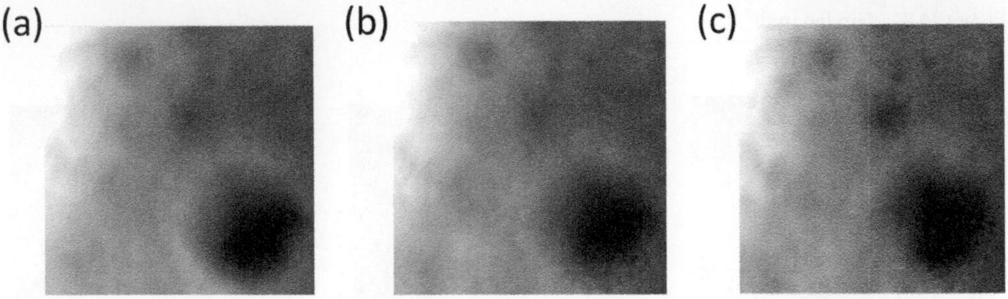

Figure 16.8. Reconstruction results for bilinear interpolation, kriging interpolation and the proposed method

The second experiment is based on lander and Navcam DEMs. The resolutions of the two DEM datasets are 0.05 m and 0.02 m, respectively (Fig. 16.7). Using two 161-x-161 datasets, we seek to reconstruct seamless DEM data (402 x 402) with a 0.02 m resolution and the same coverage as the 0.05 m DEM. The fusion results obtained using the interpolation methods and the proposed method are compared in Figure 16.8. The results show obvious visual differences; the proposed method provides more detail-enhanced DEM data with a 0.02-m resolution.

5 CONCLUSIONS

In this paper, we present an improved compressed sensing method of fusing multi-scale DEMs to produce a high-resolution DEM with extended coverage. In contrast with traditional fusion methods, the reconstructed data are generated using the supplementary information between DEMs with different resolutions and areas of coverage. In addition, grouped DEM patches rather than the whole source image are used for dictionary learning to improve the efficiency of the method. We use real DEMs generated from Chang'e-3 descent images and Navcam stereo images to validate the proposed method. Through our experiments, we reconstruct a seamless DEM with the highest resolution and the broadest spatial coverage of all of the input data. The experimental results demonstrate the feasibility of the proposed method. However, the proposed method still has some limitations. In the future, more complementary factors, such as the edginess and the noisiness of the DEM, will be taken into account to improve the accuracy of the fused data. In addition, if more training patches of different terrains from various sources are available for dictionary training, the trained model can be generalised to unseen data. We will test the suitability of other deep learning systems, such as the Generative Adversarial Network, as the generation model.

ACKNOWLEDGEMENTS

This work was supported by the National Natural Science Foundation of China (Grant Nos. 41471388, 41671458, 41771488).

REFERENCES

Boufounos, P., Kutyniok, G. & Rauhut, H. (2011) Sparse recovery from combined fusion frame measurements. *IEEE Transactions on Information Theory*, 57(6), 3864–3876.

Donoho, D. (2006) Compressed sensing. *IEEE Transactions on Information Theory*, 52(4), 1289–1306.

Jhee, H., Cho, H.C., Kahng, H.K. & Cheung, S. (2013) Multiscale quadtree model fusion with super-resolution for blocky artefact removal. *Remote Sensing Letters*, 4(4), 325–334.

Karkee, M., Steward, B.L. & Aziz, S.A. (2008) Improving quality of public domain digital elevation models through data fusion. *Biosystems Engineering*, 101(3), 293–305.

Li, R.X., Ma, F., Xu, F.L., Matthies, L.H., Olson, C.F. & Raymond, E.A. (2002) Localization of mars rovers using descent and surface-based image data. *Journal of Geophysical Research: Planets (1991–2012)*, 107(E11), 1–8.

Li, R.X., Squyres, S.W., Arvidson, R.E., Archinal, B.A., Bell, J., Cheng, Y. & Golombek, M. (2005) Initial results of rover localization and topographic mapping for the 2003 Mars Exploration Rover mission. *Photogrammetric Engineering & Remote Sensing*, 71(10), 1129–1142.

Liu, Z., Di, K., Peng, M., Wan, W., Liu, B., Li, L. & Chen, H. (2015) High precision landing site mapping and rover localization for Chang'e-3 mission. *Science China Physics, Mechanics & Astronomy*, 58(1), 1–11.

Ma, F., Di, K.C., Li, R., Matthies, L. & Olson, C. (2001) Incremental Mars rover localization using descent and rover imagery. *ASPRS 2001 Annual Conference*, St Louis, MO, USA. pp. 25–27.

Mallat, S. (1998) *A Wavelet Tour of Signal Processing*. Elsevier, London, UK.

Matthies, L., Olson, C.F., Tharp, G. & Laubach, S. (1997) Visual localization methods for Mars rovers using lander, rover and descent imagery. *Proceedings of the 4th International Symposium on Artificial Intelligence, Robotics, and Automation in Space*, Tokyo, Japan. pp. 413–418.

Papasaika, H., Kokiopoulou, E., Baltsavias, E., Schindler, K. & Kressner, D. (2011) Fusion of digital elevation models using sparse representations. *Proceedings of the 2011 ISPRS Conference on Photogrammetric Image Analysis*, Munich, Germany, 5–7 October. pp. 171–184.

Parker, T., Malin, M., Calef, F., Deen, R., Gengl, H., Golombek, M., Hall, J., Pariser, O., Powell, M., Seltten, R. & MSL Science Team (2013) Localization and contextualization of curiosity in gale crater, and other landed Mars missions. *44th Lunar and Planetary Science Conference*, Woodlands, TX, USA. Lunar and Planetary Institute. p. 2534.

Tao, W. & Qin, Z. (2011) An application of compressed sensing for image fusion. *Selected Papers from the ACM Conference on Image and Video Retrieval*, 88(18), 3915–3930.

Yang, J., Wright, J., Huang, T. & Ma, Y. (2008). Image super-resolution as sparse representation of raw image patches. In: *Computer Vision and Pattern Recognition*, Anchorage, AK, USA. pp. 1–8.

Yue, L., Shen, H., Yuan, Q. & Zhang, L. (2015) Fusion of multi-scale DEMs using a regularized super-resolution method. *International Journal of Geographical Information Science*, 29(12), 2095–2120.

Planetary Remote Sensing and Mapping – Wu et al.
© 2019 Taylor & Francis Group, London, ISBN 978-1-138-58415-0

Chapter 17

Co-registration of lunar imagery and digital elevation model constrained by both geometric and photometric information

B. Liu, X. Xin, K. Di, M. Jia and J. Oberst

ABSTRACT: The alignment of images with a digital elevation model (DEM) has many applications in the planetary mapping field. In this chapter, we propose a novel highly precise co-registration method that can achieve direct pixel-based matching between an image and a reference DEM. The DEM is first converted into a simulated image using a hill-shading technique based upon a photometric model and the image's illumination conditions. Initial matching between the simulated image and the input image is then performed based on affine scale-invariant feature transform (ASIFT). Meanwhile, the image's rational function model is established and used as a geometric constraint. Next, the tie points generated by the initial ASIFT matching are used to refine the rational function model geometric model of the image and to eliminate the gross errors of tie points in an iterative fashion. Finally, highly precise co-registration is performed by pixel-based least-squares image matching using the refined geometric model as a global geometric constraint. Two Lunar Reconnaissance Orbiter (LRO) narrow-angle camera images located at the pre-selected landing site of the Chang'E-5 mission and the SLDEM2015, a combined product of lunar orbiter laser altimetry (LOLA) and DEM generated from the Japanese Selenological and Engineering Explorer (SELENE) terrain camera images, were selected in our experiment. The results demonstrate that the proposed method can achieve effective pixel-based matching between an image and a reference DEM with a mean accuracy of 15.6 pixels in the image space and 20.95 m (0.5 pixel of the reference DEM) in the object space. Another 15 narrow-angle camera images that were evenly distributed at various latitudes were selected as an additional experiment to evaluate the matching precision of various DEM densities. The results show that the precision improved in a near-linear manner as the latitude increased. Thus, the denser the DEM, the greater the matching precision. The proposed method offers a highly precise and automatic method of matching orbiter images with DEMs.

1 INTRODUCTION

Significant numbers of lunar orbital images have been acquired by previous and ongoing lunar exploration missions, so automatic geometric processing of these images has become necessary and urgently needed for scientific and engineering applications. Owing to the precision limitations of orbit and attitude measurements, widespread spatial inconsistencies exist amongst lunar data from multiple sources, such as multiple images and laser altimetry–derived digital elevation models (DEMs). It is essential to co-register the images in a unified reference frame for orthorectification and other applications. Co-registration of images to a reference DEM can provide accurate control points for highly precise geometric processing of the images, such as geometric model refinement, instrument calibration and automatic rectification. Another application of aligned images and DEM lies in reflectance-based surface reconstruction. High-precision co-registration of images and DEMs is needed to construct a higher-resolution DEM using a coarse DEM and a high-resolution image with shape-from-shading techniques (Grumpe and Wöhler, 2011, 2014; Wu *et al.*, 2018).

Because images and DEMs represent different types of surface information, the pixel-based (area-based) image-matching method cannot be applied directly in image and DEM registration.

Co-registration between images and DEMs is mainly feature-based (i.e., craters, ridges, etc.) (Michael, 2003; Wu *et al.*, 2013). The performance of the feature-based matching method depends upon the precision of feature extraction, which is poor in featureless areas. An approach for the registration of stereo images and DEMs (or laser altimeter points) is based upon the combined block adjustment in which three-dimensional laser altimeter points are back-projected into the stereo images and used as conjugate points (Soderblom and Kirk, 2003; Wu *et al.*, 2014; Yoon and Shan, 2005). The altitude of the DEM or laser altimetry measurements is used as a constraint equation in block adjustment processing. This approach requires massive iterations because the algorithms are highly complex. Stereo images are generally needed for the co-registration process. Another approach to the registration of lunar stereo images and DEMs (or laser altimeter points) is based upon surface matching using an iterative closest point algorithm (Di *et al.*, 2012), but it is not applicable to co-registration of a single image and a reference DEM. It is noteworthy that the matching methods mentioned above do not consider the radiometric information of the image in the matching process. In a previous study, pixel-based methods for matching imagery with DEM that consider radiometric information have been proposed using earth observation satellite data (Horn and Bachman, 1978). The DEMs are simulated to images with radiometric information by illuminating an existing DEM with the image's illumination geometry information, and the traditional pixel-based matching method can then be performed. This method assumes that the DEM itself and its gradients can be considered as an illumination-invariant representation of the image (Horn, 1977). This assumption is unfounded in an area with a variety of surface features, especially in human settlements. The method was thus studied only in areas of mountainous terrain without vegetative coverage. With barren surfaces of the Moon and Mars, it will be more applicable and more widely used using the data of the Moon and Mars surfaces. Some similar studies of co-registration between altimeter tracks and images (Nefian *et al.*, 2014; Soderblom and Kirk, 2003; Soderblom *et al.*, 2002) were also based upon this assumption on the lunar or Martian surface. These methods consider images' radiometric information, but the large sampling interval of the laser altimeter provides little information to the simulated images, which limits the precision of co-registration.

Based upon our literature review of the matching method of lunar images and reference elevation data, we proposed a novel method of pixel-based co-registration of lunar images and DEMs using both geometric and photometric constraints. First, the simulated image is generated based upon the orbital images' illumination condition. Meanwhile, the rigorous sensor model and rational function model (RFM) of the orbital image is established based upon the SPICE kernels (*S*pacecraft ephemeris; *P*lanet, satellite, comet or asteroid ephemerides; *I*nstrument description kernel; *C*-Matrix pointing kernel and *E*vents kernel; NAIF, 2014). Then, Harris feature points are extracted from the simulated image and back-projected to the orbital image as the initial matching tie points. Next, highly precise co-registration based upon least-squares image matching is performed using the refined geometric model. As a result of the co-registration, the RFM geometric model of the image is refined, and highly precise tie points between the image and the reference DEM are extracted simultaneously. Lunar Reconnaissance Orbiter (LRO) narrow-angle camera (NAC) images and SLDEM2015 at various latitudes are used for experimental testing and validation of the proposed co-registration method.

2 METHODS

In our method, the image is co-registered to a reference DEM via a pixel-based matching method using both geometric and photometric constraints. Figure 17.1 outlines the framework of the proposed method. Because the DEM has much lower resolution (60 m pixel^{-1} at the equator) than the images (~1.5 m pixel^{-1}) used in our experiments, we first up-sample the DEM to the image resolution with a spline resample method. The simulated images are then generated with the hill-shading technique based upon the images' illumination conditions. To obtain the initial matching tie points

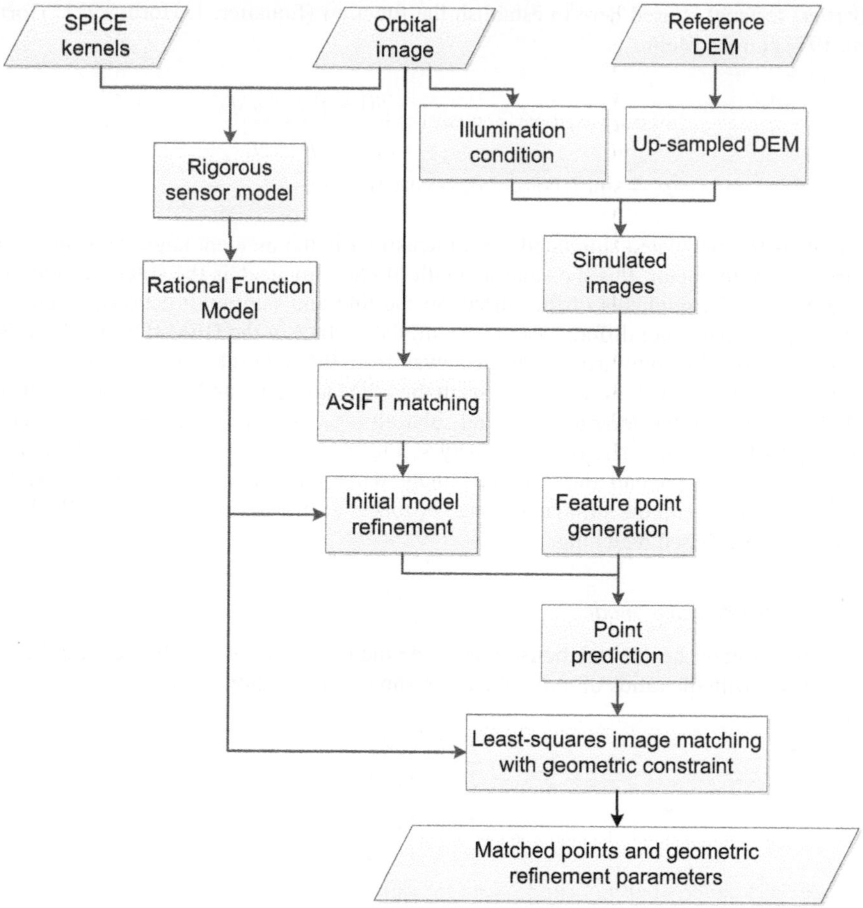

Figure 17.1. Flowchart of the proposed method

and the initial value for least-squares image matching, affine scale-invariant feature transform (ASIFT, Morel and Yu, 2009) is performed between the simulated image and the actual orbital image. Meanwhile, the rigorous sensor model of the orbital image is established (Liu *et al.*, 2017) based upon the SPICE kernels. The RFM of the image is established based upon the rigorous sensor model and is used as a geometric constraint in the co-registration process. With the tie points generated by the initial ASIFT matching, the RFM of the image is refined and the gross errors of the tie points are eliminated in an iterative manner. Next, evenly distributed feature points are generated on the simulated image and are projected to the orbital image coordinates with refined RFM. Finally, a highly precise co-registration based upon least-squares image matching is performed using the refined geometric model.

2.1 *Image simulation using DEM*

In our research, image simulation is simply based upon the hill-shading technique. The key to this method is to present the reflectance of a small, flat surface as a function of the gradient of the terrain. An idealised reflectance model (i.e., the surface is assumed to be an ideal diffuser or Lambertian surface) is used here to establish the function (Equation 1, Horn, 1981; Horn and Bachman, 1978) on the Moon:

$$\Phi(p,q) = \rho\cos(i)/\cos(e) = \frac{\rho(1+p_s p+q_s q)}{\sqrt{1+p_s^2+q_s^2}} \qquad (1)$$
$$p_s = \sin(\theta)\cot(\phi),\ q_s = \cos(\theta)\cot(\phi)$$

where $\Phi(p, q)$ is the calculated simulated pixel intensity, i is the incident angle, e is the emission angle, ρ is an 'albedo' factor, θ is the solar azimuth of each image, ϕ is the solar altitude of each image, and p, q are the gradients of the surface in the line and sample directions, which can be estimated using the first-order differences of a small, flat surface of the DEM (Horn, 1981). We set $\Phi(p, q)$ to equal zero when the surface is turned away from the light $((1 + p_s p + q_s q) < 0)$.

In this study, the p and q values for each pixel in the DEM are estimated based upon eight-neighbour grid elevations, and the solar azimuth and solar altitude used in Equation 1 can be retrieved from the image's header file. The pixel intensity in the simulated image is then calculated with the reflectance model. In the end, the simulated image will have illumination conditions similar to those of the real image, and the simulated image becomes a 'bridge' to connect the DEM and the orbital image in pixel-based matching.

2.2 *RFM and its refinement model*

The RFM, which establishes the relationship between the image-space coordinates and the object-space coordinates with the ratios of polynomials, is shown in Equation 2 (Di *et al.*, 2003):

$$r = \frac{P_1(X,Y,Z)}{P_2(X,Y,Z)}$$
$$c = \frac{P_3(X,Y,Z)}{P_4(X,Y,Z)} \qquad (2)$$

The three-order polynomials P_i (i = 1, 2, 3 and 4) take the following general form:

$$\begin{aligned} P_i(X,Y,Z) = {} & a_1 + a_2X + a_3Y + a_4Z + a_5XY + a_6XZ + a_7YZ + a_8X^2 \\ & + a_9Y^2 + a_{10}Z^2 + a_{11}XYZ + a_{12}X^3 + a_{13}XY^2 + a_{14}XZ^2 \\ & + a_{15}X^2Y + a_{16}Y^3 + a_{17}YZ^2 + a_{18}X^2Z + a_{19}Y^2Z + a_{20}Z^3 \end{aligned} \qquad (3)$$

where a_1, a_2 . . . a_{20} are the coefficients of the polynomial function P_i (i.e., rational polynomial coefficients [RPCs]). Construction of an RFM is a process of fitting the vast amount of virtual points generated based upon the rigorous sensor model (RSM) via least-squares solution (Liu *et al.*, 2016). RFM is much simpler and more independent of sensors than RSM, and RFM can fit the RSM without loss of accuracy.

Usually, the RFM of a lunar orbital image contains biases due to the limited precision of orbit and attitude measurements. The biases are reflected by back-projection error in the image space and can be corrected by co-registration of the image and a reference DEM. The affine transformation model (Liu *et al.*, 2014) in the image space, shown in Equation 4, is used to correct the biases to refine the RFM:

$$r = a_0 + a_1c' + a_2r'$$
$$c = b_0 + b_1c' + b_2r' \qquad (4)$$

where (r, c) are the measured (matched) image coordinates, (r', c') are the back-projected image coordinates of the tie points calculated from ground points using RFM and a_0, a_1, a_2, b_0, b_1 and b_2 are affine transformation parameters.

2.3 *Least-squares image matching with global geometric constraint*

The matching method presented in this paper is based upon the classic least-squares image matching method (Ackermann, 1984), which considers both geometric distortion and radiometric distortion, as shown in Equation 5:

$$g_1(x,y)+n_1(x,y)=h_0+h_1g_2(a_0+a_1x+a_2y,b_0+b_1x+b_2y)+n_2(x,y) \tag{5}$$

where g_1 and g_2 are the functions of the intensity value, n_1 and n_2 are the noises, x, y are the coordinates of the conjugate points in the image space, a_0, a_1, a_2, b_0, b_1 and b_2 are the geometric transformation (affine transformation) parameters and h_0 and h_1 are parameters of the radiometric transformation between the conjugate regions. However, the classic least-squares matching method considers only the geometric distortion in a single conjugate region and does not consider the geometric transformation amongst all matched tie points.

In this study, we present a least-squares matching method with global geometric constraints to improve the matching robustness. With the geometric information obtained in section 2.2, the geometric distortions amongst all tie points are considered with the use of RFM and the refinement model in the matching process. Equation 6 shows the function of the improved least-squares matching method:

$$g_1(x,y)+n_1(x,y)=h_0+h_1g_2(r,c)+n_2(x,y) \tag{6}$$

$$\begin{aligned} r &= a_0+a_1r'+a_2c'+\delta_r \\ c &= b_0+b_1r'+b_2c'+\delta_c \end{aligned} \tag{7}$$

where x, y are the coordinates of the tie point in the simulated image (equivalent reference DEM), r and c are the coordinates calculated via RPCs and the refined transformation parameters, δ_r and δ_c are the geometric model uncertainties or errors and r' and c' are the coordinates calculated from Equation 2. Using Equation 6, every pixel in the image patch centred at each tie point provides an error equation, and Equation 7 is used as an additional equation in the adjustment to impose a global geometric constraint. Meanwhile, δ_r and δ_c can be considered as fictitious observation equations (Equation 8):

$$\begin{aligned} F_r &= \delta_r = r-(a_0+a_1c'+a_2r') \\ F_c &= \delta_c = c-(b_0+b_1c'+b_2r') \end{aligned} \tag{8}$$

Before global least-squares image matching, the initial values of the unknowns (i.e., the geometric transformation parameters and the coordinates of the tie points) must be calculated first. The initial values of the geometric transformation parameters are calculated based upon the ASIFT matched tie points. Figure 17.2 illustrates the proposed matching method. The feature points are extracted from the simulated image based upon the Harris corner extraction method (Harris and Stephens, 1988), which is calculated from the image's radiometric information. The geographic coordinates of the feature points are calculated via the reference DEM. The feature points are back-projected onto the real orbital image based upon the initial refined RFM and are used as the initial tie points. The final tie points and the refined RFM are adjusted via an iteration procedure in least-squares image matching with global geometric constraints.

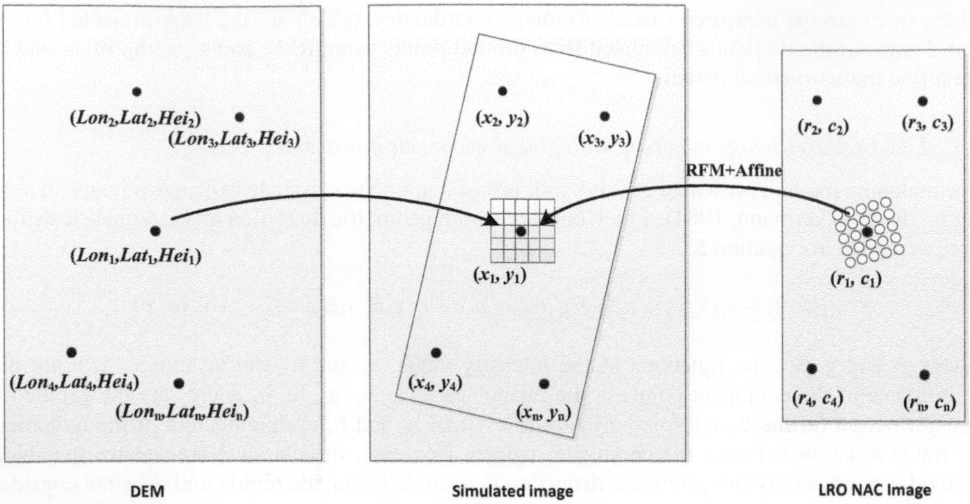

Figure 17.2. Schematic illustration of least-squares matching with global geometric constraints

3 EXPERIMENTAL ANALYSIS

3.1 *Datasets*

In our experiment, 17 LRO NAC images (two from the Chang'E-5 pre-selected landing site and 15 distributed at various latitudes) and the SLDEM2015 (Barker *et al.*, 2016) are used to test and validate the developed method. All experimental data are downloaded from the Planetary Data Systems (PDS) website (http://ode.rsl.wustl.edu/moon/). The resolution of the NAC images selected for this experiment is about 1.5 m pixel^{-1}, and the resolution of SLDEM2015 is 512 pixels per degree (for an effective resolution of ~60 m at the equator).

Figure 17.3 shows the locations of all images used in this study, with the background of an LRO wide-angle camera global image mosaic downloaded from the USGS website (https://astrogeology.usgs.gov/search/details/Moon/LRO/LROC_WAC/Lunar_LRO_LROC-WAC_Mosaic_global_100m_June2013/cub). Table 17.1 shows the image IDs and their main parameters. The solar azimuth and solar altitude of the NAC images are obtained from the headers of the PDS files.

3.2 *Simulated images and initial value of co-registration*

Seventeen simulated images are generated based upon the images' illumination information. Figure 17.4 shows two examples of the simulated images located at the Chang'E-5 pre-selected landing site based upon the method described in section 2.1. The simulated image (Fig. 7.4 (a), (c)) has similar reflectance to the real NAC image (Fig. 17.4 (b), (d)). Because we did not consider absolute radiometry, the overall tone of the simulated image is darker than that of the real image, but the pixel values of the simulated image and the real image have a nearly linear relationship. Because h_0, h_1 are used as the radiometric calibration parameters (Equation 5) to deal with the linearly radiometric distortion in the matching, the subsequent co-registration process is not affected.

In the simulated images, feature points are detected based upon the Harris detector, which computes a matrix related to the image's autocorrelation function. Meanwhile, ASIFT is performed between the simulated image and the real image to calculate the refining parameters of the RPCs. The initial values of the co-registration tie points in the images are calculated from the Harris feature points based upon the RPCs and the initial refining parameters. An obvious inconsistency

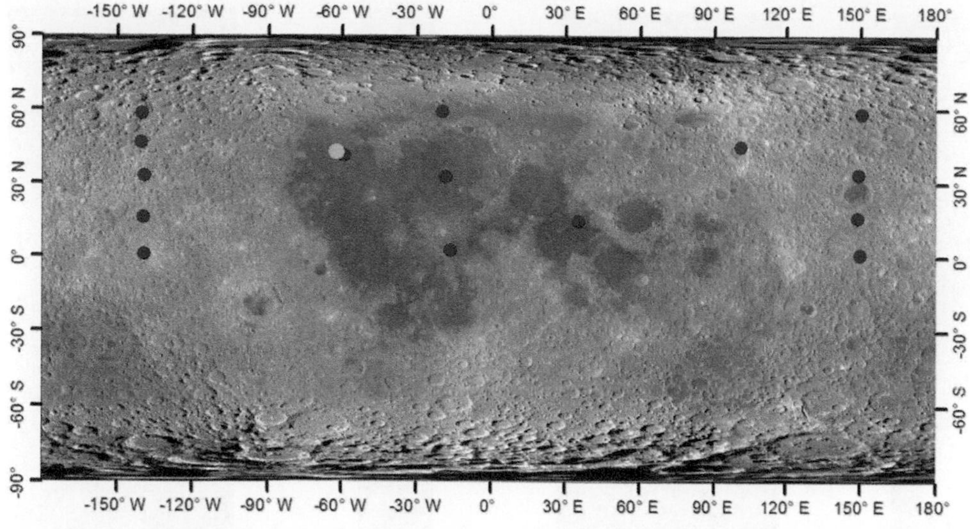

Figure 17.3. Locations of the images used in this study. Green points represent the locations of the images from the Chang'E-5 pre-selected landing site, and red points represent the locations of the images distributed at various latitudes.

Table 17.1 Main parameters of the images used in the experiment

Image ID	*Centre latitude (°)*	*Centre longitude (°)*	*Resolution (m)*	*Incidence angle (°)*	*Sub–solar azimuth (°)*
M1173478556LE	42.35	–62.3	1.43	73.89	197.68
M1188747453LE	42.35	–62.16	1.26	71.17	196.68
M1143826200LE	1.12	149.81	1.23	57.12	181.04
M1173968149LE	1.14	–139.29	1.13	61.39	181.88
M1096601220LE	2.86	–16.49	1.12	67.72	181.54
M1123363232LE	14.72	35.47	1.17	66.6	175.17
M1113197889LE	15.95	148.88	1.32	55.42	191.39
M1097401250RE	16.03	–139.41	1.39	59.84	189.38
M1157862745LE	32.58	–18.51	1.34	72.98	191.06
M1097393759LE	32.85	–139.08	1.59	63.79	198.95
M1205049446RE	33.56	149.26	1.40	61.05	202.13
M1112211163RE	41.08	–59.00	1.44	72.26	196.83
M1098172290LE	44.90	101.14	1.59	62.66	211.40
M1189253414LE	46.34	–140.54	1.49	68.45	203.51
M1158638171RE	58.34	–139.98	1.65	74.77	204.43
M1205056993RE	58.36	150.71	1.58	72.62	211.68
M1111955876LE	58.86	–19.76	1.54	79.09	199.17

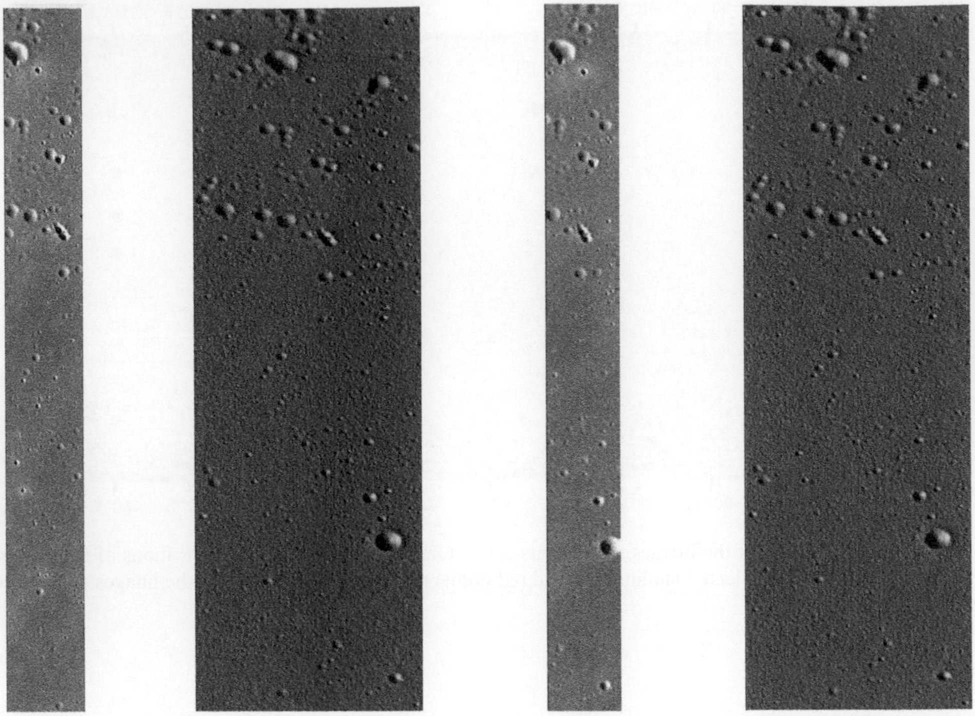

Figure 17.4. Examples of the actual LRO NAC images and simulated images. (a) NAC image (image ID: m1173478556le). (b) Simulated image with same illumination conditions as (a). (c) NAC image (image ID: m1188747453le). (d) Simulated image with same illumination conditions as (c).

exists between the reference DEM and the NAC image from the initial tie points, which are calculated from the initial refined RPCs. It is therefore necessary to co-register the image to DEM and further refine the RPC.

3.3 *Co-registration results and precision evaluation*

Two LRO NAC images are selected as examples to evaluate the precision of this method in detail. Using the method described in section 2.3, 216 and 179 feature points are extracted from these two images. The tie points with correlation coefficients higher than 0.7 are selected as the final tie points. Figure 17.5 shows the distribution of the final matched tie points. All tie points are evenly distributed over the overlapping area between the image and the DEM, so all positions of the image can be rectified well.

Two NAC orthophotos are generated based upon the initial RPCs and refinement parameters, and Figure 17.6 shows the position difference between the orthophotos and the reference DEM. Three craters are chosen as the examples from the two NAC images. The coloured part is the reference DEM, and the grayscale part of the figure is the orthophoto. A spatial inconsistency can be seen between the original orthophotos and the DEM in Figure 17.6 (a, c), whilst in Figure 17.6 (b, d), the inconsistency is eliminated with our co-registration method.

To further evaluate the co-registration precision in a quantitative manner, the matching results of ASIFT matching, classic least-squares matching and our global least-squares matching method are compared. The precision of co-registration is represented by the fitting precision (root-mean-square

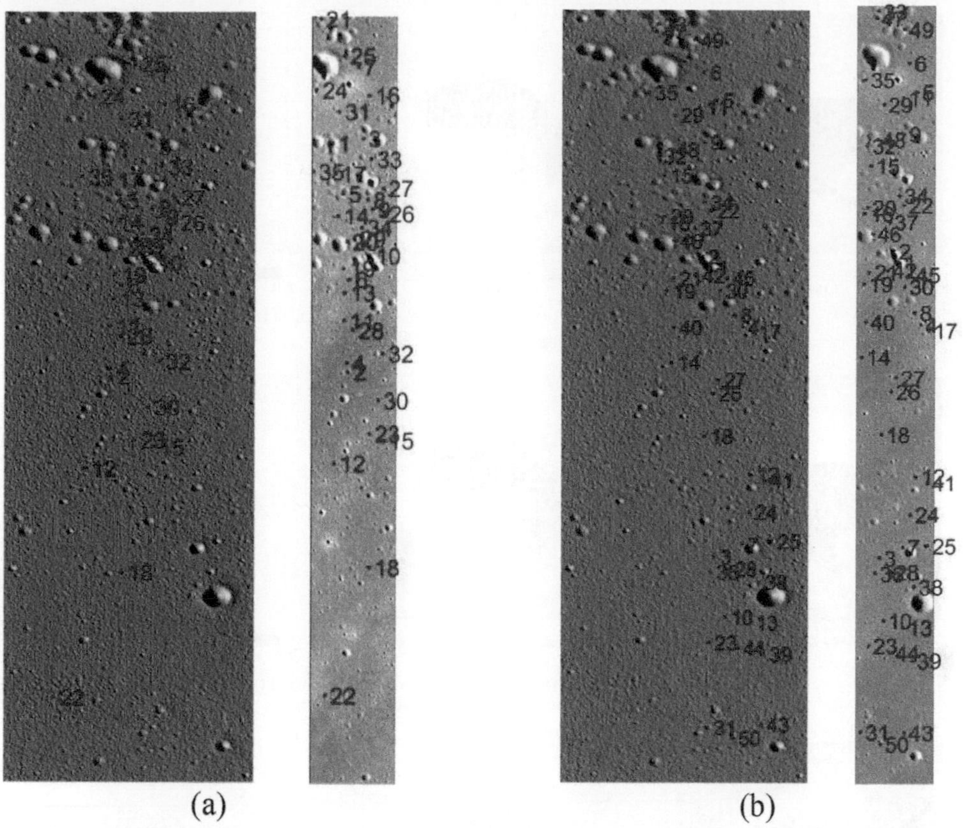

Figure 17.5. Distribution of the final matched tie points. (a) Image ID: m1173478556le. (b) Image ID: m1188747453le.

error [RMSE] in the image space) of affine transformation for RFM refinement. For the Chang'E-5 pre-selected landing site, the proposed method can reach a mean accuracy of 15.6 pixels in the image space and 20.95 m (0.5 pixel of the reference DEM) in the object space. These values show that our method is more precise than the other two methods (Table 17.2); specifically, it is 1.5 or 2 times more precise than the ASIFT method.

3.4 *Co-registration results at various latitudes*

The laser altimeter tracks are denser in higher latitude areas because of the polar orbit, so different simulated images contain different amounts of information. The higher the latitude, the greater the quantity of information the image has (see examples in Fig. 17.7). Based upon this observation, more image co-registration experiments have been performed at various latitudes. The interval of the latitude is 10°, and three images at various longitudes are chosen for each latitude.

The results of ASIFT matching, classic least-squares matching and our matching method at various latitudes are compared via the distribution of tie points and the precision of co-registration. The result given for each latitude is the average result of the three images at various longitudes. Figure 17.8 shows the number of tie points extracted from the three matching methods. The presented method extracted the most tie points, nearly twice as many as with ASIFT. Meanwhile, the distribution of the tie points extracted from ASIFT are concentrated in several craters or rocks,

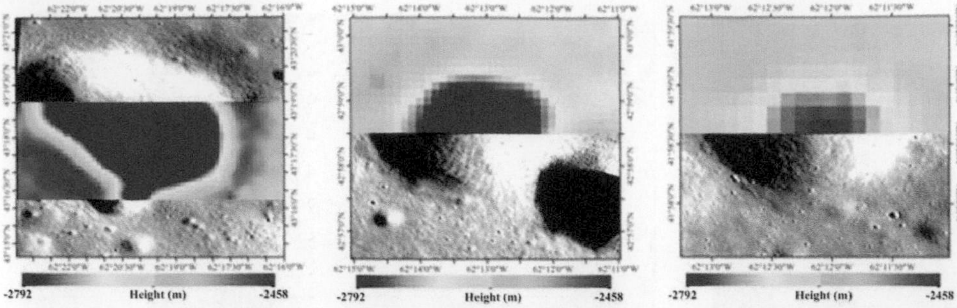

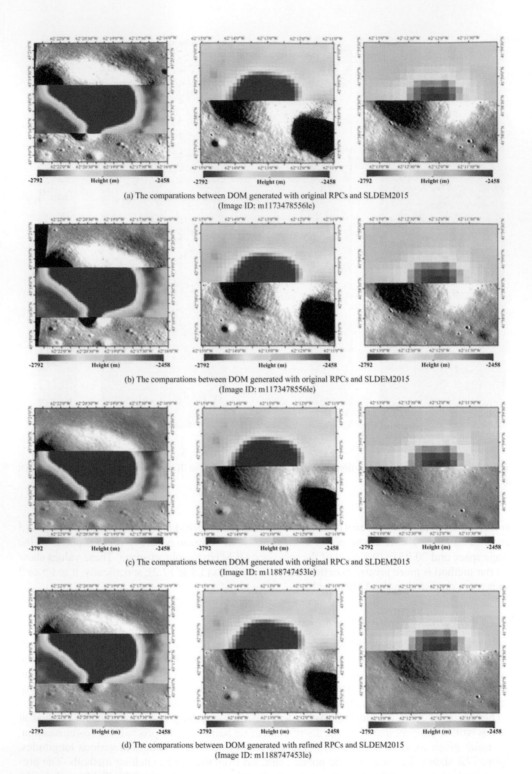

(a) The comparations between DOM generated with original RPCs and SLDEM2015
(Image ID: m1173478556le)

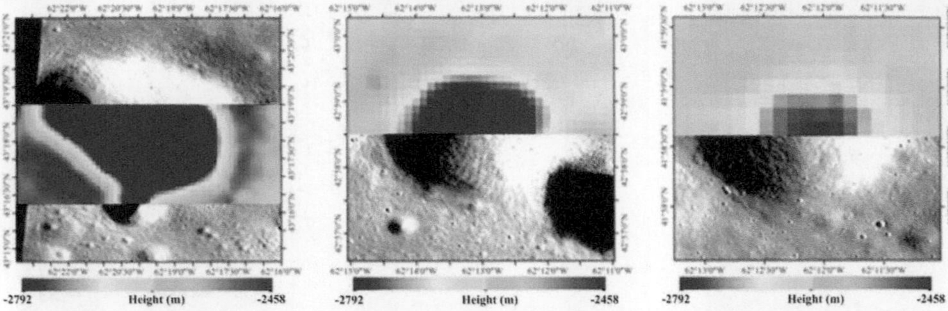

(b) The comparations between DOM generated with original RPCs and SLDEM2015
(Image ID: m1173478556le)

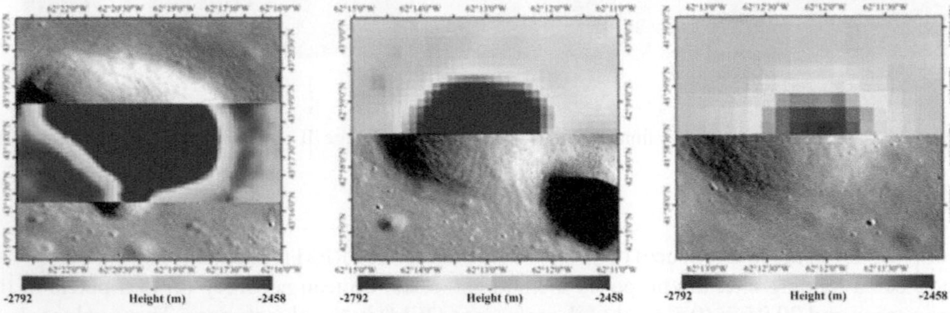

(c) The comparations between DOM generated with original RPCs and SLDEM2015
(Image ID: m1188747453le)

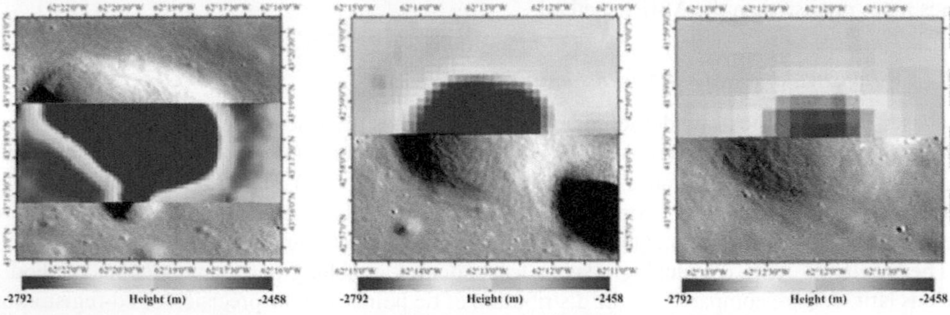

(d) The comparations between DOM generated with refined RPCs and SLDEM2015
(Image ID: m1188747453le)

Figure 17.6. Position difference between the orthophotos and the reference DEM

Table 17.2 Matching precision of different methods

Method	*Image ID*	*RMSE/X (pixels)*	*RMSE/Y (pixels)*	*RMSE/All (pixels)*
ASIFT	m1173478556le	9.532	20.980	23.044
	m1188747453le	19.863	24.036	31.181
Traditional least-squares method	m1173478556le	12.241	15.424	19.691
	m1188747453le	9.772	16.255	18.966
Our proposed method	m1173478556le	8.729	12.511	15.256
	m1188747453le	8.354	13.566	15.932

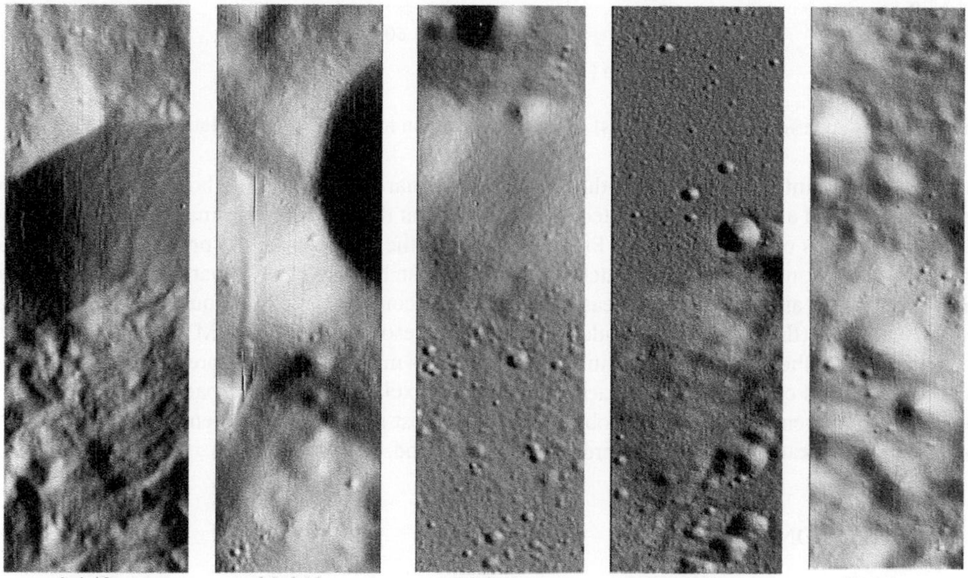

Figure 17.7. Examples of simulated images located at various latitudes

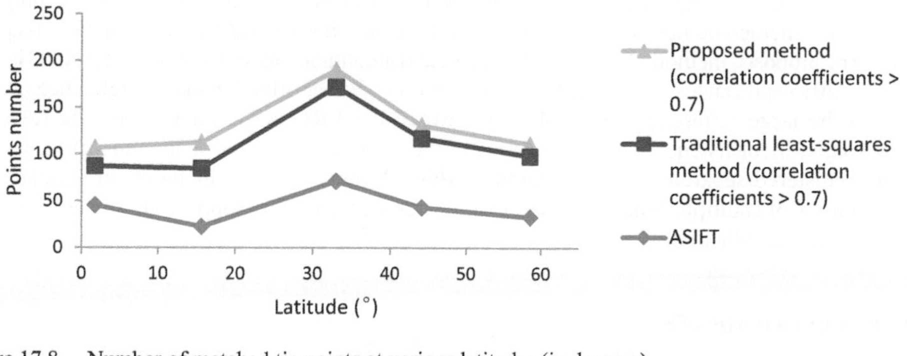

Figure 17.8. Number of matched tie points at various latitudes (in degrees)

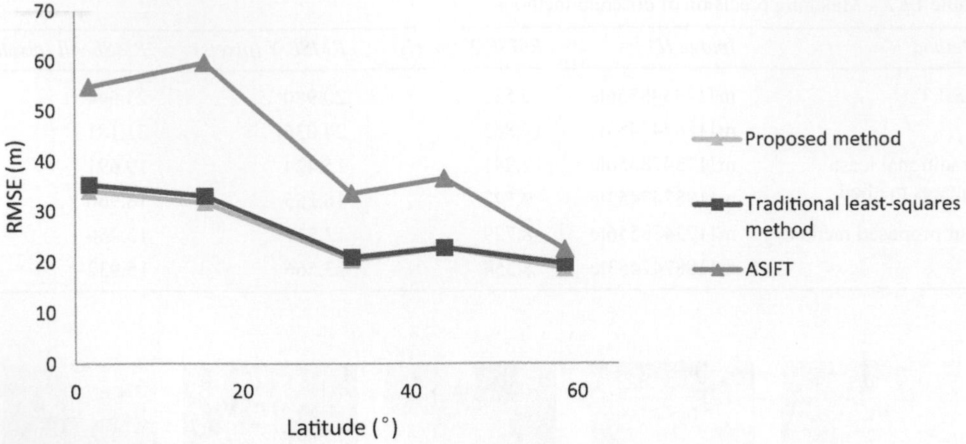

Figure 17.9. Precision (RMSE in metres) of the co-registration methods at various latitudes (in degrees)

whilst the tie points extracted from the classic least-squares matching method and the proposed matching method are evenly distributed. This ensures that the least-squares matching methods can rectify all regions of the image well. Figure 17.9 shows the precision of the presented method and that of the traditional pixel-based matching method. It can be seen clearly that the matching precision is improved as the latitude increases. This trend is consistent with the quality of the DEM at various latitudes (the higher the latitude, the greater the resolution of the DEM). Figure 17.9 shows a comparison of the precision of traditional least-squares matching and the presented method. The proposed method can reach a mean accuracy of 18.3 pixels in the image space and 25.17 m (0.5 pixel of the reference DEM) in the object space. In most regions, the presented method is more precise than the traditional least-squares matching method.

4 CONCLUSIONS

This chapter presents a novel pixel-based method using both geometric and photometric constraints for co-registration of lunar orbital image and reference DEM. The orientation of the image is also refined, which facilitates subsequent orthophoto generation. The experiments using 17 NAC images and SLDEM2015 have demonstrated the feasibility and effectiveness of the proposed method. The precision of the co-registration can reach a sub-pixel level of the reference DEM, and the proposed co-registration method outperforms both the ASIFT matching method and the traditional least-squares method. The proposed method offers a highly precise and automatic method of tying LRO NAC images to SLDEM2015 and can also be applied to co-registration of other orbital images to reference DEM.

Due to the large differences in resolution between the LRO NAC images and the reference DEMs, spatial inconsistencies may still occur in the overlapping areas of the adjacent images if they are registered separately to the reference DEM. A future study will focus on simultaneous co-registration of multiple adjacent images to a reference DEM to support automated seamless large-area orthorectification.

ACKNOWLEDGEMENTS

This study was supported by National Natural Science Foundation of China under Grants 41771490, 41671458, 41590851 and 41301528. We thank all those who worked on the Planetary Data System archive to make the LRO imagery and SLDEM2015 publicly available.

REFERENCES

Ackermann, F. (1984) Digital image correlation: performance and potential application in photogrammetry. *Photogrammetric Record*, 11(64), 429–439.

Barker, M.K., Mazarico, E., Neumann, G.A., Zuber, M.T., Haruyama, J. & Smith, D.E. (2016) A new lunar digital elevation model from the Lunar Orbiter Laser Altimeter and SELENE Terrain Camera. *Icarus*, 273, 346–355.

Di, K., Ma, R. & Li, R.X. (2003) Rational functions and potential for rigorous sensor model recovery. *Photogrammetric Engineering & Remote Sensing*, 69(1), 33–41.

Di, K., Hu, W., Liu, Y. & Peng, M. (2012) Co-registration of Chang'E-1 stereo images and laser altimeter data with crossover adjustment and image sensor model refinement. *Advances in Space Research*, 50, 1615–1628.

Grumpe, A.M. & Wöhler, C. (2011) DEM construction and calibration of hyperspectral images data using paris of radiance images. *Proceedings of the IEEE International Symposium on Image and Signal Processing and Analysis*, Dubrovnik, Croatia. pp. 609–614.

Grumpe, A.M. & Wöhler, C. (2014) Recovery of elevation from estimated gradient fields constrained by digital elevation maps of lower lateral resolution. *ISPRS Journal of Photogrammetry and Remote Sensing*, 94, 37–54.

Harris, C. & Stephens, M. (1988) A combined corner and edge detector. *Proceedings of the 4th Alvey Vision Conference*. pp. 147–151.

Horn, B.K.P. (1977) Understanding image intensities. *Artificial Intelligence*, 8, 201–231.

Horn, B.K.P. (1981) Hill shading and the reflectance map. *Proceedings of the IEEE*, 69(1), 14–47.

Horn, B.K.P. & Bachman, B.L. (1978) Using synthetic images to register real images with surface models. *Graphics and Image Processing*, 21(11), 914–924.

Liu, B., Xu B., Di K. & Jia, M. (2016) A solution to low fitting precision of planetary orbiter images caused by exposure time changing. *Proceedings of 23rd ISPRS Congress, Commission IV*, 12–19 July, Prague, Czech Republic. pp. 441–448.

Liu, B., Liu, Y.L., Di, K.C. & Sun, X.L. (2014) Block adjustment of Chang'E-1 images based on rational function model. *Remote Sensing of the Environment: 18th National Symposium on Remote Sensing of China, International Society for Optics and Photonics*, Wuhan, China, Volume 9158, 91580G.

Liu, B., Jia, M., Di, K., Oberst, J., Xu, B. & Wan, W. (2017) Geopositioning precision analysis of multiple image triangulation using LROC NAC lunar images. *Planetary and Space Science*. [Online] Available from: http://dx.doi.org/10.1016/j.pss.2017.07.016.

Michael, G.G. (2003) Coordinate registration by automated crater recognition. *Planetary and Space Science*, 51, 563–568.

Morel, J. & Yu, G. (2009) ASIFT: a new framework for fully affine invariant image comparison. *SIAM Journal on Imaging Sciences*, 2(2), 438–469.

NAIF (2014) *Lunar Reconnaissance Orbiter Camera (LROC) instrument kernel v18*. [Online] Available from: http://naif.jpl.nasa.gov/pub /naif/pds/ data/lro-l-spice-6-v1.0/lrosp_1000.

Nefian, A.V., Coltin, B. & Fong, T. (2014) Apollo metric imagery registration to Lunar Orbiter Laser Altimetry. *Lunar and Planetary Science Conference*, The Woodlands, TX, USA, Volume 45. abstract #1679.

Soderblom, L.A. & Kirk, R.L. (2003) Meter-scale 3-D models of the Martian surface from combining MOC and MOLA data. *Lunar and Planetary Science Conference*, League City, TX, USA, Volume 34. abstract #1730.

Soderblom, L.A., Kirk, R.L. & Herkenhoff, K.E. (2002) Accurate fine-scale topography for the Martian south polar region from combining MOLA profiles and MOC NA images. *Lunar and Planetary Science Conference*, League City, Texas, USA,Volume 33. abstract #1254.

Wu, B., Hu, H. & Guo, J. (2014) Integration of Chang'E-2 imagery and LRO laser altimeter data with a combined block adjustment for precision lunar topographic modeling. *Earth and Planetary Science Letters*, 391, 1–15.

Wu, B., Guo, J., Hu, H., Li, Z. & Chen, Y. (2013) Co-registration of lunar topographic models derived from Chang'E-1, SELENE, and LRO laser altimeter data based on a novel surface matching method. *Earth and Planetary Science Letters*, 364, 68–84.

Wu, B., Liu, W.C., Grumpe, A. & Wöhler, C. (2018) Construction of pixel-level resolution DEMs from monocular images by shape and albedo from shading constrained with low-resolution DEM. *ISPRS Journal of Photogrammetry and Remote Sensing*, 140, 3–19.

Yoon, J. & Shan, J. (2005) Combined adjustment of MOC stereo imagery and MOLA altimetry data. *Photogrammetric Engineering & Remote Sensing*, 71(10), 1179–1186.

Planetary Remote Sensing and Mapping – Wu et al.
© 2019 Taylor & Francis Group, London, ISBN 978-1-138-58415-0

Chapter 18

Co-registration of multiple-source DEMs for correlated slope analysis at different scales on Mars

Y. Wang and B. Wu

ABSTRACT: The surface slopes of planetary bodies must be considered when carrying out exploratory missions such as landing site selection and rover manoeuvres. Generally, high-resolution digital elevation models (DEMs) such as those generated from High Resolution Imaging Science Experiment (HiRISE) images on Mars are favoured, as they result in detailed slopes with high-fidelity terrain features. However, high-resolution datasets normally only cover small areas and are not always available, whereas lower-resolution datasets, such as those obtained using the Mars Orbiter Laser Altimeter (MOLA), provide global coverage of the Martian surface. Slopes generated from low-resolution DEMs are based on a large baseline and have been smoothed relative to the real situation. To alleviate the slope smoothness problem to carry out large-scale slope analysis of the Martian surface using low-resolution data, this chapter presents correlated slope analysis at different scales using multiple-source DEMs. DEMs from different sources often show inconsistencies due to differences in sensor configurations, data acquisition periods and production techniques. Therefore, this chapter first presents a co-registration method for multiple-source DEMs. Next, slope correlation analysis is carried out using DEMs with different resolutions, and a slope correlation function with respect to different baselines is proposed. The validity and feasibility of the slope correlation function are verified using multiple-source datasets containing HiRISE, Context Camera (CTX), High Resolution Stereo Camera (HRSC) and MOLA data. The results indicate that the proposed slope function improves the accuracy of the slopes generated from low-resolution DEMs by about 50%.

1 INTRODUCTION

The surface slopes of planetary bodies must be considered when carrying out exploratory missions and scientific research. Slope maps are widely used in tasks such as landing site selection (Braun and Manning, 2007; Golombek *et al.*, 1997, 2012; Wu *et al.*, 2014) and rover manoeuvres (Lindemann and Voorhees, 2005; Maimone *et al.*, 2007). Thus, it is vital to ensure their accuracy. Generally, slope analysis for landing site selection should be performed over a large area or even globally to identify suitable landing sites. However, most globally available DEMs operate at a low resolution (e.g., a few hundred metres). When calculating slopes, the minimum baseline (or scale) is twice the DEM resolution. Therefore, slopes generated from low-resolution DEMs have a large baseline and are smoothed relative to the real situation. As a result, the slopes calculated from low-resolution DEMs are generally smaller than in real life. To alleviate the smoothness problem resulting from a large baseline, this research conducts slope compensation through correlation analysis of slopes calculated at different scales, using Mars datasets as examples.

Topographic datasets at different resolutions have been collected from past missions exploring the Martian surface. The High-Resolution Imaging Science Experiment (HiRISE) (McEwen *et al.*, 2007) and the Context Camera (CTX) (Malin *et al.*, 2007) are two imaging systems onboard NASA's Mars Reconnaissance Orbiter. The Mars Orbiter Laser Altimeter (MOLA) (Smith *et al.*,

2001) is an instrument onboard the Mars Global Surveyor spacecraft (Albee *et al.*, 2001). The High Resolution Stereo Camera (HRSC) (Neukum and Jaumann, 2004) was the first photogrammetric stereo sensor system used in the Mars Express mission. In this research, we analyse data acquired jointly by HiRISE, the CTX, the HRSC and MOLA. The processing of the HiRISE and CTX images, including calibration and add-on components, is implemented using ISIS version 3, provided by the United States Geological Survey. Orientation parameters for the HiRISE and CTX images are retrieved from SPICE kernels by interpolating the spacecraft's trajectory and pointing vectors based on observation time. A self-adaptive triangulation-constrained matching method (Hu and Wu, 2017; Wu *et al.*, 2011, 2012; Zhu *et al.*, 2007, 2010) is used to automatically match the stereo images to obtain corresponding dense points in their overlapping region. The three-dimensional (3D) coordinates of the matched points are then obtained by photogrammetric intersection based on the image orientation parameters, from which the DEMs are interpolated. As the resolution of HiRISE images is usually 0.25–0.5 m pixel^{-1}, the corresponding HiRISE DEM spacing is 1–2 m. The resolution of CTX images is generally 6 m pixel^{-1}, and the resolution of the CTX DEMs obtained from the datasets used in this research is 20 m pixel^{-1}. The HRSC DEMs are downloaded from the Planetary Data System (PDS) archive (http://pds-geosciences.wustl.edu/missions/mars_express/hrsc.htm), and have a resolution of 50 m. The MOLA data are obtained using the Mars Global Surveyor MOLA Elevation Model (MEGDR). The resolution of the MOLA DEMs is 463 m pixel^{-1}.

However, as these DEMs are derived from different sensors used on different missions, they show numerous inconsistencies, such as translational shifts, angular rotation and scale variation, due to differences in sensor configurations, data acquisition periods and production techniques. Poole (2013) investigated the problem of mis-registration between images collected by HiRISE and the HRSC and found 100-meter offsets between them. Tao *et al.* (2014) reported that the mis-registration between DEMs derived from offsets between HiRISE and HRSC images on Mars Exploration Rover (MER) and Mars Science Laboratory (MSL) landing sites, which ranged from 100 to 200 m. Kim and Muller (2009) discovered that the DEMs generated from HiRISE and CTX images had elevation offsets ranging from several to more than 30 m.

Before conducting the correlated slope analysis, the above-mentioned problems of mis-registration between DEMs must be solved. Researchers have examined the co-registration of multiple-source topographic models for Mars topographic datasets. Anderson and Parker (2002) examined the precision registration of Mars orbiter camera (MOC) imagery and MOLA data at selected candidate landing sites. Yoon and Shan (2005) presented a combined adjustment method of processing MOC imagery and MOLA data and indicated that this method mitigated the large mis-registration between the two datasets. Spiegel (2007) developed a bundle adjustment technique for HRSC imagery, adjusting a sparse stereo point cloud to optimise its fit to a surface interpolated from MOLA data. This adjustment is now standard in the production of controlled orthorectified HRSC products. Lin *et al.* (2010) developed a co-registration process to align Mars DEMs derived from MOLA, the HRSC and HiRISE, and assessed surface matching parameters for the co-registration of multi-resolution Mars DEMs. Wang and Wu (2017) presented a bundle adjustment approach to the co-registration of DEMs derived from HiRISE and CTX images, and indicated that mis-registration was due to boresight offsets between HiRISE and CTX cameras that deviated from their default values.

This chapter reports on correlated slope analysis performed at different scales with Mars-based DEMs at various resolutions. Before conducting the correlated slope analysis, DEMs from different sources are co-registered in a common reference frame by surface matching, as described in section 2. Section 3 presents the details of the correlated slope analysis and the derivation of a slope correlation function to calibrate the slopes derived from low-resolution DEMs. In section 4, the effectiveness of the correlation function is verified based on multiple-source datasets containing HiRISE, CTX, HRSC and MOLA data. The results are discussed and conclusions drawn in the last section.

2 CO-REGISTRATION OF MULTIPLE-SOURCE DEMS THROUGH SURFACE MATCHING

Multiple-source topographic Mars datasets suffer from mis-registration problems, as they originate from different sensors onboard different platforms. A multi-feature-based surface matching method originally designed to co-register multiple-source lunar DEMs (Wu *et al.*, 2013) is extended and used in this research to co-register multiple-source Mars DEMs. It incorporates feature points, lines and surface patches through surface matching to guarantee robust surface correspondence. A combined adjustment model is used to determine seven transformation parameters (one scale factor, three rotations and three translations) based on which the multiple DEMs can be co-registered.

To co-register the multiple DEMs using point pairs, the relationship between point P on the reference DEM and point P' on the matching DEM is represented using the following equation:

$$P = sRP' + T \tag{1}$$

where s is a scale factor denoting the extent of the magnification or contraction between the two DEMs, R is a rotation matrix determined entirely by three Euler rotation angles (φ, ω and κ) and $T = (T_X, T_Y, T_Z)$ is a 3D translation vector between the two DEMs.

To co-register multiple DEMs using line pairs for surface matching, as shown in Figure 18.1, a line pair (L, L') is identified from the two DEMs. P_1 and P_2 are on line L in the reference DEM and P_3' and P_4' are on line L' in the matching DEM. If the seven transformation parameters are perfectly accurate, the corresponding points P_3' and P_4' and P_3 and P_4 should be collinear with points P_1 and P_2. Therefore, the distance from point P_3 (or P_4) to line L should be 0. A mathematical model that uses line pairs for surface matching is built based on this observation.

Assuming that a 3D line L is determined by points P_1 and P_2, it can be mathematically modelled by the intersection of two space planes, as shown in the following equation:

$$L\begin{cases} m_{11}x + m_{12}y + m_{13}z = b_1 \\ m_{21}x + m_{22}y + m_{23}z = b_2 \end{cases} \tag{2}$$

Equation 2 can be expressed in matrix form as follows:

$$MX = B \tag{3}$$

where $M = \begin{pmatrix} m_{11} & m_{12} & m_{13} \\ m_{21} & m_{22} & m_{23} \end{pmatrix}$, $B = \begin{pmatrix} b_1 \\ b_2 \end{pmatrix}$, $X = (x \;\; y \;\; z)^T$, forming a line equation.

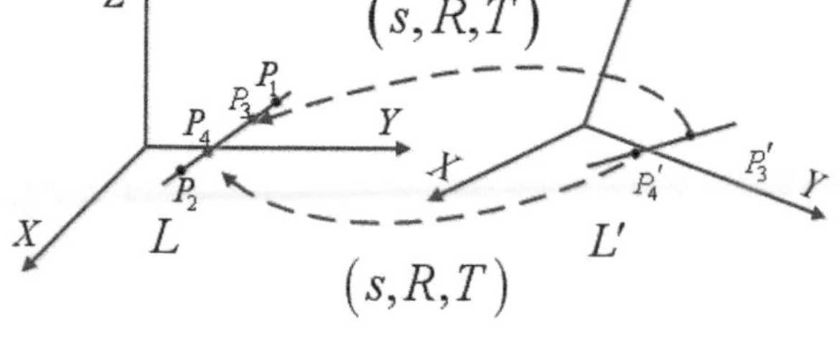

Figure 18.1. Illustration of line co-registration

For a point P' (e.g., P_3' or P_4') on line L', the distance from the corresponding point P (P_3 or P_4) to line L can be determined using the following equation:

$$d = \sqrt{(B - MP)^T \left(MM^T\right)^{-1} (B - MP)} \tag{4}$$

Ideally, the distance d is 0. The constraint $d = 0$ is then used for line correspondence in the co-registration process. For each line correspondence, two distance constraints can be generated independently, as two points are sufficient determinants of a line. Therefore, each line correspondence produces two observation equations.

To measure the correspondence between two surface patches, the normal vectors of the surface patches are used for surface matching. As shown in Figure 18.2, the relationship between a surface patch M and its corresponding surface patch M' can be determined by their surface normals n and n', using the following equation:

$$n = Rn' \tag{5}$$

For each corresponding surface (normal vector), three observation equations can be generated that are related only to the rotation matrix.

After linearising the observation equation for the corresponding points, lines and surface patches with respect to the seven unknown transformation parameters, the observation equations for the combined adjustment can be represented in matrix form as follows:

$$V = AX - L, P \tag{6}$$

where X is the unknown vector to be solved, containing the transformation parameters (s, R, T); L is the observation vector; A is the coefficient matrix containing the partial derivatives from each observation; and P is the a priori weight matrix of the observations, which reflects the measurement quality and the contribution of the observation to the final result. In particular, the process of combined adjustment using this method has the following three components, corresponding to the three types of observation:

$$\begin{aligned} V_1 &= A_1X_1 + B_1X_2 + C_1X_3 - L_1, P_1 \\ V_2 &= \qquad\quad B_2X_2 + \qquad\quad - L_2, P_2 \\ V_3 &= A_3X_1 + B_3X_2 + C_3X_3 - L_3, P_3 \end{aligned} \tag{7}$$

The first equation in Equation 7 is the observation equation for point correspondence, in which X_1 is the unknown scale parameter, X_2 represents the vector of the unknown rotation angles (φ, ω, κ)

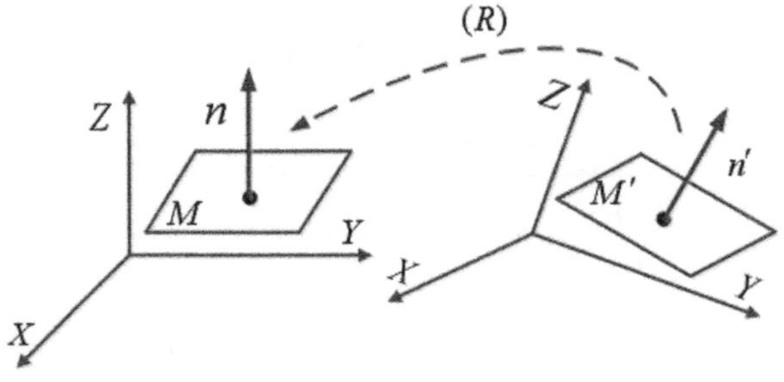

Figure 18.2. Illustration of surface patch co-registration

and X_3 is the vector of the unknown translation parameters T_X, T_Y and T_Z. The coefficient matrix A in Equation 6 is divided into A_i, B_i, and C_i (i = 1 to 3) for different types of unknown parameter. The second equation represents the observation residuals of normal vectors of the corresponding surface patches. As denoted in this equation, the surface normal is related only to the unknown rotation angles. The third equation is the observation equation for line correspondence after applying the distance constraint $d = 0$. Different weights Pi (i = 1 to 3) are assigned to different types of observation equation. Through bundle adjustment, the seven unknown transformation parameters and the scale factor are iteratively adjusted by minimising the difference between the matching surface and the reference surface, allowing different DEMs to be accurately co-registered.

3 CORRELATED SLOPE ANALYSIS USING DEMS AT DIFFERENT RESOLUTIONS

Correlated slope analysis can be performed directly using co-registered multiple-source DEMs. However, to alleviate the influence of other uncertainties on the DEMs (e.g., different data acquisition periods), this research uses DEMs with the highest resolution (i.e., DEMs derived from HiRISE images) covering different terrain types (see Fig. 18.3), and further down-samples the DEMs to lower-resolution DEMs for correlated slope analysis to derive the correlation function of slopes of different scales. The derived correlation function can then be applied to the co-registered DEMs from different sources.

3.1 *Relationship between average slope and baseline*

Seven sets of HiRISE DEMs with resolutions of 1 m pixel^{-1} are used for the correlated slope analysis. They are collected from different regions on Mars and contain different terrain types. In Figure 18.3, all of the HiRISE DEMs are visualised in the first row and their corresponding slope maps (2 m baseline for slope calculation) in the second row. Area 1, Area 3 and Area 6 are relatively flat; Area 2 and Area 7 are presentative crater areas; Area 5 is a dune area affected by some natural interactions; Area 4 is a valley floor. The slopes are calculated as follows. For each cell in the map, the maximum rate of change in value from the cell to its four neighbours is calculated. The slope map presents topographical changes in the form of slopes and in units of degrees. In the slope maps, blue represents a relatively flat area and red indicates an area with a large slope.

Low-resolution DEMs mean large baselines when calculating slopes, which usually causes smoothness; deviation is required to present the details of the real terrain. For the same area, as the resolution of a DEM decreases, the average slope will generally also decrease. Figure 18.4 shows an example from Area 2, as mentioned above. In the first row of Figure 18.4, a 1-m-pixel^{-1} HiRISE DEM is down-sampled to 20-m-pixel^{-1} and 463-m-pixel^{-1} DEMs (in greyscale) – the same resolutions as CTX DEMs and MOLA DEM. In the second row, the real CTX DEM and MOAL DEM of the same region are displayed. The slope maps (in colour) derived from these DEMs are also included in Figure 18.4. The HiRISE simulation is clearly close to the real data in terms of the slope map. An obvious trend of decrease in slope can be seen in both the simulated data and the real data.

The seven sets of data described above are used to investigate the slope correlations. The 1-m-pixel^{-1} HiRISE DEMs are gradually down-sampled to 2 m pixel^{-1}, 4 m pixel^{-1} . . . and 256 m pixel^{-1} to simulate DEMs with different resolutions. Then slope analysis is conducted based on the DEMs with different resolutions. The average slope for each dataset is illustrated in Figure 18.5. It consistently decreases with DEM resolution. Kirk *et al.* (2003, 2008) described similar findings for DEMs from different sources on Mars.

3.2 *Slope correlation function*

As slope decreases with DEM resolution, slope compensation must be conducted to better represent the real situation. The main challenge is to find the relationship between the slopes calculated from high-resolution DEMs and those from low-resolution DEMs.

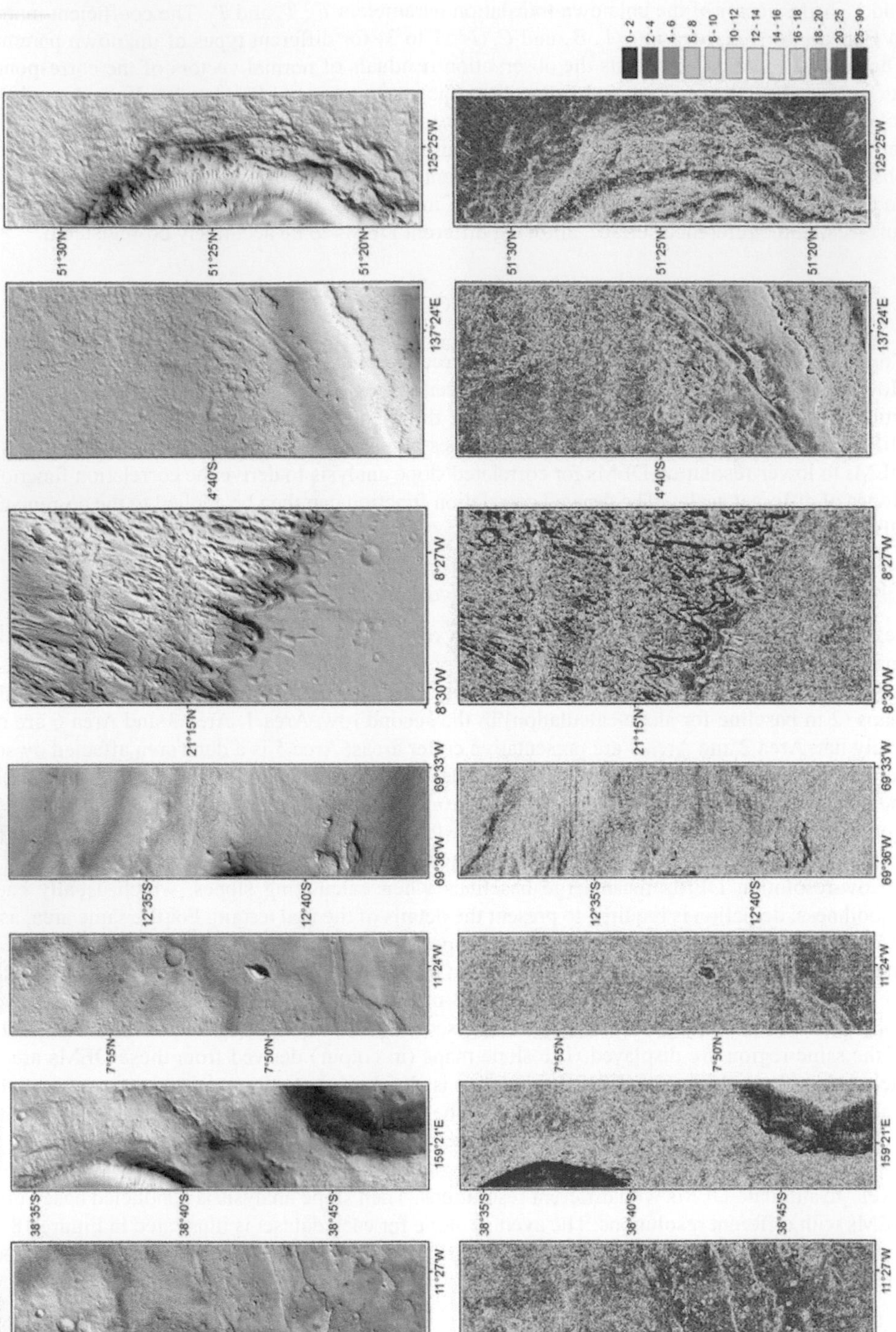

Figure 18.3. DEMs and slope maps for the seven HiRISE datasets used in the analysis

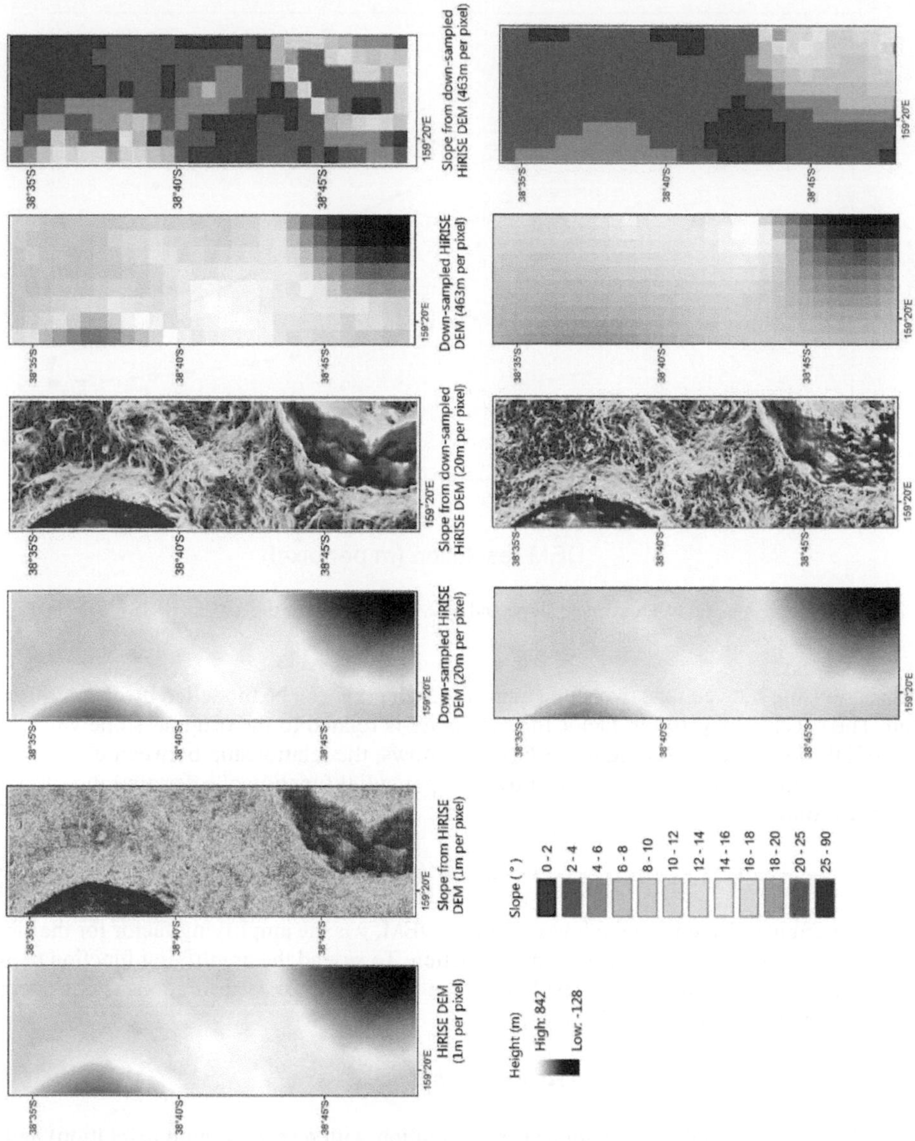

Figure 18.4. Examples of slope smoothness with decreasing resolution of DEM

Figure 18.6 shows an example of slope comparison between HiRISE DEMs with resolutions of 1 m pixel^{-1}and 256 m pixel^{-1}, respectively. First, the 1 m pixel^{-1} HiRISE DEM is down-sampled to a resolution of 256 m pixel^{-1}. Next, corresponding slope maps are calculated from the DEMs. Subsequently, the slope map from the 1-m-pixel^{-1} HiRISE DEM is down-sampled to 256 m pixel^{-1}. The approach to down-sampling the slope map is as follows. We first define a 256-m-x-256-m local window and move it across an interval of 256 m, then use the average value as the slope of the current window. We thus use a lower-resolution slope map to accurately represent the original one. Now we have two types of slope map with the same resolution, and the comparison can be conducted pixel-wise. Comparing the slope values pixel by pixel reveals

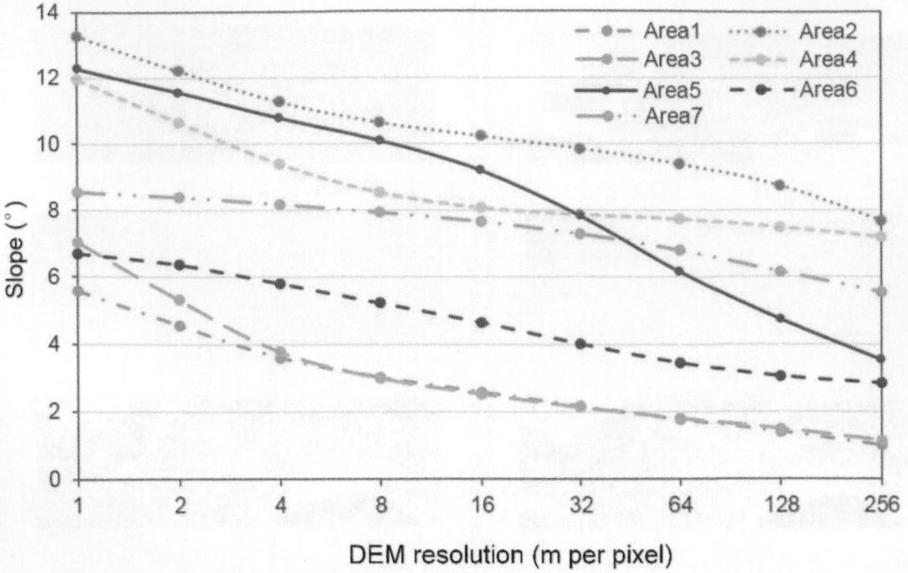

Figure 18.5. Relationships between average slope and DEM resolution for the seven HiRISE datasets

that the relationship between the two slope maps for each pixel can be modelled by an amplifying function. The specific amplifying factor for each pixel is related to the original slope value and the ratio of DEM resolution. As Figure 18.6 (right) shows, the relationship between the amplifying factor and slope value can be described by an exponential function at a certain ratio of DEM resolution, as follows:

$$y = ax^{-1} + b \tag{8}$$

where x is the original slope from the low-resolution DEM, y is the amplifying factor for the slope and a and b are the parameters of the exponent function. To extend the amplifying function model to accommodate DEMs with different resolutions, the ratio of the DEM resolutions is considered. The model can be represented as follows:

$$y = \left(alog_2(t) + b\right)x^{-1} + c \tag{9}$$

where t is the ratio of the DEM resolutions (low resolution with respect to high resolution) and a, b and c are the parameters of the function.

The seven sets of HiRISE DEMs described previously are processed. Slope comparisons similar to those described in Figure 18.6 are conducted across resolution gaps. As Figure 18.7 shows, 83,474 sets of slope pairs are obtained. The slope pairs are used to determine the parameters a, b and c. The outliers are filtered out using the bisquare method. This method minimises a weighted sum of squares, where the weight given to each data point depends on how far it is from the fitted line. Points nearer the line have larger weights. Points farther from the line have smaller weights. The adjustment process is based on the least-squares principle. The parameters a, b and c are iteratively adjusted, and the finally determined slope correlation function related to DEM resolution can be represented as follows:

$$y = \left(0.537 \times log_2(t) + 0.772\right)x^{-1} + 0.946 \tag{10}$$

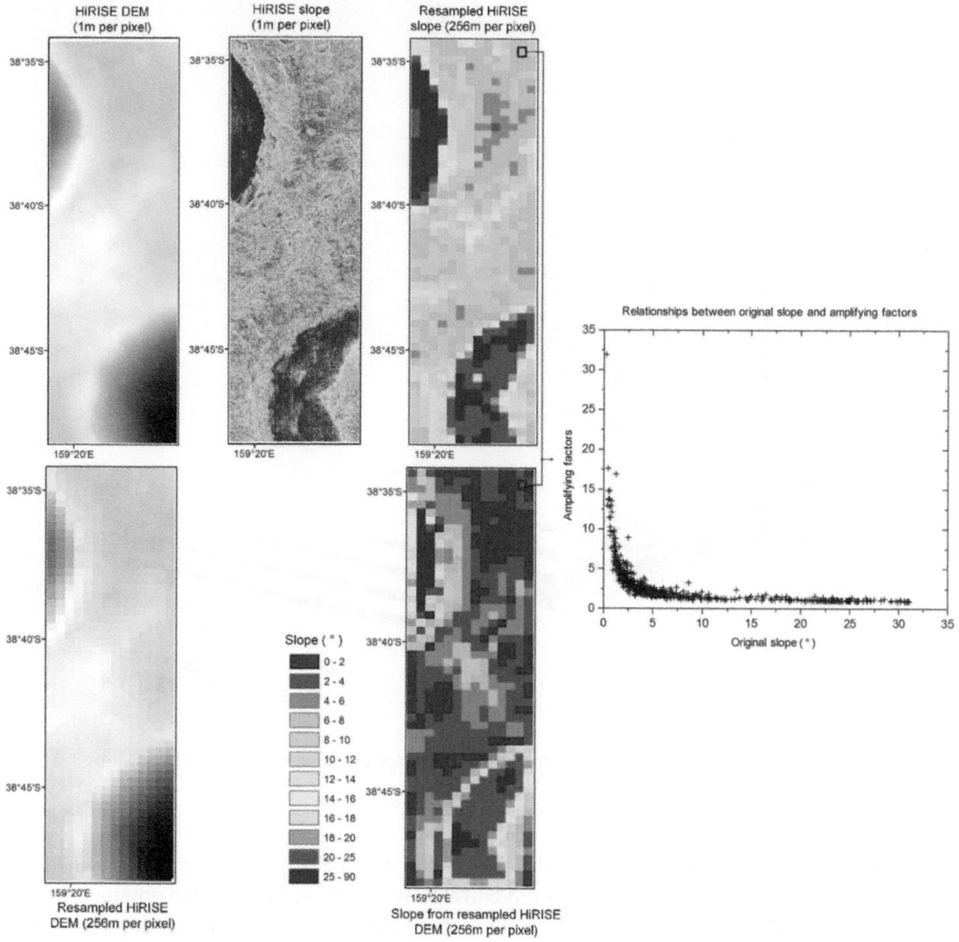

Figure 18.6. Example of slope comparison

For all of the slope pairs from the seven datasets, slopes obtained from lower-resolution DEMs are amplified by the proposed correlation function and the results are compared with those for their respective paired slopes. The root-mean-square deviation (RMSE) is 2.8 degrees, which shows that the data well fit the proposed model.

4 EXPERIMENTAL VERIFICATION

HiRISE, CTX, HRSC and MOLA datasets collected in other areas on Mars are used to verify the effectiveness of the derived slope correlation function. Comparisons are conducted between HiRISE and CTX, HiRISE and MOLA, HiRISE and HRSC, and HRSC and MOLA data, as illustrated in Figure 18.8, Figure 18.9, Figure 18.10 and Figure 18.11, respectively. At each testing site, the DEMs are co-registered using the method presented in section 2 before slope analysis and comparison are conducted.

In the verification process, each set of data contains a higher-resolution DEM and a lower-resolution DEM. The first step is to co-register the DEMs using the method presented in section 2;

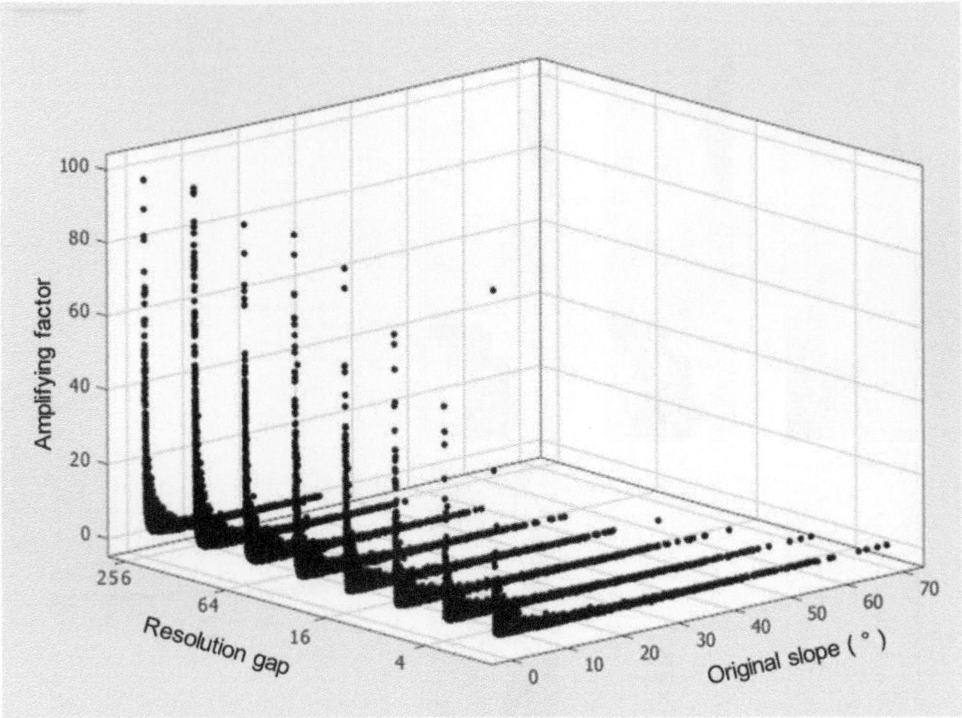

Figure 18.7. Analysis of relationships between amplifying factor (*y*), resolution gap (*t*) and slope (*x*)

next, slope maps are derived from both DEMs. Subsequently, the slopes generated from the lower-resolution DEM are amplified by the proposed correlation function. Slope maps obtained from the higher-resolution DEM are down-sampled to the same resolution as their lower-resolution counterparts for comparison. Finally, slope differences are calculated before and after amplification for comparative analysis.

The first column of Figure 18.8 shows slope maps derived from a HiRISE DEM. The resolution of the slope map is same as that of the DEM, i.e., 1 m pixel^{-1}. The second column represents a down-sampled version of the HiRISE slope map. It has the same resolution as a CTX slope map (20 m pixel^{-1}), and is regarded as a reference in the comparison. The third column shows a slope map directly derived from the CTX DEM and the fourth column shows an amplified version of this CTX slope map after applying the slope correlation function. The last two columns show the differences between the HiRISE slope map and the CTX slope map before and after slope amplification. Darker colours indicate larger slope differences and lighter colours denote smaller differences. Similarly, the verification process is demonstrated in Figure 18.9 using HiRISE data (1 m pixel^{-1}) and MOLA data (463 m pixel^{-1}); in Figure 18.10 using HiRISE data (1 m pixel^{-1}) and HRSC data (50 m pixel^{-1}); and in Figure 18.11 using HRSC data (50 m pixel^{-1}) and MOLA data (463 m pixel^{-1}). In all cases, the difference maps show that slope differences decrease after slope amplification. The RMSEs and maximum values of the slope differences are listed in Table 18.1.

Table 18.1 shows that the average slope difference is usually reduced by half after amplification. This indicates that the proposed slope correlation function works well in most cases. Table 18.1 also shows that comparing HiRISE and MOLA data, which have the largest resolution gap, yields the best results after slope amplification.

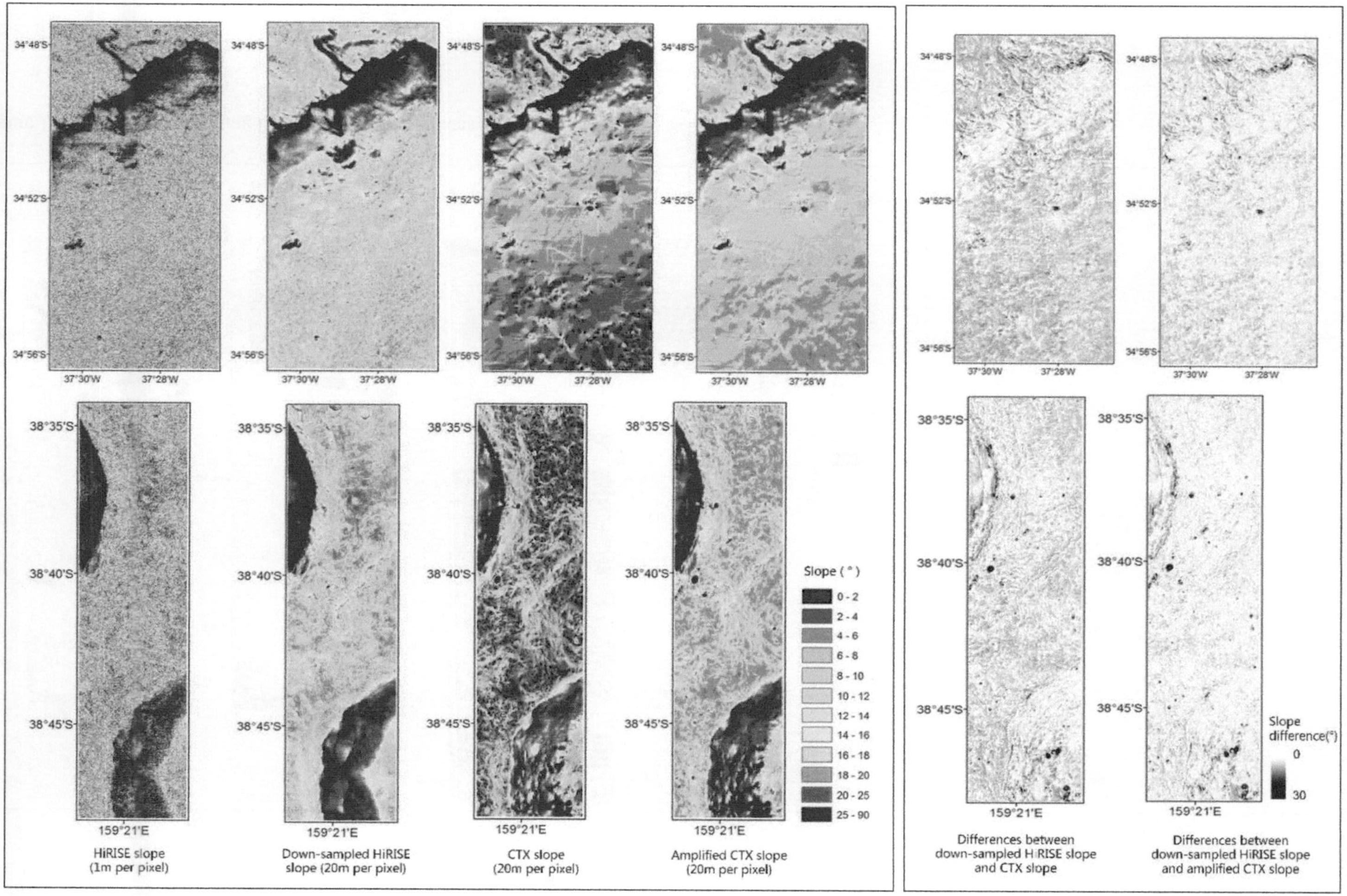

Figure 18.8. Verification using HiRISE and CTX datasets

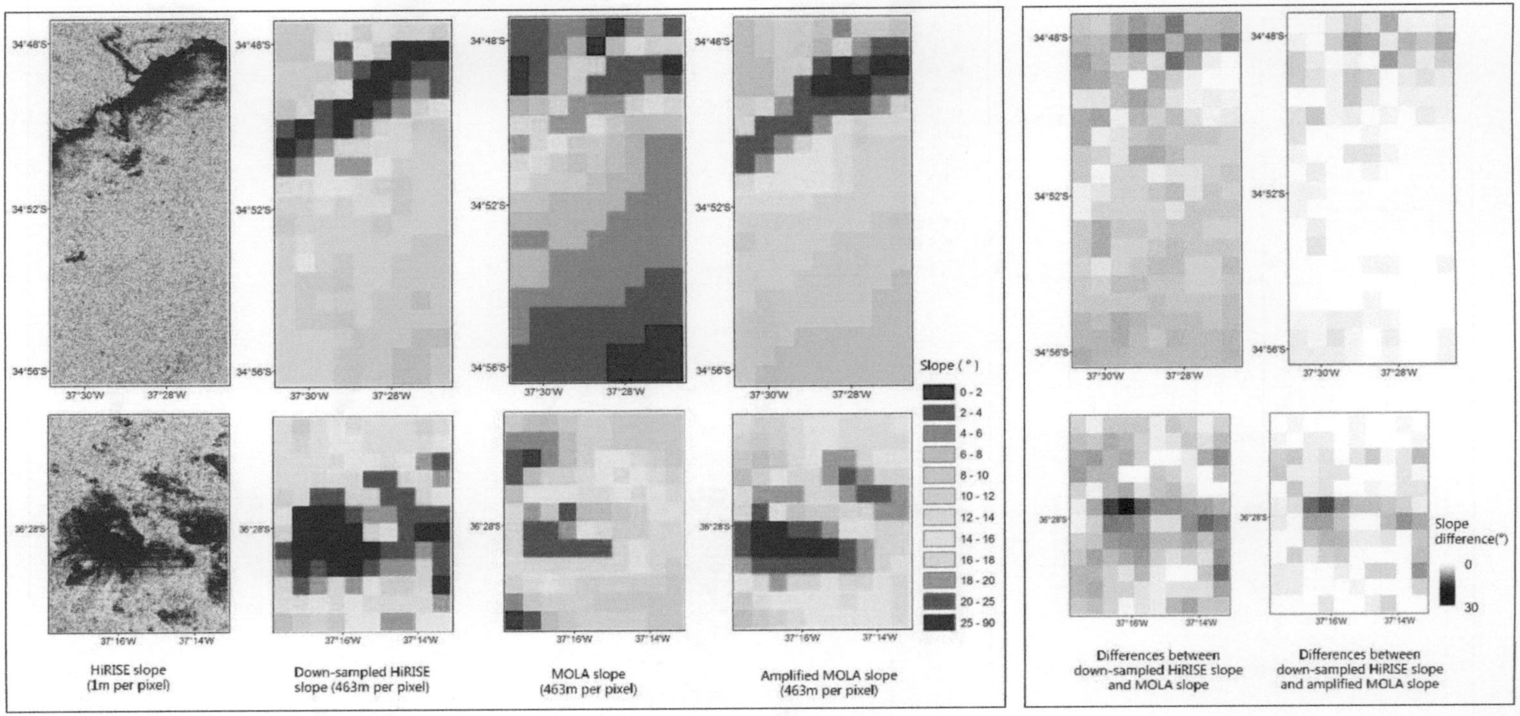

Figure 18.9. Verification using HiRISE and MOLA datasets

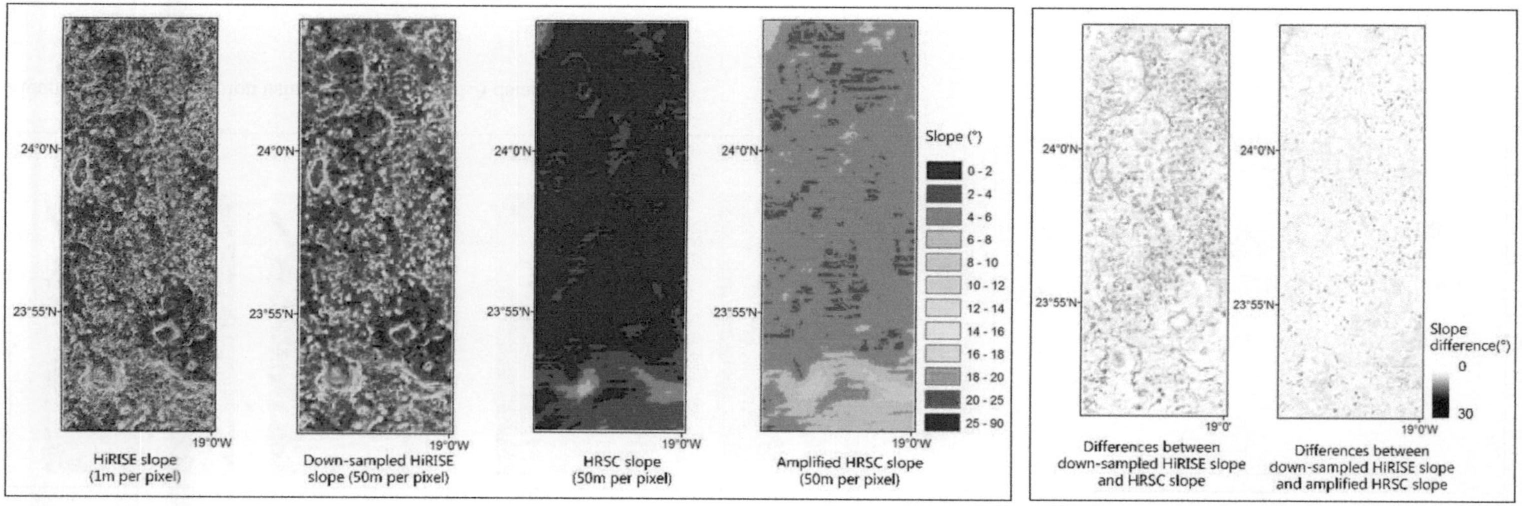

Figure 18.10. Verification using HiRISE and HRSC datasets

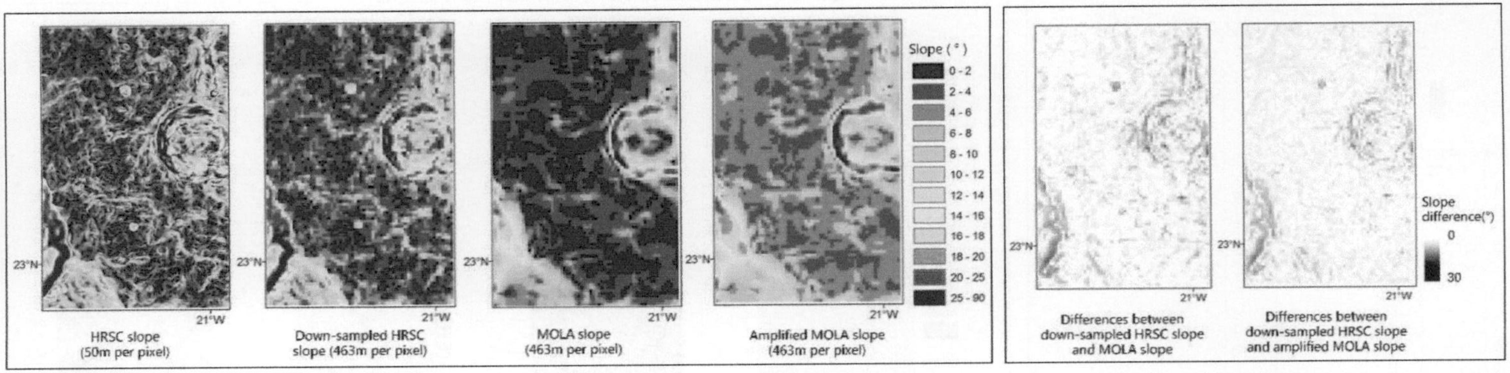

Figure 18.11. Verification using HRSC and MOLA datasets

Table 18.1 Comparison of slopes before and after amplification using the slope correlation function

		Differences before amplification		*Differences after amplification*	
		RMSE (°)	*Max. (°)*	*RMSE (°)*	*Max. (°)*
HiRISE and CTX	Region 1	6.11	64	4.99	62
	Region 2	6.32	51	4.57	50
HiRISE and MOLA	Region 1	8.19	27	4.79	21
	Region 2	6.86	16	2.82	11
HiRISE and HRSC		4.05	20	2.24	16
HRSC and MOLA		4.86	24	2.41	22

5 DISCUSSION AND CONCLUSIONS

To perform large-scale slope analysis of the Martian surface based on low-resolution data, this chapter presents a slope correlation function derived from multiple-source DEMs based on the relationship between resolution and slope. The proposed co-registration approaches effectively reduce the offsets of multiple-source DEMs. The proposed slope correlation function effectively calibrates slopes derived from low-resolution DEMs. The validity and feasibility of the proposed function are verified using multiple-source datasets containing HiRISE, CTX, HRSC and MOLA data. The experimental validation conveys the following conclusions:

1 The accuracy of slopes obtained from low-resolution DEMs is improved by the proposed slope correlation function, reducing the discrepancies between slopes obtained from low- and high-resolution DEMs.
2 In most cases, these discrepancies are reduced by about 50%. For three of the six datasets evaluated in the study, the differences are reduced to about 2 degrees.
3 Comparison of HiRISE and MOLA data, which have the largest resolution gap, yields the best result. The RMSE of slope difference decreases from 6.86 degrees to 2.82 degrees for one dataset and from about 8.19 degrees to 4.79 degrees for the other.

Although Mars is used as a research object in this study, the method is fundamentally the same for other heavenly bodies, such as the Moon and Mercury. It should be noted that this method loses validity in some extreme cases, such as a totally flat surface. However, in real-world applications, such cases rarely arise.

ACKNOWLEDGEMENTS

This work is supported by a grant from the Research Grants Council of Hong Kong (Project No. PolyU 152086/15E) and grants from the National Natural Science Foundation of China (Project No. 41671426 and Project No. 41471345). The authors thank all those who worked on the PDS archive for the Mars datasets to make the HiRISE, CTX and HRSC images and the MOLA data publicly available.

REFERENCES

Albee, A.L., Arvidson, R.E., Palluconi, F. & Thorpe, T. (2001) Overview of the Mars global surveyor mission. *Journal of Geophysical Research E*, 106, 23291–23316.

Anderson, F. & Parker, T. (2002) Characterization of MER landing sites using MOC and MOLA. Paper presented at *the 33rd Lunar and Planetary Science Conference*, 11–15 March, Pasadena, CA, USA.

Braun, R.D. & Manning, R.M. (2007) Mars exploration entry, descent, and landing challenges. *Journal of Spacecraft and Rockets*, 44, 310–323.

Golombek, M., Cook, R., Moore, H. & Parker, T. (1997) Selection of the Mars Pathfinder landing site. *Journal of Geophysical Research: Planets*, 102, 3967–3988.

Golombek, M., Grant, J., Kipp, D., Vasavada, A., Kirk, R., Fergason, R., Bellutta, P., Calef, F., Larsen, K. & Katayama, Y. (2012) Selection of the Mars Science Laboratory landing site. *Space Science Reviews*, 170, 641–737.

Hu, H. & Wu, B. (2017) Bound-constrained Multiple-image least-squares matching for multiple-resolution images. *Photogrammetric Engineering & Remote Sensing*, 83, 667–677.

Kim, J.-R. & Muller, J.-P. (2009) Multi-resolution topographic data extraction from Martian stereo imagery. *Planetary and Space Science*, 57, 2095–2112.

Kirk, R.L., Howington-Kraus, E., Redding, B., Galuszka, D., Hare, T.M., Archinal, B.A., Soderblom, L.A. & Barrett, J.M. (2003) High-resolution topomapping of candidate MER landing sites with Mars Orbiter Camera narrow-angle images. *Journal of Geophysical Research: Planets*, 108.

Kirk, R., Howington-Kraus, E., Rosiek, M., Anderson, J., Archinal, B., Becker, K., Cook, D., Galuszka, D., Geissler, P. & Hare, T. (2008) Ultrahigh resolution topographic mapping of Mars with MRO HiRISE stereo images: meter-scale slopes of candidate Phoenix landing sites. *Journal of Geophysical Research: Planets*, 113.

Lin, S.-Y., Muller, J.-P., Mills, J.P. & Miller, P.E. (2010) An assessment of surface matching for the automated co-registration of MOLA, HRSC and HiRISE DTMs. *Earth and Planetary Science Letters*, 294, 520–533.

Lindemann, R.A. & Voorhees, C.J. (2005) Mars Exploration Rover mobility assembly design, test and performance. *Proceedings of 2005 IEEE International Conference on* Systems, *Man and Cybernetics*. Waikoloa, HI, USA. pp. 450–455.

Maimone, M., Cheng, Y. & Matthies, L. (2007) Two years of visual odometry on the mars exploration rovers. *Journal of Field Robotics*, 24, 169–186.

Malin, M.C., Bell, J.F., Cantor, B.A., Caplinger, M.A., Calvin, W.M., Clancy, R.T., Edgett, K.S., Edwards, L., Haberle, R.M. & James, P.B. (2007) Context camera investigation on board the Mars Reconnaissance Orbiter. *Journal of Geophysical Research: Planets*, 112.

McEwen, A.S., Eliason, E.M., Bergstrom, J.W., Bridges, N.T., Hansen, C.J., Delamere, W.A., Grant, J.A., Gulick, V.C., Herkenhoff, K.E. & Keszthelyi, L. (2007) Mars reconnaissance orbiter's High Resolution Imaging Science Experiment (HiRISE). *Journal of Geophysical Research: Planets*, 112.

Neukum, G. & Jaumann, R. (2004) HRSC: The high resolution stereo camera of Mars Express. In: *Mars Express: The Scientific Payload*. pp. 17–35. Available from: http://sci.esa.int/mars-express/56568-esa-sp-1240-mars-express-the-scientific-payload/

Poole, W. (2013) Mars US rover traverse co-registration using multi-resolution Orbital 3D imaging datasets. Paper presented at *European Planetary Science Congress 2013*, 8–13 September, London, UK. id. EPSC2013-661, 661. Available from: http://meetings. copernicus. org/epsc2013.

Smith, D.E., Zuber, M.T., Frey, H.V., Garvin, J.B., Head, J.W., Muhleman, D.O., Pettengill, G.H., Phillips, R.J., Solomon, S.C. & Zwally, H.J. (2001) Mars Orbiter Laser Altimeter: experiment summary after the first year of global mapping of Mars. *Journal of Geophysical Research: Planets*, 106, 23689–23722.

Spiegel, M. (2007) Improvement of interior and exterior orientation of the three line camera HRSC with a simultaneous adjustment. *International Archives of Photogrammetry and Remote Sensing*, Volume 36. pp. 161–166.

Tao, Y., Muller, J., Willner, K., Morley, J., Sprinks, J., Traxler, C. & Paar, G. (2014) 3D data products and Web-GIS for Mars Rover mission for seamless visualisation from orbit to ground-level. *The International Archives of Photogrammetry, Remote Sensing and Spatial Information Sciences*, Volume 40. p. 249.

Wang, Y. & Wu, B. (2017) Investigation of boresight offsets and co-registration of HiRISE and CTX imagery for precision Mars topographic mapping. *Planetary and Space Science*, 139, 18–30.

Wu, B., Zhang, Y. & Zhu, Q. (2011) A triangulation-based hierarchical image matching method for wide-baseline images. *Photogrammetric Engineering & Remote Sensing*, 77, 695–708.

Wu, B., Zhang, Y. & Zhu, Q. (2012) Integrated point and edge matching on poor textural images constrained by self-adaptive triangulations. *ISPRS Journal of Photogrammetry and Remote Sensing*, 68, 40–55.
Wu, B., Guo, J., Hu, H., Li, Z. & Chen, Y. (2013) Co-registration of lunar topographic models derived from Chang'E-1, SELENE, and LRO laser altimeter data based on a novel surface matching method. *Earth and Planetary Science Letters*, 364, 68–84.
Wu, B., Li, F., Ye, L., Qiao, S., Huang, J., Wu, X. & Zhang, H. (2014) Topographic Modeling and analysis of the landing site of Chang'E-3 on the moon. *Earth and Planetary Science Letters*, 405, 257–273.
Yoon, J.-S. & Shan, J. (2005) Combined adjustment of MOC stereo imagery and MOLA altimetry data. *Photogrammetric Engineering & Remote Sensing*, 71, 1179–1186.
Zhu, Q., Wu, B. & Tian, Y. (2007) Propagation strategies for stereo image matching based on the dynamic triangle constraint. *ISPRS Journal of Photogrammetry and Remote Sensing*, 62, 295–308.
Zhu, Q., Zhang, Y., Wu, B. & Zhang, Y. (2010) Multiple close-range image matching based on a self-adaptive triangle constraint. *The Photogrammetric Record*, 25, 437–453.

Section VI

Planetary data management and presentation

Planetary Remote Sensing and Mapping – Wu et al.
© 2019 Taylor & Francis Group, London, ISBN 978-1-138-58415-0

Chapter 19

Status and future developments in planetary cartography and mapping

A. Naß, K. Di, S. Elgner, S. van Gasselt, T. Hare, H. Hargitai, I. Karachevtseva, E. Kersten, N. Manaud, T. Roatsch, A. P. Rossi, J. Skinner, Jr. and M. Wählisch

ABSTRACT: Planetary cartography does not only provide an extensive basis for supporting planning activities in planetary exploration, e.g., landing-site selection, orbital observations, traverse planning, but it also supports mission conduct by, e.g., observation tracking and hazard avoidance mapping. It also provides the scientific and technical basis to create science products after successful termination of a planetary mission by helping to distill data into maps. After a mission's lifetime, experiment data and eventually higher-level data such as mosaics and digital terrain models (DTMs) are stored in archives – and eventually converted into maps and higher-level data products – to form a basis for research and for new scientific and engineering studies. The complexity of such tasks increases with every new dataset that has been put on the stack of data sources. In the same way as the complexity of autonomous probes increases, tools that support these challenges also require new levels of sophistication. In planetary science, cartography and mapping share a history dating back to the roots of telescopic space exploration and are now facing new technological and organizational challenges with the rise of new missions, new global initiatives and organizations, and opening research markets. The focus of this contribution is to summarize recent activities in planetary cartography and to highlighting current issues the community is facing to identify future opportunities in this field. By this we would like to invite cartographers/researchers to join this community and to start thinking about how to jointly solve some of these challenges.

1 INTRODUCTION AND OVERVIEW

1.1 *Introduction*

As of today, hundreds of planetary maps have been produced and published during a number of different framework programs and projects. Therein, different mapping efforts exist, either on a national level or as collaboration between groups participating as investigators in mapping missions. However, coordination of such tasks does not end with the compilation and publication of a set of maps. Coordination may be considered successfully only when mapping products have been provided to upcoming generations of researchers and mappers to allow efficient re-use of a new sustainable data basis. In order to accomplish this, a mapping infrastructure, workflows, communication paths and validation tools have to be developed and made available. Related work covering these issues within planetary cartography is described by, e.g., Hare *et al.* (2017), Kirk (2016), Laura *et al.* (2017), Nass *et al.* (2017), Pędzich and Latuszek (2014), Radebaugh *et al.* (2017), and Williams (2016).

The focus of this chapter is to summarize the history and recent activities in planetary cartography across the globe and to highlight some of the issues and opportunities the community is currently facing.

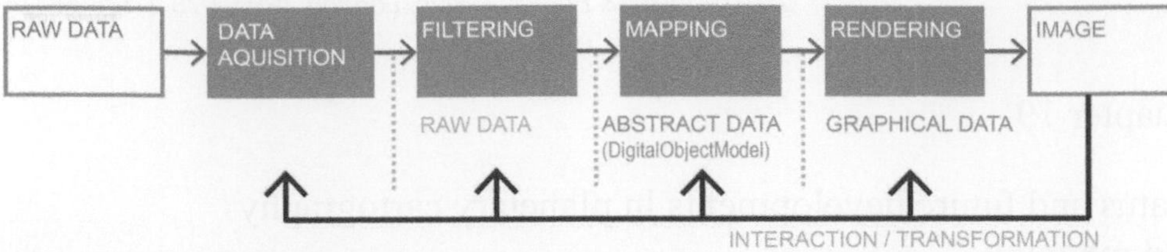

Figure 19.1. Visualization pipeline (cf. Carpendale, 2003; Haber and McNabb, 1990)

1.2 *Definition and background*

The definition of cartography and mapping has changed over the last 50 years. This has mainly been pushed by the increasing developments in information and computer technology. A concise overview of how such terms were changed over time is given by Kraak and Fabrikant (2017). The most recent one describes cartography as the "art, science and technology of making and using maps" (Strategic Plan 2003–2011 of the International Cartographic Association, http://icaci.org/strategic-plan/), and a map as "visual representation of an environment" (Kraak and Fabrikant, 2017). Thus, maps are one of the most important tools for communicating geospatial information between producers and receivers. Geospatial data, tools, contributions in geospatial sciences, and the communication of information and transmission of knowledge are matter of ongoing cartographic research. This applies to all topics and objects located on Earth or on any other body in our Solar System.

Visualization of data in general, and visualization of research data in particular, represent a simplified view of the real world, covering complex situations as well as the relationship between these (e.g. Mazza, 2009; Ware, 2004). The process to accomplish this can be divided into four major parts: (1) data pre-processing and transformation, (2) visual mapping, (3) generation of views, and (4) perception/cognition. The mapping process in Planetary Cartography is comparable to established processes commonly employed in terrestrial cartographic workflows that are described as so-called visualization pipeline (e.g., Carpendale, 2003, Figure 19.1; Haber and McNabb, 1990).

This workflow distinguishes clearly parallels to the *data-information-knowledge-wisdom* hierarchy (e.g. Ackoff, 1989), taking part in the field of information sciences and knowledge management. More recent discussions about this are shown e.g. in Rwoley (2007).

1.3 *History of planetary exploration*

Extraterrestrial mapping dates back to shortly after the invention of the telescope at the beginning of the 17th century, which marked a milestone in planetary exploration. Maps of the Moon (van Langren in 1645, Hevelius in 1647, Grimaldi and Riccioli in 1651) offered different approaches in visual and toponymic representation of an extraterrestrial landscape. Many improvements were introduced during the upcoming centuries and extraterrestrial mapping became a scientific discipline with map products similar in appearance and style as their terrestrial counterparts (Blagg and Müller, 1935; Mason and Hackman, 1961; Portee, 2013; Sadler, 1962; Shoemaker, 1961; Slipher, 1962; Whitaker *et al.*, 1963). Despite this success and many advances in the field of Earth-based telescopic observations and mapping, detailed topographic features and landforms could only be mapped from observation platforms located on spacecraft. This process started with the first set of pictures received from the far side of the Moon (Luna 3; Barabashov *et al.*, 1960) and Mars (Mariner 6; Davies *et al.*, 1970). Venus was first mapped in detail by the successful Soviet Venera probes. Results provided by landing and orbital missions (from Venera-1, 1961 to Venera-16, 1983) were published in *Atlas of Venus* that includes a series of 27 map sheets (Atlas of Venus, 1988). The first comprehensive cartographic atlas of multiple Solar System bodies based on comparative approach was published by Shingareva *et al.* (1992). China joined the countries with planetary mapping centers with the publication of several maps, atlases and globes using the images from the Chinese Chang'E lunar probe series (e.g., Compiling Committee, 2010). Global topographic data of variable resolution are now available from laser and radar altimetry, stereo photogrammetry, stereo photoclinometry for Mercury, Venus, the Moon, Mars, Ceres, Vesta, Titan, and Phobos. Following the long-employed method of replicating images using hand drawing, maps showing the topography of planets and moons used airbrush technique and manual interpretation of several sets of photographs. This was replaced by digital image mosaicking techniques in the 1990s. Parallel to terrestrial developments since the middle 1990s modern and digital mapping techniques within vector- and raster-based graphic software arose. This includes the first efforts of GIS-based data integration and mapping in planetary sciences. Since then, a few developments and approaches came up to make the usage of GI technologies more efficient for planetary mapping and cartography representation (e.g., Frigeri *et al.*, 2011; Hare and Tanaka, 2001; Hare *et al.*, 2009, 2015; Nass *et al.*, 2011; van Gasselt and Nass, 2011).

The most comprehensive review on all aspects of planetary cartography was published by Snyder (1982, 1987) and Greeley and Batson (1990). For detailed summaries on the development and evolution of planetary cartography, the reader is referred to Shevchenko *et al.* (2016) for the history of Soviet and Russian planetary cartography, and to Jin (2014) for Chinese lunar mapping results. History of planetary mapping is discussed in e.g., Kopal and Carder (1974) and Morton (2002), and recent planetary cartographic techniques and tools are reviewed in Beyer (2015) and Hare *et al.* (2017).

2 ASSOCIATED SCIENCE CENTERS, INSTITUTES, AND GROUPS

Planetary cartography has found its manifestation in governmental activities, community efforts, professional organizations, and, in recent years, private activities. That activity substantiated with the revival of planetary exploration in the early 2000s when Europe visited the Moon for the first time, and the US launched a number of exploration missions. With the success of Asian spacecraft missions to the Moon and Mars joining the global planetary exploration endeavor, planetary cartography is increasingly becoming a global collaborative effort with planetary mapping being one of its main tools to accomplish the goals.

This chapter introduces institutes and groups working in the field of planetary cartography and mapping. Some of them have a long history in planetary cartography while others represent more recent efforts. Their activities are usually organized on a national level but they are internationally related to each other through research cooperation and collaborative projects. This overview does not qualify to be complete and to list all active organizations and groups. It should provide a cross-section covering main institutions as well as groups and initiatives.

2.1 *Institutes and facilities*

In the United States, the Astrogeology Science Center (ACS) was established in Flagstaff, Arizona on 1 July 1963 as a research facility of the United State Geological Survey (USGS) (Schaber, 2005) through the efforts and requests of several geoscientists and cartographers, perhaps most notably geologist E. Shoemaker. The USGS and the NASA agreed on the benefits of a research center that focused on compiling planetary maps, developing observational instrumentation, and training both astronauts and fellow researchers. Therein, location in Flagstaff proved advantageous based on existing planetary research community, proximity to lunar-observing telescopes, as well as geologically diverse, yet highly accessible analog terrain. The USGS ASC has evolved over the past five decades in response not only to the changing needs of NASA and the planetary science community but also to the increased volume and diversity of modern, technologically advanced datasets acquired for planetary bodies. By doing so, within the Planetary Geological Mapping Program (founded by NASA and coordinated by the USGS ACS) planetary maps and cartographic products were produced which reveal topography, geology, topology, image mosaics, and more. The aim of this program is to support the international research community with high-quality peer-reviewed geologic maps of planetary bodies. To accomplish this and to have comparable and homogenous map results, the mapping process has been standardized and is coordinated from its beginning (usability of input data) up to final map layout, printing, publishing, and archiving (Tanaka *et al.*, 2011). All the resulted products are available to the international scientific community and the general public as a national resource. In order to handle these tasks effectively, i.e., to ensure that unnecessary duplication is reduced to a minimum, cooperation is critical, wherein multiple institutions and organizational bodies must cross-collaborate. Thus, the USGS ACS established cooperation with institutions like NASA (at multiple programmatic levels), European Space Agency (ESA), and Jet Propulsion Laboratory (JPL) but also community organizations such as the International Astronomical Union (IAU) and the more recently formed Mapping and Planetary Spatial Infrastructure Team (MAPSIT).

Since the launch of the Sputnik – the first artificial Earth satellite (1957) – and the outstanding flight of the first cosmonaut Yuri Gagarin (1961), planetary cartography has successfully been developing in the USSR. A new era in lunar mapping began when the world's first map of the lunar far side was published in 1960 based on *Luna 3* images (1959). This work was led by Yu.N. Lipsky in Sternberg Astronomical Institute (SAI) and by N.A. Sokolova in Central Science-Research Institute of Geodesy, Aerial Photography and Cartography (TsNIIGAiK) (Shevchenko *et al.*, 2016). At about the same time, in 1961, the Moscow State University of Geodesy and Cartography (MIIGAiK) established its Fundamental Research Laboratory at the Department of Aerial Photo Surveying. The laboratory focused on lunar image processing and mapping. In 1968, when a new Space Research Institute of the Academy of Sciences of the USSR (IKI) was officially founded, the part of MIIGAiK Laboratory was moved to IKI. Since 1999 MIIGAiK, Dresden Technical University (Germany), Eötvös Loránd University (ELTE) (Hungary), and the University of Western Ontario (Canada) have participated in the project, "Multi-language Maps of Planets and their Moons" (Shingareva *et al.*, 2005). Intense international collaborations in the course of ESA's Mars Express have resulted in a Phobos special issue (Oberst *et al.*, 2014) and *The Phobos Altas* (Savinykh *et al.*, 2015).

Nowadays, the MIIGAiK Extraterrestrial Laboratory (MExLab) is focusing on planetary geodesy and GIS-based cartography methods for Solar System bodies (Karachevtseva *et al.*, 2016a). Research topics include fundamental parameters of celestial bodies such as shape, rotational parameters, and forced librations as well as planetary coordinate systems. A variety of wall maps of Phobos and a map and globe of Mercury have also been published recently (Karachevtseva *et al.*, 2016b). One of the research branches is devoted to cartographic support for Russian landing-site selection for future missions (Luna-25, 27–28) (Kokhanov *et al.*, 2017) as well as planning of future orbital mission to the Moon (Luna-26) (Polyansky *et al.*, 2017). Using GIS and web-based technology MExLab stores results of these studies in their planetary geodatabase with web-access provided via its geoportal.

Global planetary mapping, morphologic studies of craters and work on localization in Russian of planetary nomenclature are also the domain of the Department of Lunar and Planetary Physics of SAI, and a number of collaborations have substantiated in this national context (Shevchenko *et al.*, 2016).

In Germany, the Institute of Planetary Research at the German Aerospace Center (DLR) in Berlin – among other groups in Germany – has been involved in a number of planetary cartographic topics since its establishment in the early 1990s. A series of Mars maps were compiled in the early 1990s (Hiller *et al.*, 1993; Neugebauer and Dorrer, 1996) in preparation of the Russian Mars-96 mission. Since that time the main focus of DLR's cartography has been put on generation of image mosaics and topographic maps. For the HRSC contribution for Mars Express (Jaumann *et al.*, 2007), this work has been conducted also in close collaboration with groups at universities such as the Technical University of Berlin, Freie Universität Berlin, Technical University of Dresden, Technical University of Munich or the University of Hannover. It involves production of image and topographic maps, definition of a large-scale quadrangle schema for cartographic representation of Mars (Albertz *et al.*, 2005; Lehmann *et al.*, 1997), or automatization of map-generation processes (PIMap, Gehrke *et al.*, 2006). A new set of cartographic products based on the integration of data from multiple orbits was produced and published recently (Gwinner *et al.*, 2016). Controlled images and ortho-image mosaics were also produced and published for Phobos (Wählisch *et al.*, 2010, 2014; Willner *et al.*, 2010, 2014), or the Icy Saturnian Satellites (Roatsch *et al.*, 2009). For Ceres and Vesta, global mosaics serve as base map for different atlas collections (e.g. Roatsch *et al.*, 2013, 2016).

During the last 10 years the DLR has also been focusing on GIS-based mapping and tasks of processing, analyzing, archiving, and visualizing scientific results (e.g., Deuchler *et al.*, 2004; Saiger *et al.*, 2005). Topics of interests also include standardized cartographic visualization of scientific map results in order to create homogeneous and comparable maps and data archiving products (e.g., Nass *et al.*, 2011; van Gasselt and Nass, 2011). To provide sophisticated user experiences, which satisfy scientific as well as public outreach purposes, first steps have been undertaken to set up WebMap Services and WebGL applications by using image data and maps from the Apollo-17 and LRO missions (Clever, 2014; Maslonka, 2014). For Mars, the webGIS platform *iMars* has been developed with support by the European Commission. It allows for querying spatiotemporal planetary data from the web browser (Walter *et al.*, 2018).

Driven by China's lunar exploration missions, Chang'E-1, Chang'E-2, and Chang'E-3, lunar cartography work has been undertaken by several institutions of the Chinese Academy of Sciences (CAS), including the National Astronomical Observatories of China (NAOC), Institute of Geochemistry (IGCAS), Institute of Remote Sensing and Digital Earth (RADI). NOAC established the Science and Application Center for Lunar and Deep Space Exploration in 2003. One of its major responsibilities is to produce global cartographic products from data acquired by the aforementioned Chinese lunar missions. IGCAS established the Center for Lunar and Planetary Sciences in 2005 with the focus on geological mapping using the same datasets. RADI established the Planetary Mapping and Remote Sensing laboratory in 2008. The lab works on the development of high-precision planetary mapping methods using multi-source data and has produced high-resolution topographic products of the Chang'E-3 landing site using orbital and rover images.

Researchers from the Hong Kong Polytechnic University and Macau University of Science and Technology have also been actively working on planetary mapping and related research to support China's lunar and Mars exploration missions.

Some other groups in China's universities and institutions have also been involved in relevant research of planetary mapping. Many lunar topographic products have been completed: e.g., a global image mosaic (using orbiter imagery data of Chang'E-1; Li, Liu *et al.*, 2010), a global lunar digital elevation model (DEM) (using altimetry data of Chang'E-1; Hu *et al.*, 2013; Li, Ren *et al.*, 2010; Ping *et al.*, 2009), a high-resolution global DEM and ortho-image (using stereo imagery data from Chang'E-2; Li *et al.*, 2015), a high-resolution DEM and ortho-image map of Chang'E-3 landing area (by Chang'E-2 stereo imagery and LOLA data; Wu *et al.*, 2014), high-precision DEMs and ortho-image of the Chang'E-3 landing site (by the lander's descent images and the rover's stereo images; Liu *et al.*, 2014).

As emphasized earlier, such a presentation cannot be considered complete as it only highlights some of the efforts that are being made globally. New groups in China, India, and Japan, and also Korea are developing fast. Other projects across the US, such as the JMARS/JVesta project led by Arizona State University (ASU), Vesta/MarsTrek, developed at NASA's Jet Propulsion Laboratory (JPL), play an important role internationally. And also in Europe a number of university and research institutes work on map production and aspects of cartography and coordinate international cartographic work (e.g., ELTE, Hungary; University of Oulu, Finland; University of Muenster, Germany; University of Chieti-Pescara, or University of Perugia, Italy). With key Italian instruments on board BepiColombo observing Mercury, members of several Italian scientific centers joined the planetary geologic mappers' community (Galluzzi *et al.*, 2016).

2.2 *Organizations and initiatives*

Different groups were originated by NASA to coordinate map requirements, recommend map series and standards, establish priorities for map production, monitor map distribution, and facilitate international cooperation in lunar mapping (PCWG, 1989). The last 10 years of activity within the Planetary Cartography and Geologic Mapping Working Group (PCGMWG) was largely focused on monitoring and guiding the cartographic contributions of the USGS ASC.

Naming of topographic features and the publication of the *Gazetteer of Planetary Nomenclature* are coordinated by the International Astronomical Union (IAU). It was founded in 1919 at the Constitutive Assembly of the International Research Council in Brussels, Belgium to oversee assigning names for stars, planets, moons, asteroids, comets, and surface features on them (Blaauw, 1994). The first goals were to normalize various systems used in lunar and Martian nomenclatures across different countries (Blagg and Müller, 1935). The current nomenclature database is managed by the USGS ASC on behalf of the IAU.

In 1976, the IAU established a working group on the Cartographic Coordinates and Rotational Elements of Planets and Satellites to report triennially on the preferred volumes for the parameters of the rotation rate, spin axis, prime meridian, and reference surface for planets and satellites (Archinal, 2011). This working group founded to allow the consistent data usage across many facilities, including surface exploration by robots and humans. However, the IAU's oversight does not cover other standards essential for digital mapping including common feature attributions, feature symbols, recommended mapping scales and metadata, or the documentation of the data. For the US, this role has been filled by the Federal Geographic Data Committee (FGDC) and its recommendations are generally closely adopted by the International Standards Organization (ISO). Within the FGDC (2006), feature attributes and their assigned symbols for terrestrial and also planetary digital maps are defined. For a clear understanding of planetary maps, these symbols are primarily based on the same set of attributes and symbols as used for the Earth (see e.g., Hargitai and Shingareva, 2011; Hargitai *et al.*, 2014; Nass *et al.*, 2011).

The International Society for Photogrammetry and Remote Sensing (ISPRS) working group IV/8 "Planetary Mapping and Spatial Databases" is built by the community to provide a platform for those involved in all topics of planetary cartography, such as data acquisition, processing, and information extraction from planetary remote sensing data for the mapping of celestial bodies. This also includes the evaluation and refinement of reference systems, coordinate systems, control networks, map sheet definitions, etc., and their standardization. The group organizes workshops and symposiums to exchange the developments in planetary mapping, cartography, and remote sensing, and promotes international cooperation since 1998. After the 2016 ISPRS Congress in Prague the working group has transformed to a new inter-commission working group III/II "Planetary Remote Sensing and Mapping".

The Commission on Planetary Cartography of the International Cartographic Association (ICA) was established in 1995 with the goal of disseminating products and initiating outreach and professional projects in countries where planetary cartographic materials are scarcely available or altogether absent. The commission focuses on supporting planetary cartographic projects in emerging planetary communities. Since its formation, the commission developed three multilingual outreach map series: a series edited in Dresden (Buchroithner, 1999), a Central European edition (Shingareva *et al.*, 2005), and a special series for children (Hargitai *et al.*, 2015). The commission members compiled the *Multilingual Glossary of Planetary Cartography* (Shingareva and Krasnopevsteva, 2011) and the GIS-ready *Integrated Database of Planetary Features* (Hargitai, 2016). It also developed a planetary cartographic application that can be used to compare sizes of planetary features to countries and states (Gede and Hargitai, 2015) and maintains the Digital Museum of Planetary Mapping.

The Mapping and Planetary Spatial Infrastructure Team (MAPSIT) was formulated in 2014 as a means to re-affirm that modern cartography, i.e., spatial data infrastructure (NSF, 2012), fundamentally affects all aspects of scientific investigation and mission planning, regardless of the target body of interest. MAPSIT faces tremendous challenges, not the least of which is the sheer scope of modern planetary exploration, which results in a multitude of spatial parameters related to instrument types, target body characteristics, and coordinating institutions. USGS ASC and ASI held a topical meeting in 2009 on the topic of "Geological Mapping of Mars: a workshop on new concepts and tools" (Pondrelli *et al.*, 2011). Since then, the state of the art evolved significantly, not only from an institutional (space agencies and surveys) perspective, but also in terms of technology, applications, and services. While planetary data mapping workshops have been held three times in the United States (coordinated by USGS ACS), ESA's Space Astronomy Centre (ESAC) supported its first Planetary GIS workshop in 2015 (Manaud *et al.*, 2016b). This year the 1st Workshop for "Planetary Mapping and Virtual Observatory" was organized within the VESPA (Virtual European Solar and Planetary Access) program. This workshop aims at bringing together the geologic, geospatial, and Virtual Observatory (VO) communities for bringing forward knowledge, tools, and standards for mapping the Solar System.

In addition to institutional and organizational efforts, a number of initiatives have arisen in recent years motivated not only by individuals but also by commercial entities that specialize in combining planetary (map) data with web technologies. Today, different startups and organizations offer platforms and pre-existing cartographic databases which often feature open-source mapping technologies at their core. It has made it even easier for non-GIS specialists, researchers, and data enthusiasts to visualize, manipulate, and share their data and maps on the web (Zastrow, 2015). E.g., CARTO, a company focused on web-based geospatial data visualization and analysis, collaborated with ESA to support an open-source outreach project intended to raise public awareness of ESA's ExoMars Rover mission through an interactive map of the candidate landing sites (Where On Mars?, Manaud *et al.*, 2015).

The OpenPlanetary initiative was created in 2015 (Manaud *et al.*, 2016a) providing an online framework to help collaborate on common planetary mapping and data analysis problems, on new challenges, and to create new opportunities. Furthermore, it focused on building the first Open Planetary Mapping and Social platform for researchers, educators, storytellers, and the general public (Manaud *et al.*, 2017). Also, a number of projects funded by the European Commission are directly or indirectly relevant to planetary mapping. The largest, and one of the most long-lasting efforts, is VESPA (Erard *et al.*, 2014, 2017), the EuroPlanet H2020 Research Infrastructure component that deals with accessibility and distribution of planetary science data from very diverse scientific domains, including a specific surface-mapping task (Rossi *et al.*, 2015). E-Infrastructure projects with broad Earth Science focus, such as EarthServer-2, include a Planetary Science component, PlanetServer (Baumann *et al.*, 2015; Rossi *et al.*, 2016). Rather than focusing on data searches and discovery or on-demand processing, PlanetServer primarily uses the OGC WCPS standard (Baumann, 2010) to perform real-time data analytics (Marco Figuera *et al.*, 2015; Rossi *et al.*, 2016).

In Russia, the Digital Moon project is being developed based on a new information platform that provides wide opportunities for distributed data processing and advanced functions of online communication in a synchronous virtual spatial context between remote users (Garov *et al.*, 2016) within the framework of the 3D virtual laboratory.

Lastly, few citizen-science projects with clear planetary mapping target exist. Some of them are embedded in a broader context, such as iMars (Muller *et al.*, 2016); others originate from experiment-driven effort (NASA MRO HiRISE), such as PlanetFour (Aye *et al.*, 2016), and citizen-science efforts (NASA LRO/LROC) focus on imagery mapping include Moon Zoo (Joy *et al.*, 2011).

3 CHALLENGES IN PLANETARY CARTOGRAPHY

The standardization of cartographic methods and data products is critical for accurate and precise analysis and scientific reporting. This is more relevant today than it has been ever before, as researchers have convenient access to a plethora of digital data as well as to tools to process and analyze these data products. The lifecycle of cartographic products can be short, and standardized descriptions are needed to keep track of different developments. One of our aims herein was to compartmentalize the processes of planetary cartography and to define, describe, and present the overall mapping process through its components' breakdown (Fig. 19.2).

Processes related to the INPUT compartment (Fig. 19.2) cover all aspects that allow not only to produce higher-level products but also to create a basis for their stable representation and re-usability. One of the major future issues will be to establish an international map database by digitizing analog maps and by establishing a uniform structure to describe existing data allowing them to be queried and accessed. For digital map products, a metadata description (i.e., a digital equivalent known from map legends and additional information relating to the map content), along with validation tools, and platforms capable of providing access to archiving, distribution, and querying need to be established. Standards already partially exist on a national level and some of the older higher-level map-data products are currently transferred to fit into such schemes. However, many unstandardized map products exist all around the world and are distributed across different institutes. One task will be to review such products and to establish a methodological repertoire to transfer maps, to establish a common metadata scheme, and to provide a common semantical basis.

Within the DISTILLATION processes (Fig. 19.2) core issues are concerned with the abstraction of data, the (carto)graphic visualization, and the GIS-based management of derived data. We identify three major tasks that are necessary to accomplish this: (1) the definition and setup of rules and recommendations for GIS-based mapping processes (cf. Tanaka *et al.*, 2011); (2) advocating the GIS-based implementation and distribution of international cartographic symbol standards; (3) generating generic, modular data models for GIS-based mapping, which could be used by mappers to fill in their individual mapping data and scientific results. Currently, efforts are focused on creating a template-based framework for the evaluation and optimization of existing map templates. In particular, the short lifetime of products during ongoing missions pose a considerable challenge when creating such models and putting them into operational use. Furthermore, recent work focuses on revising recommendations for cartographic symbols for geologic mapping. This encompasses critical review and updating of existing standards for planetary geologic symbols (FGDC, 2006).

OUTPUT processes (Fig. 19.2) cover all aspects of publishing and archiving mapping results in easily accessible archives using intuitive online interfaces and platforms. One method to achieve this is to incorporate already published maps along with their metadata into an accessible digital map archive. This includes digitized analog maps, digital maps, and mapping products in comparable formats, and builds on existing definitions that can benefit from existing validation tools. The

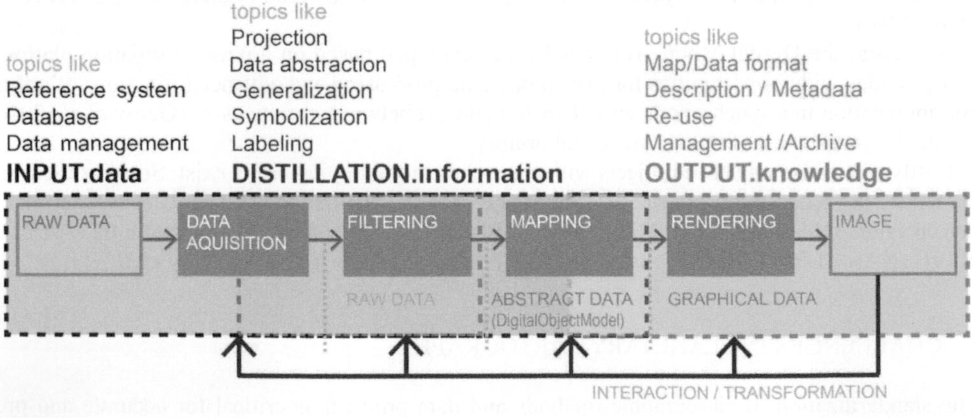

Figure 19.2. Visualization pipeline linked to the process during planetary mapping and the data-information-knowledge-wisdom hierarchy (see Section 1.2 and Fig. 19.1)

Planetary Data System (PDS, 2009), e.g., has provided a flexible toolset to accomplish parts of this task in cooperation with USGS/ACS. Existing efforts covering this topic of metadata are described in e.g., Hare (2011), Hare *et al.* (2011), and Nass *et al.* (2010). The existing archives like e.g., the PDS/NASA, the Planetary Science Archive (PSA/ESA, Besse *et al.*, 2017), or Data Archives and Transmission System (DARTS/JAXA) could be extended to include digital maps. The final issue in this part covers aspects of interoperability and exchange of map projects between different mapping and database systems. As different research institutes and individuals use different tools for mapping and data storage, procedures have to be established to allow conversions and also collaborative mapping in the future. It is the ultimate goal for planetary geologic or other thematic maps to be produced by different groups using the same principles in data collection, analysis, and display so that they are compatible.

4 CONCLUSION AND OUTLOOK

A general aim for the planetary cartography community is to develop concepts and approaches to foster future cooperation between cartographers and non-cartographers. This collaboration should focus on reducing duplication of efforts and combining limited resources in order to address technical and scientific objectives. Primarily motivated by such objectives international cross-collaborations between institutes have recently been established to provide a platform for critical discourse within organizations, as well as a constant contribution to different initiatives. Furthermore, these developments should focus on (1) identifying and prioritizing needs of the planetary cartography community along with a strategic timeline to accomplish such prioritized goals; (2) keeping track of ongoing work across the globe in the field of planetary cartography, and; (3) identifying areas of evolving technologies and innovations that deal with mapping strategies as well as output media for the dissemination and communication of cartographic results.

In addition to professional and scientific applications, planetary cartography has sufficient data resources that would enable non-planetary cartographers to produce planetary maps for the general public. Such exploration maps, e.g., the ocean floors and Antarctica, are regularly produced, but not so for extraterrestrial surfaces. Professional planetary maps, typically geologic maps, are exclusively published for scientific purposes, except for a few other cartographic products published by the USGS. Notably, some photomosaic maps available for the open public on multiple web platforms are not fully utilizing the cartographic tools and planetary datasets available. In short, collaboration and interaction between the fields of cartography (along with its tools and methods) and planetary science has not yet reached its full potential. Neither popular science books nor atlases include planetary maps, with very few exceptions, probably because knowledge about and access to planetary resources are limited. The work of planetary cartographers should serve as a bridge between those different fields, and make such datasets available for "terrestrial cartographers" in the format they can use for producing maps for the general public.

ACKNOWLEDGEMENTS

APR acknowledges support from the grant H2020-EINFRA-2014–2 no. 654367 (EarthServer-2) and H2020 Research Infrastructure EuroPlanet 2020 (EPN2020-RI) grant no. 654208.

REFERENCES

Ackoff, R.L. (1989) From data to wisdom. *Journal of Applied Systems Analysis*, 16, 3–9.

Albertz, J., Attwenger, M., *et al.* (2005) HRSC on Mars Express – photogrammetric and cartographic research. *Photogrammetric Engineering and Remote Sensing*, 71, 1153–1160.

Archinal, B.A., A'Hearn, M.F., *et al.* (2011) Report of the IAU Working Group on cartographic coordinates and rotational elements: 2009. *Celestial Mechanics and Dynamical Astronomy*, 109(2), 101–135. doi:10.1007/s10569-010-9320-4.

Atlas poverhnosti Venery (Atlas of Venus surface based on Venera 16 and Venera 17 observations). USSR, Moscow, 1988 (In Russian).

Aye, K.-M., Schwamb, M.E., *et al.* (2016) Analysis pipeline and results from the PlanetFour Citizen Science Project. Lunar and Planetary Science Conference, The Woodlands, TX, USA,, #3056.

Barabashov, N.P., Mikhailov, A.A. & Lipskiy, Y.N. (1960) *Atlas Obratnoy Storony Luny. 1960. Pod redaktsiey*. Izdatelstvo Akademii nauk SSSR, Moskva, Russia.

Baumann, P. (2010) The OGC Web Coverage Processing Service (WCPS) standard. *Geoinformatica*, 14(4), 447–479.

Baumann, P., Mazzetti, P., *et al.* (2015) Big data analytics for earth sciences: the EarthServer approach. *International Journal of Digital Earth*, 9(1), 3–29. doi:10.1080/17538947.2014.1003106.

Besse, S., Vallat, C., *et al.* (2017) The New Planetary Science Archive (PSA): exploration and discovery of scientific datasets from ESA's planetary missions. Lunar and Planetary Science Conference, The Woodlands, TX, USA, #1186.

Beyer, R.A. (2015) An introduction to the data and tools of planetary geomorphology. *Geomorphology*, 240, 137–145.

Blaauw, A. (1994) *History of IAU: The Birth and First Half-Century of the IAU*. Kluwer Academic Publishers, Dordrecht, The Netherlands. ISBN: 0-7923-2979-1.

Blagg, M.A. & Müller, K. (1935) *Named Lunar Formations*. International Astronomical Union, Percy Lund, Humphries and Co Ltd., London, GB, UK. 196 p.

Buchroithner, M.F. (1999) Mars Map: the first of the series of multilingual relief maps of terrestrial planets and their moon. *ICA Conference*, Ottawa, Canada.

Carpendale, M.S.T. (2003) Considering visual variables as a basis for information visualization. *Research Report 2001–693–16*, Department of Computer Science, University of Calgary, Calgary, Canada.

Clever, S. (2014) *LROC – Anaglyphen vom Mond: Erstellung und Visualisierung in einem Webmapping Projekt*. Ba. thesis, Beuth Hochschule für Technik Berlin.

Compiling Committee of the Chang'E-1 (2010) *The Chang'E-1 Topographic Atlas of the Moon*. NAOC and SinoMap Press, Beijing, China.

Davies, M.E., Robinson, J.C., *et al.* (1970) *Mars Chart*. US Army Topographic Command, Washington, DC, USA.

Deuchler, C., Wählisch, M. *et al.* (2004) *Combining Mars Data in GRASS GIS for Geological Mapping*. ISPRS, Istanbul, Turkey.

Erard, E., Cecconi, B., *et al.* (2014) Planetary science virtual observatory architecture. *Astronomy and Computing*, 7, 71–80.

Erard, S., Cecconi, B., *et al.* (2017) VESPA: a community-driven virtual observatory in planetary science. *Planetary and Space Science (PSS)*, 150, 65–85. doi:10.1016/j.pss.2017.05.013.

FGDC (2006) *Digital Cartographic Standard for Geologic Map Symbolization*. Federal Geographic Data Committee, Document # FGDC-STD-013-2006.

Ford, J.P., Plaut, J., *et al.* (1993) *Guide to Magellan Image Interpretation*. NASA-JPL, Pasadena, CA, USA.

Frigeri, A., Hare, T., *et al.* (2011) A working environment for digital planetary data processing and mapping using ISIS and GRASS GIS. *Planetary and Space Science (PSS)*, 59, Special Issue: Planetary Mapping, Elsevier Ltd., 1265–1272. doi:10.1016/j.pss.2010.12.008.

Galluzzi, V., Guzzetta, L., *et al.* (2016) Geology of the Victoria quadrangle (H02), Mercury. *Journal of Maps*, 12, 227–238. doi:10.1080/17445647.2016.1193777.

Garov, A.S., Karachevtseva, I.P., *et al.* (2016) Development of a heterogenic distributed environment for spatial data processing using cloud technologies. *International Archives of the Photogrammetry, Remote Sensing and Spatial Information Sciences*, 41(B4), 385–390.

Gede, M. & Hargitai, H. (2015) Country movers – an extraterrestrial geographical application. *Joint ICA Symposium on Cartography Beyond the Ordinary World*, Niteroi, Brazil. pp. 178–183.

Gehrke, S., Wählisch, M., *et al.* (2006) Generation of topographic and thematic planetary maps using the software system "PIMap". In: Mackwell, S. & Stansbery, E. (eds) *37th Annual Lunar and Planetary Science Conference*, League City, TX, USA.

Greeley, R. & Batson, R.M. (1990) *Planetary Mapping*. Cambridge University Press, Cambridge.

Gwinner, K., Jaumann, R., *et al.* (2016) The High Resolution Stereo Camera (HRSC) of Mars Express and its approach to science analysis and mapping for Mars and its satellites. *Planetary and Space Science (PSS)*, 126, 93–138.

Haber, R.B. & McNabb, D.A. (1990) Visualization idioms: a conceptual model for scientific visualization systems. In: Shriver, B., Neilson, G.M. & Rosenblum, L. (eds) *Visualization in Scientific Computing*. IEEE Computer Society Press, Los Alamitos, CA, USA. pp. 74–93.

Hare, T.M. (2011) *Standards-Based Collation Tools for Geospatial Metadata in Support of the Planetary Domain*. Ma. thesis, Northern Arizona University, Flagstaff, AZ, USA.

Hare, T.M. & Tanaka, K.L. (2001) Planetary Interactive GIS-on-the-Web Analyzable Database (PIGWAD). *ICC 2001*, Beijing, China.

Hare, T.M., Skinner, J.A. & Kirk, R.L. (2017) Cartography tools. In: Rossi, A.-P. & van Gasselt, S. (eds) *Planetary Geology*. Springer, Berlin, Germany. 540 pp.

Hare, T.M., Kirk, R.L., *et al.* (2009) Chapter 60: extraterrestrial GIS. In: Madden, M. (ed) *Manual of Geographic Information Systems*. The American Society for Photogrammetry and Remote Sensing, Bethesda, MD, USA. pp. 1199–1219.

Hare, T.M., Skinner, J.A., *et al.* (2011) FGDC geospatial metadata for the planetary domain. Lunar and Planetary Science Conference, The Woodlands, TX, USA, #1608.

Hare, T.M., Skinner, J.A., *et al.* (2015) Planetary GIS at the U.S. Geological Survey Astrogeology Science Center. *2nd Planetary Data Workshop*, 8–11 June 2015, Flagstaff, AZ, USA. LPI No. 1846, abstract #7005.

Hare, T.M., Rossi, A.P., *et al.* (2017) Interoperability in Planetary Research for Geospatial Data. *Planetary Space Science (PSS)*, 150, 36–42, doi:10.1016/j.pss.2017.04.004.

Hargitai, H. (2016) Metacatalog of planetary surface features for multicriteria evaluation of surface evolution: the integrated planetary deature database. *DPS 48/EPSC11 Meeting*, Nantes, France, abstract #426.23.

Hargitai, H. & Shingareva, K.B. (2011) Planetary nomenclature: a representation of human culture and alien landscapes. *Advances in Cartography and Giscience*, 6(4), Lecture Notes in Geoinformation and Cartography, 275–288. doi:10.1007/978-3-642-19214-2_18.

Hargitai, H., Gede, M., Zimbelman, J., *et al.* (2015) Multilingual narrative planetary maps for children. In: Robbi, S.C., Madureira Cruz, C.B. & Leal de Menezes, P.M. (eds) *Cartography – Maps Connecting the World*, Lecture Notes in Geoinformation and Cartography, Springer, Berlin, Germany. pp. 17–30.

Hargitai, H., Li, C., Zhang, Z., *et al.* (2014) Chinese and Russian language equivalents of the IAU gazetteer of planetary nomenclature: an overview of planetary toponym localization methods. *The Cartographic Journal*. 53, 1–21, doi:10.1179/1743277413.

Hiller, K., Hauber, E., *et al.* (1993) Digitale Kartenherstellung der Planetenbildkarten Olympus Mons/Planet Mars. In: *Kartograph. Nachrichten*. Heft 2 Kirschbaum Verlag, Bonn, Germany.

Hu, W., Di, K., *et al.* (2013) A new lunar global DEM derived from Chang'E-1 Laser Altimeter data based on crossover adjustment with local topographic constraint. *Planetary and Space Science (PSS)*, 87(2013), 173–182.

Jaumann, R., Neukum, G., *et al.* (2007) The High-Resolution Stereo Camera (HRSC) experiment on Mars Express: instrument aspects and experiment conduct from interplanetary cruise through the nominal mission. *Planetary and Space Science (PSS)*, 55, 928–952.

Jin, S. (2014) *Planetary Geodesy & Remote Sensing*. CRC Press, Boca Raton, FL.

Joy, K., Crawford, I., *et al.* (2011) Moon Zoo: citizen science in lunar exploration. *Astronomy & Geophysics*, 52(2), 2–10. doi:10.1111/j.1468-4004.2011.52210.x.

Karachevtseva, I.P., Kokhanov, A.A., *et al.* (2016a) Modern methodology and new tools for planetary mapping. In: Gartner, G., Jobst, M. & Huang, H. (eds) *Progress in Cartography*. Springer, Berlin, Germnay. pp. 207–227, 480 p. doi:10.1007/978-3-319-19602-2.

Karachevtseva, I.P., Kokhanov, A.A., *et al.* (2016b) Mapping of inner and outer celestial bodies using new global and local topographic data derived from photogrammetric image processing. *International Archieves of the Photogrammetry, Remote Sensings and Spatial Information Science*, Volume XLI-B4. pp. 411–415. doi:10.5194/isprs-archives-XLI-B4-411-2016. ISPRS 2016.

Kokhanov, A.A., Karachevtseva, I.P., Zubarev, A.E., Patraty, V., Rodionova, Z. & Oberst, J. (2017) Mapping of potential lunar landing areas using LRO and SELENE data. *Planetary and Space Science*. doi:10.1016/j.pss.2017.08.002 [available online August 2017].

Kirk, R. (2016) Planetary Cartography: what, how, and why begin with where. Lunar and Planetary Science Conference, The Woodlands, TX, USA, abstract #2151.

Kopal, Z. & Carder, R.W. (1974) *Mapping of the Moon*. Springer Science + Business Media, Dordrecht, The Netherlands.

Kraak, M.-J. & Fabrikant, S. (2017) Of maps, cartography and the geography of the International Cartographic Association. *International Journal of Cartography*, 23 p. doi:10.1080/23729333.2017.1288535.

Laura, J.R., Hare, T.M., *et al.* (2017) Towards a planetary spatial data infrastructure. *ISPRS International Journal of Geo-Information*, 6, 181. doi:10.3390/ijgi6060181.
Lehmann, H., Scholten, F., *et al.* (1997) Mapping a whole planet – the new topographic series 1:200 000 for Mars. *ICC 1997*, Stockholm, Sweden.
Li, C., Liu, J., *et al.* (2010) The global image of the moon obtained by the Chang'E-1: Data processing and lunar cartography. *Science China Earth Sciences*, 53(8), 1091–1102.
Li, C., Ren, X., *et al.* (2010) Laser altimetry data of Chang'E-1 and the global lunar DEM model. *Science China Earth Sciences*, 53(11), 1582–1593.
Li, C., Ren, X., *et al.* (2015) A new global and high resolution topographic map product of the Moon from Chang'E-2 image data. Lunar and Planetary Science Conference, The Woodlands, TX, USA, abstract #1638.
Liu, Z., Di K., *et al.* (2014) High precision landing site mapping and rover localization for Chang'3 mission. *Science China Physics, Mechanics and Astronomy*, 58(1), 019601.
Manaud, N., Boix, O., *et al.* (2015) Where on Mars? A Web map visualisation of the ExoMars 2018 Rover candidate landing sites. *EPSC*, Nantes, France, abstract #2015–228.
Manaud, N., Nass, A., *et al.* (2017) Open PlanetaryMap: building the first open planetary mapping and social platform for researchers, educators, storytellers, and the general public. *3rd Planetary Data Workshop 2017*, Flagstaff, AZ, USA, #7024.
Manaud, N., Rossi, A.P., *et al.* (2016a) The OpenPlanetary initiative. *DPS-EPSC*, Pasadena, CA, USA, abstract #2587836.
Manaud, N., Rossi, A.P., *et al.* (2016b) Summary and recommendations from the 2015 ESAC planetary GIS Workshop. Lunar and Planetary Science Conference, The Woodlands, TX, USA, abstract #1387.
Marco Figuera, R., Rossi, A.P., *et al.* (2015) Analyzing Lunar DTMs through web services with EarthServer/PlanetServer-2. *ISPRS Commission VI, WG VI/4 Meeting*, Berlin, Germany.
Maslonka, C. (2014) *Bestimmung und Kartierung von Stationspunkten der Traverse der Apollo-17-Landestelle auf Basis von Lunar Reconnaissance Orbiter Camera Daten*. Ma. thesis, Beuth Hochschule für Technik Berlin.
Mason, A.C. & Hackman, R.J. (1961) *Engineer Special Study of the Surface of the Moon*. Department of Interior, United States Geological Survey, Reston, VA, USA.
Mazza, R. (2009) *Introduction to Information Visualization*. Springer Verlag, London, UK.
Morton, O. (2002) *Mapping Mars: Science, Imagination, and the Birth of a World*. Picador, New York, NY, USA.
Muller, J.-P., Tao, Y., *et al.* (2016) EU-FP7-iMARS: analysis of Mars multi-resolution images using auto-coregistration data mining and crowd source techniques: processed results – a first look. *International Archives of the Photogrammetry, Remote Sensing and Spatial Information Sciences (ISPRS)*, 453–458. doi:10.5194/isprs-archives-XLI-B4-453--2016.
Nass, A., van Gasselt, S. & Jaumann, R. (2010) Map description and management by spatial metadata: requirements for digital map legend for planetary geological and geomorphological mapping. *AutoCarto, Symposium on computer-based Cartography and GIScience*, Orlando, FL, USA, abstract #1457.
Nass, A., Di, K., *et al.* (2017) Planetary cartography – activities und current status. *International Cartography Conference, Washington 2017*, Washington, DC, USA, online proceedings, abstract # 419362.
Nass, A., van Gasselt, S., *et al.* (2011) Implementation of cartographic symbols for planetary mapping in geographic information systems. *Planetary and Space Science (PSS)*, 59, Special Issue: Planetary Mapping, Elsevier Ltd., 1255–1264. doi:10.1016/j.pss.2010.08.022.
Neugebauer, G. & Dorrer, E. (1996) Experimentelle Untersuchungen zur kartographischen Darstellung der Marsoberfläche. In: *Kartograph. Nachrichten*. Heft 2 Kirschbaum Verlag, Bonn, Germany.
NSF (2012) A vision and strategy for science, engineering, and education: Cyberinfrastructure framework for 21st century. *NSF12-113*. Available from: https://www.nsf.gov/pubs/2012/nsf12113/nsf12113.pdf
Oberst, J., Zakharov, A. & Schulz, R. (eds) (2014) Phobos (special issue). *Planetary and Space Science (PSS)*, 102, Elsevier, 182 p.
PCWG (1989) The Planetary Cartography Working Group, planetary cartography in the next decade: digital cartography and emerging opportunities. *NASA TM 4092*. Available from: https://astropedia.astrogeology.usgs.gov/download/Docs/Cartography/PlanetaryCartographyInTheNextDecade.pdf
PDS (2009) Planetary data system standard reference. *Technical Report*. Jet Propulsion Laboratory, Institute of Technology, Pasadena, CA.
Pędzich, P. & Latuszek, K. (2014) Planetary cartography – sample publications, cartographic projections, new challenges. *Polish Cartographical Review*, 46(4), 388–396.

Ping, J., Huang, Q., *et al.* (2009) Lunar topographic model CLTM-s01 from Chang'E-1 laser altimeter. In: *Science in China Series G: Physics, Mechanics and Astronomy*, 52(7), 1105–1114.

Pondrelli, M., Tanaka, K., *et al.* (eds) (2011) Geological mapping of Mars. *Planetary and Space Science (PSS)*, 59(11–12). doi:10.1016/j.pss.2011.07.006.

Portee, D.S.F. (2013) *1961: USGS Astrogeology's First Published Map*. United States Geological Survey, Reston, VA, USA

Polyansky, I., Zhukov, B., Zubarev, A., *et al.* (2017) Stereo topographic mapping concept for the upcoming Luna-Resurs-1 orbiter mission. *Planetary and Space Science*. doi:10.1016/j.pss.2017.09.013.

Radebaugh, J., Thomson, B.J., *et al.* (2017) Obtaining and using Planetary Spatial Data into the future: the role of the Mapping and Planetary Spatial Infrastructure Team (Mapsit). *Planetary Science Vision 2050 Workshop 2017*, Washington, DC, USA, abstract #8084.

Roatsch, T., Jaumann, R., *et al.* (2009) Cartographic mapping of the Icy satellites using ISS and VIMS data. In: Dougherty, M., Esposito, L. & Krimigis, S. (eds) *Saturn From Cassini-Huygens*. Springer, Dordrecht, The Netherlands. pp. 763–781.

Roatsch, T., Kersten, E., *et al.* (2013) High-resolution Vesta low altitude mapping orbit atlas derived from Dawn framing camera images. *Planetary and Space Science*, 85, 293–298. doi:10.1016/j.pss.2013.06.024.

Roatsch, T., Kersten, E., *et al.* (2016) High-resolution Ceres high altitude mapping orbit atlas derived from Dawn framing camera images. *Planetary and Space Science*, 129, 103–107. doi:10.1016/j.pss.2016.05.011.

Rossi, A.P., Cecconi, B., *et al.* (2015) Planetary GIS and EuroPlanet-RI H2020. *EPSC 2015*, Nantes, France, abstract #2015-178.

Rossi, A.P., Marco Figuera, R., *et al.* (2016) Remote sensing data analytics for planetary science with Planet/EarthServer. *EGU 2016*, Vienna Austria, abstract #2016–3996.

Rwoley, J. (2007) The wisdom hierarchy: representation of the DIKW hierarchy. *Journal of Information Science*, 33(2), 163–180. doi:10.1177/0165551506070706.

Sadler, D.H. (ed) (1962) *Proceedings of the 11th General Assembly of the IAU*. 1961. IAU Trans. XIB, Academic Press, Berkeley, CA, USA. p. 234.

Saiger, P., Wählisch, M., *et al.* (2005) ArcGIS and GRASS GIS for planetary data. *1st Mars Express Science Conference*. Noordwijk, the Netherlands.

Savinykh, V.P., Karachevtseva, I.P. & Konopikhin, A.A. (eds) (2015) *The Phobos Atlas*, MIIGAiK, *Moscow*, Russia. 220 p., 43 maps (In Russian).

Schaber, G.G. (2005) The U.S. Geological Survey Branch of Astrogeology – a chronology of activities from conception through the end of project Apollo (1960–1973). *USGS Open Report 2005-1190*. Available from: https://www.hq.nasa.gov/alsj/Schaber.html

Shevchenko, V., Rodionova, Z. & Michael, G. (2016) *Lunar and Planetary Cartography in Russia*. Springer, Berlin, Germany. doi:10.1007/978-3-319-21039-1.

Shingareva, K.B. & Krasnopevsteva, B.V. (2011) A new version of the multilingual glossary of planetary cartography. *Advances in Cartography and GIScience*, 2, Series Lecture Notes in Geoinformation and Cartography, 289–295.

Shingareva, K.B., Krasnopevtseva, B.V., *et al.* (1992) *Atlas of the Earth- Type Planets and Their Satellites*. MIIGAiK, Moscow, Russia.

Shingareva, K.B., Zimbelman, J., *et al.* (2005) The realization of ICA commission projects on planetary cartography. *Cartographica*, 40(4), 105–114. doi:10.3138/3660-4078-55X1-3808.

Shoemaker, E. & Hackman, R.J. (1961) *Lunar photogeologic chart LPC 58*. Copernicus, Prototype Chart, USGS, unpublished.

Slipher, E.C. (1962) *MEC-1 Map of Mars*. United States Air Force. St. Louis, MI, USA.Snyder, J.P. (1982) Map projections used by the U.S. Geological Survey. *U.S. Geological Survey Bulletin 1532*. Available from: https://pubs.usgs.gov/bul/1532/report.pdf

Snyder, J.P. (1987) Map projections used by the U.S. Geological Survey. *U.S. Geological Survey Professional Paper 1395*. Available from: https://pubs.er.usgs.gov/publication/pp1395

Tanaka, K, Skinner, J. & Hare, T. (2011) *Planetary Geologic Mapping Handbook – 2011*. Flagstaff, AZ: USGS.

van Gasselt, S. & Nass, A. (2011) Planetary mapping: the datamodel's perspective and GIS framework. *Planetary and Space Science (PSS)*, 59, Special Issue: Planetary Mapping, Elsevier Ltd., 1231–1242. doi:10.1016/j.pss.2010.09.012.

Wählisch, M., Stooke, P.J., *et al.* (2014) Phobos and Deimos cartography. *Planetary and Space Science (PSS)*, 102, 60–73.

Wählisch, M., Willner, K., *et al.* (2010) A new topographic image atlas of Phobos. *Earth and Planetary Science Letters*, 294, 547–553.
Walter, S.H.G., Muller, J.-P., Sidiropoulos, P., Tao, Y., Gwinner, K., Putri, A.R.D., Kim, J.-R., Steikert, R., van Gasselt, S., Michael, G.G., Watson, G. & Schreiner, B. (2018) The web-based interactive Mars Analysis and Research System (iMARS Web-GIS). *Journal of Geophysical Research*, 5(7), 308–323
Ware, C. (2004) *Information Visualization – Perception for Design*, 2nd edition, Elsevier Morgan Kaufmann Publisher, San Francisco, CA, USA.
Whitaker, E.A., Kuiper, G.P., *et al.* (1963) *Rectified Lunar Atlas – Supplement Number 2 to the Photographic Lunar Atlas. Aeronaut*. University of Arizona Press, Tucson, AZ, USA.
Williams, D.A. (2016) NASA's planetary geologic mapping program: overview. *Commission IV, ISPRS, WG IV/8*. Prague, Czech Republic doi:10.5194/isprs- archives-XLI-B4-519-2016.
Willner, K., Shi, X. & Oberst, J. (2014) Phobos' shape and topography models. *Planetary and Space Science (PSS)*, 102, 51–59.
Willner, K., Oberst, J., *et al.* (2010) Phobos control point network, rotation, and shape. *Earth and Planetary Science Letters*, 294, 541–546.
Wu, B., Li, F., *et al.* (2014) Topographic modeling and analysis of the landing site of Chang'E-3 on the moon. *Earth and Planetary Science Letters*, 405(2014), 257–273.
Zastrow, M. (2015) Data visualization: Science on the map. *Nature*, 519, 120.

Planetary Remote Sensing and Mapping – Wu et al.
© 2019 Taylor & Francis Group, London, ISBN 978-1-138-58415-0

Chapter 20

The lunar sub-polar areas:

Morphometric analysis and mapping

A.A. Kokhanov, I.P. Karachevtseva, N.A. Kozlova and Zh.F. Rodionova

ABSTRACT: We apply cartographic methods to analyze lunar sub-polar areas at different scales using modern data obtained by Lunar Reconnaissance Orbiter (LRO) and Kaguya (SELENE) missions. The main focus of our study is morphometric parameters of the lunar surface at various levels of detail – from global to local, especially potential landing areas. Global and regional maps with relief features demonstrate differences between north and south lunar polar areas in general. To identify areas with presumable hazards at the local level, we applied a set of characteristics (steep slopes, high roughness, availability of fresh craters) that were used as criteria for safe landing. Based on remote sensing data and using a special algorithm developed as GIS-tool we study lunar sub-polar area from different point of view: safety of landing, engineering constraints, scientific interest. A set of maps with various parameters is an effective instrument for estimation of potential landing area at different scales: from surrounding context on global view to high-priority potential landing site.

1 INTRODUCTION

The Moon enjoyed an extensive early exploration (since 1959, launch of the "Luna-1") by orbiters and landers. However, the early lunar landings were limited to the mid-latitudes. Nowadays, it is the lunar sub-polar areas (up to 65°), especially the southern ones, which are the focus of lunar science. Remote sensing data from Lunar Prospector, which operated from lunar polar orbit during 1998–1999, have shown a high probability of hydrogen abundance near the poles (Binder, 1998). Later, abundance of hydrogen was confirmed by LCROSS experiment (Colaprete *et al.*, 2010) and mapped by results of observations from The Lunar Exploration Neutron Detector (LEND) (Sanin *et al.*, 2012) onboard the Lunar Reconnaissance Orbiter (LRO). Significant deposits are confined to the sub-polar regions (Sanin, 2015) and especially to permanently shadowed areas (PSAs). PSAs are the cold traps for volatiles and the most probable places of water ice concentration (McClanahan *et al.*, 2015).

2 "LUNA 25" MISSION AND SOUTHERN SUB-POLAR AREAS OF INTEREST

The "Luna-25" is planned for launch and landing in a lunar sub-polar area. After that the Orbiter "Luna-26" and lander "Luna-27" will be launched. The aim of the Lunar polar missions is to confirm the presence of water ice there, study lunar polar resources and lunar exosphere, and thus to begin a long-term exploration of the Moon (Khartov, 2015). For "Luna-25" some potential landing sites (PLS) were selected by Space Research Institute of Russian Academy of Science (Mitrofanov *et al.*, 2016). In our research we paid special attention to one of the high-priority landing ellipse (coordinates: 21.21°E and 68.77°S) and studied its morphometric parameters.

The search for hydrogen deposits concentration in-situ is hampered by geological and engineering factors. Taking into account this task we characterized lunar sub-polar areas by set of various characteristics: illumination and visibility as well as morphometric parameters: slopes, roughness, depth (relative depth) of craters. To present the results of study we created maps at different levels of detail, including local view one of the landing ellipse, regional view of sub-polar areas and, finally, global view of the Moon.

3 CHARACTERIZATION OF AVAILABLE DATA FOR POLAR STUDY OF THE MOON

To analyze the lunar surface we have collected all available lunar data, including orthoimages and DEMs (Table 20.1). Although remote sensing data and derived products cover most part of the lunar surface, high-resolution data are rarely available for the sub-polar regions of interest (ROI). For comprehensive characterization of ROIs we use the entire array of remote sensing data.

3.1 *WAC global mosaic and DEM GLD 100*

The global geospatial overview of the Moon is provided by an orthomosaic of LRO Wide Angle Camera (WAC) images (Robinson *et al.*, 2010). The resolution of WAC images (75 m pixel^{-1}) allows us to determine features with sizes of 500 m and more. Each of the images, included in WAC Global mosaic (100 m pixel^{-1}), has an accurate position, based on ephemerides from Lunar Orbiter Laser Altimeter (LOLA) and Gravity Recovery and Interior Laboratory (GRAIL) data and the geometric model of LROC (Speyerer *et al.*, 2016).

On the basis of WAC images a Global Lunar Digital Terrain Model (GLD100) was created, which covers 98% of the Moon (Scholten *et al.*, 2012). GLD100 reveals global and regional morphologic features (more than 1 km in size) allowing morphometric measurements, except in sub-polar areas between 90° and 85°, where data gaps were filled by LDEM (see section 3.2) based on LOLA data that had another inner accuracy: heights are accurately identified only along the altimeter tracks rather than uniformly over the area in DEM, obtained by photogrammetric methods.

3.2 *LDEM and SLDEM coverage*

LDEM is LOLA Digital Elevation Model, provided as gridded raster data (Neumann *et al.*, 2011). The gridded data "LDEM 75" covers sub-polar areas up to ±75° with nominal resolution of 30 m pixel^{-1}. The highest resolution of provided gridded DEM (5 m pixel^{-1}) has "LDEM 87.5" that covers sub-polar areas up to ±87.5°. Due to interpolation of irregularly located laser altimeter tracks the DEMs could not be used for calculation of statistical characteristics of topography for landing site characterization, based on moving window algorithms.

SLDEM 2013 – the global DEM, obtained by photogrammetric processing of SELENE Terrain Camera (TC) and supplemented by LOLA data in shadowed areas (Haruyama *et al.*, 2014). These tiles cover areas of 1°x 1° and have resolution 10 m (approximately equal to the source TC images).

Table 20.1 Remote sensing data available for the lunar study

Product	*Type*	*Pixel scale, m pixel^{-1}*	*Source*
TC Ortho map	Orthomosaic	7	Haruyama *et al.*, 2008
WAC GLOBAL	Orthomosaic	100	Scholten *et al.*, 2012
LDEM 87.5	DEM	5	Neumann *et al.*, 2011
SLDEM 2013	DEM	7	Haruyama *et al.*, 2014
LDEM 75	DEM	30	Neumann *et al.*, 2011
GLD 100	DEM	100	Scholten *et al.*, 2012

Note: All abbreviations are explained in the text below.

According to estimates by Barker *et al.* (2016) effective resolution of the model is ~100 m. Unfortunately, vertical quantization is rather coarse (1 m), and internal vertical precision is limited (~10 m).

Both vertical and horizontal accuracy of TC-derived topography is lower than its internal precision due to imperfect knowledge of the Kaguya orbit, errors in camera pointing, imperfect knowledge of focal length, distortion, as well as the spacecraft jitter, which is especially harmful since TC is a push-broom camera (Kreslavsky *et al.*, 2016).

4 MORPHOMETRIC AND HYPSOMETRIC MAPPING OF SUB-POLAR AREAS

4.1. *Global roughness mapping by GLD100*

Global level supposes to measure landscape features more than 1 km in size and estimate morphometric and geologic context of the landing area in general. Topographic roughness shows that the relative height variations have been calculated for analysis of surface entire sub-polar area. To select smooth areas at the global level, we estimate roughness by a specially developed method using an interquartile range of second derivative of the relief (Laplacian), providing scale-defined stable results (Kreslavsky *et al.*, 2013). With this method and DEM GLD100, both sub-polar areas were characterized by Laplacian based on automated tool implemented for ArcGIS software (Kokhanov *et al.*, 2013).

Roughness was calculated in a round moving window with radius 23 pixels (the minimal window size, allowing to calculate the interquartile range without loss of information). The final map resolution is 3.8 km pixel^{-1}. The created map (Fig. 20.1) demonstrates lunar relief parameters with a special look to the Moon, showing differences between the presentation of main features with roughness calculation (left) and topographic information (right). Topographic roughness calculated as Laplasian demonstrates most typical landforms on global view. More attention is paid to the south sub-polar area (main look) as a basic focus of Russian future missions. North lunar sub-polar area is shown at two insets.

The map includes description of Laplacian method and its possibilities in comparison with a visual analysis of the image (at bottom) and can be used for educational purposes and for public outreach. It demonstrates landing sites of past missions in equatorial areas (Luna and Apollo) as well as landing areas proposed for future Russian missions in sub-polar regions. The map provides possibilities to keep a historic view on Moon exploration and to show the main differences between past missions in lunar maria and modern missions planned to more cratered regions.

4.2. *Regional hypsometric mapping by LDEM 75*

To characterize studied sub-polar regions of the Moon at the regional level, we have created a topographic map (1:1 600 000) using LDEM 75 (Fig. 20.2). The special hypsometric color ramp was developed for this map, which highlighted structures of crater floors and rims. The permanently shadowed areas (Mazarico *et al.*, 2011) are shown on the map. Comparing roughness, the hypsometric map shows differences between Moon polar areas in more detail.

Peculiarities of heights distribution are demonstrated by histograms and hypsographic curves (Figs. 20.3, 20.4). The minimal height point (–5207 m) of the northern sub-polar area is situated at the bottom of the Milankovic crater, maximal height point (2977 m) – on the crater Ricco rim. The lowest point of southern sub-polar area is situated in the De Forest crater: –7735 m, the highest, 7027 m – on the crater Nobile rim.

Maxima on histogram of northern sub-polar area fall on -1.5–0.5 km. Mean height is –824 m, median of heights is –758 m. Local minimum at –3 km is caused by old flat-bottomed craters: Rozhdestvenskiy, Nansen, Hermitte, Plaskett, Milankovic and its satellite craters.

The southern sub-polar area is partially occupied by South Pole-Aitken basin. This explains the bigger heights range: southern area is crossed by basins rim, and a big part of the territory is slope of the biggest lunar crater. Mean height of southern sub-polar area is –1339 m, median of heights

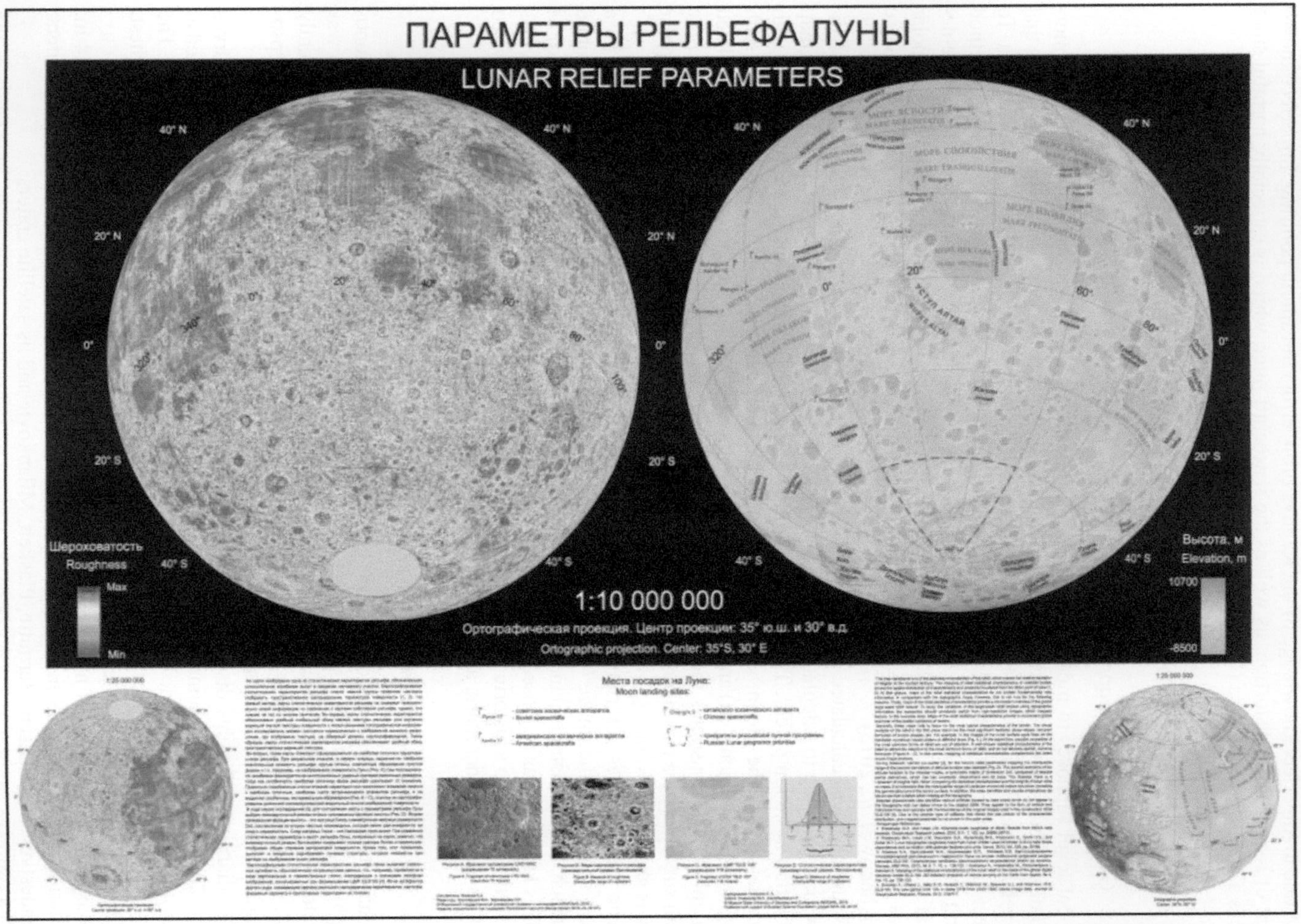

Figure 20.1. Printed version of bilingual map of lunar relief parameters (published by MIIGAiK, 2016), original size is 841 x 594 mm²

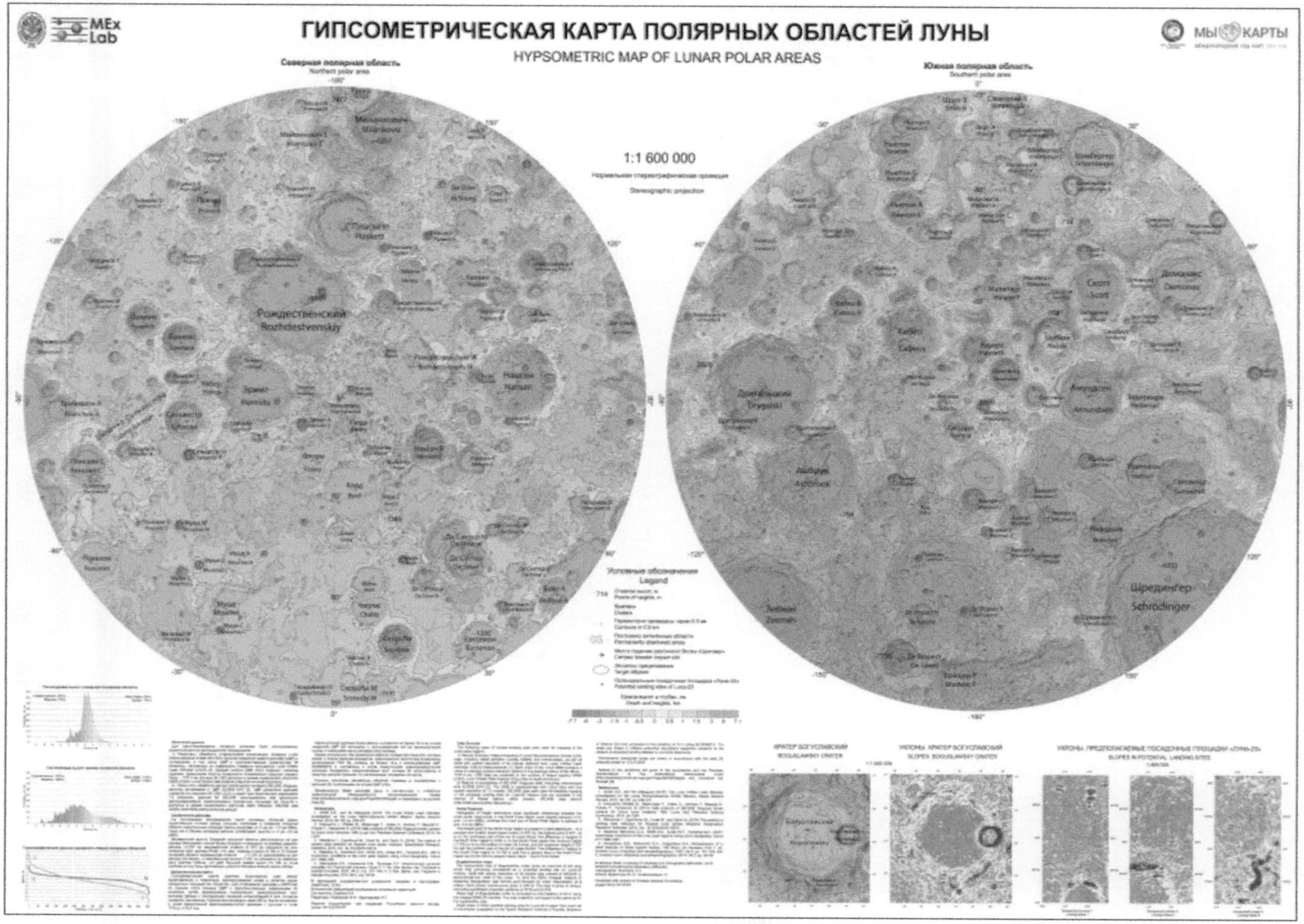

Figure 20.2. Printed version of bilingual hypsometric map of lunar sub-polar areas (published by MIIGAiK, 2016), original size is 1189 x 841 mm^2

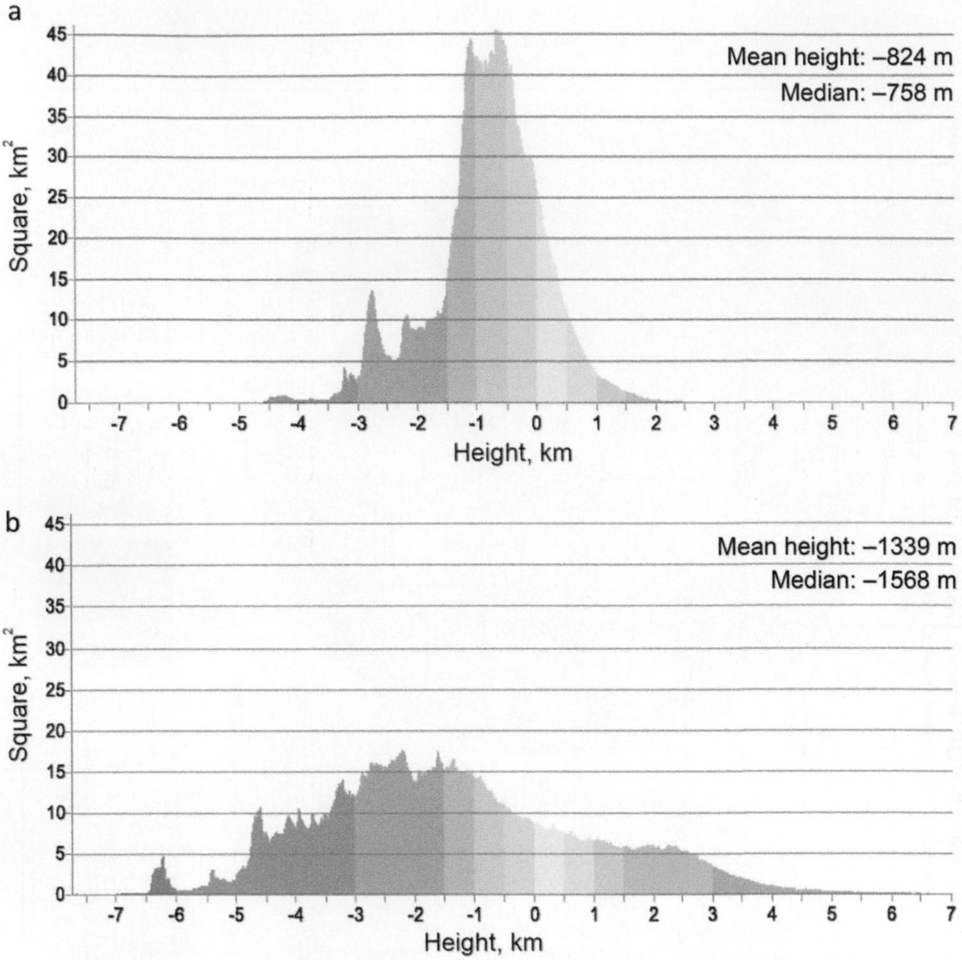

Figure 20.3. Histograms of heights: in northern (a) and southern (b) sub-polar areas

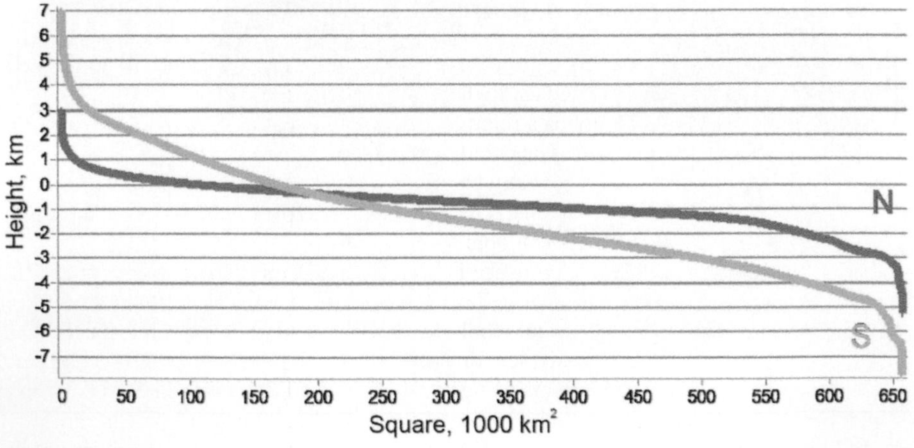

Figure 20.4. Hypsographic curves of lunar sub-polar areas: N (dark curve) for northern, S (light curve) for southern

is –1568 m. Thus the northern and southern sub-polar areas represent two types of mega-relief. The northern area is more flattish, while the southern area is occupied by slope of South Pole–Aitken basin.

4.3. *Mapping of crater density distribution*

Identification of craters as main relief features at celestial bodies is very useful for geology analysis. Crater measurements are a frequent component of the geological interpretation of surface age estimates by means of accumulated crater populations (Michael and Neukum, 2010). Various crater catalogues have been used to investigate morphological differences of planetary units at different scales, such as study of lunar maria on the local level with small craters 20–100 m, using LRO data (Basilevsky *et al.*, 2014) or mapping Moon polar area on the global level (Rodionova *et al.*, 2013) with big craters 10–600 km (Fig. 20.5, top), derived from morphological catalogue based on results of previous lunar mission that includes 15,000 objects (Rodionova *et al.*, 1987). Recently created global morphological catalogue with middle-size craters 1–10 km in diameter (Kozlova *et al.*, 2017) complements the catalogue (1987) and provides a new look in Moon sub-polar area. Big and middle craters were identified using Global orthomosaic WAC GLOBAL (Scholten *et al.*, 2012) with resolution 100 m px^{-1}. The maps, created by craters catalogues, show low density of medium-size craters (Fig. 20.5, middle) in regions with high density of big craters (Fig. 20.5, top).

5 MAPPING OF POTENTIAL LANDING SITES AT LOCAL LEVEL

To study PLS considering direct hazards for the spacecraft landing we used sufficient data: SLDEM, derived from SELENE (Kaguya) images, and LOLA data. It provides the possibility to characterize potential landing sites at the highest available spatial resolution (10 m $pixel^{-1}$ against 30 m $pixel^{-1}$ by LDEM at –68° latitude).

5.1 *GIS-tool for landing site selection*

For secure contact of landers with the surface it is necessary to conduct a comprehensive study of surface characteristics within selected landing sites. Based on criteria proposed for safety landing (Mitrofanov *et al.*, 2016) a special algorithm for estimation various parameters of surface was developed (Fig. 20.6). Using the Model Builder application the algorithm was integrated into ArcGIS (www.arcgis.com/features/index.html) as spatial tool for selection site more suitable for landing. To test tool the high-priority landing, ellipse 1 was chosen from 12 proposed ROIs (Mitrofanov *et al.*, 2016); see other suggested ellipses on Figure 20.10.

At the beginning the developed algorithm creates a map of slope distribution and diagram of slopes less than 7°. Slopes are calculated using the method suggested by Horn (1981). Then roughness is computed in two ways. Firstly as a relative topographic position (Tagil and Jennes, 2008), separating cratered and smoother areas and indicating crater edges, it is more suitable at local and regional scales, providing a map of relatively smooth areas. Another method based on Laplacian index (see Chapter 4.1) adequately characterizes the most typically occurring land forms and excludes extreme relief formations on the global level.

As fresh small craters are more dangerous for landing, to avoid this hazard, a catalogue of small craters should be created. As usual visual geology analysis is applied to estimate the degree of crater degradation. If DEMs with sufficient resolution which could be threatening for the lander's footing (up to 1–2 m $pixel^{-1}$) are available, for example, from LROC NAC stereo image processing, it allows detection of fresh craters using relative depth (depth/Diameter, d/D). Automatic selection of objects with depth to diameter ratio more than 0.15, as established in various lunar researches, for example (Basilevsky *et al.*, 2014), provides identification of the most dangerous fresh craters as part of the developed algorithm.

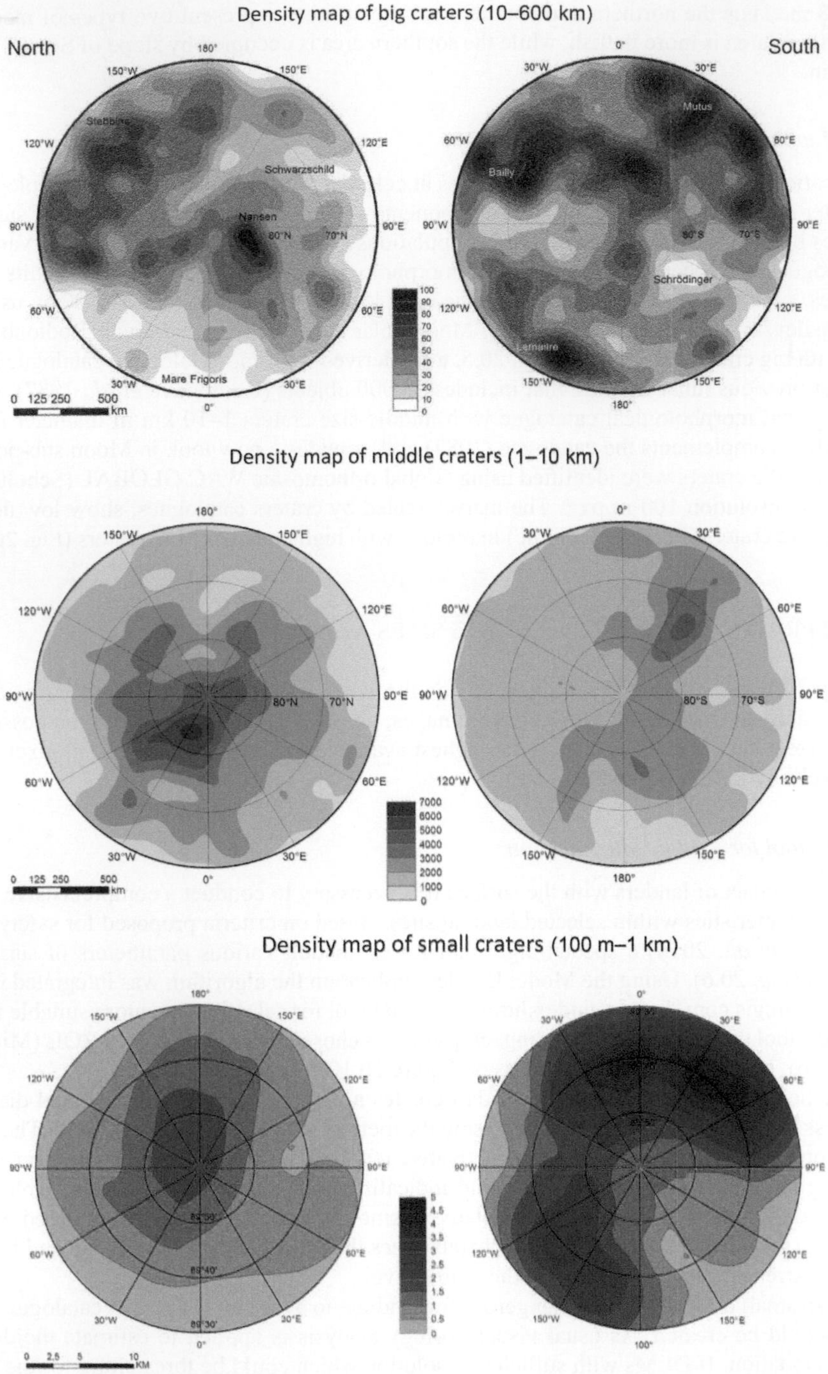

Figure 20.5. Crater density maps of Moon sub-polar area showing: number of big craters (10–600 km) per unit area, normalized to 100,000 sq. km calculated in neighborhoods of radius 150 km (top); number of middle craters (1–10 km), normalized to 100,000 sq. km calculated in neighborhoods of radius 150 km (middle); number of small craters (100 m–1 km) normalized to 700 sq. km calculated in neighborhoods of radius 15 km (bottom)

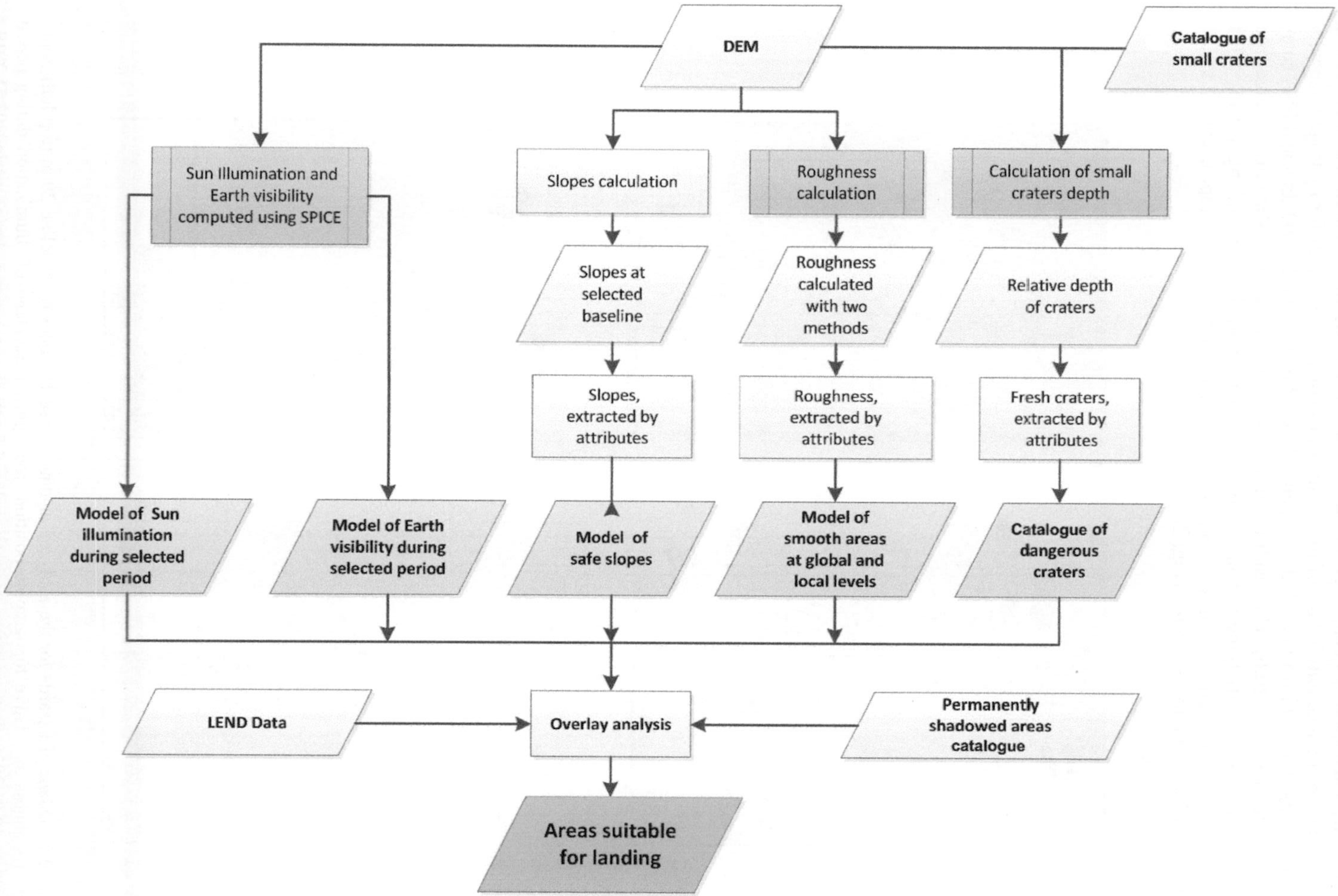

Figure 20.6. Automated GIS-algorithm developed for landing site characterization (*blue blocks – specially developed instruments, green blocks – derived data product*).

Based on SPICE tools (https://naif.jpl.nasa.gov/pub/naif/toolkit_docs/), the visibility of Sun and Earth was calculated using specially developed software "Compute Illumination" (Zubarev *et al.*, 2016). To support main scientific task of future lunar missions – the search for hydrogen deposits concentration in-situ, and also to exclude permanently shadowed area, LEND data and PSA catalogue are subtracted (Fig. 20.6). The result of automated analysis are maps showing the area free of potential hazards and satisfying various factors for scientific, geological and engineering tasks (Fig. 20.7). Then overlay analysis in ArcGIS provides overlap zone with safe slopes, smooth areas, good illumination and Earth visibility (Fig. 20.8).

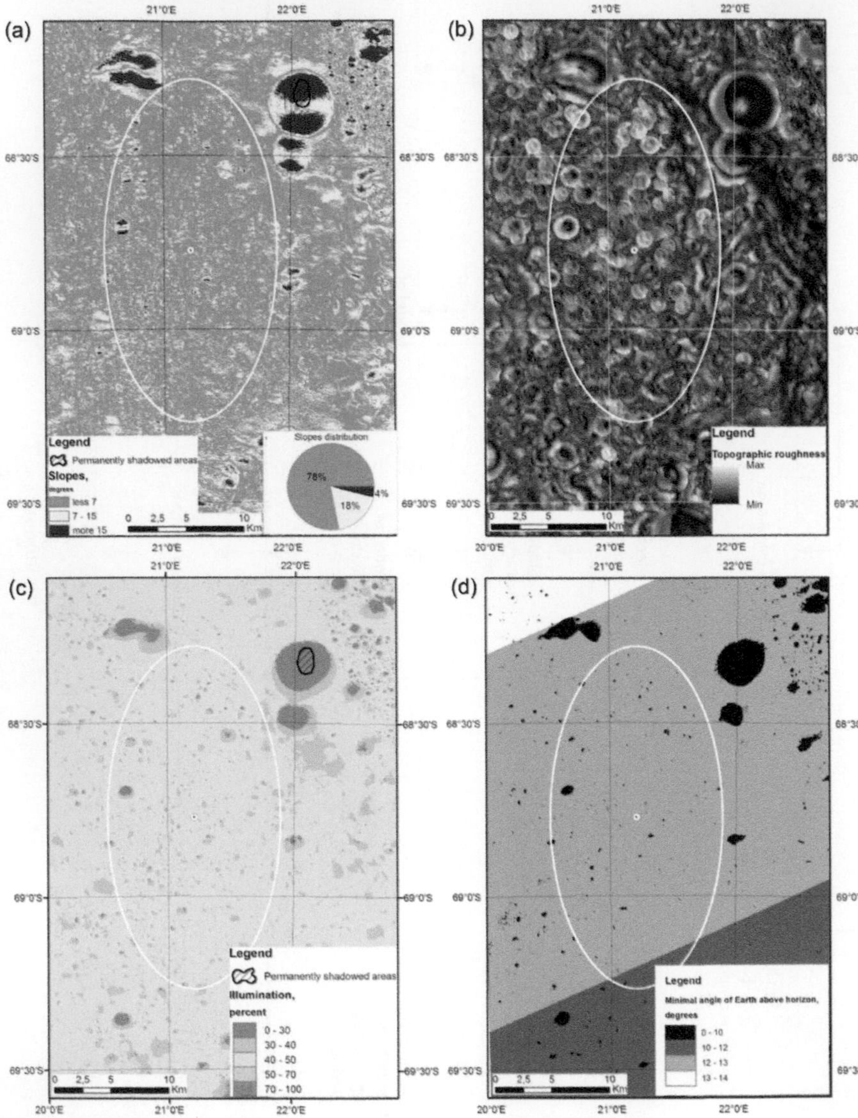

Figure 20.7. Maps of hazards for the analyzed landing ellipse 1: slopes on baseline 20 m (top left); topographic roughness as relative topographic position (top right); percentage of illumination during period 03.11.2019–30.01.2020 (bottom left); minimal angle of Earth above horizon during period 03.11.2019–30.01.2020 (bottom right)

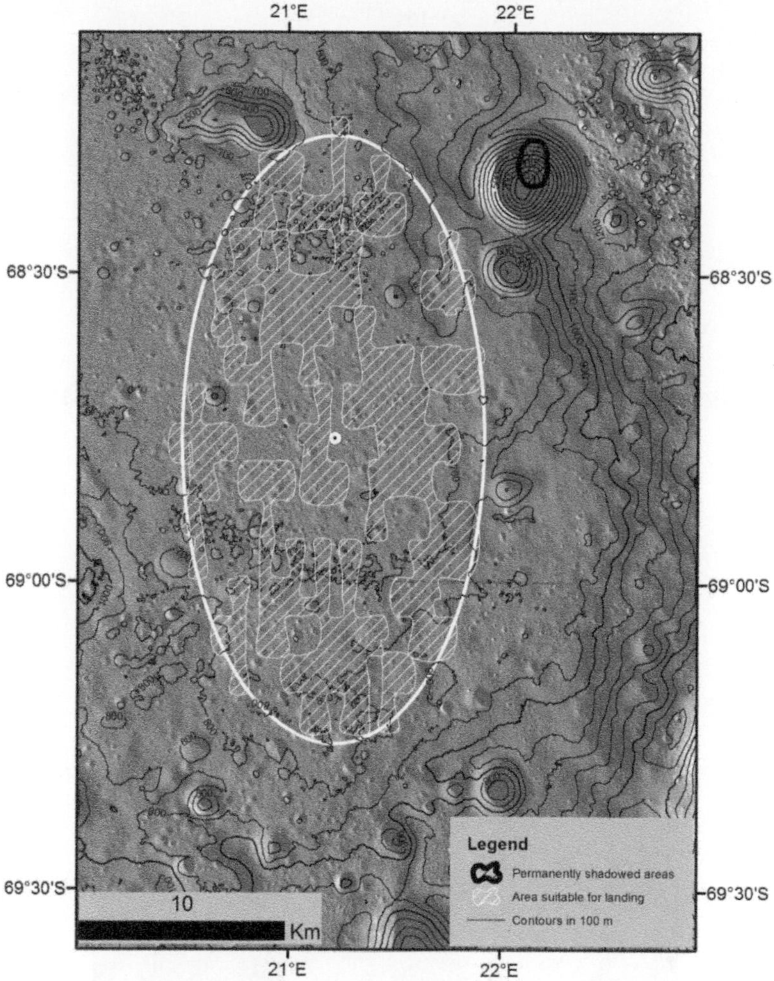

Figure 20.8. Topographic map of the landing ellipse 1 with results of overlay analysis. Background: Kaguya images TCO_MAP_02_S66E018S69E021SC, TCO_MAP_02_S66E021S69E024SC, TCO_MAP_02_S69E018S72E021SC, TCO_MAP_02_S69E018S72E021SC

5.2 *Mapping of the landing ellipse*

As a result of application of the algorithm, a set of maps characterizing the relief parameters of the studied surface, topographic profiles and safe areas within PLS was created. Maps of hazards inside PLS show critical areas for landing of spacecraft values: steep slopes, a rugged territory, low illumination, low angle of Earth above horizon. The map diagram shows (Fig. 20.7, top left) that 78% of the landing ellipse is safe in terms of the slope criterion, 93% is illuminated more than 40% of period 03.11.2019–30.01.2020 (Fig. 20.7, bottom right), 51% of territory is suitable by all criteria (Fig. 20.8). The central part of selected area as a more potential point was characterized in more detail: inspection of created small crater catalogue reveals that there are no fresh craters larger than 70 m in diameter inside this area (Fig. 20.9).

To consider the hazards of the loss of Sun and Earth visibility during the proposed period of the future mission Luna-25, we used LDEM and created cumulative maps at regional level

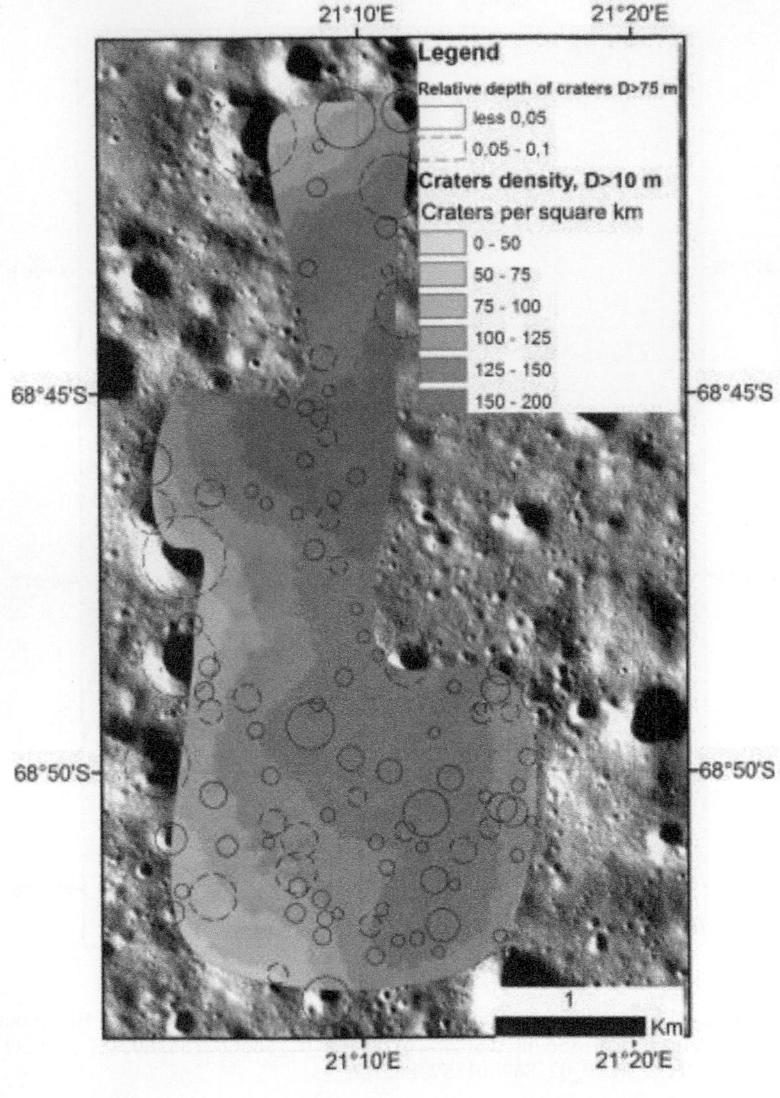

Figure 20.9. Map of small crater density (10 m to 1 km in diameter) at the central area of landing ellipse 1

(Fig. 20.10) to cover all ROIs – 12 ellipses were selected as a potential landing area. Maps show that Sun and Earth visibility are limited on south-oriented slopes of craters of the proposed landing area, 84% of the displayed territory is illuminated during 40% and more of period 03.11.2019–30.01.2020.

6 CONCLUSIONS

Using LRO and Selene (Kagyua) data we carried out morphometric analysis and mapping of the lunar surface. A set of maps for sub-polar areas, including proposed landing sites, was created. Maps at various scales are useful in providing a comparative geographic description for orientation,

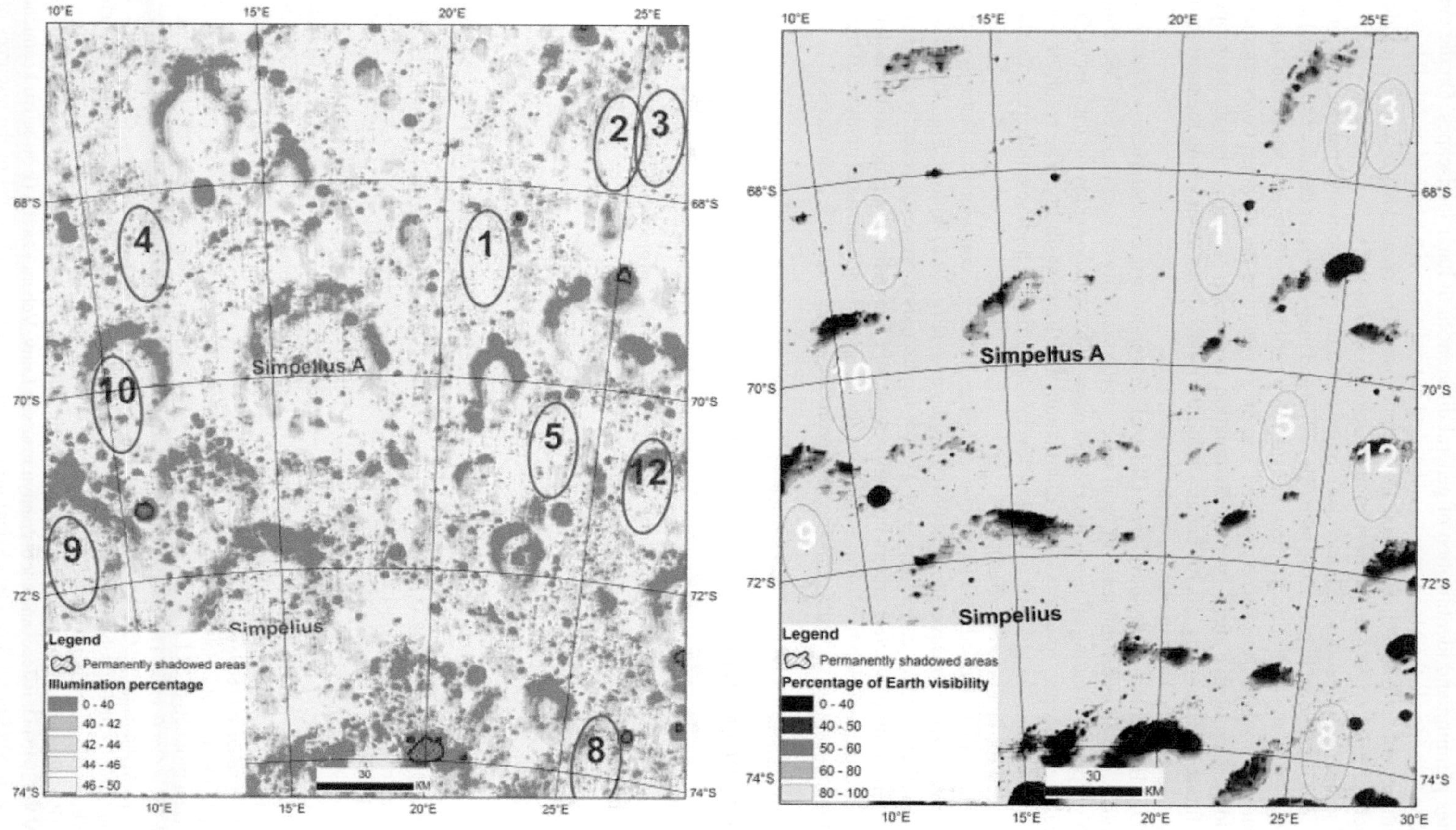

Figure 20.10. Cumulative maps of illumination (left) and Earth visibility (right) during period 03.11.2019–30.01.2020 for area of PLS based on LDEM

flight routes planning and identification of general differences in lunar landscapes. On the eve of new Russian missions to the Moon, maps can be used not only for scientific and practical tasks, but also for public outreach, as well as for educational purposes. The created maps will be included in the new school *Atlas of the Solar System* being developed in Russia.

Global maps show features of lunar mega-relief and their morphometric parameters for both sub-polar areas. Regional hypsometric map (Fig. 20.2) and maps of crater density for big and small craters (Fig. 20.5, top and bottom) show considerable differences between south and smoother north-polar areas, while craters with a size of 1 to 10 km more in the northern polar region (Fig. 20.5, middle). Cumulative maps of Sun illumination and Earth visibility during the proposed period of future missions (Fig. 20.10) demonstrated the possibilities of choosing the most suitable places for all supposed landing regions in south-polar area.

Using DEM and the specially developed GIS-tool, morphometric characteristics of surface can be calculated, as well as the influence of local relief on the visibility of Earth and Sun. We concentrated at high-priority landing ellipse 1 and created maps of hazards for studying dangerous factors at the local scale: slopes, roughness, Sun illumination and Earth visibility. The result of overlay analysis outlines zones in ellipse 1 with a low level of hazards and high scientific priority that are more preferred for landing (Fig. 20.8).

Nevertheless, main hazards are small craters inside PLSs. To consider this factor we created small crater catalogues for several parts of Moon polar area (Fig. 20.5, bottom; Fig. 20.9). As there is a great number of small craters inside of 17 new landing sites proposed recently (Sanin *et al.*, 2017), it is impossible to detect craters manually at each PLS in a limited time. So the task of our next work is further development of an automatic tool for spatial and statistical analysis of the lunar surface at local scales as an indirect method to predict the areas most free from craters.

ACKNOWLEDGEMENTS

This work was carried out at MIIGAiK Extraterrestrial Laboratory (MExLab). Kozlova N.A. was supported by Russian Foundation for Basic Research (RFBR), project No.16–37–00323, in the part of creation of middle-size lunar crater catalogue.

REFERENCES

Barker, M., Mazarico, E., Neumann, G., Zuber, M., Haruyama, J. & Smith D. (2016) A new lunar digital elevation model from the Lunar Orbiter Laser Altimeter and SELENE Terrain Camera. *Icarus*, 273, 346–355.

Basilevsky, A.T., Kreslavsky, M.A., Karachevtseva, I.P. & Gusakova, E.N. (2014) Morphometry of small impact craters in the Lunokhod-1 and Lunokhod-2 study areas. *Planetary and Space Science*, 92, 77–87. http://dx.doi.org/10.1016/j.pss.2013.12.016.

Binder, A.B. (1998) Lunar prospector: overview. *Science*, 281(4), 1475–1476.

Colaprete, A., Schultz, P., Heldmann, J., Wooden, D., Shirley, M., Ennico, K., Hermalyn, B., Marshall, W., Ricco, A., Elphic, R.C., Goldstein, D., Dummy, D., Bart, G.D., Asphaug, E., Korycancky, D., Landis, D. & Sollitt, L. (2010) Detection of water in the LCROSS ejecta plume. *Science*, 330(60033), 463–458. doi:10.1126/science.1186986.

Haruyama, J., Ohtake, M., Matsunaga, T., Otake, H., Ishihara, Y., Masuda, K., Yokota, Y. & Yamamoto, S. (2014) Data products of SELENE (Kaguya) terrain camera for future lunar missions. *45th Lunar and Planetary Science Conference*, The Woodlands, TX, USA, 17–21 March, abstract #1304.

Haruyama, J., Matsunaga, T., Ohtake, M., Morota, T., Honda, C., Yokota, Y., Torii, M., Ogawa, Y. & Lism Working Group. (2008) Global lunar-surface mapping experiment using the Lunar Imager/Spectrometer on SELENE. *Earth, Planets, and Space*, 60, 243–255.

Horn, B.K.P. (1981) Hill shading and the reflectance map. *Proceedings of the IEEE*, 69(1), 14–47.

Khartov, V.V. (2015) Ot issledovaniya k osvoeniyu resursov Luny. Vchera i zavtra (k 50-letiyu kosmicheskoi deyatel'nosti NPO imeni S.A. Lavochkina) (From research to the development of the resources of The

Moon. Yesterday and tomorrow [to the 50th anniversary of Space activity of NPO Lavochkin]). *Vestnik NPO im. S.A. Lavochkina*, 3, pp. 8–14 (in Russian).

Kokhanov, A.A., Kreslavsky, M.A., Karachevtseva, I.P. & Matveev, E.V. (2013) Mapping of the statistical characteristics of the lunar relief on the basis of the global digital elevation. *Current Problems in Remote Sensing of the Earth from Space*, 10(4), 136–153 (in Russian).

Kozlova, N.A., Zavyalov, I.Yu. & Kolenkina, M.M. (2017) Morphometric catalogue of lunar craters 1–10 km in diameter. *8th Moscow Solar System Symposium (8M-S3)*, Space Research Institute (IKI), Moscow, Russia, 8MS3-PS-14.

Kreslavsky, M.A., Zubarev, A.E. & Karachevtseva, I.P. (2016) Advanced global topographic mapping of the Moon: an important objective of upcoming lunar orbital missions. *7th Moscow Solar System Symposium*, Space Research Institute (IKI), Moscow, Russia, 7MS3-MN-24.

Kreslavsky, M.A., Head, J.W., Neumann, G.A., Rosenburg, M.A., Aharonson, O., Smith, D.E. & Zuber, M.T. (2013) Lunar topographic roughness maps from Lunar Orbiter Laser Altimeter (LOLA) data: scale dependence and correlation with geologic features and units. *Icarus*, 226, 52–66.

Mazarico, E., Neumann, G.A., Smith, D.E., Zuber, M.T. & Torrence, M.H. (2011) Illumination conditions of the lunar polar regions using LOLA topography. *Icarus*, 211, 1066–1081.

McClanahan, T.P., Mitrofanov, I.G., Boynton, W.V., Chin, G., Bodnarik, J., Droege, G., Evans, L.G., Golovin, D., Hamara, D., Harshman, K., Litvak, M., Livengood, T.A., Malakhov, A., Mazarico, E., Milikh, G., Nandikotkur, G., Parsons, A., Sagdeev, R., Sanin, A., Starr, R.D., Su, J.J. & Murray, J. (2015) Evidence for the sequestration of hydrogen-bearing volatiles towards the Moon's southern pole-facing slopes. *Icarus*, 255, 88–99.

Michael, G.G. & Neukum, G. (2010) Planetary surface dating from crater size-frequency distribution measurements: partial resurfacing events and statistical age uncertainty. *Earth and Planetary Science Letters*, 294(3–4), 223–229.

Mitrofanov, I., Djachkova, M., Litvak, M. & Sanin, A. (2016) The method of landing sites selection for Russian lunar lander missions. *Geophysical Research Abstracts*, EGU 2016, Vienna, Austria, Volume 18, abstract #EGU2016-10018.

Neumann, G.A., Smith, D.E. & Scott, S. (2011) Lunar Reconnaissance Orbiter – Lunar Orbiter Laser Altimeter. *ArchiveVolume – Software Interface Specification. Version 2.5.*

Robinson, M.S., Brylow, S.M., Tschimmel, M., Humm, D., Lawrence, S.J., Thomas, P.C., Denevi, B.W., Bowman-Cisneros, E., Zerr, J., Ravine, M.A., Caplinger, M.A., Ghaemi, F.T., Schaffner, J.A., Malin, M.C., Mahanti, P., Bartels, A., Anderson, J., Tran, T.N., Eliason, E.M., McEwen, A.S., Turtle, E., Jolliff, B.L. & Hiesinger, H. (2010) The Lunar Reconnaissance Orbiter Camera (LROC) instrument overview. *Space Science Reviews*, 150, 81–124.

Rodionova, Zh.F., Karlov, A.A. & Skobeleva, P.P. (1987) *Morphological Catalog of Craters of the Moon.* Moscow Lomonosov University, Moscow, Russia. p. 173 (in Russian).

Rodionova, Zh.F., Lazarev, E.N. & Karachevtseva, I.P. (2013) Comparison of impact crater populations in the lunar polar regions. *The Fourth Moscow Solar System Symposium*, Moscow, Russia.

Sanin, A.B. (2015) Water maps of the polar Moon: LEND data after 6 years on the lunar orbit. *6th* Moscow *Solar System Symposium*, Space Research Institute (IKI), Moscow, Russia, 6MS3-MN10.

Sanin, A.B., Mitrofanov, I. G, Litvak, M.L., Malakhov, A., Boynton, W.V., Chin, G., Droege, G., Evans, L.G., Garvin, J., Golovin, D.V., Harshman, K., McClanahan, T.P., Mokrousov, M.I., Mazarico, E., Milikh, G., Neumann, G., Sagdeev, R., Smith, D.E., Star, R.D. & Zuber, M.T. (2012) Testing lunar permanently shadowed regions for water ice: LEND results from LRO. *Journal of Geophysical Research*, 117, E00H26.

Sanin, A.B., Mitrofanov, I., *et al.* (2017) Potentially interesting landing sites around the South pole of the Moon. *8th Moscow Solar System Symposium*, Space Research Institute (IKI), Moscow, Russia, 8MS3-PG-08.

Scholten, F., Oberst, J., Matz, K.-D., Roatsch, T., Wählisch, M., Speyerer, E.J. & Robinson, M.S. (2012) GLD100: The near-global lunar 100 m raster DTM from LROC WAC stereo image data. *Journal Geophysical Research*, 117, E00H17.

Speyerer, E.J., Wagner, R.V., Robinson, M.S., Licht, A., Thomas, P.C., Becker, K., Anderson, J., Brylow, S.M. & Humm, D.C. (2016) Pre-flight and on-orbit geometric calibration of the Lunar Reconnaissance Orbiter Camera. *Space Science Reviews*, 200, 357–392. doi:10.1007/s11214-014-0073-3.

Tagil, S. & Jenness, J. (2008) GIS-based automated landform classification and topographic, landcover and geologic attributes of landforms around the Yazoren Polje. *Turkey Journal of Applied Science*, 8, 910–921.

Zubarev, A.E., Nadezhdina, I.E., Kozlova, N.A., Brusnikin, E.S. & Karachevtseva, I.P. (2016) Special software for planetary image processing and research. *The International Archives of the Photogrammetry, Remote Sensing and Spatial Information Science*, XLI-B4. pp. 529–536.

Planetary Remote Sensing and Mapping – Wu et al.
© 2019 Taylor & Francis Group, London, ISBN 978-1-138-58415-0

Chapter 21

Geoportal of planetary data: Concept, methods and implementations

I. P. Karachevtseva, A. S. Garov, A. E. Zubarev, E. V. Matveev, A. A. Kokhanov and A. Yu. Zharkova

ABSTRACT: To store results of planetary image processing and spatial surface analysis we are creating a Geoportal of Planetary Data. The first pilot version of the information system as a 2D WebGIS was based on Phobos data processed in frame of preparation for Phobos-Grunt mission. Starting firstly with a small Martian satellite, we have extended our system and now it is being developed as geodetic and cartographic node that collects scientific spatial products for the Moon, Mercury, Mars, Enceladus (Saturn system) and Jupiter's Galilean satellites Io, Europe, Ganymede and Calisto (http://cartsrv.mexlab.ru/geoportal/). In a huge data flow, it becomes impossible to exploit the whole obtained data for scientific research. To solve this problem, it is proposed to update the current version of Geoportal with new possibilities.

New Geoportal will allow full and effective exploitation of data for planetary research with web access and interactive online tools for high-level analysis of planetary data. The web platform is organized as an online 3D WebGIS portal. The web portal and applications have a unified basic interface, implemented using HTML markup. Based on the new software architecture several cross-platform solutions are developed that can be applied for information system "Digital Moon" proposed to integrate various collections of lunar data. It is of great importance because the Moon is a main focus of the Russian space program. The proposed system is related to the possibility of the comprehensive use of lunar data in a unified software environment, providing a qualitatively new level of planetary research where expert analysis and automated data processing complement each other. Some executable applications that have been created using new software architecture and developed infrastructure are presented.

1 INTRODUCTION

In the last decade, there has been a sharp increase in the amount of data obtained by planetary missions, as well as the rapid growth of existing planetary data archives, for example, a lot of spacecraft from different countries have been exploring the Moon lately. The data volume obtained so far by Lunar Reconnaissance Orbiter (LRO), launched in 2009, is impressive: more than 1.3 million images from the onboard Narrow Angle Camera (LROC NAC) (Robinson *et al.*, 2010), which corresponds to more than 600 terabytes (Keller *et al.*, 2016). Another LRO device – Lunar Orbiter Laser Altimeter (LOLA) – transmitted about 6.8 billion measurements to the Earth (Smith *et al.*, 2017).

Access to planetary data for scientific investigations requires a solution to the problems of data storage and archiving, as well as distribution, visualization and analysis (Oosthoek *et al.*, 2014). Providing to international groups the ability to search and access the data via the Internet (this is especially important for scientific cooperation in planetary sciences) requires the specialized online tools, capabilities for distributed processing, usage of commonly used standards, specialized both for planetary research and for organization of online access.

Nowadays, in order to produce high-level scientific data products, there is a need to update software and processing tools, as well as to save processing scenarios to reaffirm applying the same methods of analysis. These may become possible in specially developed systems, such as, for example, virtual observatories (Erard *et al.*, 2018). Also, various web services are to be organized to provide access to planetary data, including time-dimensional visualization to study dynamics of phenomena to a better understanding of the surface formation processes (e.g., Walter *et al.*, 2017). Specialized tools for geologic studies morphometric measurements and morphological analysis, auxiliary settings, joint processing and analysis and interpretation of data at different scales are to be developed.

GIS technologies become essential for direct planning of future planetary missions: for the landing site selection on the lunar surface for the Luna-25 lander (Djachkova *et al.*, 2017), European spacecraft to Mercury BepiColombo (http://sci.esa.int/bepicolombo/) or the international project (with the participation of Russia) EXOMARS (www.esa.int /Our_Activities/Space_Science/ Four_candidate_landing_sites_for_ExoMars_2018). The use of GIS methods and crowdsourcing in planetary science also works towards educational goals helping to attract future researchers and popularize space science, e.g. i-Mars project (www.i-mars.eu/). But also they are important for the solution of fundamental problems in planetary geology, providing new capabilities for data mining, search for new craters and the determination of the speed of crater degradation (Muller and Tao, 2016). Another issue in modern planetary sciences is the problem of integration of heterogeneous data, both new and archival. This problem becomes even more urgent considering the increasing array of new planetary information. It can be solved during photogrammetric processing by methods of co-registration (Sidiropoulos and Muller, 2016).

For decision of these tasks we suggest a concept of distributed communication environment using new software architecture (Garov *et al.*, 2016). The proposed methodological approach is based on ensuring the possibility of a comprehensive use of heterogeneous data within a single unified information environment. This is of great importance for establishing planetary research at a qualitatively new level, where expert analysis and automated approaches to data processing are complementary and are based on the use of existing standards used in planetary studies: the International Astronomical Union cartographic standards that support unification of the coordinate systems (Archinal *et al.*, 2011); the new standard PDS4 (Crichton *et al.*, 2013; Sarkissian *et al.*, 2016), developed to improve the storage and access to data archived in a Planetary Data System (PDS) format, as well as the Open Geospatial Consortium (OGS) web standards created for access and extracting spatial Earth-based information but also valuable for planetary data (Hare *et al.*, 2017).

2 SOFTWARE ARCHITECTURE

To update the current version of Geoportal we use the developed unified distributed communication software architecture for processing of spatial data which integrates web-, desktop- and mobile platforms and combines volunteer computing model and public cloud possibilities. The main idea is to create a flexible working environment for research groups, which may be scaled according to required data volume and computing power, while keeping infrastructure costs at minimum. It is based upon the "single window" principle, which combines data access via geoportal functionality, processing possibilities and communication between researchers. Based on this approach it will be possible to organize the research and representation of results on a new technology level, which provides more possibilities for immediate and direct reuse of research materials, including data, algorithms, methodology and components. The new software environment is targeted at remote scientific teams, and will provide access to existing spatial distributed information for which we suggest implementation of a user interface as an advanced front-end.

Based on an innovative software environment the developed planetary information system is being updated now as 3D WebGIS that will support online spatial data processing, analysis and three-dimensional data visualization. To implement the updated version of Geoportal as a specific interactive online scientific laboratory for planetary research, we offer several software solutions that provide: (1) exchange of scientific knowledge based on saving of processing scenarios; (2) organization of the collective workspace, which will allow joint processing of spatial data by remote scientific groups; (3) synchronous broadcasting of the spatial context in telecommunication mode, which allows organizing online discussions and workshops. The developed software architecture has the following characteristics:

- Cross-platform C++ code, provided as set of modules, which can be compiled to server or end-user (client) application, the latter being desktop or in-browser app (built using Emscripten toolchain);
- Message-based data exchange between modules via typed data channels abstraction; support of several network protocols for data exchange, such as inproc, IPC, TCP, WebSocket. WebRTC (Data channels); rejection of using single request/response (RPC) model in favor of more rich model with support of distributed requests with parallel execution and prolonged data collection, usage of push notification;
- Google Protocol Buffers (Protobuf) have been selected as a standard for serialization of messages between modules. This selection was due to a number of reasons: compactness (Protobuf is a binary protocol), support for the schema (possibility of formal message description without locking to concrete programming language), i.e. possibility of public API creation, existence of bindings to wide range of programming languages (which thus can be used for implementation of functional modules);
- Unified 2D/3D graphic interface is implemented using modern graphic API based on shaders (OpenGL/WebGL) with supplied common set of instruments which operate in both modes (2D/3D);
- Application user interface (UI), such as menu system, forms, is based on HTML and architecturally can be described as hybrid MVVM. The specialty of this UI is that JavaScript is not directly used, as well as no one of existing JavaScript frameworks/libraries (such as Angular, React, or VueJS). These libraries make development of UI easier, providing the so-called data-binding mechanism between presentation (View), implemented as HTML markup, and ViewModel, which usually prepares required data in client JavaScript code. However, specifics of 3D WebGIS application implies that besides representing data in HTML in textual form, we need to transfer a large amount of related data to GPU for rendering of 2D/3D scenes, and thus we also need data binding for that (especially if we are going to manipulate them visually using tools or read data back). Following MVVM paradigm the 2D/3D scene structure in code is representing yet another ViewModel where output graphic canvas represents View. Therefore we decided to use single data-binding mechanics, implemented in 3D-engine, which have been extended and modified to provide HTML generation, including form support (for user input parameters). As directives that guide HTML generation, we use a set of control attributes taken from VueJS, however data binding occurs with code written in C++ (to be correct with DataProviders, implemented in C++). Using this approach we succeeded in getting rid of the necessity to carry full JS runtime (e.g. NodeJS) together with its own system of packages and dependences for desktop application.

Using this architecture and changing packaging, different program solutions can be created. Packaging options vary from single tool/body thematic application to full-blown data portal with rich capabilities. While at the development stage we are more inclined to opt for the first option with plans for further integration. However, even single-instrument thematic application needs to provide some core WebGIS functionality such as fetching and visualization of context spatial data. The current developer version of whole Geoportal is available at the following link (http://cartsrv.mexlab.ru/portal.v3.dev).

The overall architecture of a Planetary Data Geoportal, which is being developed as 3D WebGIS, is depicted in Figure 21.1.

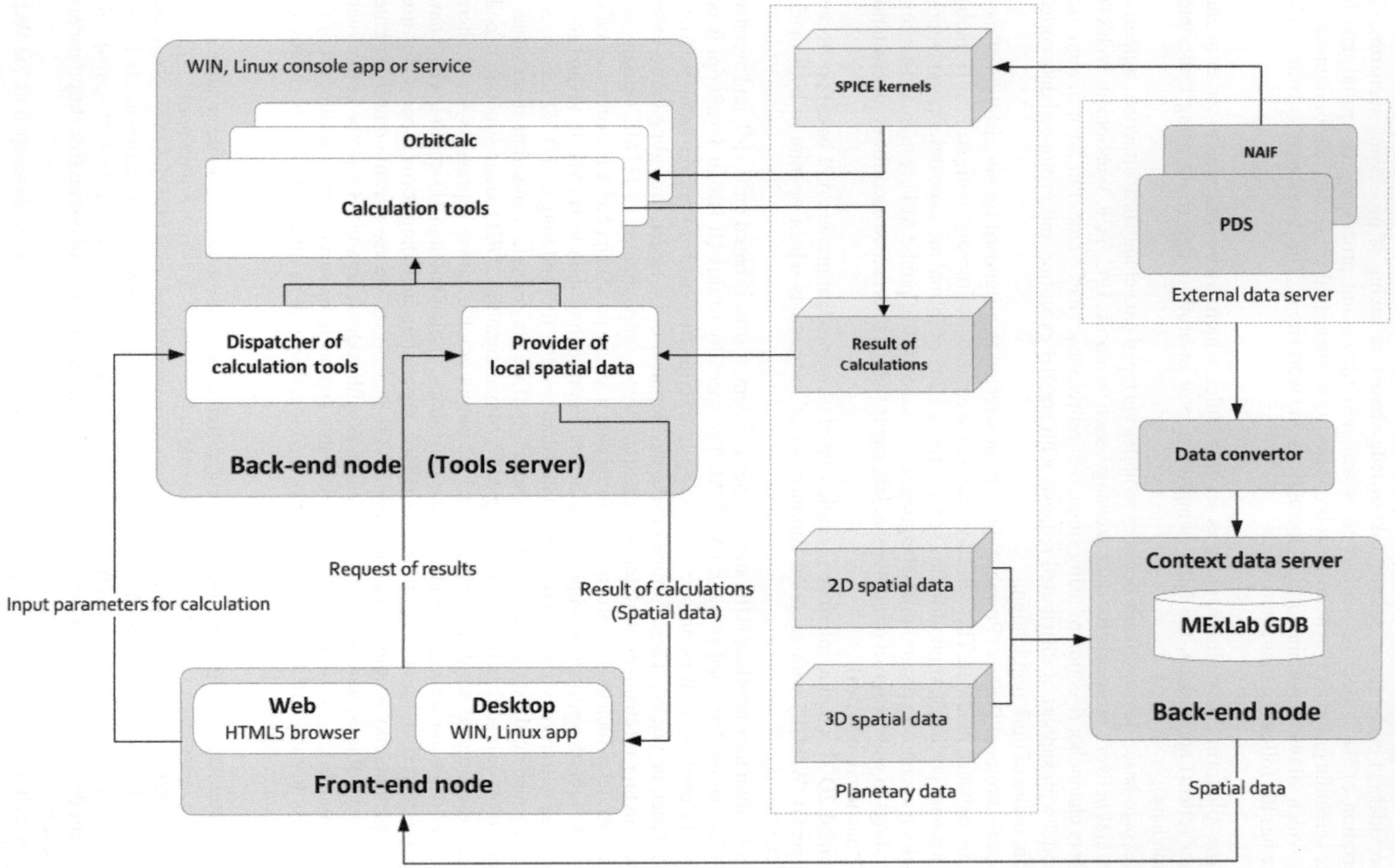

Figure 21.1. Software architecture of Planetary Data Geoportal illustrated on thematic application "ObitCalc" (see for details paragraph 3.1)

Structurally Geoportal represents a network of multiple nodes (Fig. 21.1) where every node belongs to one of three types: (1) a context data server back-end, which supplies existing context data; (2) tools server back-end, which provides processing (orbit calculation in case of "OrbitCalc"), data modeling and update procedures; (3) a client front-end, running in a browser (or a desktop) and providing a user with an interface and visualization of spatial data.

2.1 *Context data server back-end*

Context data server back-end node serves WebGIS basic data needs. It supplies front-end with available spatial data presented as raster (geo-referenced grids or images transformed to one of supported projections) and vector layers. The data are taken from MExLab geodatabase (GDB), which is in turn taking the data from external sources using Data Converter. The currently implemented version of context data server node is built using new architecture described above, but functionally represents a program wrapper over previously developed back-end of MIIGAiK Planetary Data Geoportal (Karachevtseva *et al.*, 2014) and utilizes its GDB and application servers. Back-end uses MS SQL Server as data storage layer. Application server has been developed in-house using. Net platform and supports OGC WFS and WMTS specifications.

Currently we are working on a new version of Context data server that will support operation with planetary data storages, in particular PDS, using transparent indexing of metadata and providing the ability to access and use selected data. This version of the server uses the basic structure of PDS online catalogs (PDS Data Volumes). Supporting current initiatives to standardize access to planetary data, a recent review of which is presented in (Hare *et al.*, 2017), we plan to provide interoperability not only with PDS archives, but to other online data sources (portals) using updated and re-implemented OGC protocol set.

2.2 *Tools server back-end*

The structure and functionality of the tools server back-end is described on the sample of "OrbitCalc" application (see Fig. 21.1 and section 3.1):

- Dispatcher module accepts from user a set of input parameters (see Fig. 21.3 below) for calculation such as orbiter's initial position and speed vectors, camera parameters (for example, fields of view), shooting period and interval between shots and region of interest for shooting coverage presented as rectangle area (bounding box). The module packs these parameters, forming a calculation task that is transferred to one of the workers and thus initiates calculation.
- Local vector data server (provider) takes from calculation module the resulting array of data, containing orbit position and corresponding projected coverage of the image (footprint), and transforms them into set consisting of two vector layers – 3D orbit vector layer and 2D surface footprints vector layer, and makes the set available to the front-end.
- Local raster server/tile-server (provider) takes resulting coverage data from the calculation module, forms the coverage raster containing the image overlap count as the raster's data. After that it performs slicing of the whole coverage into tiles of different levels of detail, compresses the tiles to adapt them for web transfer (the tile generation can be done using GPU-acceleration or by just software rendering using osmesa) and announces the new raster layer to the front-end.

2.3 *Front-end*

Front-end provides the user interface to system and can be implemented as in-browser or standalone desktop application, featuring the same appearance. When developing the front-end we followed traditional GIS interface layout familiar to users, which consists of an adjustable content menu with layer information, including legend, main graphical working area where spatial data are visualized and operated upon with respect to observer position and number of service windows. Core services supported in working area include: separate and joint visualization of layers, multiple object selection, pixel data retrieval, transparency, legend adjustment and availability in both 2D/3D modes. They operate using the same graphic API, where GPU builds a graphic scene consisting of reused blocks of quad geometries which are transformed dynamically using OpenGL Shading Language (GLSL) shaders and "stitched" for seamless surface.

To provide performance 3D visualization, we used a fast extensible graphics "engine" written in C++ programming language, which runs natively in the case of a desktop client or cross-compiled to ASM.JS or WebAssembly for an in-browser app.

Front-end can be extended to include new thematic tools which can be operating locally (e.g., as filters) or remotely using different thematic tool servers, e.g., for landing site selection or terrain modeling. Mapping functionality of Geoportal with new tools will provide data presentation with cartographic quality: multi-scale visualization of planetary data and labels on-demand, extended cartographic design (different hypsometric scales with various steps, new types of symbols for presentation of unique features on planetary surface, etc.), support of various types of cartographic projections (conformal, equal area and equidistant), etc. Proposed concept provides not only quality data presentation and wide dissemination of planetary research, but a more flexible approach suitable for users: with interactive spatial tools for analysis and dynamic cartographic design, it will be possible to create high-quality online maps for better understanding of processes on planetary surfaces.

3 APPLICATIONS

Using the proposed software architecture, we are developing various applications for the processing, analysis and three-dimensional visualization of spatial data for presentation of scientific results. Accomplishment of the software architecture and further development of special tools for thematic tasks can form a kind of "crystallization center" – a collection point for research and projects in the field of geosciences, ensuring the next stage in the evolution of GIS systems with the organization of free Internet access to the results of thematic processing.

3.1 *Image shooting planning*

The exploration of the lunar sub-polar regions is pivotal to the Russian Lunar program: lander Luna-25 (Luna-Glob) planned to launch to 2019; then will start orbital spacecraft Luna-26 (Luna-Resurs-1) scheduled for launch in 2020. An important task of the orbiter mission is the high-resolution topographic mapping of the descent areas of the next Russian Luna landers (Luna-27–29). Stereo imaging in the Luna-Resurs-1 mission will be performed with the Lunar Stereo Topographic Camera (LSTC), currently under development in the Space Research Institute of the Russian Academy of Sciences (Polyansky *et al.*, 2017). Image shooting planning is part of mission preparation to provide requirements for further stereo processing and the creation of digital terrain models for topographic mapping.

To support image shooting planning we have developed web instrument "OrbitCalc" (http://cartsrv.mexlab.ru/orbitcalc/). The application (Fig. 21.1) is based on an algorithm for the calculation of a spacecraft orbit's parameters and the coverage of images using SPICE libraries and kernels. The calculation module of "OrbitCalc" is written in C ++ and compiled as a DLL library, which is accessed by the tools server.

In the current version of the web application, the calculation of orbits and imaging parameters for the Moon and Mercury is realized (Fig. 21.2). NASA's orbital missions Lunar Reconnaissance Orbiter (LRO) and MESSENGER provided data for testing the developed algorithm using real orbit parameters of these spacecrafts (Figs. 21.3–21.4). "OrbitCalc" supports calculation and visualization of the three-dimensional coordinates of the spacecraft orbit not only for current or already

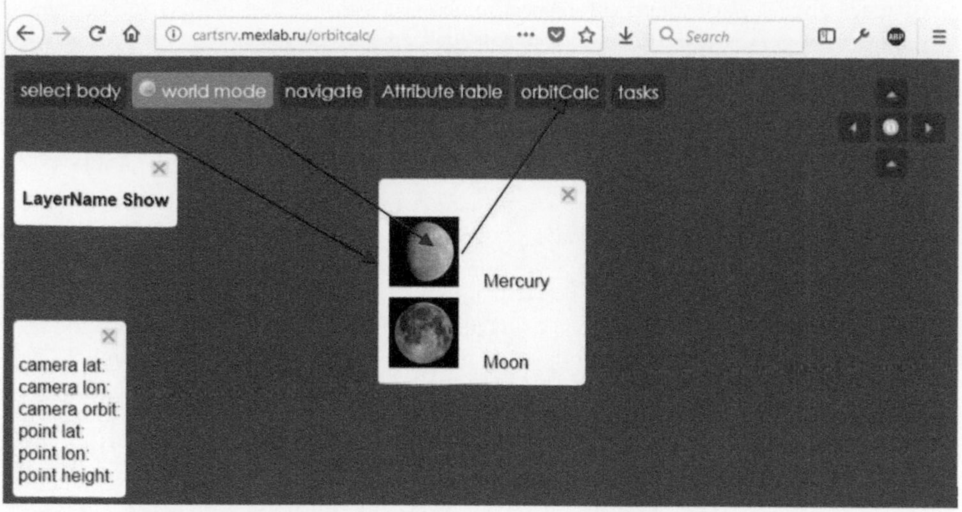

Figure 22.2. "OrbitCalc" user interface: select planetary object and visualization method – 2D (map mode) or 3D-View (world mode)

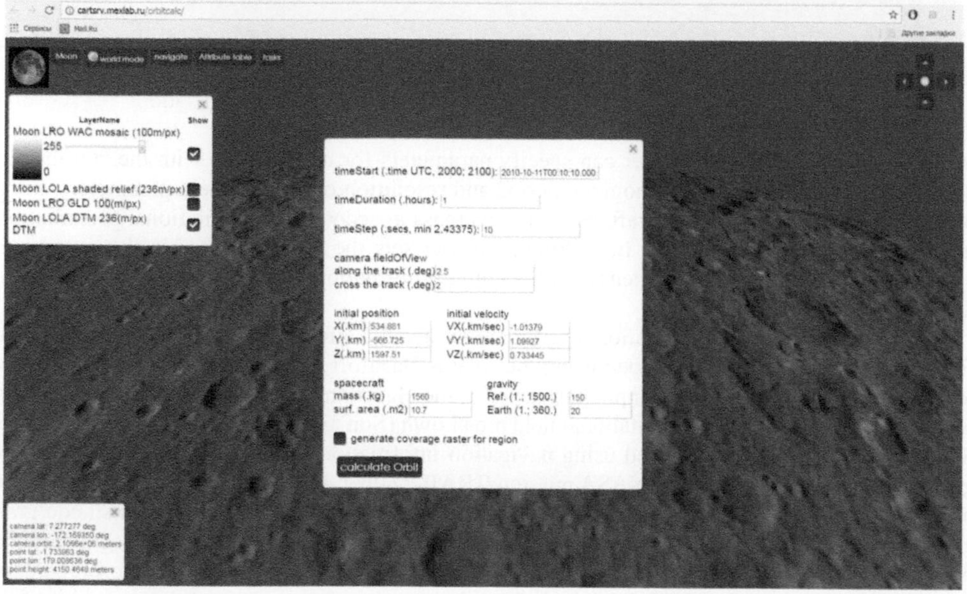

Figure 21.3. "OrbitCalc" user interface: enter initial parameters for calculation of spacecraft orbit and image coverage

completed missions for the selected time in the past, but for modeling using new camera parameters at times in the future, e.g., to planning imaging by LSTC for Luna-26 (Fig. 21.5).

When the interactive web application is loading, after the user selects the celestial body of interest in the interface menu with the SelectBody command (the list of available bodies is specified in the configuration file), the application sends a broadcast request to all registered services with a request for information about raster/vector layers available for the selected body, as well

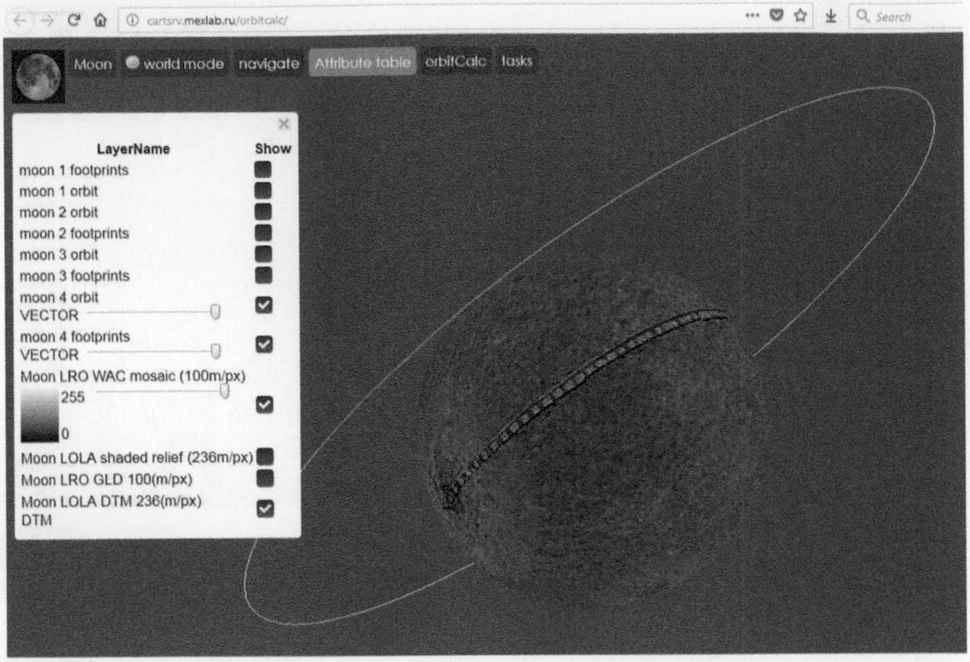

Figure 21.4. Example of orbit and image footprint modeled using web application "OrbitCalc"

as available server tools. The user can specify parameters for calculations with the "OrbitCalc" command, including the initial moment, period and resolution of data, camera parameters, initial orbits and velocity of the spacecraft, as well as its mass to account for gravitational perturbations and the effect of solar pressure. In addition, the user sets the coverage area in the geographic coordinate system and the discreteness of the output information, e.g., step of the regular grid in degrees (Fig. 21.3).

To take into account the gravitational perturbations and the effect of solar pressure, an algorithm for integrating orbits was developed using the Adams-Bashfort method of the fifth orders (Montenbruck and Gill, 2001). As an input parameter of the algorithm, a celestial body is specified for which the parameters of the gravitational field are known (Sun and Mercury, Venus, Mars, Jupiter, Saturn). The algorithm was tested using navigation information of the LRO, because the Moon gravity now is well known from NASA mission GRAIL (Zuber *et al.*, 2013). For testing the model of the gravitational field of the Moon up to 100 orders (http://pds-geosciences.wustl.edu/grail/grail-l-lgrs-5-rdr-v1/grail_1001/shadr/) was used, as well as a model of the Earth's gravitational field up to 20 orders (https://geod.ru/projects/data/gao2012/). From the comparison orbit parameters of LRO calculated in "OrbitCalc" with the reference data (parameters of the real orbit of the LRO), an accuracy about 50 m was obtained in the interval of 24 hours, which allows predicting the motion of the spacecraft according to the requirements of future Russian orbital mission Luna-26.

After the calculations, the server informs the user about the presence of the results in the form of new vector and raster layers, e.g., Moon footprints, Moon orbits (Fig. 21.4), that can be visualized in overlay mode with available base map layers, for example, global mosaic and DEM derived from LRO (Scholten *et al.*, 2012). Since the calculation of the orbit and image cover can take a long time, the user is informed of the current progress for the calculation instance.

Using the "OrbitCalc" computing algorithm, calculations were performed to estimate the coverage of the Moon surface by LSTC image footprints for the future Russian orbital mission Luna-26. "OrbitCalc" provides an estimation of overlapping between images, which is required in digital

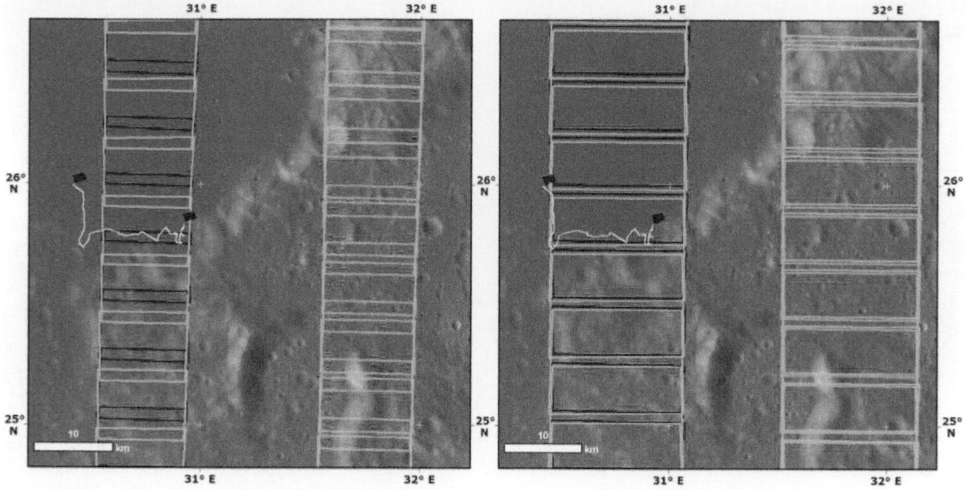

Figure 21.5. Examples of modeled image footprints from two adjacent 50-km (left) and 70-km (right) orbits: The footprints of the forward camera of LSTC system are selected by red colors and the backward camera by blue colors (Polyansky *et al.*, 2017)

photogrammetry to create DEMs. Stereo image footprints modeled by "OrbitCalc" covering the Luna-21 landing site and the Lunokhod-2 track are shown in an enlarged view on Figure 21.5 (background is LRO WAC mosaic).

3.2 *Three-dimensional visualization and terrain modeling*

We implemented the created software architecture and developed Geoportal infrastructure to demonstrate the global characteristics of the terrestrial planets and other celestial bodies in three-dimensional form. One application is the virtual morphometric globes (http://cartsrv.mexlab.ru/virtualglobe/), which show some terrain parameters in raster format for the Earth, Mars and the Moon (Florinsky and Filippov, 2017). For globes the set of the local, nonlocal and combined morphometric variables were preliminarily computed from DEMs using algorithms described in (Florinsky, 2016). Recently the same parameters were also calculated for Mercury (Florinsky, 2018) based on global DEM (with resolution of about 665 m pixel^{-1} at the equator) obtained from MESSENGER images (Becker *et al.*, 2016). The study of Mercury using the latest MESSENGER data is very important for comparative analysis of the Moon and Mercury's surfaces. We better understand the characteristics of the Moon when exploring the features of Mercury because both celestial bodies – atmosphereless and heavily cratered – have very similar relief.

Morphometric parameters can be calculated in different ways, and it depends on the tasks. For example, earlier global topographic roughness of the Moon was calculated based on Lunar Orbiter Laser Altimeter (LOLA) data (Kreslavsky *et al.*, 2013). We also estimated a global lunar roughness, but using *Laplacian* or interquartile range of the second derivative of heights (Kokhanov *et al.*, 2013). It was successfully implemented to characterize and mapping of the Moon sub-polar area (Kokhanov *et al.*, 2019). To compare the Moon surface with Mercury's surface, in this morphological study besides Laplacian obtained from global MESSENGER DEM mentioned above, additionally we used DEM on Mercury H-6 quadrangle (Kuiper) with resolution ~222 m pixel^{-1} (Preusker *et al.*, 2017). It was processed by another method – as *Relative topographic position* (RTP), which is identified each pixel with respect to its local neighborhood (Jenness, 2006).

Both parameters of terrain roughness characterize the surface at different scales: Laplacian works better on the global level and gives the average notion of planetary relief showing old hidden

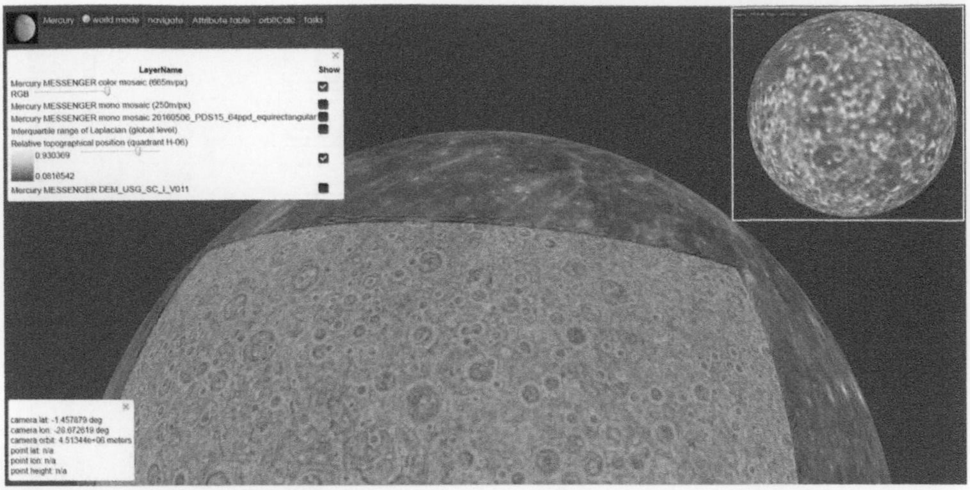

Figure 21.6. The Mercury morphometric parameters in 3D-View of Geoportal (world mode): Relative topographical position for H-6 quadrangle (Kuiper) and global Laplacian (top right corner)

structures that are not visible on images, and RTP excellently distinguishes the crater rims and plains (concave/convex objects) at a detailed level (Fig. 21.6). Results of Mercury terrain modeling are useful for identifying types of landscapes, prevailing geomorphological processes and defining boundaries of surface types as well as for comparative analysis with the Moon. Comparison of Mercury's surface with lunar relief parameters allows detection of similar patterns or differences in the evolution of the bodies (Kreslavsky and Head, 2015).

Mercury global Laplacian and RTP for H-6 quadrangle (Kuiper) pre-calculated using ArcGIS tools (Kokhanov *et al.*, 2016 and Jenness, 2006, correspondingly) and available on Geoportal via the Context Data Server, as well as the Moon global Laplacian derived from GLD-100 (Kokhanov *et al.*, 2013) and topographic roughness of Mercury's north sub-polar area obtained from the Mercury Laser Altimeter (Kreslavsky *et al.*, 2014). In spite of terrain modeling of planetary surface still being carried out on the server from the various DEM, the architecture and infrastructure of Geoportal provide not only interactivity of 3D-visualization, but also will support online calculation. Using a cross-compiled C++ engine in a browser gives the possibility to inject into it custom processing algorithms which calculate terrain parameters. Moreover, we can achieve unification of parallel (shader) and sequential (C++) parts of an algorithm by using GLM math library. As a result a set of filters will be developed for online terrain modeling and calculations of morphometric parameters based on DEM.

3.3 *External plug-ins of Geoportal*

In addition to proposed applications for research tasks, specialized online services and telecommunication capabilities provide effective organization of the education process using the obtained scientific results as methodical aids (digital data products, morphological maps, ready scenarios for thematic data processing). These possibilities can be integrated to Geoportal as extended plug-ins.

3.3.1 Translation of spatial context in teleconferencing mode

Based on adaptation of existing software and network libraries, a plug-in for communication of the scientific community and educational purposes (online lectures and presentations) was created (Fig. 21.7). Incorporation of teleconferencing mode provides online user collaboration in a single software environment using joint spatial context. We built the plug-in following architectural

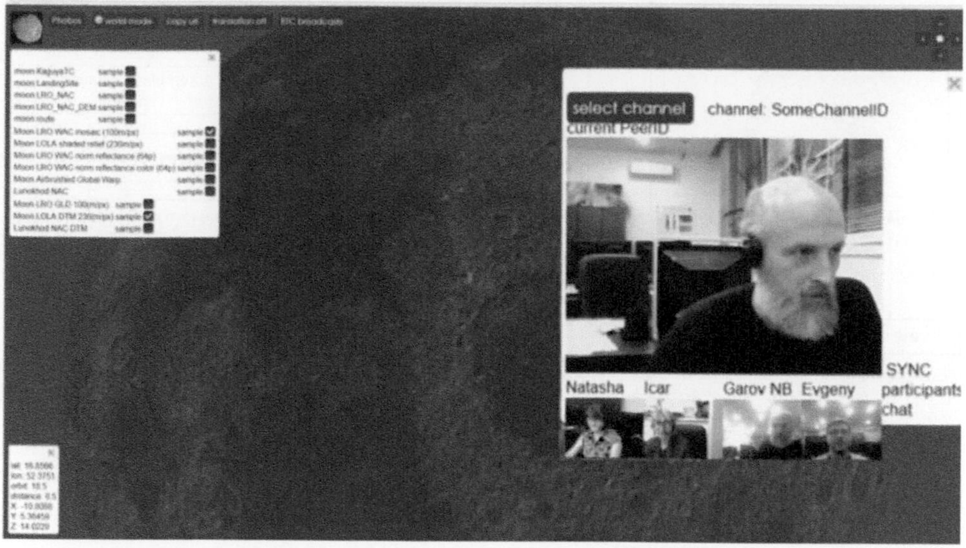

Figure 21.7. Online communication within a single software environment based on joint user three-dimensional spatial context

guidelines described in section 2. Additional design decisions, which are specific for this plug-in, are as follows:

- Restriction of using HTTP, XHR in favor of WS/WSS and WebRTC (HTTP is used to load the user interface).
- Scripting support – running external scripts and animation written in a simple language (e.g., Lua). It will be possible to transmit a static (set of layers/position of the observer) and dynamic (position script/animation) context through URL or via message.
- Teleconference regime (including video/audio broadcasting) – The presence of the scripting and WebRTC use for communications provides an opportunity to implement such a regime in a peer-to-peer mode. Scripting provides context synchronization between the speaker and other participants.

Implementation of this plug-in provides domain aware GIS-specific telecommunication solution and scenario recording to portal users. Using this solution, we can obtain better collaboration and learning experience, providing a level of interaction which cannot be reached using existing general-purpose communication software.

3.3.2 Intelligent search of external planetary sources

The necessity of a special tool for the intellectualization of search services is caused by the huge volume and diversity of information accumulated in various scientific fields, and in particular, in planetary research. The functionality of the searching web service is implemented via the spatial and semantic context of planetary data on the online map (Fig. 21.8).

Neural network methods were used to intellectualize the search algorithm. Neural network is extremely effective for search operations on the Internet (Ferreira, 2006). Also, neural network methods provide quick learning even on partial data, filter "noises" of entry parameters and easily adjust to changes in these parameters (Graves *et al.*, 2014). With neural networks, the list of interrelated entry parameters which are coherent with a user's request is created. The list is dynamic:

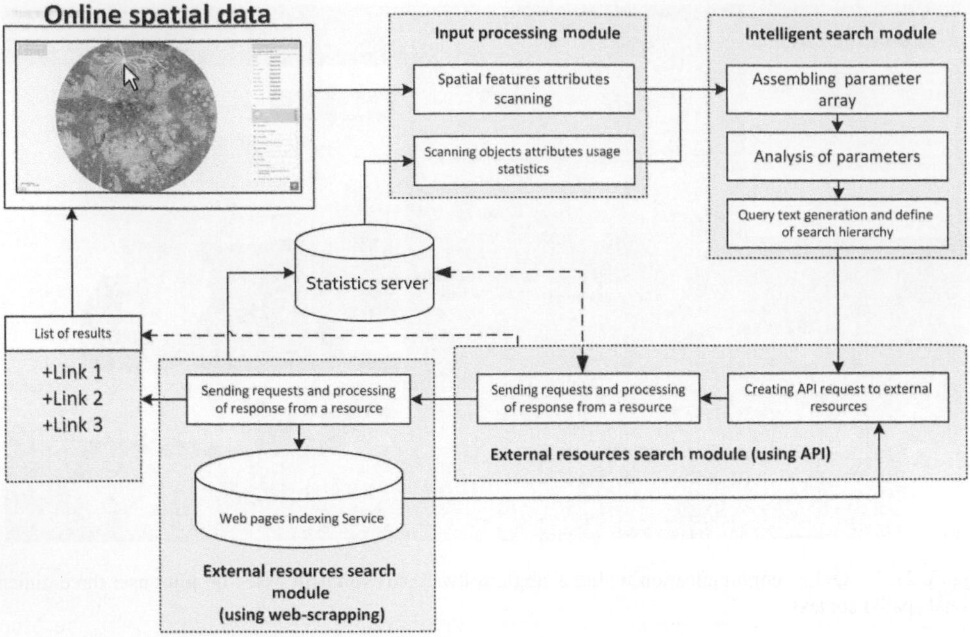

Figure 21.8. An overall scheme for the web-based intelligent search service using spatial and semantic context of planetary data

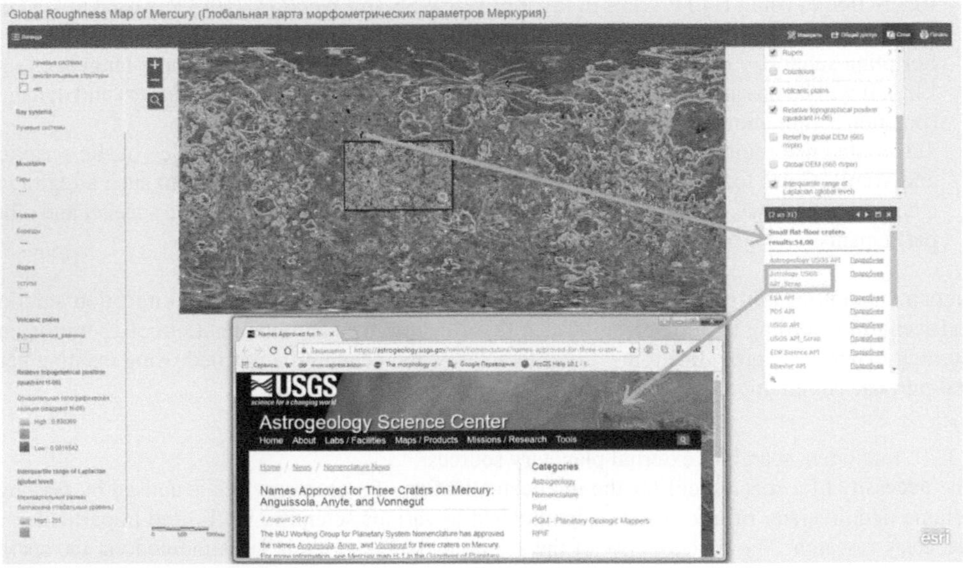

Figure 21.9. Example of online map of Mercury's morphometric parameters with integrated web service created for intelligent search of planetary information in external resources

the algorithm adjusts itself through machine learning methods and provides a user with more personalized data for every new request.

Various sources that provide access to the planetary information are selected and analyzed, such as NASA Planetary Data System (PDS), newswires of space agencies, including ROSCOSMOS and other foreign agencies, international publishers (Elsevier, Springer, EDP Sciences, etc.). To create the service, besides the API of selected external sources, we use web-scraping for thematic indexing of web pages. Search using API provides the user with all found information regardless of thematic and time request. The combination of methods provides access to the newest information, including data which better satisfy the user's request.

The created specialized web service was tested with different interactive maps created using ArcGIS online tools (e.g., http://bit.ly/2H78Qq9) and provides data search taking into account the spatial and semantic context of planetary data (see results on Fig. 21.9). The developed algorithm will be integrated into user interface of Geoportal to support the possibility of intelligent search of planetary information on external sources.

4 SUMMARY

The intensity of scientific lunar exploration at the present stage has led to the fact that the volume of obtained data has grown rapidly and continues to grow. However, in most cases only raw data from onboard instruments are published by spacecraft teams, leading to opportunities for wide further research based on joint processing and analysis of the remote sensing data. The Moon exploration is essential for the Russian space program. During next few years several launches are planned: the landing missions Luna-Glob (Luna-25, 27, 28) to the sub-polar areas (Zeleny, 2012) and an orbital mission Luna-Resource (Luna-26) (Petrukovich *et al.*, 2016), which is aimed at a global topographic mapping of the Moon. So, we also expect large amount of data: for example, a spacecraft in the quasi-polar circular orbit at an altitude of 50–100 km above the surface makes 12 flights around the Moon during the Earth's day, obtaining about 5000 images per day or about 150,000 images per month, assuming a typical lunar data transfer rate (Polyansky *et al.*, 2017).

Remote sensing data of the Moon, like any spatial planetary data, include arrays of space images and digital relief models, which are characterized by diversity, significant volumes and high complexity. Processing and analysis of such data requires specialized approaches, including online processing, various methods of categorizing and data integration, as well as joint analysis of heterogeneous information obtained from different sources. To solve these problems we have proposed the concept of "Digital Moon" for the integration of existing information as well as new data to be obtained in future and developed software architecture of Geoportal, including solutions for storage, archiving, joint processing, access, distribution, visualization and analysis of lunar data that can be used for implementation of an advanced front-end for the Russian segment of the planetary archive to collect and manage data from future missions (Batanov *et al.*, 2017).

ACKNOWLEDGEMENT

A.Yu. Zharkova was supported by Russian Foundation for Basic Research (RFBR), project № 18–35–00334.

REFERENCES

Archinal, B.A., A'Hearn, M.F., Bowell, E., Conrad, A., Consolmagno, G.J., Courtin, R., Fukushima, T., Hestroffer, D., Hilton, J.L., Krasinsky, G.A., Neumann, G., Oberst, J., Seidelmann, P.K., Stooke, P., Tholen, D.J., Thomas, P.C. & Williams, I.P. (2011) Report of the IAU Working Group on cartographic coordinates and rotational elements: 2009., *Celestial Mechanics and Dynamical Astronomy*, 109(2), 101–135. doi:10.1007/s10569-010-9320-4.

Batanov, O., Nazarov, V., Korotkov, F., Markov Ya., Konoplev, V., Melnik, A., Tretiakov, A. & Mischenko, A. (2017) Russian Science Ground Segment and IKI activities on receiving ExoMars 2016 science data. *6th European Ground System Architecture Workshop (#ESAW2017) at the European Space Operations Centre, 2017*, Darmstadt, Germany, 20–21 June.

Becker, K.J., Robinson, M.S., Becker, T.L., Weller, L.A., Edmundson, K.L., Neumann, G.A., Perry, M.E. & Solomon, S.C. (2016) First global digital elevation model of Mercury., *47th Lunar and Planetary Conference*, 2016, The Woodlands, TX, USA, 21–25 March, abstract #1903.

Crichton, D.J., Sarkissian, A., Hughes, J.S., Heather, D., Martinez, S., Beebe, R., Neakrase, L.D.V., Yu, Y. & Krishna, B.G. (2013) Towards an international planetary data standard based on PDS4., *44th Lunar and Planetary Science Conference*, 2013, The Woodlands, TX, USA, 21–25 March, abstract #1815.

Djachkova, M.V., Litvak, M.L., Mitrofanov, I.G. & Sanin, A.B. (2017) Selection of Luna-25 landing sites in the South Polar Region of the Moon. *Solar System Research*, 51(3), 185–195.

Erard, S., Cecconi, B., Le Sidaner, P., Rossi, A.P., Capria, M.T., Schmitt, B., Géno, V., André, N., Vandaele, A.C., Scherf, M., Hueso, R., Määttänen, A., Thuillot, W, Carr, B., Achilleos N, Marmo, C., Santolik, O., Benson, K. & Fernique, P. (2018) VESPA: a community-driven virtual observatory in planetary science. *Planetary and Space Science*, 150, 65–85. doi:10.1016/j.pss.2017.05.013.

Ferreira, C. (2006) Designing neural networks using gene expression programming. In: Abraham, A., de Baets, B., Köppen, M. & Nickolay, B. (eds) *Applied Soft Computing Technologies: The Challenge of Complexity*. Springer-Verlag, Berlin, Germany. pp. 517–536.

Florinsky, I.V. (2016) *Digital Terrain Analysis in Soil Science and Geology*. 2nd edition. Elsevier, Academic Press, Amsterdam, The Netherlands. p. 486.

Florinsky, I.V. (2018) Multiscale geomorphometric modeling of Mercury. *Planetary and Space Science*, 151, 56–70. doi:10.1016/j.pss.2017.11.010.

Florinsky, I.V. & Filippov, S.V. (2017) A desktop system of virtual morphometric globes for Mars and the Moon. *Planetary and Space Science*, 137, 32–39. doi:10.1016/j.pss.2017.01.005.

Garov, A.S., Karachevtseva, I.P., Matveev, E.V., Zubarev, A.E. & Patratiy, V.D. (2016) Development of heterogenic distributed environment for spatial data processing using cloud technologies. *The International Archives of the Photogrammetry, Remote Sensing and Spatial Information Sciences*, XLI-B4. pp. 385–390, XXIII ISPRS Congress, 12–19 July 2016, Prague. doi:10.5194/isprs-archives-XLI-B4-385-2016/.

Graves, A., Wayne, G. & Danihelka, I. (2014) Neural turing machines. *arXiv preprint arXiv:1410.5401*.

Hare, T.M., Rossi, A.P., Frigeri, A. & Marmo, Ch. (2017) Interoperability in planetary research for geospatial data analysis. *Planetary and Space Science*, 150, 36–42. doi:10.1016/j.pss.2017.04.004.

Jenness, J. (2006) Topographic Position Index (TPI) v 1.2. [Online]. *Jenness Enterprices*. Available from: www.jennessent.com/downloads/TPI_Documentation_online.pdf. [Accessed: 20 April 2018].

Karachevtseva, I.P., Oberst, J., Zubarev, A.E., Nadezhdina, I.E., Kokhanov, A.A., Garov, A.S., Uchaev, D.V., Uchaev Dm.V., Malinnikov, V.A. & Klimkin, N.D. (2014) The Phobos information system. *Planetary and Space Science*, 102, 74–85.

Keller, J.W., Petro, N.E. & Vondrak, R.R. (2016) The Lunar Reconnaissance Orbiter Mission – six years of science and exploration at the Moon. *Icarus*, 273, 2–24. doi:10.1016/j.icarus.2015.11.024.

Kokhanov, A.A., Karachevtseva, I.P., Kozlova, N.A. & Rodionova, Zh.F. (2019) The Lunar sub-polar areas: morphometric analysis and mapping. *Planetary Remote Sensing and Mapping, ISPRS Book Series*, Section VI: Planetary Data Management and Presentation, Chapter 20, CRC Press/Balkema, Leiden, The Netherlands..

Kokhanov, A.A., Kreslavskiy, M.A., Karachevtseva, I.P. & Matveev, E.N. (2013) Mapping of the statistical characteristics of the lunar relief on the basis of the global digital elevation model GLD-100. *Sovremennye problemy distantsionnogo zondirovaniya Zemli iz kosmosa (Current Problems in Remote Sensing of the Earth from Space)*, 10(4), 136–153 (In Russian).

Kokhanov, A.A., Bystrov, A.Y., Kreslavsky, M.A., Matveev, E.V. & Karachevtseva, I.P. (2016) Automation of morphometric measurements for planetary surface analysis and cartography. *The International Archives of the Photogrammetry, Remote Sensing and Spatial Information Sciences*, XLI-B4. pp. 431–433, 2016. doi:10.5194/isprs-archives-XLI-B4-431-2016.

Kreslavsky, M.A. & Head, J.W. (2015) A thicker regolith on Mercury. *46th Lunar and Planetary Science Conference*, The Woodlands, TX, USA, 16–20 March, abstract #1246.

Kreslavsky, M.A., Head, J.W., Neumann, G.A., Zuber, M.T. & Smith, D.E. (2014) Kilometer-scale topographic roughness of Mercury: Correlation with geologic features and units, *Geophysical Research Letters*, 41. doi:10.1002/2014GL062162.

Kreslavsky, M.A., Head, J.W., Neumann, G.A., Rosenburg, M.A., Aharonson, O., Smith, D.E. & Zuber, M.T. (2013) Lunar topographic roughness maps from Lunar Orbiter Laser Altimeter (LOLA) data: scale dependence and correlation with geologic features and units. *Icarus*, 226, 52–66. doi:10.1016/j.icarus.2013.04.027.

Montenbruck, O. & Gill, E. (2001) Book review: satellite orbits: models, methods and applications/Springer. *The Observatory*, 121(1162), 182.

Muller, J.-P. & Tao, Y. (2016) EU-FP7-iMARS: analysis of Mars multi-resolution images using auto-coregistration data mining and crowd source techniques: processed results – a first look. *The International Archives of the Photogrammetry, Remote Sensing and Spatial Information Sciences*, XLI-B4. pp. 453–458. doi:10.5194/isprs-archives-XLI-B4-453-2016.

Oosthoek, J.H.P, Flahaut, J., Rossi, A.P., Baumann, P., Misev, D., Campalani, P. & Unnithan, V. (2014) PlanetServer: innovative approaches for the online analysis of hyperspectral satellite data from Mars. *Advances in Space Research*, 53(12), 1858–1871. doi:10.1016/j.asr.2013.07.002.

Petrukovich, A., Zelenyi, L., Anufrejchik, K., Korablev, O., Mitrofanov, I. & Polyansky, I. (2016) Russian Lunar orbiter mission. *7th Moscow Solar System Symposium*, Space Research Institute (IKI), Moscow, Russia, 7MS3-MN-20.

Polyansky, I., Zhukov, B., Zubarev, A., Nadejdina, I., Brusnikin, E., Oberst, J. & Duxbury, T. (2017) Stereo topographic mapping concept for the upcoming Luna-Resurs-1 orbiter mission. *Planetary and Space Science*. doi:10.1016/j.pss.2017.09.013.

Preusker, F., Oberst, J., Stark, A., Matz, K.-D., Gwinner, K. & Roatsch, T. (2017) High resolution topography from MESSENGER orbital stereo imaging – the southern hemisphere. *EPSC*, Riga, Latvia, Vol. 11, EPSC2017-591.

Robinson, M.S., Brylow, S.M., Tschimmel, M., Humm, D., Lawrence, S.J., Thomas, P.C., Denevi, B.W., Bowman-Cisneros, E., Zerr, J., Ravine, M.A., Caplinger, M.A., Ghaemi, F.T., Schaffner, J.A., Malin, M.C., Mahanti, P., Bartels, A., Anderson, J., Tran, T.N., Eliason, E.M., McEwen, A.S., Turtle, E., Jolliff, B.L. & Hiesinger, H. (2010) Lunar Reconnaissance Orbiter Camera (LROC) instrument overview. *Space Science Reviews*, 150(1–4), 81–124. doi:10.1007/s11214-010-9634-2.

Sarkissian, A., Gopala Krishna, B., Crichton, D.J., Beebe, R., Yamamoto, Y., Arviset, C., Di Capria, M.T., Mickaelian, A.M. & The International Planetary Data Alliance (IPDA): overview of the activities. *Astronomical Surveys and Big Data*, 505, 29–34.

Scholten, F., Oberst, J., Matz, K.-D., Roatsch, T., Wählisch, M., Speyerer, E.J. & Robinson, M.S. (2012) GLD100: the near-global lunar 100 m raster DTM from LROC WAC stereo image data. *Journal Geophysical Research*, (117), E00H17.

Sidiropoulos, P. & Muller, J.-P. (2016) Batch co-registration of Mars high-resolution images to HRSC MC11-E mosaic. *The International Archives of the Photogrammetry, Remote Sensing and Spatial Information Sciences*, XLI-B4. pp. 491–495, XXIII ISPRS Congress, 12–19 July 2016, Prague.

Smith, D.E., Zuber, M.T., Neumann, G.A., Mazarico, E., Lemoine, F.G., Head Iii, J.W., Lucey, P.G., Aharonson, O., Robinson, M.S., Sun, X., Torrence, M.H., Barker, M.K., Oberst, J., Duxbury, T.C., Mao, D., Barnouin, O.S., Jha, K., Rowlands, D.D., Goossens, S., Baker, D., Bauer, S., Gläser, P., Lemelin, M., Rosenburg, M., Sori, M.M., Whitten, J. & McClanahan, T. (2017) Summary of the results from the lunar orbiter laser altimeter after seven years in lunar orbit. *Icarus*, 283, 70–91.

Walter, S.H.G., Steikert, R., Schreiner, B., Muller, J.-P., van Gasselt, S., Sidiropoulos, P. & Lanz-Kroechert, J. (2017) The iMars webGIS – space-time queries and dynamic time series of single images. *Lunar Planetary Science Conference*, Volume XLVIII, The Woodlands, TX, USA, abstract #1066.

Zeleny, L. (2012) Lunar program of Russia for 2011–2020 and 2020–2025, potential Cooperation. *The 3rd Moscow International Solar System Symposium*, Moscow, Russia,.

Zuber, M.T., Smith, D.E., Watkins, M.M., Asmar, S.W., Konopliv, A.S., Lemoine, F.G., Melosh, H.J., Neumann, G.A., Phillips, R.J., Solomon, S.C., Wieczorek, M.A., Williams, J.G., Goossens, S.J., Kruizinga, G., Mazarico, E., Park, R.S. & Yuan, D.-N. (2013) Gravity field of the Moon from the Gravity Recovery and Interior Laboratory (GRAIL) mission. *Science*, 339(6120), 668–671. doi:10.1126/science.1231507.

ISPRS book series

1. Advances in Spatial Analysis and Decision Making (2004)
   Edited by Z. Li, Q. Zhou &W. Kainz
   ISBN: 978-90-5809-652-4 (HB)

2. Post-Launch Calibration of Satellite Sensors (2004)
   Stanley A. Morain & Amelia M. Budge
   ISBN: 978-90-5809-693-7 (HB)

3. Next Generation Geospatial Information: From Digital Image Analysis to Spatiotemporal Databases (2005)
   Peggy Agouris & Arie Croituru
   ISBN: 978-0-415-38049-2 (HB)

4. Advances in Mobile Mapping Technology (2007)
   Edited by C. Vincent Tao & Jonathan Li
   ISBN: 978-0-415-42723-4 (HB)
   ISBN: 978-0-203-96187-2 (E-book)

5. Advances in Spatio-Temporal Analysis (2007)
   Edited by Xinming Tang,Yaolin Liu, Jixian Zhang &Wolfgang Kainz
   ISBN: 978-0-415-40630-7 (HB)
   ISBN: 978-0-203-93755-6 (E-book)

6. Geospatial Information Technology for Emergency Response (2008)
   Edited by Sisi Zlatanova & Jonathan Li
   ISBN: 978-0-415-42247-5 (HB)
   ISBN: 978-0-203-92881-3 (E-book)

7. Advances in Photogrammetry, Remote Sensing and Spatial Information Science. Congress Book of the XXI Congress of the International Society for Photogrammetry and Remote Sensing, Beijing, China, 3–11 July 2008 (2008)
   Edited by Zhilin Li, Jun Chen & Manos Baltsavias
   ISBN: 978-0-415-47805-2 (HB)
   ISBN: 978-0-203-88844-5 (E-book)

8. Recent Advances in Remote Sensing and Geoinformation Processing for Land Degradation Assessment (2009)
   Edited by Achim Röder & Joachim Hill
   ISBN: 978-0-415-39769-8 (HB)
   ISBN: 978-0-203-87544-5 (E-book)

9. Advances inWeb-based GIS, Mapping Services and Applications (2011)
   Edited by Songnian Li, Suzana Dragicevic & Bert Veenendaal
   ISBN: 978-0-415-80483-7 (HB)
   ISBN: 978-0-203-80566-4 (E-book)

10. Advances in Geo-Spatial Information Science (2012)
    Edited byWenzhong Shi, Michael F. Goodchild, Brian Lees &Yee Leung
    ISBN: 978-0-415-62093-2 (HB)
    ISBN: 978-0-203-12578-6 (E-book)

11. Environmental Tracking for Public Health Surveillance (2012)
    Edited by Stanley A. Morain & Amelia M. Budge
    ISBN: 978-0-415-58471-5 (HB)
    ISBN: 978-0-203-09327-6 (E-book)

12. Spatial Context: An Introduction to Fundamental Computer Algorithms for Spatial Analysis (2016)
    Christopher M. Gold
    ISBN: 978-1-138-02963-7 (HB)
    ISBN: 978-1-4987-7910-4 (E-book)

Printed and bound by CPI Group (UK) Ltd, Croydon, CR0 4YY
23/10/2024
01778380-0001